FOOD MICROBIOLOGY

FOOD MICROBIOLOGY

Second Edition

W. C. FRAZIER

Professor Emeritus of Bacteriology

University of Wisconsin

McGRAW-HILL BOOK COMPANY

NEW YORK   ST. LOUIS   SAN FRANCISCO   TORONTO   LONDON   SYDNEY

PREFACE TO THE SECOND EDITION

Since the writing of the first edition of this textbook considerably more information has become available on the microbial content of foods and on their preservation and spoilage. New methods of handling and processing foods have created new microbiological problems. Catering and mechanical vending of foods are on the increase, as are the uses of partially or completely precooked foods and of freeze-dried foods. Irradiation of potatoes and bacon with gamma rays is now permitted, and it probably will be allowed soon for more foods. Definite rules have been formulated concerning chemical preservatives permitted in foods. Biological hazards in food are receiving an increased amount of attention, as are suggested microbiological standards for foods.

All this new information demands the expansion of the size of this volume. However, in the interest of keeping the new edition down to textbook size, most additions have been made as brief as practicable, and some sections of the first edition have been shrunk or even eliminated.

In response to suggestions, a few changes have been made in the arrangement of material. Individual chapters deal with the contamination, preservation, and spoilage of each important class of food, and a chapter has been added on chemical changes produced by microorganisms in constituents of foods.

The author regrets that he has been unable to give acknowledgment and thanks to all the numerous authors whose reports have been the bases for the facts presented in this textbook. He is especially grateful to the many authors from whose publications the data in Tables 10–1, 16–3, 16–4, and 26–2 were gathered.

The author wishes to express his thanks for criticism and advice from those who read parts of the manuscript: Dr. E. M. Foster, Dr. K. B. Raper, and Dr. Elizabeth McCoy of the Department of Bacteriology of the University of Wisconsin. Also, thanks for suggestions regarding the revisions are due Dr. Carl S. Pederson of Cornell University and Dr. John C. Ayres of Iowa State University.

W. C. FRAZIER

PREFACE TO THE FIRST EDITION

The purpose of *Food Microbiology* is to condense into a volume of modest size the basic principles of food microbiology, together with illustrations of these principles, in such a form that the book can serve as a college textbook or as an aid to workers in the fields related to the food industries. An attempt has been made to summarize and digest material for the reader and to avoid referring to and quoting long lists of reports from various workers. Although hundreds of scientific papers on food microbiology have been consulted during the preparation of the manuscript, reference has been made, whenever possible, to adequate books and review articles, each of which gives an extensive bibliography, and to individual articles only when books and reviews are not available. The summarizing references should be especially useful to the student, who can and will consult only a limited number of sources.

Each of the main subjects treated in this book is worthy of a separate volume, but limitations of space have not permitted the inclusion of all of the material that each specialist, in his enthusiasm for his field, might wish to see. The author has tried to avoid giving undue attention to any single phase of food microbiology. It has been found necessary, of course, to include a limited amount of food technology, enough for an understanding of the microbiology of some foods.

The subject matter in *Food Microbiology* has been divided into six main parts: Part One on the microorganisms important in food microbiology, their characteristics and their entrance into foods; Part Two on the preservation of foods, including basic methods and their application to specific foods; Part Three on the general principles concerned in the spoilage of foods and the spoilage of specific foods; Part Four on foods and enzymes produced by microorganisms; Part Five on foods in relation to disease, with emphasis on food poisonings and infections; and Part Six on food sanitation, control, and inspection.

By the time this volume is published new methods and discoveries will have been reported. Nevertheless, the basic principles of food microbiology as presented here should still apply.

The author wishes to express his thanks for criticism and advice from those who read parts of the manuscript: Dr. E. M. Foster, Dr. K. B. Raper, Dr. J. B. Wilson, and Dr. Elizabeth McCoy of the Department of Bacteriology of the University of Wisconsin, and Dr. H. J. Peppler of the Red Star Yeast and Products Company.

W. C. FRAZIER

CONTENTS

PREFACE TO THE SECOND EDITION   v

PREFACE TO THE FIRST EDITION   vii

PART ONE
MICROORGANISMS IMPORTANT IN FOOD MICROBIOLOGY   1

1 MOLDS   2
General Characteristics. Classification and Identification.
Molds of Industrial Importance.

2 YEASTS AND YEASTLIKE FUNGI   25
General Characteristics. Classification and Identification.
Yeasts of Industrial Importance.

3 BACTERIA   36
General Characteristics. Genera Important in Food Bacteriology.
Groups Important in Food Bacteriology.

4 CONTAMINATION OF FOODS FROM NATURAL SOURCES   63
From Green Plants and Fruits. From Animals. From Sewage. From
Soil. From Water. From Air. Contamination during Handling and
Processing.

PART TWO
PRINCIPLES OF FOOD PRESERVATION AND SPOILAGE   73

5 GENERAL PRINCIPLES OF FOOD PRESERVATION—
ASEPSIS, REMOVAL, ANAEROBIC CONDITIONS   74
Methods. Principles. Asepsis. Removal. Anaerobic Conditions.

6 PRESERVATION BY USE OF HIGH TEMPERATURES   82
Heat Resistance. Heat Penetration. Thermal Processes. Canning.

7 PRESERVATION BY USE OF LOW TEMPERATURES   109
Growth at Low Temperatures. Lethal Effects. Chilling. Freezing.

8 PRESERVATION BY DRYING   122
Methods. Pretreatments. Procedures after Drying. Microbiology.

9 PRESERVATION BY PRESERVATIVES   131
Additives. Added Inorganic. Added Organic. Developed.

10 PRESERVATION BY RADIATIONS AND BY PRESSURE   146
Heating Radiations. Ultraviolet Rays. Ionizing Radiations.
Mechanical Pressure.

11 GENERAL PRINCIPLES UNDERLYING SPOILAGE   160
Fitness of Food. Causes of Spoilage. Numbers and Kinds of Microorganisms. Growth.

12 CHEMICAL CHANGES CAUSED BY MICROORGANISMS   175
Nitrogenous Organic Compounds. Nonnitrogenous Organic
Compounds.

PART THREE
CONTAMINATION, PRESERVATION, AND SPOILAGE OF DIFFERENT
KINDS OF FOODS   179

13 CONTAMINATION, PRESERVATION, AND SPOILAGE OF CEREALS AND
CEREAL PRODUCTS   180
Contamination. Preservation. Spoilage of Grains, Meals, Flours,
Bread, Cakes, Macaroni, and Tapioca.

14 CONTAMINATION, PRESERVATION, AND SPOILAGE OF SUGARS AND
SUGAR PRODUCTS   192
Contamination. Preservation. Spoilage of Sucrose, Maple Sap, Honey,
and Candy.

15 CONTAMINATION, PRESERVATION, AND SPOILAGE OF VEGETABLES
AND FRUITS   201
Contamination. Preservation of Vegetables and Products; Sauerkraut,
Pickles, Olives. Preservation of Fruits and Products. Spoilage of Raw
Vegetables and Fruits, Juices, Fermented Products.

16 CONTAMINATION, PRESERVATION, AND SPOILAGE OF MEATS AND
MEAT PRODUCTS   252
Contamination. Preservation: Heat, Cold, Rays, Drying, Preserva-
tives. Spoilage: General Principles; Types of Spoilage; Spoilage of
Different Kinds of Meats.

17 CONTAMINATION, PRESERVATION, AND SPOILAGE OF FISH AND OTHER SEAFOODS   283

Contamination. Preservation: Heat, Cold, Drying, Preservatives. Spoilage: Factors Influencing; Evidences; Causative Bacteria; Spoilage of Special Kinds of Fish and Seafoods.

18 CONTAMINATION, PRESERVATION, AND SPOILAGE OF EGGS   296

Contamination. Preservation: Removal, Heat, Cold, Drying, Preservatives. Spoilage: Defects and Changes, Bacterial Rots, Fungal Rots.

19 CONTAMINATION, PRESERVATION, AND SPOILAGE OF POULTRY   310

Contamination. Preservation: Removal, Heat, Cold, Preservatives. Spoilage.

20 CONTAMINATION, PRESERVATION, AND SPOILAGE OF MILK AND MILK PRODUCTS   318

Contamination. Preservation: Asepsis, Removal, Heat, Cold, Drying, Preservatives, Rays. Spoilage: Milk and Cream, Condensed and Dry Products, Frozen Desserts, Butter, Fermented Products.

21 MISCELLANEOUS FOODS   350

Contamination. Fatty Foods. Salad Dressings. Essential Oils. Bottled Beverages. Spices. Condiments. Salt. Nutmeats.

22 SPOILAGE OF HEATED CANNED FOODS   357

Causes. Grouping on pH. Types of Spoilage. Canned Meats and Fish.

PART FOUR
FOODS AND ENZYMES PRODUCED BY MICROORGANISMS   369

23 PRODUCTION OF CULTURES FOR FOOD FERMENTATIONS   370
Culture Maintenance and Preparation. Bacterial Cultures. Yeast Cultures. Mold Cultures.

24 FOOD FERMENTATIONS   380
Bread. Malt Beverages. Wines. Distilled Liquors. Vinegar. Tea, Coffee, Cacao, Vanilla, and Citron. Oriental Fermented Foods.

25 FOODS AND ENZYMES FROM MICROORGANISMS   416
Microorganisms as Food. Fats. Vitamins. Other Substances. Enzymes.

PART FIVE
FOODS IN RELATION TO DISEASE   433

26   FOOD POISONINGS AND INFECTIONS   434
Chemicals. Poisonous Plants and Animals. Food Poisoning. Food
Infections. Other Infections. Trichinosis.

27   INVESTIGATION OF FOOD-BORNE DISEASE OUTBREAKS   463
Diseases. Objectives. Personnel. Materials and Equipment. Field
Investigation. Laboratory Testing. Interpretation and Application of
Results. Preventive Measures.

PART SIX
FOOD SANITATION, CONTROL, AND INSPECTION   475

28   MICROBIOLOGY IN FOOD PLANT SANITATION   476
Water. Waste Treatment and Disposal. Microbiology of Food
Product. Food Vending. Standards. Employee Health.

29   MICROBIOLOGICAL LABORATORY METHODS   492
Purposes of Tests. Sampling. Kinds of Tests. Agencies Recommend-
ing Methods. Eating and Drinking Utensils.

30   FOOD CONTROL   502
Enforcement and Control Agencies. Federal Food, Drug, and Cos-
metic Act. Food Additives Amendment. Meat Inspection Act.
Poultry Products Inspection Act. Federal Inspection and Grading
Service. State and Municipal Food Laws.

APPENDIX: MICROBIOLOGICAL STANDARDS   509

INDEX   515

FOOD MICROBIOLOGY

PART ONE

MICROORGANISMS IMPORTANT
IN FOOD MICROBIOLOGY

The food microbiologists must become acquainted with the microorganisms important in foods, at least to the extent that will enable him to identify the main types encountered, so that he can make use of what is known about their characteristics and can compare his results with those of other workers. Part 1 makes an attempt to outline briefly the identification and classification of food molds, because these organisms are not studied much in the usual beginning course in microbiology. There is some discussion of the classification of yeasts, but no attempt has been made to cover determinative bacteriology, a subject to which the student usually receives an adequate introduction in his first course in bacteriology.

The contamination of foods with microorganisms is discussed only briefly in this first section because this subject will be considered repeatedly and in more detail in following sections of the textbook.

CHAPTER ONE

MOLDS

Mold growth on foods, with its fuzzy or cottony appearance, sometimes colored, is familiar to everyone; and usually a moldy or "mildewed" food is considered unfit to eat. While it is true that molds are concerned in the spoilage of many kinds of foods, special molds are useful in the manufacture of certain kinds of foods or of ingredients of foods. Thus, some kinds of cheese are mold-ripened, e.g., blue, Roquefort, Camembert, Brie, Gammelost, etc., and molds are used in the making of Oriental foods such as soy sauce, miso, sonti, and others that will be discussed later. Molds have been grown as food or feed, and are employed to produce products used in foods, such as amylase for bread-making or citric acid used in soft drinks.

The food bacteriologist should be able to identify the genera of molds important in foods and be able to recognize some of the common species. It is the purpose of the following discussion to give only enough information to enable preliminary identification of common genera of molds without special regard for their place in botanical classification and to teach how to favor or inhibit mold growth as desired.

GENERAL CHARACTERISTICS OF MOLDS

The term **mold** is a common one applied to certain multicellular, filamentous fungi whose growth on foods usually is readily recognized by its fuzzy or cottony appearance. The main part of the growth commonly appears white but may be colored or dark or smoky. Colored spores are typical of mature mold plants of some kinds and give color to part or all of the growth.

Morphological characteristics

The morphology, that is, the form and structure, of molds, as judged by their macroscopic and microscopic appearance, is used in their identification and classification.

HYPHAE AND MYCELIUM. The mold plant consists of a mass of branching, intertwined filaments called **hyphae** (singular **hypha**), and the whole mass of these hyphae is known as the **mycelium.** The hyphae may be **submerged,** or growing within the food, or **aerial,** or growing into the air above the food. Hyphae also may be classed as **vegetative,** or growing, and hence concerned chiefly with the nutrition of the mold, or **fertile,** concerned with the production of reproductive parts. In most molds the fertile hyphae are aerial, but in some molds they may be submerged. The hyphae of some molds are full and smooth, but the hyphae of others are characteristically thin and ragged. A few kinds of molds produce **sclerotia** (singular **sclerotium**), which are tightly packed masses of modified hyphae, often thick-walled, within the mycelium. These sclerotia are considerably more resistant to heat and other adverse conditions than the rest of the mycelium and for this reason may be important in some processed food products.

Microscopic examination of mold hyphae reveals characteristics useful in the identification of genera. Molds are divided into two groups: **septate,** that is, with cross walls dividing the hypha into cells; and **nonseptate,** with the hyphae apparently consisting of cylinders without cross walls. The nonseptate hyphae have nuclei scattered throughout their length and are considered multicellular. The hyphae of most molds are clear, but some are dark or smoky. Hyphae may appear uncolored and transparent on microscopic examination but colored when large masses of hyphae are viewed macroscopically.

Septate hyphae increase in length by division of the tip cell (apical growth) or of cells within the hypha (intercalary growth), the type of growth being characteristic of the kind of mold. Division of the nuclei distributed throughout nonseptate hyphae is accompanied by an increase in the length of filaments.

Special mycelial structures or parts aid in the identification of molds. Examples are the rhizoids, or "holdfasts," of *Rhizopus* and *Absidia,* the foot cell in *Aspergillus,* and the dichotomous, or Y-shaped, branching in *Geotrichum,* all of which will be described later.

REPRODUCTIVE PARTS OR STRUCTURES. Molds can grow from a transplanted piece of mycelium, but there is seldom an opportunity for this to occur. Reproduction of molds is chiefly by means of asexual spores. Some molds also form sexual spores. Such molds are termed "perfect" and are classified as either *Oömycetes* or *Zygomycetes* if nonseptate, or *Ascomycetes* or *Basidiomycetes* if septate, as contrasted with "imperfect" molds, the *Fungi Imperfecti* (typically septate), which have only asexual spores.

ASEXUAL SPORES.   The asexual spores of molds are produced in large numbers and are small, light, and resistant to drying. They are readily spread through the air to alight and start new mold plants where conditions are favorable. The three principal types of asexual spores are (1) **conidia** (singular **conidium**) (Figure 1–9); (2) **arthrospores,** or **oidia** (singular **oidium**) (Figure 1–15); and (3) **sporangiospores** (Figure 1–5). Conidia are cut off, or bud, from special fertile hyphae called **conidiophores,** and usually are in the open, that is, not enclosed in any container, as contrasted to the sporangiospores, which are in a **sporangium** (plural **sporangia**), or sac, at the tip of a fertile hypha, the **sporangiophore.** Arthrospores are formed by fragmentation of a hypha, so that the cells of the hypha become arthrospores. Examples of these three kinds of spores will be given in the discussion of important genera of molds. A fourth kind of asexual spore, the **chlamydospore,** is formed by many species of molds when a cell here and there in the mycelium stores up reserve food, swells, and forms a thicker wall than that of surrounding cells. This chlamydospore, or resting cell, can withstand unfavorable conditions better than ordinary mold mycelium and later, under favorable conditions, can grow into a new mold plant.

The morphology of the asexual spores is helpful in the identification of genera and species of molds. Sporangiospores differ in size, shape, and color. Conidia not only vary in these respects, but also may be smooth or roughened and one-, two-, or many-celled.

Also helpful in the identification of molds is the appearance of the fertile hyphae and the asexual spores on them. If sporangiospores are formed, points to be noted are whether the sporangiophores are simple or branched, the type of branching, and the size, shape, color, and location of the sporangia. The swollen tip of the sporangiophore, the **columella,** which usually projects into the sporangium, assumes shapes typical of species of mold. Conidia may be borne singly on conidiophores or in spore heads of differing arrangement and complexity. A glance at the general appearance of a spore head often is sufficient for identification of the genus. Some molds have conidia in chains, squeezed off one by one from a special cell, a **sterigma** (plural **sterigmata**) or **phialide,** at the tip of the conidiophore. Other molds have irregular masses of conidia, which cut off from the tip of the conidiophore without evident sterigmata. These masses of conidia may be loosely or tightly packed or even enslimed. The conidia of some molds bud from the conidiophore and continue to multiply by budding; they appear yeastlike.

SEXUAL SPORES.   The molds which can produce sexual spores are classified on the basis of the manner of formation of these spores and the type produced. The nonseptate molds (*Phycomycetes*) that produce **oöspores** are termed *Oömycetes*. These molds are mostly aquatic forms and uncommon in foods. The oöspores are formed by the union of a small male gamete and a large female gamete. The *Zygomycetes* form **zygospores** by the union of the tips of two hyphae which often appear similar and which may come from the same mycelium or from different mycelia. Both oöspores and zygospores are covered by a tough wall and can survive drying for long periods. The *Ascomycetes* (septate) form sexual spores known as **ascospores,** which are formed after the union of two cells from the same mycelium or from two separate mycelia. The ascospores, resulting from cell division after conjugation, are in an **ascus** or sac, with usually eight spores per ascus. The asci may be single or may be grouped within a covering called an **ascocarp,** formed by branching and intertwining adjacent hyphae. The *Basidiomycetes,* which include most mushrooms, plant rusts, smuts, etc., form a fourth type of sexual spore, the basidiospore, but will not be discussed because they are relatively unimportant in food microbiology.

Cultural characteristics

The gross appearance of a mold plant growing on a food often is sufficient to indicate its genus. Some molds are loose and fluffy; others are compact. Some look velvety on the upper surface, some dry and powdery, and others wet or gelatinous. Some molds are restricted in size, while others seem limited only by the food or container. Definite zones of growth in the plant distinguish some molds, e.g., *Aspergillus niger.* Pigments in the mycelium, red, purple, yellow, brown, gray, black, etc., are characteristic, as are the pigments of masses of asexual spores: green, blue-green, yellow, orange, pink, lavender, brown, gray, black, etc. The appearance of the reverse side of a mold plant on an agar plate may be striking, like the opalescent blue-black or greenish-black color of the underside of *Cladosporium.*

Physiological characteristics

The physiological characteristics of molds will be reviewed only briefly here and will be discussed in more detail later.

MOISTURE REQUIREMENTS. In general most molds require less available moisture than most yeasts and bacteria, although notable exceptions will be cited in Chapter 11, where moisture requirements of molds, yeasts, and bacteria will be compared. Tolerance of solutes may be expressed in terms of **water activity** (a_w), which is the vapor pressure of the solution (of solutes in water in most foods), divided by the vapor pressure of the solvent (usually water). The a_w for pure water would be 1.00, and for a 1.0 molal solution of the ideal solute the a_w would be 0.9823. The a_w would be in equilibrium with a relative humidity (R.H.) of the atmosphere about the food 100 times as large, if the R.H. is expressed as a percentage. A relative humidity about a food that would correspond to an a_w lower than that of the food would tend to dry the surface of the food; conversely, if the R.H. were higher than that corresponding to the a_w, the latter would be increased at the surface of the food.

Water is made unavailable in various ways:

1. Solutes and ions tie up water in solution. Therefore an increase in the concentration of dissolved substances such as sugars and salts is in effect a drying of the material. Not only is water tied up by solutes, but water tends to leave the microbial cells by osmosis if there is a higher concentration of solute outside the cells than inside.
2. Hydrophilic colloids (gels) make water unavailable. As little as 3 to 4 percent agar in a medium may prevent bacterial growth by leaving too little available moisture.
3. Water of crystallization or hydration is usually unavailable to microorganisms. Water itself, when crystallized as ice, no longer can be used by microbial cells. The a_w of water-ice mixtures (vapor pressure of ice divided by vapor pressure of water) decreases with a drop in temperature below 0 C. The a_w values of pure water are 1.00 at 0 C, 0.953 at -5 C, 0.907 at -10 C, 0.864 at -15 C, 0.823 at -20 C, and so on. In a food, as more ice is formed, the concentration of solutes in the unfrozen water is increased, lowering its a_w.

The reduction of a_w by a solute depends primarily on the total concentration of dissolved molecules and ions, each of which is surrounded by water molecules held more or less firmly. The solution then has a lower freezing point than pure water and a lower vapor pressure. The organisms must compete with these particles for water molecules. The decrease in vapor pressure for an ideal solvent follows Raoult's law: the vapor pressure of the solution relative to that of the pure solvent is equal to the mole fraction of the solvent; that is, $p/p_0 = n_2/n_1 + n_2$, where p and p_0 are the vapor pressures of the solution and solvent and n_1 and n_2 are the number of moles of solute and solvent, respectively. Although a_w varies with temperature, the variations are only slight within the range of temperatures permitting microbial growth. Variations in temperature increase in importance with increasing concentrations of solutes and increasing effects on ionization.

Each microorganism has a maximal, optimal, and minimal a_w for growth. This range depends upon factors discussed below. As the a_w is reduced below the optimal level, there is a lengthening of the lag phase of growth, a decrease in the rate of growth, and a decrease in the amount of cell substance synthesized—changes that vary with the organism and with the solute employed to reduce a_w.

Factors that may affect a_w requirements of microorganisms include: (1) The kind of solute employed to reduce the a_w. For many organisms, especially molds, the lowest a_w for growth is practically independent of the kind of solute used. Other organisms, however, have lower limiting a_w values with some solutes than with others. Potassium chloride, for example, usually is less toxic than sodium chloride, and it in turn is less inhibitory than sodium sulfate. (2) The nutritive value of the culture medium. In general, the better the medium is for growth, the lower will be the limiting a_w. (3) Temperature. Most organisms have the greatest tolerance to low a_w at about optimal temperatures. (4) Oxygen supply. Growth of aerobes takes place at a lower a_w in the presence of air than in its absence, and the reverse is true of anaerobes. (5) pH. Most organisms are more tolerant of low a_w at pH values near neutrality than in acid or alkaline media. (6) Inhibitors. The presence of inhibitors narrows the range of a_w for growth of microorganisms.

Methods for the control of a_w are (1) equilibration with controlling solutions, (2) determination of the water-sorption isotherm for the food, or (3) addition of solutes.

Freezing-point determinations may be employed to establish a_w values of liquid foods. A rough estimate of the a_w value of a solid food can be obtained by means of filter papers impregnated with certain salts and dried. The papers are left in a closed space with the food and indicate a_w by their wet or dry state. The dew-point method measures the relative humidity of air in equilibrium with the food.

Molds differ considerably among themselves as to optimal a_w and range of a_w for germination of the asexual spores. The range for spore germination is greater at temperatures near the optimum for germination and in a better culture medium. The minimal a_w for spore germination has been found to be as low as 0.62 for some molds, and as high as 0.93 for others (e.g., *Mucor, Rhizopus,* and *Botrytis*). Each mold also has an optimal a_w and range of a_w for growth. Examples of optimal a_w are 0.98 for an *Aspergillus* sp., 0.995 to 0.98 for a *Rhizopus* sp., and 0.9935 for a *Penicillium* sp. The a_w would have to be below 0.62 to stop all chances for mold growth, although an a_w below 0.70 inhibits most molds causing food spoilage; and an a_w below 0.94 inhibits molds like *Rhizopus,* and below 0.85 inhibits *Aspergillus* spp. The reduction of the a_w below the optimum for a mold delays spore germination and reduces the rate of growth and therefore can be an important factor in food preservation. Many of the molds can grow in media with an a_w approaching 1.00 (pure water).

An approximate limiting total moisture content of a given food for mold growth can be estimated, and therefore it has been claimed that below 14 to 15 percent total moisture in flour or some dried fruits will prevent or greatly delay mold growth.

TEMPERATURE REQUIREMENTS. Most molds would be considered **mesophilic,** i.e., able to grow well at ordinary temperatures. The optimal temperature for most molds is around 25 to 30 C (77 to 86 F), but some grow well at 35 to 37 C (95 to 98.6 F) or above, e.g., *Aspergillus* species, and some at still higher temperatures. A number of molds are **psychrophilic,** i.e., they grow fairly well at temperatures of freezing or just above, and some can grow slowly at temperatures below freezing. Growth has been reported at as low as —5 to —10 C (23 to 14 F). A few are **thermophilic,** i.e., they have a high optimal temperature.

OXYGEN AND pH REQUIREMENTS. Molds are aerobic, i.e., they require oxygen for growth; this is true at least for the molds growing on foods. Most molds can grow over a wide range of hydrogen-ion concentration (pH 2 to 8.5), but the majority are favored by an acid pH.

FOOD REQUIREMENTS. Molds in general can utilize many kinds of foods, ranging from simple to complex. Most of the common molds possess a variety of hydrolytic enzymes, and some are grown for their amylases, pectinases, proteinases, and lipases.

INHIBITORS. Compounds inhibitory to other organisms are produced by some molds, such as penicillin from *Penicillium chrysogenum* and clavacin from *Aspergillus clavatus.* Certain chemical compounds are **mycostatic,** inhibiting the growth of molds (sorbic acid, propionates, and acetates are examples), or are specifically **fungicidal,** killing molds.

Initiation of growth of molds is slow compared to that of bacteria or yeasts, so that when conditions are favorable for all these organisms, molds usually lose out in the competition. After mold growth is under way, however, it may be very rapid.

CLASSIFICATION AND IDENTIFICATION OF MOLDS

Molds are plants of the division *Thallophyta,* having no roots, stems, or leaves, and of the subdivision *Fungi,* devoid of chlorophyll. They belong to the *Eumycetes,* or true fungi, and are subdivided further as shown in Figure 1–1.

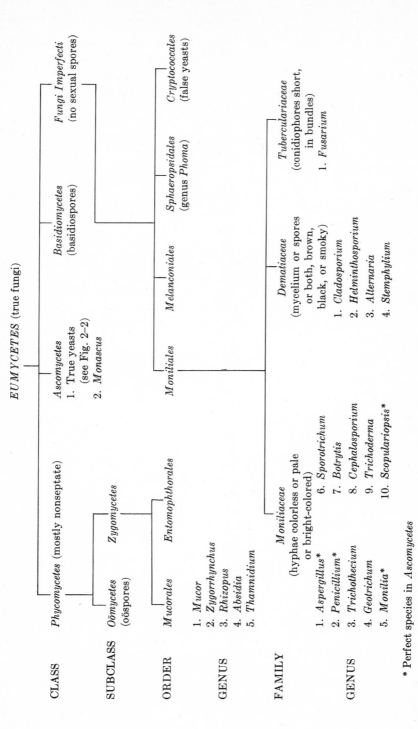

Figure 1–1. Partial classification of *Eumycetes*, or true fungi.

9

Review of the previous discussion and of Figure 1–1 will indicate that the following criteria are used chiefly for differentiation and identification of molds:

1. Hyphae septate or nonseptate.
2. Mycelium clear or dark (smoky).
3. Mycelium colored or colorless.
4. Whether sexual spores are produced and the type: oöspores, zygospores, or ascospores.
5. Type of asexual spores: sporangiospores, conidia, or arthrospores (oidia).
6. Characteristics of the spore head:
   a. Sporangia: size, color, shape, and location.
   b. Spore heads bearing conidia: single conidia, chains, budding conidia, or masses; shape and arrangement of sterigmata or phialides; gumming together of conidia.
7. Appearance of sporangiophores or conidiophores: simple or branched, and if branched the type of branching. Size and shape of columella at tip of sporangiophore. Whether conidiophores are single or in bundles.
8. Microscopic appearance of the asexual spores, especially of conidia: shape, size, color; smooth or rough; one-, two-, or many-celled.
9. Presence of special structures (or spores): stolons, rhizoids, foot cells, apophysis, chlamydospores, sclerotia, etc.

Examples of these differential characteristics will be given in the description of genera of molds to follow.

MOLDS OF INDUSTRIAL IMPORTANCE

Genera of molds of industrial importance, especially in foods, will be described briefly, many of them only enough to enable differentiation from other commonly encountered genera.

I. Molds of the class *Phycomycetes* (all food molds in this class are non-septate).
   A. Subclass *Oömycetes:* Sexual spores are oöspores.
      1. Order *Saprolegniales* (water molds)
         a. Genus *Saprolegnia. S. parasitica* causes a fungal growth on fish. Asexual reproduction is by sporangia and motile zoöspores.
      2. Order *Peronosporales*
         a. Genus *Pythium.* Some species cause rots of vegetables and some are pathogens of plant roots. Sporangia give forth motile zoöspores.
   B. Subclass *Zygomycetes:* Sexual spores are zygospores.
      1. Order *Mucorales:* Asexual sporangiospores in sporangia.
         a. Genus *Mucor* (Figure 1–2). Mucors are involved in the spoilage of some foods and in the manufacturing of others. A widely distributed species is *M. racemosus. M. rouxii* is used in the "Amylo"

process for saccharification of starch, and mucors aid in the ripening of some cheeses (e.g., Gammelost) and the making of certain Oriental foods. Distinguishing characteristics are (1) Nonseptate. (2) Sporangiophores which form on all aerial parts of mold and may be simple or branched. (3) Columella round, cylindrical, or pear-shaped. (4) Spores smooth, regular. (5) Zygospore suspensors equal in size (Figure 1–3). (6) No stolons, rhizoids, or sporangioles (small sporangia containing a few spores).

b. Genus *Zygorrhynchus.* These soil molds are similar to *Mucor* except that the zygospore suspensors are markedly unequal in size and arise from neighboring branches of the same hypha (Figure 1–4).

c. Genus *Rhizopus* (Figures 1–5, 1–6). *R. nigricans,* the so-called "bread mold," is very common and is involved in the spoilage of many foods: berries, fruits, vegetables, bread, etc. Distinguishing characteristics of *Rhizopus* are (1) Nonseptate. (2) Has stolons and rhizoids, often darkening in age. (3) Sporangiophores arise at the

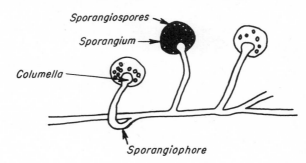

Figure 1–2. *Mucor.*

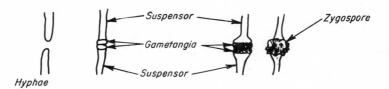

Figure 1–3. Zygospore formation in *Mucor.*

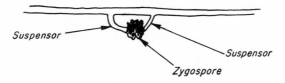

Figure 1–4. Zygospore formation in *Zygorrhynchus.*

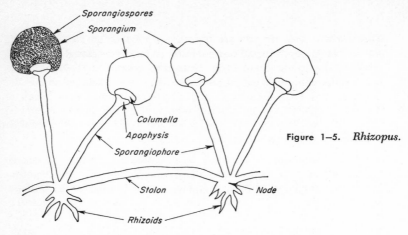

Figure 1–5. *Rhizopus.*

Figure 1–6. Photomicrograph of *Rhizopus nigricans.*
Note rhizoids and sporangiophores arising from node,
and sporangia at ends of sporangiophores (×200).
(*Courtesy of Harper & Row, Publishers, Incorporated.
From W. B. Sarles et al., Microbiology: General and
Applied, 2nd ed., copyright, 1956, New York.*)

nodes, where rhizoids also are formed. (4) Sporangia are large and usually black. (5) Hemispherical columella and cup-shaped **apophysis** (base to the sporangium). (6) Abundant, cottony mycelium which may fill the container, e.g., a petri dish. (7) No sporangioles.

d. Genus *Absidia*. Similar to *Rhizopus,* except that sporangiophores arise from the internodes and sporangia are small and pear-shaped (Figure 1–7).

e. Genus *Thamnidium* (Figure 1–8). *T. elegans* is found on meat in chilling storage. Distinguishing characteristics are (1) Nonseptate. (2) Sporangiophores with large sporangia at tip and lateral clusters of sporangioles near the base. The sporangioles are like miniature sporangia, each with two to twelve or more spores, and are borne on many-branched outgrowths from near the base of the sporangiophore.

Simple Key to Genera of *Mucorales* Described

  I. Sporangioles present ....*Thamnidium*
 II. Sporangioles absent
    A. Rhizoids and stolons formed
       1. Sporangiophores arise at nodes ....*Rhizopus*
       2. Sporangiophores arise at internodes ....*Absidia*
    B. Rhizoids and stolons not formed
       1. Zygospore suspensors equal in size ....*Mucor*
       2. Zygospore suspensors unequal in size ....*Zygorrhynchus*

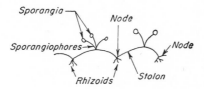

Figure 1–7. Diagram of *Absidia,* showing location of sporangiophores and rhizoids.

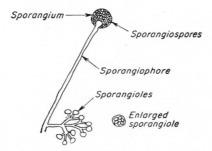

Figure 1–8. *Thamnidium.*

II. Septate molds

    A. Class *Fungi Imperfecti:* No sexual spores (for convenience, a few molds forming sexual spores are included in the following discussion).

        1. Order *Moniliales:* Conidiophores free, arising irregularly from the mycelium.

           a. Family *Moniliaceae:* Mycelium clear and colorless or palely or brightly colored.

                (1) Genus *Aspergillus* (Figures 1–9, 1–10). The aspergilli are very widespread. Many are involved in the spoilage of foods, and some are useful in the preparation of certain foods. Raper and Fennell list eighteen groups of aspergilli and recognize 132 species, but only a few will be mentioned here. The *A. glaucus* group, with *A. repens* as an important species, is often involved in food spoilage. The molds grow well in high concentrations of sugar and salt and hence in many foods of low moisture content. Conidia of this group are some shade of green, and ascospores are in asci within yellow to reddish perithecia. Some authors include these molds in the *Ascomycetes* and the genus *Eurotium,* a name reserved for those members having a perfect (sexual) stage.

                The *A. niger* group, with *A. niger* as a leading species, is widespread and may be important in foods. The spore-bearing heads are large, tightly packed, and globular and may be black, brownish-black, or purple-brown. The conidia are rough, with bands of pigment. Many strains have sclerotia, colored from buff to gray to blackish. Selected strains are used for the commercial production of citric and gluconic acids, and a variety of enzyme preparations.

                The *A. flavus-oryzae* group includes molds important in the making of some Oriental foods and the production of enzymes, but

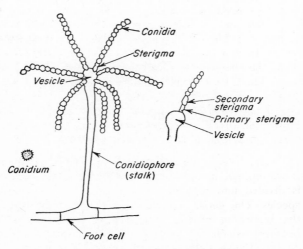

Figure 1–9. Diagram of a simple *Aspergillus.*

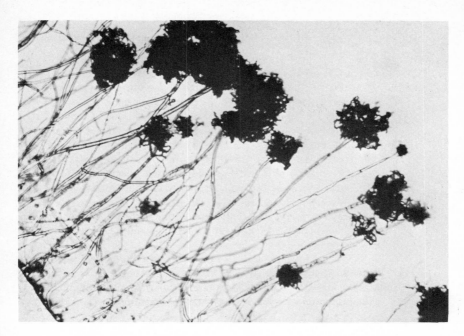

Figure 1–10. Photomicrograph of *Aspergillus niger*, showing fruiting heads bearing chains of conidia (×200). (*Courtesy of Harper & Row, Publishers, Incorporated. From W. B. Sarles et al., Microbiology: General and Applied, 2nd ed., copyright, 1956, New York.*)

molds in this group often are involved in the spoilage of foods. Conidia give various yellow to green shades to the spore heads, and dark sclerotia may be formed.

The distinguishing characteristics of *Aspergillus* are (*a*) Septate, branched mycelium, usually uncolored; the submerged part is vegetative and the aerial part mostly fertile. (*b*) Colonies often zonate. (*c*) The conidiophore or stalk septate or nonseptate and arising from a **foot cell** (see Figure 1–9), a special mycelial cell which has become enlarged and thick-walled. The conidiophore swells into a vesicle at the end, bearing sterigmata from which the conidia are cut off. (*d*) Sterigmata or phialides simple or compound and colored or uncolored. (*e*) Conidia in chains—green, brown, or black more commonly than other colors. (*f*) Some species grow well at 37 C (98.6 F) or above.

(2) Genus *Penicillium* (Figure 1–11). This is another genus that is widespread in occurrence and important in foods. The genus is divided into groups and subgroups, and there are numerous species. The genus is divided into large groups on the basis of the branching of the spore-bearing heads, or penicilli (little brushes), which may be simple with one penicillus (Figure 1–12), double (Figure 1–12, sketch 2), or complex (Figure 1–12, sketches

3 and 4), and the latter may be symmetrical (Figure 1–12, sketch 3), or asymmetrical (Figure 1–12, sketch 4). Most species important in foods are complex and asymmetrical. One type forms **coremia,** or compact bundles of conidiophores, as in *P. expansum,* the blue-green-spored mold that causes soft rots of fruits. Other important species are *P. digitatum,* with olive- or yellowish-green conidia, causing a soft rot of citrus fruits; *P. italicum,* called the "blue contact mold," with blue-green conidia, also rotting citrus fruits; *P. camemberti,* with grayish conidia, useful in the ripening of Camembert cheese; and *P. roqueforti* (Figure 1–13), with bluish-green conidia, aiding in the ripening of blue cheeses, e.g., Roquefort. A few species form asci with ascospores in cleistothecia, and a few exhibit sclerotia and therefore have caused trouble in canned acid foods.

The distinguishing characteristics of *Penicillium* are (*a*) Septate, branching mycelium which usually is uncolored. (*b*) Septate, aerial conidiophores that are perpendicular to and walled off from the submerged hyphae from which they arise. They may be

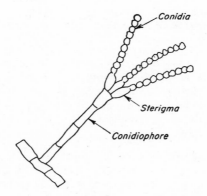

Figure 1–11. Diagram of a simple *Penicillium.*

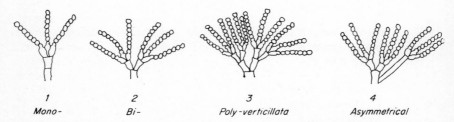

1	2	3	4
Mono-	Bi-	Poly-verticillata	Asymmetrical

Figure 1–12. Diagram showing different types of spore heads of *Penicillium* molds: (1) Monoverticillata; (2) Biverticillata (also symmetrical); (3) Polyverticillata (symmetrical); (4) Polyverticillata (asymmetrical).

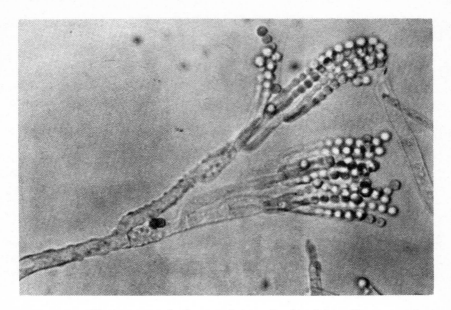

Figure 1–13. Photomicrograph showing fruiting heads of *Penicillium roqueforti* (×1,200). (*Courtesy of K. B. Raper.*)

branched or unbranched. (*c*) Brushlike spore-bearing heads, with sterigmata or phialides borne in clusters and essentially in one plane. A chain of conidia arises from each sterigma as the result of the conidia being cut off, one by one, from the tip of the sterigma. (*d*) Conidia of most species are green when young, but later may turn brownish.

(3) Genus *Trichothecium*. The common species, *T. roseum* (Figure 1–14), is a pink mold which grows on wood, paper, fruits such as apples and peaches, and vegetables such as cucumbers and cantaloupes. This mold is easily recognized by the clusters of two-celled conidia at the ends of short, erect conidiophores. Conidia have a nipplelike projection at the point of attachment, and the smaller of the two cells of each conidium is at this end.

(4) Genus *Geotrichum* (*Oöspora* or *Oidium*) (Figure 1–15). This genus is included with the yeastlike fungi by some writers and with the molds by others. Species may be white, yellowish, orange, or red, with the growth appearing first as a firm, feltlike mass that later becomes soft and creamy. *Geotrichum candidum* (*Oöspora lactis*), often called the "dairy mold," gives white to cream-colored growth. The hyphae are septate and in common species are dichotomously branched. The asexual spores are arthrospores (oidia) which may appear rectangular if from submerged hyphae and oval if from aerial hyphae.

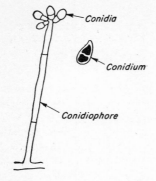

Figure 1–14. *Trichothecium.*

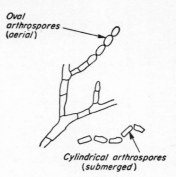

Figure 1–15. *Geotrichum.*

(5) Genus *Neurospora* (*Monilia*) (Figure 1–16). This genus has been described under various names because of the confusion concerning its classification, but most mycologists believe that it should be classed among the perfect molds (producing sexual spores), and they call the genus *Neurospora*. *Neurospora* (*Monilia*) *sitophila*, the most important species in foods, sometimes is termed the "red bread mold" because its pink, loose-textured growth often occurs on bread. It also grows on sugar-cane bagasse and on various foods. The perfect, or ascosporogenous, stage seldom is seen.

Distinguishing characteristics of *N. sitophila* are (*a*) Septate mycelium, which later may break up into cells. (*b*) Loose network of aerial, long-stranded mycelium. (*c*) Aerial hyphae bearing many ovate, pink to orange-red, budding conidia, which form branched chains and are found near the top of the plant.

(6) Genus *Sporotrichum*. Among the saprophytic species is *S. carnis* (Figure 1–17), found growing on chilled meats, where it causes "white spot."

Distinguishing characteristics are (*a*) Clear, septate mycelium. (*b*) Colonies cream-colored, first wet and later dry and powdery.

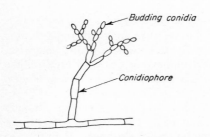

Figure 1–16. *Neurospora* (*Monilia*).

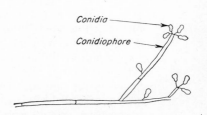

Figure 1–17. *Sporotrichum.*

(*c*) Conidia mostly small and pear-shaped (round or oval in some species), arising laterally or at the tips of conidiophores from all parts of the mycelium. Conidia often single, but may be in small clusters.

(7) Genus *Botrytis*. Only one species is important in foods, *B. cinerea* (Figure 1–18). It causes a disease of grapes but may grow saprophytically on many foods.

Distinguishing characteristics are (*a*) Woolly, pale, dirty-brown, septate mycelium. (*b*) Fairly long, stiff conidiophores which are branched irregularly at the end and bear conidia in grapelike bunches. (*c*) Small, ovate conidia. (*d*) Dirty-green sclerotia which later turn black.

(8) Genus *Cephalosporium* (Figure 1–19). *C. acremonium* is a common species.

Distinguishing characteristics are (*a*) Septate mycelium. (*b*) Short, erect, nonseptate conidiophores arising at random from fertile hyphae. (*c*) Colorless, elongate or elliptical conidia, arising from the tip of the conidiophore, which also exudes a sticky fluid. Conidia, held together by this fluid, form into small balls.

(9) Genus *Trichoderma*. *T. viride* (Figure 1–20) is a common species. The mature mold plant is bright-green because the balls of green conidia are glued together, and tufts of white hyphae (sterile) stick up well above the conidiophores.

Distinguishing characteristics are (*a*) Septate mycelium. (*b*) Many-branched, septate conidiophores, the final branch being a sterigma, which cuts off spherical to slightly ovate, bright-green conidia into slimy balls.

(10) Genus *Scopulariopsis*. *S. brevicaulis* (Figure 1–21) is a common species. This genus may be confused with *Penicillium*, for both have brushlike penicilli and chains of spores cut off from the sterigmata, but the conidia of *Scopulariopsis* are never green.

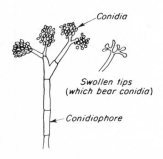

Figure 1–18. *Botrytis* (*cinerea*).

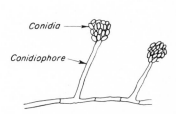

Figure 1–19. *Cephalosporium*.

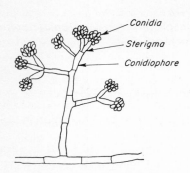

Figure 1–20. *Trichoderma.*

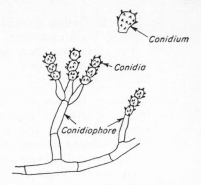

Figure 1–21. *Scopulariopsis.*

Conidiophores may be branched or unbranched in *Scopulariopsis,* and simple or complex, and the branching usually is irregular. The spore-bearing heads may vary from complex, branching systems with penicilli to single sterigmata arising from short branches of aerial hyphae. The spores are distinctive in microscopic appearance and are not green but commonly yellowish-brown; they are lemon-shaped, thick-walled, spiny, and pointed at one end, with a thick ring at the opposite end. Colonies are brownish and cottony.

(11) Genus *Pullularia* (Figure 1–22). Ovate, hyaline conidia (blastospores or buds from preexisting cells) borne as lateral buds on all parts of the mycelium. Colonies are pale and slimy and yeastlike when young, becoming mycelial and dark and leathery in age. *P. pullulans* is a common species.

Simple Key to Genera of *Moniliaceae* Described

I. Single-celled conidia
  A. Conidia cut off in chains from sterigmata
    1. Sterigmata arising from all parts or the upper part of surface of vesicle (swollen tip) of conidiophore. Foot cell . . . . *Aspergillus*
    2. Sterigmata in clusters, spore head brushlike. . .*Penicillium*
    3. Sterigmata single on a conidiophore or on irregularly branched conidiophore to form brushlike heads
    Scopulariopsis
  B. Budding conidia; form treelike heads . . . .*Neurospora*
  C. Conidia often borne singly on conidiophore or in small clusters on all parts of mycelium . . . .*Sporotrichum*
  D. Conidia borne as single lateral buds on mycelium. .*Pullularia*
  E. Conidia borne in clusters at tip of conidiophore
    1. In loose, grapelike bunches on irregularly branched conidiophore . . . .*Botrytis*

2. Conidia gummed together in balls
    a. Unbranched conidiophores ....*Cephalosporium*
    b. Many-branched conidiophores ....*Trichoderma*
II. Two-celled conidia ....*Trichothecium*
III. Arthrospores (oidia) ....*Geotrichum*

b. Family *Dematiaceae:* Mycelium dark or smoky (black, gray, brown, or olive). Usually conidia also are dark. Mass of mycelium appears dark, but individual hyphae may be very light in color when examined microscopically.

(1) Genus *Cladosporium. C. herbarum* (Figure 1–23) is a leading species. These dark molds cause "black spot" on a number of foods, on cellar walls, etc. Colonies of *C. herbarum* are restricted in growth and are thick, velvety, and olive- to gray-green, and the reverse side of the plant is a striking opalescent blue-black to greenish-black.

Distinguishing characteristics are (*a*) Septate and dark mycelium. (*b*) Spore heads consisting of large, treelike clusters of dark conidia, similar, except for pigment, to those of *Neurospora.* (*c*) Dark, ovate, budding conidia which are one-celled when young and may be two-celled when old.

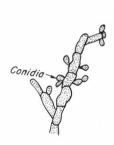

Figure 1–22. *Pullularia.*

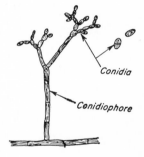

Figure 1–23. *Cladosporium.*

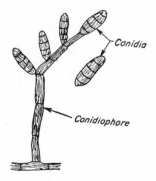

Figure 1–24. *Helminthosporium.*

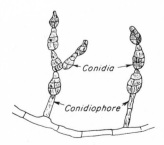

Figure 1–25. *Alternaria.*

(2) Genus *Helminthosporium* (Figure 1–24). Species of this genus are, for the most part, plant pathogens, but may grow saprophytically on vegetable materials.

Distinguishing characteristics are (*a*) Septate and usually dark mycelium. (*b*) Dark, long, wormlike, septate conidia, usually containing four to six cells, and borne singly or in small bunches, sometimes like bunches of bananas, at the end of short conidiophores.

(3) Genus *Alternaria*. Molds of this genus are common causes of the spoilage of foods. *A. citri* (rotting citrus fruits), *A. tenuis*, and *A. brassicae* are common species. The mass of mycelium usually is dirty gray-green, but hyphae often look nearly colorless under the microscope. The brown, many-celled conidia (Figure 1–25) are in a chain on the conidiophore.

Distinguishing characteristics are (*a*) Dirty gray-green, loose, woolly, septate mycelium. (*b*) Large, ovate to inverted club-shaped, greenish-brown to dark-brown, multicellular conidia, with both cross and longitudinal or diagonal walls. (*c*) Conidiophore bearing a chain of conidia, with the blunt end of each conidium toward the mycelium and the opposite end more beaked. Conidiophores branched or unbranched.

(4) Genus *Stemphylium* (Figure 1–26). This, too, is a common genus. The conidia are dark and multicellular but have fewer cross walls than those of *Alternaria* and are rounded at both ends.

Distinguishing characteristics are (*a*) Septate mycelium. (*b*) Very dark, ovate conidia that are multicellular but have few septa, which are both cross and longitudinal. (*c*) Conidia in small clusters on short conidiophores arising from trailing hyphae.

Simple Key to Genera of *Dematiaceae* Described

I. Budding, one-, and two-celled conidia . . . . *Cladosporium*
II. Many-celled conidia
    A. Conidia cylindrical, elongate, transversely septate; produced singly on conidiophore . . . . *Helminthosporium*
    B. Beaked conidia in chain borne on conidiophore; both cross and longitudinal septa in conidia . . . . *Alternaria*
    C. Conidia rounded at both ends; in clusters on conidiophore; both cross and longitudinal septa in conidia . . . . *Stemphylium*

c. Family *Tuberculariaceae:* Short conidiophores in bunches or bundles, forming cushion-shaped aggregates.

(1) Genus *Fusarium*. Molds of this genus often grow on foods. The species are very difficult to identify, and the appearance of growth is variable. The chief distinguishing characteristic is the several-celled, sickle-shaped macroconidium (Figure 1–27) that may be colored, but never is dark. One-celled, ovoid, or oblong

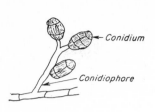

Figure 1—26. *Stemphylium.*

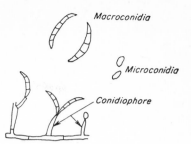

Figure 1—27. *Fusarium.*

microconidia usually also are present; they may be borne singly or in chains.

d. Family *Cryptococcaceae:* Yeastlike fungi or false yeasts; reproduced by budding; mycelium present or lacking. Genera: *Candida, Cryptococcus* (see Chapter 2).

e. Family *Rhodotorulaceae:* Mostly orange- to red-pigmented nonsporing (false) yeasts. Genus: *Rhodotorula.*

2. Order *Melanconiales:* Do not have free conidiophores, but rather a layer of closely compacted conidiophores. Important genera in foods, chiefly fruits, are *Colletotrichum* and *Gleosporium* (causing anthracnoses in plants) and *Pestalozzia* (causing stem-end rots of vegetables and fruits).

3. Order *Sphaeropsidales:* Conidia are in flask-shaped receptacles called **pycnidia.**

a. Genus *Phoma:* Cause of rots in beets, tomatoes. Conidia are one-celled and borne in dark pycnidia.

b. Genus *Diplodia:* Cause of rots, especially stem-end rots, of vegetables and fruits. Conidia are colored and two-celled.

B. Class *Ascomycetes:* Sexual spores are ascospores. Some genera of the *Ascomycetes* have been discussed with the *Fungi Imperfecti* for convenience, for example, *Neurospora* and *Eurotium.* Other genera significant in foods include:

1. Genus *Endomyces:* Yeastlike fungi, forming mycelium and arthrospores. Some species rot fruits.

2. Genus *Monascus:* Colonies of *M. purpureus* are thin and spreading and reddish or purple in color. Found on dairy products and on Chinese red rice (ang-khak).

3. Genus *Sclerotinia.* Some species cause rots of vegetables and fruits, where they are present in the conidial stage. The lemon-shaped conidia are in chains, with a "plug" separating conidia.

In the preceding discussion of molds important in foods it has been deemed advisable to omit mention of genera that only occasionally grow on specific foods. They will be mentioned with those foods. For the descriptions of these genera, reference should be made to text and reference books on fungi listed at the end of the chapter.

REFERENCES

BARNETT, H. L. 1955. Illustrated genera of imperfect fungi. Burgess Publishing Company, Minneapolis.

BESSEY, E. A. 1950. Morphology and taxonomy of fungi. McGraw-Hill Book Company, Blakiston Division, New York.

FOSTER, J. W. 1949. Chemical activities of fungi. Academic Press Inc., New York.

FUNDER, S. 1961. Practical mycology. 2nd rev. ed. Broggers Boktr. Forlag, Oslo, Norway.

GILMAN, J. C. 1945. A manual of soil fungi. Iowa State College Press, Ames, Iowa.

RAPER, K. B., and DOROTHY I. FENNELL. 1965. The genus *Aspergillus*. The Williams & Wilkins Company, Baltimore.

RAPER, K. B., and C. THOM. 1949. A manual of the penicillia. The Williams & Wilkins Company, Baltimore.

SCOTT, W. J. 1957. Water relations of food spoilage microorganisms. Advances Food Res. 7:83-127.

SKINNER, C. E., C. W. EMMONS, and H. M. TSUCHIYA. 1947. Henrici's molds, yeasts, and actinomycetes. 2nd ed. John Wiley & Sons, Inc., New York.

SMITH, G. 1960. An introduction to industrial mycology. 5th ed. Edward Arnold (Publishers) Ltd., London.

WOLF, F. A., and F. T. WOLF. 1947. The fungi. John Wiley & Sons, Inc., New York. 2 vol.

CHAPTER TWO

YEASTS AND YEASTLIKE FUNGI

Yeasts are as difficult to define as molds. In Henrici's *Molds, Yeasts, and Actinomycetes,* yeasts are defined as "true fungi whose usual and dominant growth form is unicellular." Some of the molds in their conidial stage are like budding yeasts, and some yeasts have a mycelial stage. An example of a genus that sometimes is listed with the molds and sometimes with the yeasts is *Geotrichum.*

Yeasts may be useful or harmful in foods. Yeast fermentations are involved in the manufacture of foods like bread, beer, wines, vinegar, and surface-ripened cheese, and yeasts are grown for enzymes and for food. Yeasts are undesirable when they cause spoilage of sauerkraut, fruit juices, sirups, molasses, honey, jellies, meats, wines, beer, and other foods.

Little attempt will be made to describe genera of yeasts so that they can be identified. Instead the general characteristics of yeasts will be discussed, the important genera listed, and yeasts of industrial importance mentioned.

GENERAL CHARACTERISTICS OF YEASTS

As with most plants, yeasts are classified botanically chiefly on their morphological characteristics, although their physiological ones are more important to the food microbiologist.

Morphological characteristics

The morphological characteristics of yeasts are determined by microscopic examination.

FORM AND STRUCTURE. The form of yeasts may be spherical to ovoid, lemon-shaped, pear-shaped, cylindrical (see Figure 2–1), triangular, or even elongated into a false or true mycelium. They also differ in size. Visible parts of the structure are the cell wall, cytoplasm, water vacuoles, fat globules, and granules which may be metachromatic, albuminous, or starchy. Special staining is necessary to demonstrate the nucleus.

REPRODUCTION. Most yeasts reproduce asexually by multilateral or polar budding (Figure 2–1), a process in which some of the protoplasm bulges out the cell wall; the bulge grows in size and finally walls off as a new yeast cell. In some yeasts, notably some of the film yeasts, the bud appears to grow from a tubelike projection from the mother cell. Replicated nuclear material is divided between the mother and daughter cells. A few species of yeasts reproduce by fission, and one by a combination of fission and budding.

Sexual reproduction of "true" yeasts (*Ascomycetes*) results in the production of ascospores, the yeast cell serving as the ascus. The formation of ascospores follows conjugation of two cells in most species of true yeasts, but some may produce ascospores without conjugation, followed by conjugation of ascospores or small daughter cells. The usual number of spores per ascus and the appearance of the ascospores are characteristic of the kind of yeast. The ascospores may differ in color, in smoothness or roughness of their walls, and in their shape (round,

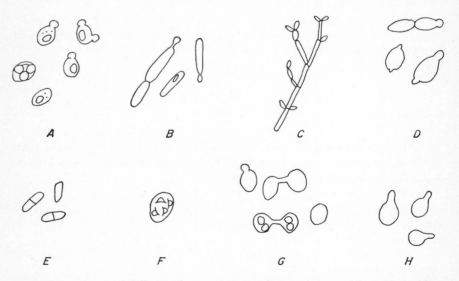

Figure 2–1.   Yeasts of different shapes: (A) *Saccharomyces cerevisiae* with budding cells and one ascus with four ascospores; (B) *Candida* yeast with elongated cells; (C) *Candida* showing pseudomycelium; (D) apiculate (lemon-shaped) yeast; (E) *Schizosaccharomyces*, multiplying by fission; (F) *Hansenula* with ascospores shaped like derby hats; (G) *Zygosaccharomyces* showing conjugation and ascus with four ascospores; (H) flask-shaped yeasts.

oval, reniform, bean- or sickle-shaped, Saturn- or hat-shaped, hemispherical, angular, fusiform, or needle-shaped).

"False" yeasts, which produce no ascospores or other sexual spores, belong to the *Fungi Imperfecti*. Cells of some yeasts become chlamydospores by formation of a thick wall about the cell.

Cultural characteristics

The appearance of massed yeast growth is not, for the most part, useful in the identification of yeasts, although growth as a film on the surface of liquid media suggests an oxidative or film yeast, and production of a carotenoid pigment indicates the genus *Rhodotorula*. However, the appearance of the growth is important when it causes colored spots on foods. It is difficult to tell yeast colonies from bacterial ones on agar plates; the only certain way is by means of microscopic examination of the organisms. Most young yeast colonies are moist and somewhat slimy, but may appear mealy; and most colonies are whitish, but some are cream-colored or pink. Some colonies change little with age, but others become dry and wrinkled.

Yeasts are oxidative, fermentative, or both. The oxidative yeasts may grow as a film or scum on the surface of a liquid and then are termed *film yeasts*. Fermentative yeasts usually grow throughout the liquid.

Physiological characteristics

Although species of yeasts may differ considerably in their physiology, those of industrial importance have enough physiological characteristics in common to permit generalizations, provided that it is kept in mind that there will be exceptions to every statement made.

Most of the commonly encountered yeasts grow best with a plentiful supply of available moisture. But since many yeasts grow in the presence of greater concentrations of solutes, like sugar or salt, than do most bacteria, it can be concluded that these yeasts require less moisture than the majority of bacteria. Most yeasts require more moisture than do molds, however. On the basis of water activity (a_w) supporting growth (see Chapter 1), yeasts may be classified as ordinary if they do not grow in high concentrations of solutes—that is, in a low a_w—and as osmophilic if they do. Lower limits of a_w for ordinary yeasts tested thus far range from 0.88 to 0.94. Specific examples of minimal a_w are 0.94 for a beer yeast, 0.90 for a yeast from condensed milk, and 0.905 for a bakers' yeast. By contrast, osmophilic yeasts have been found growing slowly in media with an a_w as low as 0.62 to 0.65 in sirups, although some osmophilic yeasts are stopped at about 0.78 in both salt brine and sugar sirup. Each yeast has its own characteristic optimal a_w and range of a_w for growth for a given combination of environmental conditions. These a_w values will vary with the nutritive properties of the substrate, pH, temperature, availability of oxygen, and presence or absence of inhibitory substances.

The range of temperature for growth of most yeasts is, in general, similar to that for molds, with the optimum around 25 to 30 C (77 to 86 F) and the maximum about 35 to 47 C (98.6 to 116.6 F). Some kinds can grow at 0 C (32 F) or less. The growth of most yeasts is favored by an acid reaction in the vicinity of pH 4 to 4.5, and they will not grow well in an alkaline medium unless adapted to it. Yeasts grow best under aerobic conditions, but the fermentative types can grow anaerobically, although slowly.

In general, sugars are the best food for energy for yeasts, although oxidative yeasts, e.g., the film yeasts, oxidize organic acids and alcohol. Carbon dioxide produced by bread yeasts accomplishes the leavening of bread; and alcohol made by the fermentative yeasts is the main product in the manufacture of wines, beer, industrial alcohol, and other products. The yeasts also aid in the production of flavors or "bouquet" in wines.

Nitrogenous foods utilized vary from simple compounds like ammonia and urea to amino acids and polypeptides. In addition, yeasts require accessory growth factors.

Yeasts may change in their physiological characteristics, especially the true, or ascospore-forming, yeasts, which have a sexual method of reproduction. These yeasts can be bred for certain characteristics, or may mutate to new forms. Most yeasts can be adapted to conditions which previously would not support good growth. Illustrative of different characteristics within a species is the large number of strains of *Saccharomyces cerevisiae* suited to different uses, e.g., bread strains, beer strains, wine strains, and high-alcohol-producing strains or varieties.

CLASSIFICATION AND IDENTIFICATION OF YEASTS

Yeasts belong to the division *Fungi* and phylum *Eumycetes* (Figure 2–2). True yeasts are in the class *Ascomycetes* (a few are *Basidiomycetes*), and the false, or asporogenous, yeasts are in the class *Fungi Imperfecti*. Further subdivisions are shown in Figure 2–2.

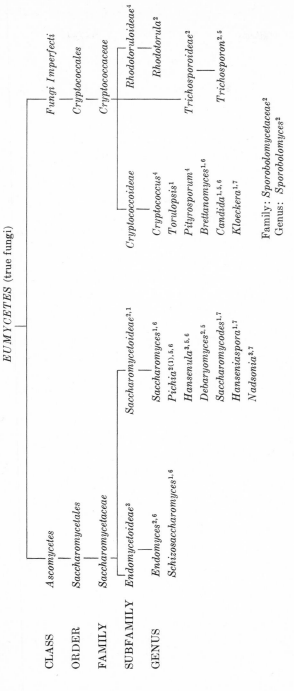

Figure 2–2. Classification of yeasts to include genera found in foods.

EUMYCETES (true fungi)

CLASS	Ascomycetes	Fungi Imperfecti
ORDER	Saccharomycetales	Cryptococcales
FAMILY	Saccharomycetaceae	Cryptococcaceae

SUBFAMILY: Endomycetoideae[3]   Saccharomycetoideae[2,1]   Cryptococcoideae   Rhodotoruloideae[4]

GENUS:
Endomyces[2,6]
Schizosaccharomyces[1,6]

Saccharomyces[1,6]
Pichia[2(1),5,6]
Hansenula[3,5,6]
Debaryomyces[2,5]
Saccharomycodes[1,7]
Hanseniaspora[1,7]
Nadsonia[3,7]

Cryptococcus[4]
Torulopsis[1]
Pityrosporum[4]
Brettanomyces[1,6]
Candida[1,5,6]
Kloeckera[1,7]

Rhodotorula[2]

Trichosporoideae[2]
Trichosporon[2,5]

Family: Sporobolomycetaceae[2]
Genus: Sporobolomyces[2]

[1] Fermentative
[2] Oxidative
[3] Fermentative and oxidative
[4] Nonfermenting

[5] May form pellicle
[6] True or pseudomycelium
[7] Apiculate (lemon-shaped)

The principal bases for the identification and classification of genera of yeasts are as follows:

1. Whether or not ascospores are formed.
2. If they are spore-forming—
   a. The method of production of ascospores:
      (1) Produced without conjugation of yeast cells (parthenogenetically). Spore formation may be followed by (a) conjugation of ascospores or (b) conjugation of small daughter cells.
      (2) Produced after isogamic conjugation (conjugating cells appear similar).
      (3) Produced by heterogamic conjugation (conjugating cells differ in appearance).
   b. Appearance of ascospores: shape, size, and color. Most spores are spheroidal or ovoid, but some have odd shapes, such as those of most species of *Hansenula,* which look like derby hats (Figure 2–1, F).
   c. The usual number of ascospores per ascus: one, two, four, or eight.
3. Appearance of vegetative cells: shape, size, color, inclusions.
4. Method of asexual reproduction:
   a. Budding.
   b. Fission.
   c. Combined budding and fission.
   d. Arthrospores (oidia).
5. Production of a mycelium, pseudomycelium, or no mycelium.
6. Growth as a film over surface of a liquid (film yeasts) or growth throughout medium.
7. Color of macroscopic growth.
8. Physiological characteristics (used primarily to differentiate species or strains within a species):
   a. Nitrogen and carbon sources.
   b. Vitamin requirements.
   c. Oxidative or fermentative: film yeasts are oxidative; other yeasts may be fermentative or fermentative and oxidative.
   d. Lipolysis, urease activity, acid production, or formation of starchlike compounds.

YEASTS OF INDUSTRIAL IMPORTANCE

True yeasts (class *Ascomycetes*)

Most yeasts used industrially are ascomycetes, and most are in the genus *Saccharomyces.* The term "wild yeast" is applied to any yeast other than the one being used or encouraged. Thus a yeast employed in one process could be a wild yeast in another. Most of the troublesome wild yeasts are asporogenous, or false, yeasts.

Genus *Schizosaccharomyces.* These yeasts, which reproduce asexually by fission and form four or eight ascospores per ascus after isogamic conjugation, have been found in tropical fruits, molasses, soil, honey, and elsewhere. A common species is S. *pombe*.

GENUS *Saccharomyces*.   Cells of these yeasts may be round, ovate, or elongated and may form pseudomycelium. Reproduction is by multipolar budding or by ascospore formation which may follow conjugation or may develop from diploid cells when these represent the vegetative stage. The ascospores, one to four per ascus, are usually round or ovate. The leading species, S. *cerevisiae*, is employed in many food industries, with special strains being used for the leavening of bread (Figure 2–3), as top yeasts for ale, for wines, and for the production of alcohol, glycerol, and invertase. **Top yeasts** clump together during growth, collect carbon dioxide, and are buoyed up to the surface of the fermenting liquid, from which they can be skimmed. **Bottom yeasts** do not clump but settle to the bottom of the liquid following the period of growth and activity. S. *cerevisiae* var. *ellipsoideus* is a high-alcohol-yielding variety used to produce industrial alcohol, wines, and distilled liquors. S. *carlsbergensis*, a bottom yeast, is used in making beer. S. *fragilis*

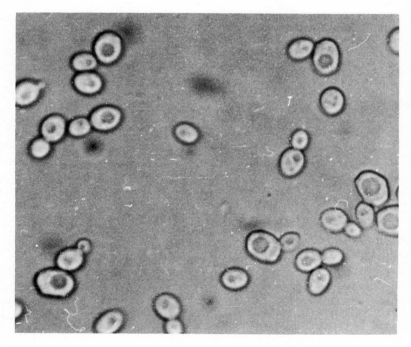

Figure 2–3. Photomicrograph of budding *Saccharomyces cerevisiae* (bread yeasts) (×1,500). (*Courtesy of Universal Foods Corporation.*)

and S. *lactis*, because of their ability to ferment lactose, may be important in milk or milk products. S. *rouxii* and S. *mellis* are osmophilic.

GENUS *Zygosaccharomyces*. Some workers consider this a subgenus
of *Saccharomyces*. These yeasts are notable for their ability to grow
in high concentrations of sugar (hence are termed **osmophilic**), and
are involved in the spoilage of honey, sirups, and molasses and in the
fermentation of soy sauce and some wines. *Z. nussbaumeri* grows in
honey.

GENUS *Pichia*. These oval to cylindrical yeasts may form pseudomycelium. Ascospores are round or hat-shaped, and there are one to
four per ascus. A pellicle is formed on liquids, e.g., *P. membranaefaciens*
grows a pellicle on beers or wines.

GENUS *Hansenula*. These yeasts resemble *Pichia* in appearance
but they are usually more fermentative, although some species form
pellicles. Ascospores are hat- or Saturn-shaped.

GENUS *Debaryomyces*. These round or oval yeasts form pellicles
on meat brines. Ascospores have a warty surface. *D. kloeckeri* grows
on cheese and sausage.

GENUS *Hanseniaspora*. These lemon-shaped (apiculate) yeasts grow
in fruit juices. *Nadsonia* yeasts are large and lemon-shaped.

False yeasts (class *Fungi Imperfecti*)

GENUS *Torulopsis*. These round to oval, fermentative yeasts with
multilateral budding cause trouble in breweries and spoil various foods.
T. sphaerica ferments lactose and may spoil milk products. Other species
can spoil sweetened condensed milk, fruit-juice concentrates, and acid
foods.

GENUS *Candida*. These yeasts form pseudo- or true hyphae, with
abundant budding cells or blastospores, and may form chlamydospores.
Many form films and can spoil foods high in acid and salt. *C. utilis*
is grown for food and feed. *C. krusei* has been grown with dairy starter
cultures to maintain activity and increase longevity of the lactic acid
bacteria. Lipolytic *C. lipolytica* can spoil butter and oleomargarine.

GENUS *Brettanomyces*. These ogive- or arch-shaped yeasts produce high amounts of acid, and are involved in the late fermentation

of Belgian lambic beer and English beers. They also are found in French wines. *B. bruxellansis* and *B. lambicus* are typical species.

GENUS *Kloeckera.* These are imperfect apiculate or lemon-shaped yeasts. *K. apiculata* is common on fruits and flowers and in the soil.

GENUS *Trichosporon.* These yeasts bud and form arthrospores. They grow best at low temperatures and are found in breweries and on chilled beef. *T. pullulans* is a common species.

GENUS *Rhodotorula.* These red, pink, or yellow yeasts may cause discolorations on foods, e.g., colored spots on meats or pink areas in sauerkraut.

Groups of yeasts

Film yeasts (Figure 2–2), in the genera *Pichia, Hansenula, Debaryomyces, Candida,* and *Trichosporon,* grow on the surface of acid products like sauerkraut and pickles (Figure 2–4), oxidize the organic acids, and enable less acid-tolerant organisms to continue the spoilage. *Hansenula* and *Pichia* tolerate high levels of alcohol and may oxidize it in alcoholic

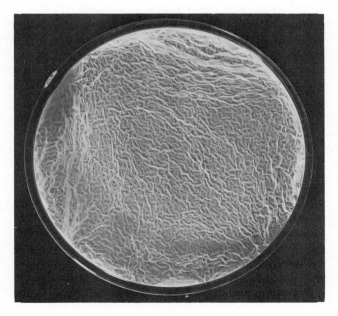

Figure 2–4. Wrinkled growth of typical film yeast from pickle fermentation. (*Courtesy of J. L. Etchells and I. D. Jones, Glass Packer. 30(5):358–360. 1951.*)

beverages. *Pichia* species are encouraged to grow on Jerez and Arbois wine, to which they are supposed to impart distinctive flavors and esters. *Debaryomyces* is very salt-tolerant and can grow on cheese brines with as much as 24 percent salt. The film yeasts produce little or no alcohol from sugars.

Apiculate or *lemon-shaped yeasts,* in *Saccharomycodes, Hansenia-spora, Nadsonia,* and *Kloeckera,* are considered objectionable in wine fermentations because they give off-flavors, low yields of alcohol, and high volatile acid.

Osmophilic yeasts grow well in an environment of high osmotic pressure, i.e., in high concentrations of sugars, salts, or other solutes. It will be recalled that minimal a_w values for growth of ordinary yeasts range from 0.88 to 0.94, whereas some osmophilic yeasts grow slowly in sirups at an a_w as low as 0.62 to 0.65, and others at about 0.78 in both brines and sirups. Most highly sugar-tolerant yeasts have been classified as species of *Zygosaccharomyces* (although now listed by some as *Saccharomyces rouxii* and *S. mellis*), causing spoilage of dry fruits, concentrated fruit juices, honey, maple sirup, and other high sugar solutions.

Salt-tolerant yeasts grow in curing brines, salted meats and fish, soy sauce, miso paste, and tamari sauce. Most salt-tolerant of the film yeasts are species of *Debaryomyces,* which grow on curing brines and on meats and cucumbers in them, as does *Saccharomyces rouxii,* which can grow as a film on brine. Yeasts in various other genera (*Torulopsis, Brettanomyces,* and others) also grow in brines. Yeasts grow in soy sauce with its high content of salt (about 18 percent). *Saccharomyces rouxii* is considered of great importance in production of alcohol and flavor, but species of *Torulopsis, Pichia, Candida,* and *Trichosporon* also may grow. Film-forming *S. rouxii* and *Pichia* are sometimes involved in the spoilage of soy sauce. Similar yeasts are involved in miso production, but kinds will vary as the salt concentration is varied between 7 and 20 percent.

From the foregoing discussion it can be gathered that yeasts often are placed in the following groups that have industrial significance but little relationship to botanical classification: the alcohol yeasts, apiculate yeasts, film yeasts, osmophilic yeasts, food yeasts, lactose-fermenting yeasts, etc.

REFERENCES

Cook, A. H. (*Ed.*) 1958. The chemistry and biology of yeasts. Academic Press Inc., New York.

ETCHELLS, J. L., T. A. BELL, and I. D. JONES. 1953. Morphology and pigmentation of certain yeasts from brines and the cucumber plant. Farlowia 4:265-304.

HENRICI, A. T. 1941. The yeasts: genetics, cytology, variation, classification and identification. Bacteriol. Revs. 5:97–179.

LINDEGREN, C. C. 1949. The yeast cell, its genetics and cytology. Educational Publishers, St. Louis.

LODDER, J., and N. J. W. KREGER-VAN RIJ. 1952. The yeasts, a taxonomic study. Interscience Publishers (Division of John Wiley & Sons, Inc.), New York.

LODDER, J., and N. J. W. KREGER-VAN RIJ. 1954, 1955. Classification and identification of yeasts. Lab. Prac. 3:483–490; 4:20–24, 53–57.

MRAK, E. M., and H. J. PHAFF. 1948. Yeasts. Annu. Rev. Microbiol. 2:1–46.

ONISHI, H. 1963. Osmophilic yeasts. Advances Food Res. 12:53–94.

SCOTT, W. J. 1957. Water relations of food spoilage microorganisms. Advances Food Res. 7:83–127.

SKINNER, C. E., C. W. EMMONS, and H. M. TSUCHIYA. 1947. Henrici's molds, yeasts, and actinomycetes. 2nd ed. John Wiley & Sons, Inc., New York.

SMITH, G. 1960. An introduction to industrial mycology. 5th ed. Edward Arnold (Publishers) Ltd., London.

CHAPTER THREE

BACTERIA

Determinative bacteriology (classification and identification of bacteria) is beyond the scope of this book and should be reviewed in textbooks on the subject. However, some of the morphological, cultural, and physiological characteristics of bacteria will be mentioned briefly because of their relationship to the preservation or spoilage of foods.

MORPHOLOGICAL CHARACTERISTICS IMPORTANT IN FOOD BACTERIOLOGY

One of the first steps in the identification of bacteria in a food would be microscopic examination to ascertain the shape, size, aggregation, structure, and staining reactions of the bacteria present. Characteristics of special significance might be:

Encapsulation

The presence of capsules (Figure 3–1) or slime could account for sliminess or ropiness of a food. In addition, capsules serve to increase the resistance of bacteria to adverse conditions such as heat or chemicals. Capsule-forming bacteria may produce much slimy material under some conditions and little or none under others. The factors that influence growth, to be discussed in following paragraphs, will influence slime production.

Formation of endospores

Bacteria of the genera *Bacillus* and *Clostridium* form endospores (Figure 3–2), but other rod forms do not, nor do the cocci encountered in foods. Spores of different bacterial species or even different strains vary widely in their resistance to heat and other adverse conditions. In general, however, the bacterial spores are considerably more resistant to heat, chemicals, and other destructive agencies than are the vegetative cells.

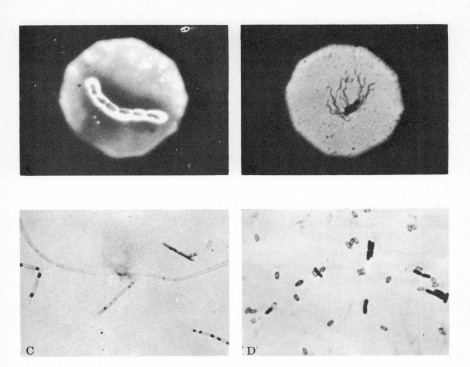

Figure 3–1. Bacterial structures: (*A*) capsules; (*B*) flagella; (*C*) granules in *Lactobacillus bulgaricus;* (*D*) spores of *Bacillus subtilis.* (*Courtesy of P. R. Elliker.*)

According to some, sporulation takes place in mature cells in the late logarithmic phase when nutrients become depleted and products accumulate. It is induced by special chemical compounds that bring about an increase in DNA and then cause spore formation, and is favored by a narrow range of pH, presence of oxygen for aerobes and its absence for anaerobes, a narrower temperature range than for growth, the presence of certain metallic ions, especially Mn^{++}, the absence of inhibitors such as fatty acids, and a good supply of glucose and available nitrogen. During the process, cell protein is converted to spore protein, and special enzymes are formed. Complexes are also formed, such as dipicolinic acid (DPA), glucosamine, and muramic acid.

Germination is favored, in general, by conditions that are favorable to growth of the vegetative cells, but it may take place under conditions that do not permit such growth, e.g., at low temperatures. It is triggered by mixtures of amino acids, e.g., L-alanine, adenosine, L-cysteine, or L-valine, by Mg^{++} and Mn^{++} ions, by glucose, by dipicolinic acid plus Ca^{++} ions, and by heat shocking or heat activation which activates dormant enzymes. The optimal temperature and time for heat shocking depends upon the kind of spore, the heat-treatment being greater for

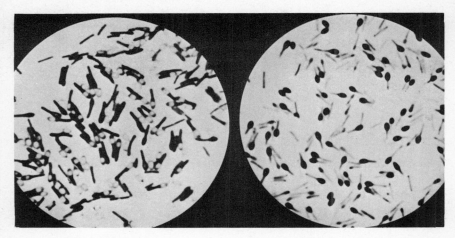

Figure 3–2. Photomicrograph showing terminal endospores in a rod. The spores are clear in the left-hand field and stained in the right-hand one (×1,330) (*From J. Nowak.*)

spores of thermophiles, for example, than for those of mesophiles. Germination is inhibited by sorbic acid at an acid pH, by some divalent cations, by starch, and by oleic and linoleic acids.

"Dormancy" of spores has been defined as delayed germination (and outgrowth) under conditions favorable for it. The spores fail to germinate, however, probably because of unfavorable conditions such as inhibitors in the environment or lack of essential nutrients, e.g., amino acids. Some spores may germinate but fail to grow out, and some may have been damaged by heat, rays, or other agent, so that they need a more complex or specialized medium for growth than did their ancestors. Delayed germination of spores from a few days to many months has been reported, e.g., "dormancy" for from a few days to 3 or 4 months with spores of *Bacillus megaterium* and for 15 days to 72 months with spores of *Clostridium botulinum*.

Formation of cell aggregates

It is characteristic of some bacteria to form long chains and of others to clump under certain conditions. It is more difficult to kill all bacteria in intertwined chains or sizable clumps than to destroy separate cells.

CULTURAL CHARACTERISTICS IMPORTANT
IN FOOD BACTERIOLOGY

Bacterial growth in and on foods often is extensive enough to make the food unattractive in appearance or otherwise objectionable.

Pigmented bacteria cause discolorations on the surfaces of foods; films may cover the surfaces of liquids; growth may make surfaces slimy; or growth throughout the liquids may result in undesirable cloudiness or sediment.

PHYSIOLOGICAL CHARACTERISTICS IMPORTANT IN FOOD BACTERIOLOGY

The bacteriologist is concerned with the growth and activity of bacteria (and other microorganisms) in foods and with the accompanying chemical changes. These changes include hydrolyses of complex carbohydrates to simple ones, of proteins to polypeptides, amino acids, and ammonia or amines, and of fats to glycerol and fatty acids. Oxidation-reduction reactions which are utilized by the bacteria to obtain energy from foods (carbohydrates, other carbon compounds, simple nitrogen-carbon compounds, etc.) yield products such as organic acids, alcohols, aldehydes, ketones, and gases. These changes will be discussed in more detail in chapters to follow. A knowledge of the factors that favor or inhibit the growth and activity of bacteria is essential to an understanding of the principles of food preservation and spoilage.

Factors influencing bacterial growth

The chief environmental factors that influence the growth of bacteria are food, moisture, temperature, hydrogen-ion concentration, oxidation-reduction potential, and presence of inhibitory substances. While each of these factors is important, it is the combination of all of them that determines which microorganism will grow and how rapidly it will grow, and what changes will be produced and at what rate.

Food. Each kind of bacterium (or other microorganism) has a definite range of food requirements. For some species that range is wide, and growth takes place in a variety of substrates, as is true for coliform bacteria; but for others, e.g., many of the pathogens, the range is narrow, and the organisms can grow in only a limited number of kinds of substrates. Thus, bacteria differ in the foods that they can utilize for energy: some can use a variety of carbohydrates, e.g., the coliform bacteria and *Clostridium* species, and others only one or two (many *Pseudomonas* species), while some can use other carbon compounds like organic acids and their salts, alcohols, and esters (*Pseudomonas* species). Some can hydrolyze complex carbohydrates, although others cannot. Likewise, the nitrogen requirements of bacteria such as *Pseudomonas* species may be satisfied by simple compounds such as ammonia or nitrates; or more complex compounds like amino acids, peptides, or proteins may be utilized or even required, as is true for the lactics. Bacteria also vary in their need for vitamins or accessory growth factors; some (*Staphylococcus aureus*) synthesize part and others (*Pseudomonas* or *Escherichia coli*) all of the factors needed, and still others must have them all furnished (the lactics and many pathogens). It should be emphasized that, in general, the better the medium is for an organism, the wider will be the ranges of temperature, pH, and a_w over which growth can take place.

MOISTURE. In general, bacteria require more available moisture than do yeasts or molds. It is the amount of *available moisture* and not the total moisture which determines the lower limit of moisture for growth.

Most bacteria grow well in a medium with a water activity, or a_w (see Chapter 1), approaching 1.00 (at 0.995 to 0.998, for example); that is, they grow best in low concentrations of sugar or salt, although there are notable exceptions that will be mentioned later. Culture media for most bacteria contain not more than 1 percent of sugar and 0.85 percent of sodium chloride ("physiological salt solution"); as little as 3 to 4 percent sugar and 1 to 2 percent salt may inhibit some bacteria. The optimal a_w and the lower limit for growth vary with the bacterium, as well as with food, temperature, pH, and the presence of oxygen, carbon dioxide, and inhibitors, and are lower for bacteria able to grow in high concentrations of sugar or salt. Examples of reported lower limits of a_w for growth of some food bacteria are 0.97 for *Pseudomonas*, 0.96 for *Achromobacter*, 0.96 for *Escherichia coli*, 0.95 for *Bacillus subtilis*, 0.945 for *Aerobacter aerogenes*, 0.86 for *Staphylococcus aureus*, and 0.95 for *Clostridium botulinum*. Other bacteria will grow with the a_w below 0.90. These figures would be different under other conditions of growth than those used in obtaining the values. Some optimal a_w figures reported for food bacteria are 0.99 to 0.995 for *Staphylococcus aureus* and *Salmonella* spp., 0.995 for *Escherichia coli*, and 0.982 for *Streptococcus faecalis*.

TEMPERATURE. Each bacterium (or yeast or mold) has an **optimal temperature**, at which it grows best, a **minimal temperature**, which is the lowest one at which growth will occur, and a **maximal temperature**, which is the highest one for growth. The term **psychrophilic (cryophilic)** will be applied here to bacteria which grow well at refrigerator temperatures, although actually this term should imply a low optimal temperature. Psychrophiles grow at a relatively rapid rate at 0 C (32 F), according to one definition. The term **psychrotrophic** has been applied to bacteria able to grow at refrigerator temperatures (below 10 C or 50 F). Most of these organisms would be considered mesophilic on the basis of their optimal temperatures. Some bacteria have been reported able to grow at temperatures as low as —5 to —7 C (23 to 19.4 F). Bacteria with an optimal temperature between 20 and 45 C (68 and 113 F) are **mesophilic,** and those with an optimum above 45 C (usually 55 C or 131 F or higher) are termed **thermophilic.** Bacteria may be **obligate or facultative thermophiles.** Small differences in the temperature at which a food is held may result in the growth of entirely different microorganisms and hence in different changes in the food.

HYDROGEN-ION CONCENTRATION.   Hydrogen-ion concentration, usually expressed as pH, often determines the kind of bacterium to grow in a food and the changes produced. Each organism has its own optimal, minimal, and maximal pH for growth. Most bacteria grow best at a pH near neutrality, but some are favored by an acid reaction, and a few kinds can grow in fairly acid or alkaline media.

OXIDATION-REDUCTION POTENTIAL.   On the basis of their processes of respiration, bacteria are classified as **aerobic** if they require uncombined oxygen, **anaerobic** if they do not require free oxygen and grow better in its absence, and **facultative** if they grow either with or without free oxygen. **Microaerophilic** bacteria require a definite but small amount of free oxygen. Oxidizing or reducing substances in a medium will tend to poise it at a level favorable for aerobic or anaerobic bacteria, respectively. Thus KNO_3 in cured meat would poise it at a more oxidized level than that of meat without nitrate; but if the nitrate were reduced to nitrite, the meat would be more reduced, i.e., at a lower oxidation-reduction potential. Exposure of a food to free oxygen (air) results in a rise in the oxidation-reduction potential at the surface and will affect the interior at a rate that will depend upon the rate of penetration of the oxygen and the poising power of the food.

INHIBITORY SUBSTANCES.   Products produced by bacteria during growth will, in time, slow down or stop that growth, and may be inhibitory to the multiplication of other organisms. Natural foods may contain compounds that inhibit some organisms more than others, for example, the benzoic acid in cranberries. Inhibitory substances added during the processing of foods may check the growth of most microorganisms or at least of undesirable ones, for instance, propionates added to bread to inhibit molds and rope bacteria.

BACTERIOPHAGES

Bacteriophages are viruses that cause disease of bacteria, resulting in disintegration of the cells. To date, most food industries have not had trouble with phage development, the exception being the dairy industry, where phages may attack some of the starter bacteria used to produce lactic acid during the manufacture of cheese and cultured buttermilk. Phage has been encountered in some of the fermentation industries, e.g., in the production of streptomycin and of acetone and butyl alcohol.

GENERA OF BACTERIA IMPORTANT IN FOOD BACTERIOLOGY

The discussion to follow will emphasize the characteristics of genera of bacteria that make them important in foods and will give less attention to the characteristics used in their classification and identification. The classification given in "Bergey's Manual of Determinative Bacteriology," seventh edition, 1957, will be followed.

All food bacteria are included in the class *Schizomycetes* and most of them in the orders *Pseudomonadales* and *Eubacteriales*. Of the "higher bacteria," some in the orders *Actinomycetales* and *Chlamydobacteriales* will be mentioned. Most of the bacteria of importance in foods are in the following families: *Pseudomonadaceae, Spirillaceae, Achromobacteraceae, Enterobacteriaceae, Micrococcaceae, Brevibacteriaceae, Lactobacillaceae, Propionibacteriaceae, Corynebacteriaceae* and *Bacillaceae*. Another family, *Brucellaceae*, contains some of the human pathogens that may contaminate foods.

Family *Pseudomonadaceae*

This family includes the genus *Pseudomonas*, species of which are important spoilage bacteria, the genus *Acetobacter*, containing the vinegar bacteria, the genus *Photobacterium*, of luminescent bacteria, and the genus *Halobacterium*, of obligate halophiles.

Pseudomonas.   A number of species of *Pseudomonas* (Figure 3–3) can cause food spoilage. These bacteria are Gram-negative, usually motile, asporogenous rods. The motile forms have one flagellum or a group of flagella at one end or both ends of the rod, distinguishing them from motile *Achromobacter* species, in which the flagella are scattered over the surface of the cell. It is probable that spoilage of foods

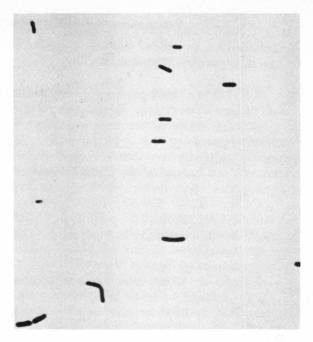

Figure 3–3. Photomicrograph of *Pseudomonas fluorescens* (×2,000).

often has been attributed wrongly to *Achromobacter* rather than to *Pseudomonas* because the type of flagellation has not been ascertained.

Characteristics of some of the *Pseudomonas* species that make them important in foods are (1) their ability to utilize a large variety of noncarbohydrate carbon compounds for energy and their inability to use most carbohydrates; (2) their ability to produce a variety of products that affect flavor deleteriously; (3) their ability to use simple nitrogenous foods; (4) their ability to synthesize their own growth factors or vitamins; (5) the proteolytic and lipolytic activity of some species; (6) their aerobic tendencies, enabling them to grow rapidly and produce oxidized products and slime at the surfaces of foods, where heavy contamination is most likely; (7) their ability to grow well at low (refrigeration) temperatures; and (8) pigment production by some species, such as the greenish fluorescence by pyoverdin of *Pseudomonas fluorescens*, and white, cream-colored, reddish, brown, or even black (*P. nigrifaciens*) colors of other species.

On the other hand, the pseudomonads are limited by a fairly high a_w (0.97 to 0.98), are readily killed by heat and rays, grow poorly if oxygen is not readily available, are not especially resistant to drying, and grow poorly or not at all above 42 C.

The pseudomonads often are added to foods by water, soil, utensils, and equipment, and their presence is undesirable.

Acetobacter. These acetic acid–producing bacteria oxidize ethyl alcohol to acetic acid, and other organic compounds to other oxidation products. Some species can oxidize acetic acid and others cannot. They are motile or nonmotile, Gram-negative to Gram-variable, asporogenous, aerobic rods, usually catalase-positive. Some workers maintain that the acetic acid bacteria do not belong in the *Pseudomonadaceae* and can be divided into two genera: *Acetobacter,* which contains organisms that oxidize ethanol (to $CO_2 + H_2O$), lactate (to carbonate) and various amino acids, use the citric acid cycle, have peritrichous flagella if motile, and have a lactaphilic nutrition; and *Acetomonas,* which contains organisms that do not oxidize ethanol, lactate, or amino acids, do not use the citric acid cycle, have polar flagella if motile, and are glycophilic. Characteristics that make the acetic acid bacteria important are (1) their ability to oxidize ethanol to acetic acid, making them useful in vinegar manufacture and harmful in alcoholic beverages; (2) their strong oxidizing power, which may result in the oxidation of the desired product, acetic acid, by undesirable species or by desirable species under unfavorable conditions; this oxidizing power may be useful, as in the oxidation of D-sorbitol to L-sorbose in the preparation of ascorbic acid by synthetic methods (Chapter 25); and (3) excessive sliminess of some species, e.g., *Acetobacter xylinum,* that clog vinegar generators.

Photobacterium. The genus includes coccobacilli and rods that are luminescent. *P. phosphoreum,* for example, has been known to cause phosphorescence of meats and fish. They are not widespread.

Halobacterium. Rod-shaped bacteria of this genus (for example, *H. salinarium*) are obligate halophiles and are usually chromogenic. They may grow and cause discolorations on foods high in salt, such as salt fish.

Family *Spirillaceae*

The only genus of this family encountered to any extent in foods is *Vibrio,* which has been found in salt and in fish and meat brines.

Family *Achromobacteraceae*

Three genera of this family, *Alcaligenes, Achromobacter,* and *Flavobacterium,* enter foods and may grow there. These bacteria are small- to medium-sized rods that are Gram-negative and aerobic to facultative.

If motile, the cells have flagella scattered over their surface. Carbohydrates usually are not attacked, but if so, the action of the bacteria is feeble. A number of species are psychrophilic.

Alcaligenes. As the name suggests, an alkaline reaction usually is produced in the medium of growth. *A. viscolactis* (*viscosus*) causes ropiness of milk, and *A. metalcaligenes* gives a slimy growth on cottage cheese. These organisms come from manure, feeds, soil, water, and dust.

Achromobacter. These unpigmented water or soil bacteria often are confused with *Pseudomonas,* as has been previously indicated. It seems probable that the pseudomonads are more often the cause of spoilage of meats, fish, poultry, and eggs than *Achromobacter,* but the latter probably is involved in the production of slimy growth on foods at times.

Flavobacterium. The yellow- to orange-pigmented species of this genus may cause discolorations on the surface of meats and be involved in the spoilage of shellfish, poultry, eggs, butter, and milk. Some of these organisms are psychrotrophic and have been found growing on thawing vegetables. These organisms tend to clump.

Bacteria of these three genera are, for the most part, unable to grow at much above 37 C, have a fairly high limiting a_w, are readily killed by heat, and grow best aerobically.

Family *Enterobacteriaceae*

In this family the genera *Escherichia, Aerobacter, Klebsiella, Paracolobactrum, Erwinia, Serratia, Proteus, Salmonella,* and *Shigella* are of interest. The family includes Gram-negative, asporogenous, rod-shaped bacteria that grow well on and in artificial media. The first four genera are important saprophytes in foods, and the last three genera consist chiefly of pathogens.

Coliform bacteria. Bacteria in the genera *Escherichia, Aerobacter* (*Klebsiella, Enterobacter*), and *Paracolobactrum* are included in the coliform or coli-aerogenes group, and the organisms are termed the coliform bacteria or coliforms. They are short rods that are defined in standard methods for the examination of both water and milk as "all aerobic and facultative anaerobic, Gram-negative, non-spore-forming bacteria which ferment lactose with gas formation." The leading species of coliform bacteria are *Escherichia coli* (Figure 3–4) and *Aerobacter aerogenes* (or *Klebsiella aerogenes* for the nonmotile form and *Enterobacter aerogenes* for the motile). Because *E. coli* is considered to be primarily

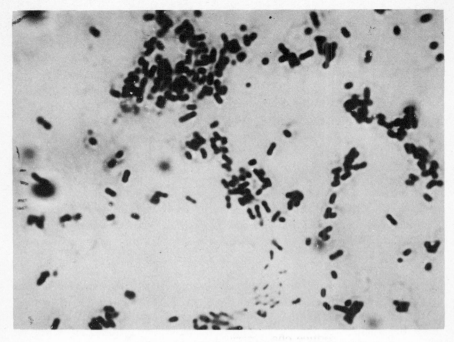

Figure 3–4. Photomicrograph of *Escherichia coli* (×2,000).

of intestinal origin, whereas *A. aerogenes* is commonly of plant origin (although occasionally from the intestines), methods for distinguishing between the two species have been developed. A typical *E. coli* produces more acid in a glucose broth as indicated by methyl red indicator, forms indole but not acetoin (acetylmethylcarbinol), yields carbon dioxide and hydrogen in the ratio of 1:1, and cannot utilize citrate as a sole source of carbon. *A. aerogenes* produces less acid, forms acetoin but not indole, yields carbon dioxide and hydrogen in the ratio of about 2:1, and utilizes citrate as a sole source of carbon. It also usually produces more gas than *E. coli* and therefore is the more dangerous gas-producing organism in cheese, milk, and other foods. Both species ferment sugars to yield lactic acid (more from *E. coli*), ethanol, acetic and succinic acids, carbon dioxide, and hydrogen. A number of coliform bacteria are intermediate between *E. coli* and *A. aerogenes* in their characteristics. Some that ferment lactose slowly or not at all are in the genus *Paracolobactrum*.

The coliform bacteria are, in general, undesirable in foods, for their presence in some foods—water and oysters for example—is considered to be indicative of sewage contamination and hence of the possible presence of enteric pathogens; and growth in foods results in their spoilage.

Some of the characteristics that make the coliform bacteria important in food spoilage are (1) their ability to grow well in a variety of substrates and to utilize a number of carbohydrates and some other organic compounds as food for energy and a number of fairly simple nitrogenous compounds as a source of nitrogen; (2) their ability to synthesize most of the necessary vitamins; (3) the ability of the group to grow well over a fairly wide range of temperatures, from below 10 C (50 F) to about 46 C (114.8 F); (4) their ability to produce considerable amounts of acid and gas from sugars; (5) their ability to cause off-flavors, often described as "unclean" or "barny"; and (6) the ability of A. aerogenes to cause sliminess or ropiness of foods.

Erwinia. The species of this genus are plant pathogens that cause necrosis, galls, wilts, or soft rots in plants and therefore damage the plants and vegetable or fruit products from them. *E. carotovora* will be mentioned as the cause of a market disease called bacterial soft rot (Chapter 15). The organisms are motile, Gram-negative rods.

Serratia. Bacteria of this genus are small, aerobic, Gram-negative, motile rods that produce characteristic red pigments and therefore cause red discolorations on the surfaces of foods. *S. marcescens* is the species most commonly encountered.

Proteus. Bacteria of this genus are straight, Gram-negative, motile, mesophilic rods. They have been found to be involved in the spoilage of meats, seafood, and eggs, and sometimes give a putrefactive odor. The presence of these bacteria in large numbers in unrefrigerated foods has made them suspects as causes of food poisoning. Acid and gas are produced from sugars. *P. vulgaris* is a common species.

Salmonella. Species of these enteric pathogens (e.g., S. *enteritidis*) may grow in foods and cause food infections, or may be commonly only transported by foods.

Shigella. Species of *Shigella*, causing bacillary dysenteries, may be transported by foods.

Family *Brucellaceae*

This family is mentioned only because some of the pathogenic bacteria in it may be transmitted through foods: *Pasteurella tularensis* from squirrels or rabbits can cause human tularemia, or *Brucella* species from food animals can cause brucellosis.

Family *Micrococcaceae*

The important genera in this family are *Micrococcus* and *Staphylococcus,* although *Sarcina* may be encountered occasionally in foods.

Micrococcus. The micrococci are spherical cells arranged in irregular masses but never in packets. Most of the species prominent in foods are Gram-positive, aerobic, and catalase-positive. They have an optimal temperature around 25 to 30 C (77 to 86 F) and grow well on ordinary laboratory culture media. Otherwise, it is difficult to generalize about their characteristics, which may differ considerably from species to species. Characteristics that make various groups of micrococci important in foods are (1) some species can utilize ammonium salts or other simple nitrogenous compounds as sole source of nitrogen; (2) most species can ferment sugars with the production of moderate amounts of acid; (3) some are acid-proteolytic (*M. freudenreichii*) (Figure 3–5); (4) some are very salt-tolerant and hence able to grow at relatively low levels of available moisture; these grow in meat-curing brines, brine tanks, etc.; (5) many are thermoduric, that is, survive the pasteurization treatment given market milk (*M. varians*); (6) some

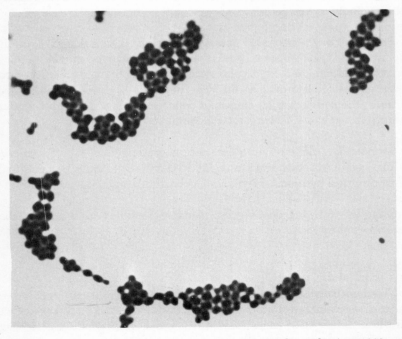

Figure 3–5.   Photomicrograph of *Micrococcus freudenreichii* (×1,000).

are pigmented and discolor the surfaces of foods on which they grow;
M. *flavus* is yellow, for example, and M. *roseus* is pink; and (7) some
of the micrococci can grow fairly well at temperatures around 10 C
(50 F) or below.

Micrococci are widespread in occurrence in nature but have been
isolated most often from dust and water. They often are found on inadequately cleaned and sanitized food utensils and equipment.

Staphylococcus. The Gram-positive staphylococci grow singly, in
pairs, in tetrads, or in irregular, grapelike clusters. The most important
species, *S. aureus*, usually gives yellow to orange growth, although it
may be white on occasion. The species requires an organic source of
nitrogen and is facultative in oxygen requirements. Many of the beta
hemolytic coagulase-positive strains are pathogenic, and some produce
an enterotoxin to cause food poisoning. *S. epidermidis* is considered
parasitic and *S. lactis* saprophytic.

Family *Brevibacteriaceae*

The genus of this family of chief interest is *Brevibacterium,* which
includes a number of species formerly placed in the genus *Bacterium.*
Bacteria of the genus are asporogenous, Gram-positive, motile or nonmotile rods, some distinctly rod-shaped and others almost coccoid. This
is a heterogenous group of bacteria which differ in their ability to attack
carbohydrates, their oxygen requirements, etc. Two pigmented species,
Brevibacterium linens and *B. erythrogenes,* may be important in the
surface smears of certain cheeses, e.g., brick or Limburger, where they
give orange to red pigmentation and aid in ripening.

Family *Lactobacillaceae*

The very important food bacteria in this family are called the lactic
acid bacteria, or "lactics." Members of the family are long or short rods
or cocci that divide like rods in one plane only, and are nonmotile,
microaerophilic, Gram-positive, and, for the most part, catalase-negative.
They give poor growth on most ordinary laboratory culture media, with
very little surface growth on any medium. They need complex foods:
various vitamins, an array of amino acids or certain peptides for nitrogenous food, and a fermentable carbohydrate for energy (most species).
They ferment sugar to chiefly lactic acid, if they are **homofermentative,**
plus small amounts of acetic acid, carbon dioxide, and trace products;
or if they are **heterofermentative,** they produce appreciable amounts
of volatile products, including alcohol, in addition to lactic acid. The

optical type of lactic acid produced is characteristic of the organism and the medium.

The most important characteristic of the lactic acid bacteria is their ability to ferment sugars to lactic acid. This may be desirable in making products such as sauerkraut or cheese but undesirable in the spoilage of wines. Because they form acid rapidly and commonly in considerable amounts, they usually eliminate for the time being much of the competition from other microorganisms.

The cocci in this family are in the tribe *Streptococceae,* and the rods in *Lactobacilleae.* The three genera of lactic cocci important in foods are *Streptococcus, Pediococcus,* and *Leuconostoc.*

Streptococcus. The cocci in this genus are characteristically in pairs, in short chains, or in long chains, depending upon the species and the conditions of growth, and all are homofermentative. The streptococci may be classified serologically by a precipitin reaction into Lancefield groups with capital-letter designations (A, B, C, D, etc.), but ordinarily the streptococci important in foods are divided into four groups, the pyogenic, viridans, lactic, and enterococcus groups.

The **pyogenic** group (pus-producing) includes species of pathogenic streptococci, of which S. *agalactiae,* a cause of mastitis in cows, and S. *pyogenes,* a cause of human septic sore throat, scarlet fever, and other diseases, are representatives that have been encountered in raw milk. The pyogenic streptococci cannot grow at 10 C (50 F) or 45 C (113 F).

The **viridans** group includes S. *thermophilus,* a coccus important in cheeses made by cooking the curd at high temperatures and in certain fermented milks such as yoghurt, and S. *bovis,* which comes from cow manure and saliva and, like S. *thermophilus,* is thermoduric and therefore is counted in the plating of pasteurized milk. These species can grow at 45 C (113 F) but not at 10 C (50 F).

The **lactic** group contains the important dairy bacteria, S. *lactis* (Figure 3–6) and S. *cremoris,* which can grow at 10 C (50 F) but not at 45 C (113 F). These bacteria are used as starters for cheese, cultured buttermilk, and some types of butter, along with *Leuconostoc* species, and S. *lactis* often is concerned in the souring of raw milk. These lactic bacteria tolerate no more than 2 to 4 percent salt and therefore are not concerned in the lactic fermentation of brined vegetables. Some sources of the lactics are green plants, feeds, silage, and utensils.

The **enterococcus** group consists of S. *faecalis,* S. *faecium,* and S. *durans.* S. *faecalis* and S. *faecium* resemble each other, but can be distinguished by physiological tests. S. *faecalis* is usually the more heat-resistant and comes more from human sources, whereas S. *faecium* has

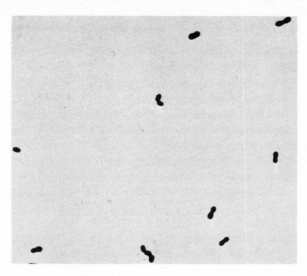

Figure 3–6. Photomicrograph of *Streptococcus lactis* (×1,000).

been reported to be more from plant sources. *S. faecalis* var. *liquefaciens* is an acid-proteolytic variety of *S. faecalis*, and *S. faecalis* var. *zymogenes* is a beta hemolytic variety. The two varieties formerly were called *S. liquefaciens* and *S. zymogenes*, respectively. *S. faecalis* and *S. faecium* are more commonly encountered in foods than *S. durans*. Bacteria of this group can grow both at 10 C (50 F) and at 45 C (113 F). The enterococci have several characteristics in common that make them unusual streptococci: (1) they are thermoduric, readily surviving the pasteurization treatment of milk or even more heating; (2) they tolerate 6.5 percent and more of salt; (3) they can grow at the alkaline pH of 9.6; and (4) they can grow over a wide range of temperatures, some multiplying at as low as 5 to 8 C (41 to 46.4 F) and most of them at as high as 48 to 50 C (118.4 to 122 F). *S. faecalis* and the closely related *S. faecium* sometimes survive the heat-treatment given canned hams and are potential spoilage organisms. *S. faecalis* also has been found growing on bacon. As indicated by the name, the enterococci come from the intestinal tracts of man and animals and are sometimes used as indicator organisms of fecal contamination of foods. They also can survive in dairy products and may contaminate utensils and equipment.

Pediococcus. Only one accepted species has been encountered in foods to any extent, *P. cerevisiae.* The cocci occur singly, in pairs or short chains, or in tetrads (division in two planes), and are Gram-positive, catalase-negative, and microaerophilic. They are homofermentative, fermenting sugars to yield 0.5 to 0.9 percent acid, mostly lactic, and they grow fairly well in salt brines up to 5.5 percent and poorly in concentrations of salt up to about 10 percent. Their range of temperatures for growth is from about 7 C to 45 C (44.6 to 113 F), but 25 to 32 C (77 to 89.6 F) is best. The characteristics that make the organism important in foods have been mentioned: salt tolerance, acid production, and temperature range, especially the ability to grow at cool temperatures. Pediococci have been found growing during the fermentation of brined vegetables and have been found responsible for the spoilage of alcoholic beverages, e.g., beer, where their production of diacetyl is undesirable.

Leuconostoc. This genus, called *Betacoccus* by Orla-Jensen, contains the heterofermentative lactic streptococci, which ferment sugar to lactic acid plus considerable amounts of acetic acid, ethyl alcohol, and carbon dioxide. The ability of two species, *L. dextranicum* and *L. citrovorum,* to ferment citric acid of milk and produce the desirable flavoring substance, diacetyl (the more reduced acetoin and 2,3-butanediol also are produced), and to stimulate the lactic streptococci has led to their inclusion in the so-called "lactic starter" for buttermilk, butter, and cheese. *Streptococcus diacetilactis* is similar to *L. dextranicum,* but produces more gas.

Some of the characteristics of *Leuconostoc* species that make them important in foods are their: (1) production of diacetyl and other flavoring products; (2) tolerance of salt concentrations such as are present in sauerkraut and dill-pickle fermentations, permitting *L. mesenteroides* to carry on the first part of the lactic fermentation; (3) ability to initiate fermentation in vegetable products more rapidly than other lactics or other competing bacteria and to produce enough acid to inhibit nonlactics; (4) tolerance of high sugar concentrations—up to 55 to 60 percent for *L. mesenteroides*—permitting the organism to grow in sirups, liquid cake and ice-cream mixes, etc.; (5) production of considerable amounts of carbon dioxide gas from sugars, leading to undesirable "openness" in some cheeses, to spoilage of foods high in sugars (sirups, mixes, etc.), and to leavening in some breads; (6) heavy slime production in media containing sucrose. This is a desirable characteristic for the production of dextran (Chapter 25) but a hazard in materials high in sucrose, as in the production of sucrose from sugar cane or beets.

The habitat of this genus is the surface of plants.

Lactobacillus. The rod-shaped lactic acid bacteria of food in the tribe *Lactobacilleae* are in the genus *Lactobacillus.* The lactobacilli are rods that usually are long and slender, and form chains in most species

(Figure 3–7). They are microaerophilic, catalase-negative, and Gram-positive, and ferment sugars to yield lactic acid as the main product. Some species can produce diacetyl. It is convenient to divide the lactobacilli into the homofermentative and heterofermentative species or to divide them into those that grow best at 37 to 45 C (98.6 to 113 F) or above and those whose optimal temperature is about 28 to 32 C (82.4 to 89.6 F). The homofermentative lactobacilli with optimal temperatures of 37 C (98.6 F) or above include *L. caucasicus, L. bulgaricus, L. helveticus, L. lactis, L. acidophilus, L. thermophilus,* and *L. delbrueckii. L. fermenti* is the chief example of a heterofermentative lactobacillus growing well at the higher temperatures. The homofermentative lactobacilli with lower optimal temperatures (about 30 C, or 86 F) include *L. casei, L. plantarum,* and *L. leichmannii,* and heterofermentative species in the subgenus *Saccharobacillus* are *L. brevis, L. buchneri, L. pastorianus, L. hilgardii,* and *L. trichodes.* All of the above species

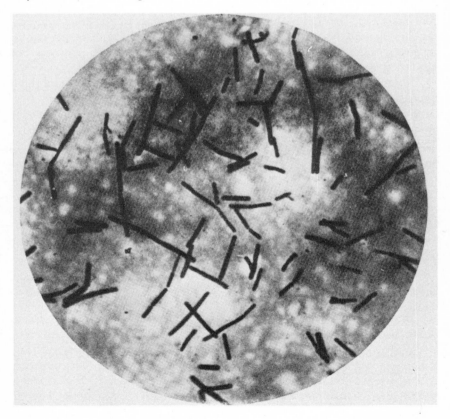

Figure 3–7.   Photomicrograph of *Lactobacillus bulgaricus* (×1,000).

except *L. delbrueckii, L. leichmannii, L. hilgardii, L. trichodes,* and some strains of *L. brevis* ferment lactose with the production of lactic acid and therefore may be of importance in the dairy industries. Chief sources of the lactobacilli are plant surfaces, manure, and dairy products.

Characteristics that make the lactobacilli important in foods are (1) their ability to ferment sugars with the production of considerable amounts of lactic acid, enabling their use in the production of fermented plant and dairy products or the manufacture of industrial lactic acid, but resulting in the deterioration of some products, e.g., wine or beer; (2) production of gas and other volatile products by heterofermentative species, sometimes with damage to the quality of the food, as with *L. fermenti* growing in Swiss cheese or *L. hilgardii* or *L. trichodes* in wines; (3) their inability to synthesize most of the vitamins they require, making them unable to grow well in foods poor in vitamins, but making them useful in assays for the vitamin content of foods; and (4) the heat resistance, or thermoduric properties, of most of the high-temperature lactobacilli, enabling them to survive pasteurization or other heating processes, such as that given the curd in the manufacture of Swiss and similar cheeses.

Species of *Lactobacillus* different from the above-named have been found growing in refrigerated meats, but only a few names for these lactobacilli have been suggested, e.g., *L. viridescens* for one causing greening of sausage and *L. salimandus* for one growing in sausage. These lactobacilli are exceptional because of their ability to grow at low temperatures.

OTHER LACTIC ACID BACTERIA.   Other genera than those discussed contain bacteria that produce chiefly lactic acid during the fermentation of sugars but are not included with the lactics because of other characteristics. Certain *Bacillus* species or strains (*B. cereus, B. stearothermophilus*) or *Escherichia coli* are examples.

Family *Propionibacteriaceae*

Bacteria of the genus *Propionibacterium* may be found in foods. These bacteria are small, nonmotile, Gram-positive, asporogenous, catalase-positive, anaerobic to aerotolerant rods that often are coccoid and sometimes chain. They ferment lactic acid, carbohydrates, and polyalcohols to propionic and acetic acids and carbon dioxide. In Swiss cheese certain species (e.g., *Propionibacterium shermanii*) ferment the lactates to produce the gas that aids in the formation of the holes, or eyes, and also contribute to the flavor. Pigmented propionibacteria can cause color defects in cheese.

Family *Corynebacteriaceae*

Bacteria of this family, sometimes called coryneforms, often are banded or beaded with metachromatic granules and may assume odd shapes.

Microbacterium. Bacteria of this genus are important because of their resistance to adverse conditions and their use in production of vitamins. They are small, nonmotile, Gram-positive, asporogenous, catalase-positive, aerobic, homofermentative, lactic acid–producing rods, which sometimes produce palisade arrangements. *M. lacticum* is the species usually encountered. Microbacteria are very resistant to heat for non-spore-forming bacteria, surviving pasteurization of milk readily and even temperatures of 80 to 85 C (176 to 185 F) for 10 min. They therefore are among the thermodurics that give high counts in pasteurized dairy products, such as market milk and dry milk. Their range of temperatures for growth is 15 to 35 C (59 to 95 F) and their optimum is about 30 C (86 F). Therefore, in plating for these organisms, incubation of plates should be at below 35 C (95 F), preferably at about 30 C (86 F).

Corynebacterium. This genus is of interest in food bacteriology chiefly because of the diphtheria organism, *C. diphtheriae,* which may be transported by foods. *C. pyogenes* causes mastitis in cows. *C. bovis,* with the slender, barred, or clubbed rods characteristic of the genus, has been blamed for the rancidity of cream. Corynebacteria are found on fresh fish and in water and ice.

Arthrobacter. Some *Arthrobacter* cultures can grow at refrigerator temperatures (5 C or 41 F) and therefore are counted as psychrophiles. These soil organisms are inert in most foods.

Family *Bacillaceae*

This family consists of the two genera of spore-forming rods, *Bacillus* and *Clostridium.* The resistance of bacterial spores to heat and other adverse conditions makes them a problem in canning and other methods of preservation.

Bacillus. The endospores of species of this aerobic to facultative genus usually do not swell the rods in which they are formed. Different species may be mesophilic or thermophilic, actively proteolytic, moderately proteolytic, or nonproteolytic, gas-forming or not, and lipolytic or not. In general the spores of the mesophiles, e.g., *B. subtilis,* are less heat-resistant than spores of the thermophiles. Spores of the obligate thermophiles, e.g., *B. stearothermophilus,* are more resistant than those of facultative thermophiles, e.g., *B. coagulans.* The actively proteolytic species usually also have a bacterial rennin which will sweet-curdle milk; *B. cereus* is such a species. The two chief acid- and gas-forming species, *B. polymyxa* and *B. macerans,* sometimes are termed "aerobacilli." Many of the mesophiles can form acid from glucose or other sugar, but usually form only a small amount that often is neutralized by ammonia produced from the nitrogenous food. The thermophilic flat sour bacteria that spoil canned vegetables can produce considerable amounts of lactic acid from sugar, and such a culture, e.g., *B. coagulans,* may be employed for the manufacture of lactic acid. The soil is an important source of *Bacillus* species.

Clostridium. The endospores of species of this genus of anaerobic to microaerophilic bacteria usually swell the end or middle of the rods in which they are formed. All species are catalase-negative. Many species actively ferment carbohydrates with the production of acids, usually including butyric, and gases, usually carbon dioxide and hydrogen. Different species may be mesophilic or thermophilic and proteolytic or nonproteolytic. *C. thermosaccharolyticum* is an example of a saccharolytic obligate thermophile; this organism causes gaseous spoilage of canned vegetables. Putrefaction of foods often is caused by mesophilic, proteolytic species, such as *C. lentoputrescens* or *C. putrefaciens.* The violent disruption of the curd in milk by *C. perfringens* or similar species results in a "stormy fermentation" of milk; and the lactate-fermenting *C. tyrobutyricum (C. butyricum)* (Figure 3–8) is a cause of late gas in cured cheese. The soil is the primary source of *Clostridium* species, although they also may come from bad silage, feeds, and manure.

Order *Actinomycetales*

This order of higher bacteria contains species of the genus *Streptomyces* which can cause undesirable flavors and appearance when growing on foods; musty or earthy odors and tastes from these organisms may be absorbed by nearby foods when growth of the actinomycetes is near at hand. These aerobic higher bacteria grow to form a much-branched mycelium and bear conidia in chains. The tubercle bacilli

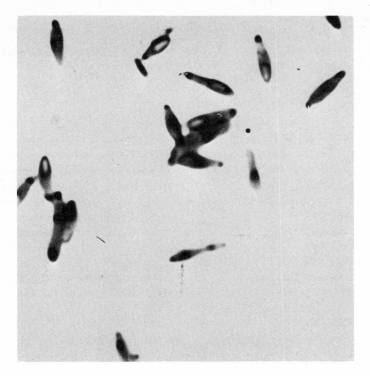

Figure 3-8. Photomicrograph of sporulating *Clostridium tyrobutyricum* (*C. butyricum*) (×1,300).

that cause tuberculosis are in the genus *Mycobacterium* in this order. Foods, especially raw milk, are means of spread of tuberculosis.

Order *Chlamydobacteriales*

This order of higher bacteria contains the iron bacteria, including the important genera *Crenothrix* and *Clonothrix*, which deposit ferric hydroxide in or on their sheaths and cause trouble in water supplies, where they may be responsible for unattractive appearance, bad odors, or even clogging of pipes. Species of the genus *Gallionella* (order *Pseudomonadales*) also are important iron bacteria and may give more trouble than *Clonothrix* or *Crenothrix*.

GROUPS OF BACTERIA IMPORTANT IN FOOD BACTERIOLOGY

Bacteria important in foods often are grouped on the basis of one characteristic that they have in common, without regard for their systematic classification. It is obvious that some bacterial species might be included in two or more of these artificial groups. Examples of the commonly employed groupings will be presented.

Lactic acid-forming bacteria, or lactics

This most important group includes the bacteria in the *Lactobacillaceae* that yield lactic acid as their main product in the fermentation of sugars. In the discussion of members of this family it was noted that the lactics are subdivided into cocci and rods and into the homofermentative lactics, which produce mostly lactic acid and only small amounts of volatile products such as acetic acid and carbon dioxide from sugars, and the heterofermentative lactics, which produce considerable amounts of acetic acid, alcohol, and carbon dioxide in addition to lactic acid. They also are divided into high-temperature lactics, with an optimal temperature of 37 C (98.6 F) or above, and low-temperature lactics, which grow best at about 30 C (86 F). Sometimes the lactics are classified as plant, or cereal, lactics, which do not ferment lactose, or as dairy lactics, which do; but the division between the groups is not definite enough to make the separation useful. Another group that grows in meats held at low temperatures has been called the "meat lactics." Examples of bacteria in these various subdivisions have been given in the discussion of genera in the family *Lactobacillaceae*. The lactics are mostly in the genera *Streptococcus, Leuconostoc, Pediococcus, Lactobacillus,* and *Microbacterium*.

Acetic acid-forming bacteria, or acetics

The bacteria that form acetic acid as a main product, usually by the oxidation of ethyl alcohol, are mostly in the genera *Acetobacter* and *Acetomonas*.

Butyric acid-forming bacteria, or butyrics

Most bacteria of this group are spore-forming anaerobes of the genus *Clostridium*.

Propionic acid-forming bacteria, or propionics

Most bacteria of this group are in the genus *Propionibacterium*, although propionic cocci have been reported.

Proteolytic bacteria

This is a heterogeneous group of actively proteolytic bacteria which produce extracellular proteinases, so termed because the enzymes diffuse outside the cells. All bacteria have proteinases within the cell, but only a limited number of kinds have extracellular proteinases. The proteolytic bacteria may be divided into those that are aerobic or facultative and may be spore-forming or not, and those that are anaerobic and spore-forming. *Bacillus cereus* is an aerobic, spore-forming, proteolytic bacterium, *Pseudomonas fluorescens* is non-spore-forming and aerobic to facultative, and *Clostridium sporogenes* is spore-forming and anaerobic. Many of the species of *Clostridium, Bacillus, Pseudomonas,* and *Proteus* are proteolytic. Some bacteria, termed **acid-proteolytic**, carry on an acid fermentation and proteolysis simultaneously. *Streptococcus faecalis* var. *liquefaciens* and *Micrococcus caseolyticus* are acid-proteolytic. Some bacteria are putrefactive, that is, they decompose proteins anaerobically to produce foul-smelling compounds such as hydrogen sulfide, mercaptans, amines, indole, and fatty acids. Most proteolytic species of *Clostridium* are putrefactive, as are some species of *Proteus, Pseudomonas,* and other genera of non-spore-formers. Putrefaction of split products of proteins also can take place.

Lipolytic bacteria

This is a heterogeneous group of bacteria which produce **lipase**, an enzyme which catalyzes the hydrolysis of fats to fatty acids and glycerol. Many of the aerobic, actively proteolytic bacteria also are lipolytic. *Pseudomonas fluorescens,* for example, is strongly lipolytic. *Pseudomonas, Achromobacter, Alcaligenes, Serratia,* and *Micrococcus* are genera that contain lipolytic species.

Saccharolytic bacteria

These bacteria hydrolyze disaccharides or polysaccharides to simpler sugars. A limited number of kinds of bacteria are **amylolytic**, that is, possess amylase to bring about the hydrolysis of starch outside the cell. *Bacillus subtilis* and *Clostridium butyricum* are amylolytic. Few kinds of bacteria can hydrolyze cellulose. Species of *Clostridium* sometimes are classified as proteolytic ones that may or may not attack sugars, or saccharolytic ones that attack sugars but not proteins. *C. lentoputrescens* is proteolytic but ordinarily does not ferment carbohydrates, whereas *C. butyricum* is nonproteolytic but ferments sugars.

Pectolytic bacteria

Pectins are complex carbohydrates that occur in vegetables and fruits. The mixture of pectolytic enzymes, called **pectinase**, may be responsible for softening of plant tissues or loss of gelling power of fruit juices. Species of *Erwinia, Bacillus,* and *Clostridium* (as well as of molds) may be pectolytic.

Intestinal bacteria

The coliform bacteria and the enterococci, previously discussed, often are used as indicator organisms for contamination of foods with fecal matter. Tests for *Clostridium* also have been employed. These intestinal bacteria also are spoilage organisms.

Thermophilic bacteria, or thermophiles

These bacteria, with an optimal temperature at least above 45 C (113 F) but usually 55 C or above, are important in foods held at high temperatures. *Bacillus* species causing flat sour spoilage of canned foods and *Clostridium thermosaccharolyticum* causing gaseous spoilage are examples. *Lactobacillus thermophilus* is an obligately thermophilic lactic acid bacterium.

Psychrophilic bacteria, or psychrophiles

These bacteria, which grow well at temperatures not far above freezing, are important in refrigerated foods. Psychrophilic bacteria are found chiefly in the genera *Pseudomonas, Achromobacter, Flavobacterium,* and *Alcaligenes,* although the genera *Micrococcus, Lactobacillus, Aerobacter, Arthrobacter,* and others may contain psychrophilic species.

Halophilic bacteria, or halophiles

Truly **halophilic** bacteria require certain minimal concentrations of dissolved sodium chloride for growth. The salt requirements for optimal growth is lower for **moderately halophilic** bacteria (5 to 20 percent) than for **extremely halophilic** bacteria (20 to 30 percent). Some that grow best in media with 2 to 5 percent salt are called slightly halophilic. Other bacteria are **salt-tolerant,** that is, they can grow with or without salt. Halophilic and halotolerant bacteria may be important in highly salted foods and in salt brines. Such bacteria are found in the genera *Halobacterium, Sarcina, Micrococcus, Pseudomonas, Vibrio, Pediococcus,* and *Achromobacter.*

Osmophilic or saccharophilic bacteria

Strictly speaking, **osmophilic** bacteria are those that grow best in high concentrations of sugar, but most bacteria called osmophiles are merely **sugar-tolerant,** e.g., species of *Leuconostoc.*

Bacteria of food poisonings and infections

Bacteria causing food poisoning include enterotoxin-producing forms such as *Staphylococcus aureus* (*Micrococcus pyogenes* var. *aureus*) and *Clostridium botulinum*. Food infections are caused by bacteria, such as certain species of *Salmonella* and *Streptococcus*, that are able to grow in foods and infect consumers of the foods.

Other pathogenic bacteria

Foods can carry pathogenic bacteria in the same manner that they are carried by other things, such as clothing, books, money, doorknobs, etc. Organisms causing the dysenteries may be spread by foods, as may those causing brucellosis, tuberculosis, and other diseases.

Pigmented bacteria

Colors produced by pigmented bacteria growing on or in foods range through the visible spectrum and also include black and white. Examples will be numerous in a later discussion of spoilage of foods. All species in some genera are pigmented, as in *Flavobacterium* (yellow to orange) and *Serratia* (red). Pigmented species are found in many genera; many species of *Micrococcus* are pigmented, for example. Also pigmented varieties occur within some species, e.g., the rust-colored *Lactobacillus plantarum* var. *rudensis* that discolors cheese.

Slime- or rope-forming bacteria

Examples of these bacteria already have been given: *Alcaligenes viscolactis* (*viscosus*) and *Aerobacter aerogenes* causing ropiness of milk, *Leuconostoc* species producing slime in sucrose solutions, and slimy surface growth of various bacteria occurring on foods. Some of the species of *Streptococcus* and *Lactobacillus* have varieties that make milk slimy or ropy. A micrococcus makes curing solutions for meats ropy. Strains of *Lactobacillus plantarum* and other lactobacilli may cause ropiness in various fruit, vegetable, and grain products, e.g., in cider, sauerkraut, and beer. In addition, special species of bacteria in a number of the genera previously discussed may be responsible for sliminess or ropiness.

Gas-forming bacteria

Many kinds of bacteria produce such small amounts of gas and yield it so slowly that it ordinarily is not detected. This sometimes is true of the heterofermentative lactics, although under other conditions gas evolution is evident. Among the genera that contain gas-forming bacteria are *Leuconostoc, Lactobacillus* (heterofermentative), *Propionibacterium, Escherichia, Aerobacter, Proteus, Bacillus* (the aerobacilli), and *Clostridium.* Bacteria of the first three genera produce only carbon dioxide, and those of the other genera yield both carbon dioxide and hydrogen.

Still other groupings of food bacteria could be made, such as those producing bitter flavors, those causing phosphorescence, etc.

REFERENCES

BREED, R. S., E. G. D. MURRAY, and N. R. SMITH. 1957. Bergey's manual of determinative bacteriology. 7th ed. The Williams & Wilkins Company, Baltimore.

CAMPBELL, L. L., and H. O. HALVORSON (*Eds.*) 1965. Spores III. American Society for Microbiology, Ann Arbor, Mich.

CLIFTON, C. E. 1957. Introduction to bacterial physiology. McGraw-Hill Book Company, New York.

FLANNERY, W. L. 1956. Current status of knowledge of halophilic bacteria. Bacteriol. Revs. 20:49–66.

GUNSALUS, I. C., and R. Y. STANIER (*Eds.*) 1960, 1962, 1963. The bacteria: Vol. I, Structure; Vol. II, Metabolism; Vol. IV, The physiology of growth. Academic Press Inc., New York.

HALVORSON, H. O. (*Ed.*) 1958. Spores. Amer. Inst. Biol. Sci. Publ. 5. Washington, D.C. (Symp.)

HALVORSON, H. O. (*Ed.*) 1961. Spores II. Burgess Publishing Company, Minneapolis.

MOSSEL, D. A. A. 1962. Significance of microorganisms in foods, p. 157–201. *In* J. C. Ayers, A. A. Kraft, H. E. Snyder, and H. W. Walker (*Eds.*) Chemical and biological hazards in food. The Iowa State University Press, Ames, Iowa.

OGINSKI, EVELYN L., and W. W. UMBREIT. 1959. An introduction to bacterial physiology. 2nd ed. W. H. Freeman and Company, San Francisco.

PELCZAR, M. J., and R. D. REID. 1965. Microbiology. 2nd ed. McGraw-Hill Book Company, New York.

PORTER, J. R. 1946. Bacterial chemistry and physiology. John Wiley & Sons, Inc., New York.

SARLES, W. B., W. C. FRAZIER, J. B. WILSON, and S. G. KNIGHT. 1956. Microbiology, general and applied. 2nd ed. Harper & Row, Publishers, Incorporated, New York.

SCOTT, W. J. 1957. Water relations of food spoilage microorganisms. Advances Food Res. 7:83–127.

CHAPTER FOUR

CONTAMINATION OF FOODS
FROM NATURAL SOURCES

Growing plants carry a typical flora of microorganisms on the surfaces of their parts and in addition may become contaminated from outside sources. Animals likewise have a typical surface flora plus an intestinal one, give off organisms in excretions and secretions, and also become contaminated from outside sources. Plants and animals with parasitic disease, of course, carry the pathogen causing the disease. The inner, healthy tissues of plants and animals, however, have been reported to contain few living microorganisms or none. Microorganisms from various natural sources are indicated in Table 4–1.

CONTAMINATION FROM GREEN PLANTS AND FRUITS

The natural surface flora of plants varies with the plant but usually includes species of *Pseudomonas, Alcaligenes, Flavobacterium, Achromobacter,* and *Micrococcus,* and coliforms and lactic acid bacteria. Lactic acid bacteria include *Lactobacillus brevis* and *plantarum, Leuconostoc mesenteroides* and *dextranicum,* and *Streptococcus faecium* and *faecalis. Bacillus* species, yeasts, and molds also may be present. The numbers of bacteria will depend upon the plant and its environment and may range from a few hundred or thousand per square centimeter of surface to millions. The surface of a well-washed tomato, for example, may show 400 to 700 microorganisms per square centimeter, while an unwashed tomato would have several thousand. Outer tissue of unwashed cabbage might contain 1 to 2 million microorganisms per gram, but washed and trimmed cabbage 200,000 to 500,000. The inner tissue of the cabbage, where the surface of the leaves would harbor primarily the natural flora, contains fewer kinds and lower numbers, ranging from a few hundred to 150,000 per gram. Exposed surfaces of plants become contaminated from soil, water, sewage, air, and animals, so that microorganisms from these sources are added to the natural flora. Whenever conditions for growth of natural flora and contaminants are present, increases in numbers of special kinds of microorganisms will take place.

TABLE 4-1. Bacterial contamination of foods from natural sources

Source	Pseudomonas	Achromobacter	Flavobacterium	Alcaligenes	Escherichia	Aerobacter	Proteus	Micrococcus	Sarcina	Streptococcus	Leuconostoc	Lactobacillus	Corynebacterium	Arthrobacter	Chromobacterium	Bacillus	Clostridium	Streptomyces
Plants	+	+	+	+		+		+		+	+	+			+	+	+	+
Animals*	+	+	+		+	+		+	+	+	+	+		+		+	+	+
Manure†			+	+	+		+			+		+		+		+	+	
Poultry	+	+	+	+	+	+	+	+		+						+		
Fish:																		
Slime‡	+	+	+	+			+	+	+					+		+		
Gut	+	+	+		+											+	+	
Water	+	+	+			+	+	+		+					+	+		
Ice	+	+						+						+		+		
Soil§	+	+	+	+	+	+	+	+	+	+	+	+	+	+	+	+	+	+

Genera

* Also *Staphylococcus.*
† Also *Propionibacterium.*
‡ Also *Vibrio, Serratia.*
§ Also *Acetobacter.*

This occurs most following harvesting. It will be discussed later. Some fruits have been found to contain viable microorganisms in their interior. Normal, healthy tomatoes have been shown to contain *Pseudomonas,* coliforms, *Achromobacter, Micrococcus* and *Corynebacterium,* and yeasts have been found within undamaged fruits. Organisms also have been found in healthy root and tuber vegetables.

CONTAMINATION FROM ANIMALS

The natural surface flora of meat animals usually is not as important as the contaminating microorganisms from their intestinal tracts and from their hides, hoofs, and hair. These usually contain not only large numbers of organisms from soil, manure, feed, and water, but also important kinds of spoilage organisms. Feathers and feet of poultry carry heavy contamination from similiar sources. Occasionally pathogenic organisms capable of causing human disease may come from animals, e.g., *Salmonella* from poultry or meats.

Animals, from the lowest to the highest forms, contribute their wastes and finally their bodies to the soil and water and to plants growing there. Little attention has been paid to the direct contamination of food plants from this source, except insofar as coliform bacteria or enterococci may be added. Insects and birds cause mechanical damage to fruits and vegetables, introduce microorganisms, and open the way for microbial spoilage. Milk aseptically drawn from the cow already contains bacteria from the interior of the udder. Animal manure may be a source of coliforms, lactics (*Streptococcus faecium, S. faecalis,* and other enterococci, *S. bovis, S. thermophilus,* and lactobacilli), propionics, bacilli, clostridia, alkali-formers, inert bacteria, *Micrococcus.* and *Arthrobacter.*

CONTAMINATION FROM SEWAGE

When untreated domestic sewage is used to fertilize plant crops, there is a likelihood that raw plant foods may be contaminated with human pathogens, especially with those causing gastrointestinal diseases. The use of "night soil" as a fertilizer still persists in some parts of the world but is rare in the United States. In addition to the pathogens, coliform bacteria, anaerobes, enterococci, and other intestinal bacteria can contaminate the foods from this source. Natural waters contaminated with sewage contribute their microorganisms to shellfish, fish, and other seafood.

Treated sewage going onto soil or into water also contributes microorganisms, although it should contain smaller numbers and fewer pathogens than raw sewage.

CONTAMINATION FROM SOIL

The soil contains the greatest variety of microorganisms of any source of contamination. Whenever the microbiologist searches for new kinds of microorganisms or new strains for special purposes, he usually turns first to the soil. Not only numerous kinds of microorganisms but also large total numbers are present in fertile soils, ready to contaminate the surfaces of plants growing on or in them and the surfaces of animals roaming over the land. Soil dust is whipped up by air currents, and soil particles are carried by running water to get into or onto foods. No attempt will be made to list the microorganisms important in food microbiology that could come from the soil, but it can be stated with certainty that nearly every important microorganism can come from soil. Especially important are various molds and yeasts and species of the bacterial genera *Bacillus, Clostridium, Aerobacter, Escherichia, Micrococcus, Alcaligenes, Achromobacter, Flavobacterium, Chromobacterium, Pseudomonas, Proteus, Streptococcus, Leuconostoc,* and *Acetobacter,* as well as some of the higher bacteria such as the actinomycetes and the iron bacteria.

Modern methods of food handling usually involve the washing of the surfaces of foods and hence the removal of much of the soil from those surfaces, and care is taken to avoid contamination by soil dust.

CONTAMINATION FROM WATER

Natural waters contain not only their natural flora but also microorganisms from soil and possibly from animals or sewage. **Surface** waters in streams or pools and **stored** waters in lakes and large ponds vary considerably in their microbial content, from many thousands per milliliter after a rainstorm to the comparatively low numbers that result from self-purification of quiet lakes and ponds or of running water. **Ground** waters from springs or wells have passed through layers of rock and soil to a definite level, and hence most of the bacteria, as well as the greater part of other suspended material, have been removed. Bacterial numbers in these waters may range from a few to several hundred bacteria per milliliter.

The kinds of bacteria in natural waters are chiefly species of *Pseudomonas, Chromobacterium, Proteus, Archromobacter, Micrococcus, Bacillus, Streptococcus* (enterococci), *Aerobacter,* and *Escherichia.* Bacteria of the last three genera probably are contaminants rather than part of the natural flora. These bacteria in the water surrounding fish and other sea life establish themselves on the surfaces and in the intestinal tracts of the sea fauna.

The food microbiologist is interested in two aspects of water bacteriology: (1) public health aspects and (2) economic aspects. From the public-health point of view the water used about foods should be absolutely safe to drink, as shown by freedom from sewage contamination. The usual test for possible sewage contamination is the presumptive test for coliform bacteria. When dilutions of the water are cultured in fermentation or Durham tubes of lactose broth at 35 to 37 C, the production of acid and gas is a positive presumptive test indicating the probable presence of coliform bacteria. The test can be confirmed and completed by methods described in "Standard Methods for the Examination of Water and Wastewater"; and *Escherichia coli,* which is considered more often of intestinal origin, can be distinguished from *Aerobacter aerogenes,* which is found on plant surfaces and in soils more often than in intestinal contents. Some control laboratories run plate counts and presumptive tests on the water at regular intervals and chlorinate more heavily at the first sign that something may be wrong. Chlorination of drinking water is practiced when there is any doubt about the sanitary quality of the water, the amount of chlorine finally present ranging from 0.025 to 2 or more parts of available chlorine per million parts of water, depending on the composition of the water and the amount of contamination.

From the economic point of view, a water with agreeable chemical and bacteriological characteristics is desired for use in connection with the food being handled or processed. The water should have an acceptable taste, odor, color, clarity, chemical composition, and bacterial content and should be available in sufficient volume at a desired temperature and be uniform in composition. Desirable chemical composition is affected by hardness and alkalinity as well as by content of organic matter, iron, manganese, and fluorine.

As has been stated, the water used about foods should meet the bacteriological standards for drinking water and should be acceptable from the sanitary as well as the economic viewpoint. Usually, however, water is more important from the standpoint of the kinds of microorganisms it may introduce into or onto foods than of the total numbers. Contamination may come from water used as an ingredient, for washing foods, for cooling heated foods, and for manufactured ice for preserving foods. For each food product there will be certain microorganisms to be feared especially. The gas-forming coliform bacteria may enter milk from cooling-tank water and cause trouble in cheese made from the milk. Anaerobic gas-formers may enter foods from soil-laden water. Cannery cooling water often contains coliform and other spoilage bacteria that enter canned foods during cooling through minute defects in the seams or seals of the cans. This water commonly is chlorinated, but there are reports that a chlorine-resistant flora can build up in time. Bacteria causing ropiness of milk, e.g., *Alcaligenes viscolactis* (*viscosus*) and *Aerobacter aerogenes,* usually come from water, as do slime-forming species of *Achromobacter, Alcaligenes,* and *Pseudomonas,* which cause trouble in cottage cheese. The bacterium causing the surface taint of butter, *Pseudomonas putrefaciens,* comes primarily from water. The iron bacteria, whose sheaths contain ferric hydroxide, may gum up an entire water supply and are difficult to eliminate. The bacterial flora of ice, crushed to be applied to fish or other food, consists mostly of *Corynebacterium, Achromobacter, Flavobacterium, Pseudomonas,* and cocci.

It is evident from the preceding discussion that it is important to select a location with a good water supply when establishing a plant for handling or processing foods. And it often is necessary to treat the water to make it of satisfactory chemical and bacteriological quality. Water supplies should be protected against sewage pollution. They may be purified by sedimentation in reservoirs or lakes, by filtration through sand or finer filters, or by chlorination, ultraviolet irradiation, or boiling. Only partial purification is likely to result from sedimentation. Efficient filtration greatly reduces the microbial content, but filters sometimes can be a source of contamination of the water with undesirable bacteria. Thus the filters for water for soft drinks have been found sometimes to add large numbers of coliform bacteria. Treatment of water with ultraviolet rays has been used on water for soft drinks. Water may be treated with silver ions by the "catadyn" process mentioned in Chapter 9.

CONTAMINATION FROM AIR

Contamination of foods from the air may be important for sanitary as well as economic reasons. Disease organisms, especially those causing respiratory infections, may be spread among employees by air, or the food product may become contaminated. Total numbers of microorganisms in a food may be increased from the air, especially if the air is being used for aeration of the product, as in growing bread yeast, although the numbers of organisms introduced by sedimentation from air usually are negligible. Spoilage organisms may come from air, as may those interfering with food fermentations. Mold spores from air may give trouble in cheese, meat, sweetened condensed milk, and sliced bread and bacon.

Sources of microorganisms in air

Air does not contain a natural flora of microorganisms, for all that are present have come there by accident and usually are on suspended solid materials or in moisture droplets. Microorganisms get into air on dust or lint, dry soil, spray from streams, lakes, or oceans, droplets of moisture from coughing, sneezing, or talking, from growths of sporulating molds on walls, ceilings, floors, foods, and ingredients, and from sprays or dusts from food products or ingredients. Thus the air around a plant manufacturing yeast usually is high in yeasts, and the air of a dairy plant may contain bacteriophage, or at least the starter bacteria being used there.

Kinds of microorganisms in air

The microorganisms in air have no opportunity for growth but merely persist there, and the kinds that are most resistant to desiccation will live the longest. Mold spores, because of their small size, resistance to drying, and the large numbers per mold plant, are usually present in air. Many mold spores do not water-wet readily and therefore are less likely to sediment from humid air than particles that wet readily. It is possible for any kind of bacterium to be suspended in air, especially on dust particles or in moisture droplets, but some kinds are more commonly found than others in undisturbed air. Cocci usually are more numerous than rod-shaped bacteria, and bacterial spores are relatively uncommon in dust-free air. Yeasts, especially asporogenous chromogenic ones, are found in most samples of air. Of course, whenever dusts or sprays of various materials are carried up into the air, the microorganisms characteristic of those suspended materials will be present: soil organisms from soil and dust, water organisms from water spray, plant organisms from feed or fodder dust, etc.

Numbers of microorganisms in air

The numbers of microorganisms in air at any given time depend upon a number of factors including amount of movement, sunshine, humidity, location, and amount of suspended dust or spray. Numbers vary from less than one per cubic foot at a mountain top to thousands in very dusty air. Individual microorganisms and those on suspended dust or in droplets settle out in quiet air, and, conversely, moving air brings organisms up into it. Therefore, numbers of microorganisms in air are increased by air currents caused by movements of people, by ventilation, and by breezes. Direct rays from the sun kill microorganisms suspended in air and hence reduce numbers. Dry air usually contains more organisms than similar air in a moist condition. Rain or snow removes organisms from the air, so that a hard, steady rainfall may practically free the air of organisms.

The influence of location of air on its microbial content depends chiefly on the factors just discussed. The air over a crop-bearing field would contain fewer organisms than the same area when newly plowed. City air usually contains more organisms than country air on a quiet day. Ocean air would have fewer organisms than air over land. The air of an empty room would contain fewer microorganisms than the air of the same room occupied by people, and factory air might run still higher in count. Always the amounts and kinds of dusts or sprays entering the air will influence numbers and kinds of microorganisms present.

Treatment of air

It has been pointed out that numbers of microorganisms in air may be reduced under natural conditions by sedimentation, sunshine, and washing by rain or snow. Removal of microorganisms from air by artificial means may involve these principles or those of filtration, chemical treatment, heat, or electrostatic precipitation. Most used of these methods is filtration through fibers of various sorts, e.g., cotton, Fiberglas, asbestos, etc., or activated carbon. The fibers are replaced periodically or sterilized with heat or a gas. Washing by means of a water spray or by bubbling air through water is not efficient and seldom is used by itself. Chemical treatments of air are finding increasing use. Sending. air through chemical solutions has been employed, but the use of aerosols or finely divided fogs of chemicals is more successful. Suggested chemicals for such treatment include di- or triethylene glycol, propylene glycol, hypochlorites, formaldehyde, and o-p-benzyl phenols. Passage of air through tunnels lined with ultraviolet lamps or installation of these lamps in a room or over an area where contamination from air is feared is used in some places in the food industries, as noted in Chapter 10. Electrostatic precipitation of dust particles and microorganisms from air also has been accomplished successfully. Heat-treatment of air at very high temperatures is successful but expensive.

After the microorganisms have been removed from air, precautions must be taken to prevent their reentrance. Positive pressure in rooms keeps outside air away. Filters in ventilating or air-conditioning systems prevent the spread of organisms from one part of a plant to another, and ultraviolet-irradiated air locks at doors reduce numbers of organisms carried in by workers.

Sampling and analysis of air

Measured volumes of air are sampled and numbers of microorganisms are expressed per cubic foot, yard, liter, etc. A simple, qualitative method is to expose open petri plates containing a nutrient agar to the air for a stated period, after which the plates are closed and incubated and numbers of colonies are counted. For counting organisms, usually a metered quantity of air is impinged onto a solid medium

or bubbled through a liquid which then is plated, or passed through
an electrostatic precipitator, followed by plating.

CONTAMINATION DURING HANDLING AND PROCESSING

The contamination of foods from the natural sources just discussed
may take place before the food is harvested or gathered, or may occur
also during handling and processing of the food. Additional contamination may come from equipment coming in contact with foods, from
packaging materials, and from personnel. The processor attempts to
cleanse and "sanitize" equipment to reduce such contamination, and
to employ packaging materials that will minimize contamination. The
term sanitize is used here rather than "sterilize," because although an
attempt is made to sterilize the equipment, that is, free it of all living
organisms, sterility seldom is attained. Contamination during handling
and processing of individual kinds of foods will be discussed in later
chapters on those foods.

REFERENCES

ALEXANDER, M. 1961. Introduction to soil microbiology. John Wiley & Sons,
    Inc., New York.
American Public Health Association. 1965. Standard methods for the examina-
    tion of water and wastewater. 12th ed. American Public Health Association,
    New York.
COLLINS, VERA G. 1964. The fresh water environment and its significance
    in industry. J. Appl. Bacteriol. 27:143–150.
EMPEY, W. A., and W. J. SCOTT. 1939. Investigations on chilled beef: I,
    Microbial contamination acquired in the meatworks. Council Sci. Ind. Res.
    [Australia] Bull. 126.
GAINEY, P. L., and T. H. LORD. 1952. Microbiology of water and sewage.
    Prentice-Hall, Inc., Englewood Cliffs, N.J.
GREGORY, P. H. 1961. The microbiology of the atmosphere. Interscience Pub-
    lishers (Division of John Wiley & Sons, Inc.), New York.
HAINES, R. B. 1937. Microbiology in the preservation of animal tissues. Gt.
    Brit., Dep. Sci. Ind. Res., Food Invest. Spec. Rep. 45.
HELDMAN, D. R., T. I. HEDRICK, and C. W. HALL. 1964. Air-borne micro-
    organism population in food packaging areas. J. Milk Food Technol.
    27:245–251.
JENSEN, L. B. 1943. Bacteriology of ice. Food Res. 8:265–272.
JOHNSTON, MARIAN M., and MILDRED J. KAAKE. 1935. Bacteria on fresh fruit.
    Amer. J. Public Health 25:945–947.
MOULTON, F. R. (Ed.) 1942. Aerobiology. American Association for the
    Advancement of Science, Washington, D.C.

MUNDT, J. O. 1961. Occurrence of enterococci: bud, blossom, and soil studies. Appl. Microbiol. 9:541–544.

PRESCOTT, S. C., C.-E. A. WINSLOW, and MAC H. McCRADY. 1946. Water bacteriology. 6th ed. John Wiley & Sons, Inc., New York.

SAMISH, ZDENKA, R. ETINGER-TULCZYNSKA, and MARIAN BICK. 1963. The microflora within the tissue of fruits and vegetables. J. Food Sci. 28:259–266.

SYKES, G., and D. V. CARTER. 1954. The sterilization of air. J. Appl. Bacteriol. 17:286–294.

WILLINGALE, JOAN M., and C. A. E. BRIGGS. 1955. The normal intestinal flora of the pig: II, Quantitative bacteriological studies. J. Appl. Bacteriol. 18:284–293.

WILSSENS, A., and R. BUTTIAUX. 1958. Les bactéries de la flore fécale de la vache saine. Ann. Inst. Pasteur 94:332–340.

WÖLLER, H. 1929. Ueber die epiphytische Bakterienflora gesunder grüner Pflanzen. Centbl. Bakt., II Abt. 79:173–177.

PART TWO

PRINCIPLES OF FOOD PRESERVATION AND SPOILAGE

An attempt is made in this section to outline the principles, especially those of a microbiological nature, involved in the various methods of food preservation and in the spoilage of foods resulting from failure of these preservative methods.

Improvements in methods of food preservation and transportation, made mostly in comparatively recent times, have made possible the successful feeding of heavy populations in countries unable to raise all their own food and in areas in and about large cities. As a result of the improved methods of preservation and transportation our diet has become more varied and better balanced, perishable foods have been made available the year round instead of only seasonally, the preparation of meals has been made easier, and foods in general are being produced in a cleaner and more sanitary manner than previously.

CHAPTER FIVE

GENERAL PRINCIPLES OF FOOD PRESERVATION— ASEPSIS, REMOVAL, ANAEROBIC CONDITIONS

Foods for human consumption may be divided into eight main groups, four of plant and four of animal origin, and several lesser groups. The eight main classes of foods are:

Foods from Plants	*Foods from Animals*
Cereals and cereal products	Meats and meat products
Sugar and sugar products	Poultry and eggs
Vegetables and vegetable products	Fish and other seafood
Fruits and fruit products	Milk and milk products

To the list of foods of plant origin could be added: spices and other flavoring materials, nutmeats, and fungi grown for food (yeasts, molds, mushrooms, etc.). Sodium chloride is a mineral food, a flavoring material, an essential nutrient, and a chemical preservative. Some foods may be fortified with minerals, e.g., iron and calcium compounds added to flour. Some of the coloring and flavoring materials used in foods are synthetic. Vitamins usually are present in foods, but may be added or consumed separately, after chemical synthesis or production by microorganisms.

Most kinds of food are readily decomposed by microorganisms unless special methods are used for their preservation.

METHODS OF FOOD PRESERVATION

The chief methods of food preservation are:

1. Asepsis, or keeping out microorganisms.
2. Removal of microorganisms.
3. Maintenance of anaerobic conditions, e.g., in a sealed, evacuated container.
4. Use of high temperatures.
5. Use of low temperatures.

6. Drying: this includes the tying up of water by solutes, hydrophilic colloids, etc.
7. Use of chemical preservatives, either developed by microorganisms or added.
8. Irradiation.
9. Mechanical destruction of microorganisms, e.g., by grinding, high pressures, etc. (not used industrially).
10. Combinations of two or more of the above methods. Only rarely is a single method effective, and usually several are combined. For example, canned foods are preserved by heat-processing them in an evacuated, sealed can. When preservative methods are combined, the required intensity of each usually is reduced to less than that for preservation by one agency alone. When benzoate or sorbate is added to fruit juices, less heat is required for sterilization of these products. If salt, sugar, and vinegar are all added to catchup, pickles, or relishes, each can be used at a lower concentration than if only one were added. Foods previously irradiated with gamma rays or treated with the antibiotic tylosin require less heat for their sterilization than foods not so treated. Numerous other examples will be found in chapters to follow.

PRINCIPLES OF FOOD PRESERVATION

In accomplishing the preservation of foods by the various methods, the following principles are involved:

1. Prevention or delay of microbial decomposition:
   a. By keeping out microorganisms (asepsis)
   b. By removal of microorganisms, e.g., by filtration
   c. By hindering the growth and activity of microorganisms, e.g., by low temperatures, drying, anaerobic conditions, or chemicals
   d. By killing the microorganisms, e.g., by heat or radiations
2. Prevention or delay of self-decomposition of the food:
   a. By destruction or inactivation of food enzymes, e.g., by blanching
   b. By prevention or delay of purely chemical reactions, e.g., prevention of oxidation by means of an antioxidant
3. Prevention of damage because of insects, animals, mechanical causes, etc. This subject is beyond the scope of this textbook.

The methods used to control the activities of the microorganisms usually are effective against enzymatic activity in the food or chemical reactions. Methods like drying or use of low temperatures, however, permit autodecomposition to continue unless special precautions are taken. For example, most vegetables are blanched (heated) to inactivate their enzymes before being frozen.

Delay of microbial decomposition

Many of the commonly employed methods of food preservation depend not on the destruction or removal of microorganisms but rather on delay in the initiation of growth and hindrance to growth once it has begun.

GROWTH CURVE OF MICROBIAL CULTURES. Whenever microorganisms are added to a food and conditions are favorable, the organisms will begin to multiply and will pass through a succession of phases. When counts of organisms are made periodically and the results are plotted with logarithms of numbers of organisms per milliliter as ordinates and time units as abscissas, a **growth curve** is obtained as diagramed in Figure 5-1. This curve ordinarily is divided into phases as indicated in the figure: (1) the initial **lag phase** (A to B), during which there is no growth or even a decline in numbers; (2) the phase of **positive acceleration** (B to C), during which the rate of growth is continuously increasing; (3) the **logarithmic** or **exponential phase** of growth (C to D), during which the rate of multiplication is most rapid and is constant; (4) the phase of **negative acceleration** (D to E), during which the

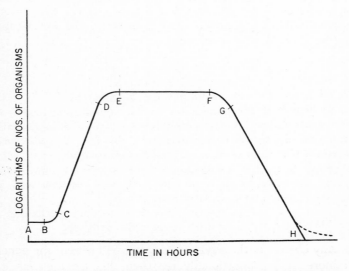

Figure 5-1. Growth curve of microorganism. A to B, lag phase; B to C, phase of positive acceleration; C to D, logarithmic or exponential phase; D to E, phase of negative acceleration; E to F, maximal stationary phase; F to G, accelerated death phase; and G to H, death phase.

rate of multiplication is decreasing; (5) the **maximum stationary phase** (E to F), where numbers remain constant; (6) the **accelerated death phase** (F to G); and (7) the **death phase** or phase of decline (G to H), during which numbers decrease. With many bacteria (or other microorganisms) the numbers do not decrease at a fixed rate to zero as indicated by the unbroken line in the figure, but taper off very gradually as low numbers are approached, as shown by the broken line, and a few viable cells remain for some time.

APPLICATIONS TO FOOD PRESERVATION. Especially important in food preservation (i.e., prevention of spoilage) is the lengthening, as much as possible, of the lag phase and the phase of positive acceleration, often combined and called the lag phase. This can be accomplished in different ways: (1) By the introduction of as few spoilage organisms as possible, that is, by reducing the amount of contamination; for the fewer organisms there are present, the longer will be the lag phase. (2) By avoiding the addition of actively growing organisms (from the logarithmic phase of growth). Such organisms might be growing on unclean containers, equipment, or utensils that come in contact with foods. (3) By one or more unfavorable environmental conditions: unfavorable food, moisture, temperature, pH, or O-R potential, or presence of inhibitors. The greater the numbers of these conditions that are unfavorable, the longer will be the delay of the initiation of growth. (4) By actual damage to organisms by processing methods such as heating or irradiation. Thus, for example, bacteria or their spores subjected to sublethal heat-treatments have been found to require a better culture medium for growth than the unheated organisms. Often a combination of methods for delaying the initiation of growth is enough to give a food the desired storage life.

From the growth curve, the generation time of the organisms, that is, the time that elapses between the formation of a daughter cell and its division into two new cells, can be calculated. The generation time will be shortest during the logarithmic phase of growth, and its length will vary with the environmental conditions during growth, such as the type of food, its pH, temperature, oxidation-reduction potential, available moisture, and presence of inhibitors. The generation time shortens as conditions become more favorable and lengthens as they become less favorable. Any change in the environment that will extend the generation time will more than proportionally lengthen the keeping time of the food. A drop in temperature, for example, will lengthen the generation time and hence the keeping time. If we start with a single cell, and if it divides every 30 min there will be about a million cells in 10 hr, but only about 1,000 cells if the generation time is 60 min, and only 32 cells if it is 120 min. This emphasizes the importance of avoiding contamination of food with microorganisms that are in the logarithmic phase of growth, for when their generation time is the shortest, the lag phase will be brief, and multiplication will proceed at its most rapid rate.

Prevention of microbial decomposition

Microbial decomposition of foods will be prevented if all spoilage organisms are killed (or removed) and recontamination is prevented. Merely stopping the multiplication of microorganisms, however, does not necessarily prevent decomposition, for viable organisms or their enzymes may continue to be active. As will be pointed out in later chapters, the killing of microorganisms by most agencies is easier with smaller initial numbers present than with larger numbers, which reemphasizes the importance of contamination. Especially important is the introduction or building up of forms markedly resistant to the lethal agency being employed, of heat-resistant bacterial spores when foods are to be heat-processed, or of organisms resistant to irradiation when ionizing radiations are to be applied. Vegetative cells of organisms in their logarithmic phase of growth are least resistant to lethal agencies, and they are more resistant in their late lag or their maximum stationary phase of growth.

ASEPSIS

In nature there are numerous examples of asepsis, or keeping out microorganisms, as a preservative factor. The inner tissues of healthy plants and animals usually are free from microorganisms, and if any are present, they are unlikely to initiate spoilage. If there is a protective covering about the food, microbial decomposition is delayed or prevented. Examples of such coverings are the shells of nuts, the skins of fruits and vegetables, the husks of ear corn, the shells of eggs, and the skin, membranes, or fat on meat or fish. It is only when the protective covering has been damaged or decomposition has spread from the outer surface that the inner tissues are subject to decomposition by microorganisms.

In the food industries an increasing amount of attention is being given to the prevention of the contamination of foods, from the raw material to the finished product. The food technologist is concerned with the "load" of microorganisms on or in a food and considers both kinds and numbers of organisms present. The *kinds* are important in that they may include dangerous spoilage organisms, those desirable in a food fermentation, or even pathogenic (disease-producing) microorganisms. The *numbers* of microorganisms are important because the more spoilage organisms there are present, the more likely will be the spoilage of the food, the more difficult its preservation, and the more likely the presence of pathogens. The "load" may be the result of contamination, of growth of organisms, or of both. The quality of many kinds of foods is judged partly by the numbers of microorganisms present. The following are some examples of the importance of aseptic methods in food industries.

Packaging of foods is a widely used application of asepsis. The covering may range from a loose carton or wrapping, which prevents primarily contamination during handling, to the hermetically sealed container of canned foods, which, if tight, protects the processed contents from contamination by microorganisms.

In the dairy industry, contamination with microorganisms is avoided as much as is practicable in the production and handling of market milk and milk for other purposes, and the quality of the milk is judged by its bacterial content.

In the canning industry the load of microorganisms determines the heat process necessary for the preservation of a food, especially if the contamination introduces heat-resistant spoilage organisms, such as spore-forming thermophiles that may come from equipment; and the sealed can prevents recontamination after the heat-treatment.

In the meat-packing industry sanitary methods of slaughter, handling, and processing reduce the "load" and thus improve the keeping quality of the meat or meat products.

In the industries involving controlled food fermentation, e.g., in cheese making, the fewer the competing organisms present in the fermenting material the more likely will be the success of the fermentation.

REMOVAL OF MICROORGANISMS

For the most part the removal of microorganisms is not very effective in food preservation, but under special conditions it may be helpful. Removal may be accomplished by filtration, centrifugation (sedimentation or clarification), washing, or trimming.

Filtration is the only successful method for the complete removal of organisms, and its use is limited to clear liquids. The liquid is filtered through a previously sterilized "bacteriaproof" filter made of asbestos pads, sintered glass, diatomaceous earth, unglazed porcelain, membrane pads, or similar material, and the liquid is forced through by positive or negative pressure. This method has been used successfully with fruit juices, beer, soft drinks, wine, and water.

Centrifugation, or sedimentation, generally is not very effective, in that part of, but not all, the microorganisms are removed. Sedimentation is used in the treatment of drinking water, but is insufficient by itself. When centrifugation (clarification) is applied to milk for cheese making, the main purpose is not to remove bacteria but to take out other suspended materials, although centrifugation at high speeds removes most of the spores.

Washing of raw foods can be helpful in their preservation but may be harmful under some conditions. The washing of cabbage heads or cucumbers prior to their fermentation into sauerkraut and pickles, respectively, removes most of the soil microorganisms on the surface, and in this way increases the proportion that the desirable lactic acid bacteria make of the total flora. The washing of fresh fruits and vegetables removes soil organisms that might be resistant to the heat process during the canning of these foods. Obviously the removal of organisms and of food for them from equipment coming into contact with foods, followed by a germicidal treatment of the apparatus, is an essential and effective procedure during the handling of all kinds of foods. Washing of foods may be dangerous if the water adds spoilage organisms or increases the moisture so that growth of spoilage organisms is encouraged.

Trimming away spoiled portions of a food or discarding spoiled samples is important from the standpoint of food laws and may be helpful in food preservation. Although large numbers of spoilage organisms are removed in this way, heavy contamination of the remaining food may take place. Trimming of the outer leaves of cabbage heads is recommended for the manufacture of sauerkraut.

MAINTENANCE OF ANAEROBIC CONDITIONS

A preservative factor in sealed, packaged foods may be the anaerobic conditions in the container. A complete fill, evacuation of the unfilled space (the head space in a can), or the replacement of the air by carbon dioxide or by an inert gas like nitrogen will bring about anaerobic conditions. Spores of some of the aerobic spore-formers are especially resistant to heat and may survive in canned food but be unable to germinate or grow in the absence of oxygen. Production of carbon dioxide during fermentation and accumulation at the surface will serve to make conditions anaerobic there and prevent the growth of aerobes.

An example is the water seal on grape must being fermented to wine, where a light pressure of carbon dioxide is maintained.

REFERENCES

ASCHEHOUG, V., and R. VESTERHUS. 1941. The microbiology of food preservation. Zentralbl. Bakteriol. Parasitenk., II Abt. 104:169–185.

BATE-SMITH, E. C., and T. N. MORRIS (*Eds.*) 1952. A symposium on quality and preservation of foods. Cambridge University Press, London.

DESROSIER, N. W. 1963. The technology of food preservation. Rev. ed. Avi Publishing Co., Inc., Westport, Conn.

FELLERS, C. R. 1955. Food preservation, chap. 11. *In* F. C. Blanck (*Ed.*) Handbook of food and agriculture. Reinhold Publishing Corporation, New York.

LECHTMAN, S. C., and O. FANNING (*Eds.*) 1957. The future of food preservation. Midwest Research Institute, Kansas City, Mo.

PIKE, M. 1964. Food science and technology. John Murray (Publishers), Ltd., London.

PRESCOTT, S. C., and B. E. PROCTOR. 1937. Food technology. McGraw-Hill Book Company, New York.

SARLES, W. B., W. C. FRAZIER, J. B. WILSON, and S. G. KNIGHT. 1956. Microbiology, general and applied, chap. 24. 2nd ed. Harper & Row, Publishers, Incorporated, New York.

TANNER, F. W. 1944. Microbiology of foods. 2nd ed. Garrard Press, Champaign, Ill.

VON LOESECKE, H. W. 1942. Outlines of food technology. Reinhold Publishing Corporation, New York.

CHAPTER SIX

PRESERVATION BY USE OF HIGH TEMPERATURES

The killing of microorganisms by heat is supposed to be due to coagulation of the proteins and especially to the inactivation of enzymes required for metabolism. The heat-treatment necessary to kill organisms or their spores varies with the kind of organism, its state, and the environment during heating. Depending upon the heat-treatment employed, only part of the vegetative cells, most or all of the cells, part of the bacterial spores, or all of them may be killed. The heat-treatment selected will depend upon the kinds of organisms to be killed, other preservative methods to be employed, and the effect of heat on the food.

FACTORS AFFECTING HEAT RESISTANCE
(THERMAL DEATH TIME)

Cells and spores of microorganisms differ widely in their resistance to high temperatures. Some of these differences are the result of factors that can be controlled, but others are characteristic of the organisms and cannot always be explained. There are differences in heat resistance within a population of cells or spores, as is illustrated by the frequency-distribution curve in Figure 6-1. A small number of cells have low resist-

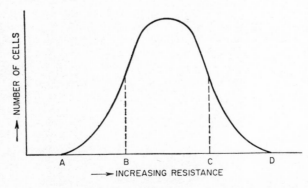

Figure 6–1. Frequency-distribution curve showing heat resistance of cells in a culture.

ance (points A–B); most of the cells have a medium resistance (points B–C); and a small number have high resistance (points C–D). Conditions of growth may favor one or the other of these groups, and, by selection, cultures that are more or less heat-resistant than usual may be produced.

Certain factors are known to affect the heat resistance of cells or spores and must be kept in mind when microorganisms are compared and when heat-treatments for the destruction of an organism are considered. The chief known factors are as follows:

1. The temperature-time relationship. The time for killing cells or spores under a given set of conditions decreases as the temperature is increased. This is illustrated in Table 6–1 by the results of Bigelow and Esty with

TABLE 6–1. Effect of temperature of heating on time needed to kill spores of flat sour bacteria

Temperature, C	Time to destroy all spores, min
100	1,200
105	600
110	190
115	70
120	19
125	7
130	3
135	1

115,000 spores of flat sour bacteria per milliliter in corn juice at pH 6.1. Analogous results would be obtained if vegetative cells were heated at different lethal temperatures.

2. Initial concentration of spores (or cells). The more spores or cells present, the greater is the heat-treatment necessary to kill all of them. Bigelow and Esty heated spores of a thermophile from spoiled canned food in corn juice at pH 6.0 at 120 C, with the results in Table 6–2.

TABLE 6–2. Effect of initial numbers of spores on time required to kill them

Initial concentration of spores, nos/ml	Time required to kill spores, min
50,000	14
5,000	10
500	9
50	8

3. Previous history of the vegetative cells or spores. The conditions under which the cells have been grown and spores have been produced, and their treatment thereafter will influence their resistance to heat.
   a. Culture medium. The medium in which growth takes place is especially important. The effect of the nutrients in the medium, their kind and amount, varies with the organism, but, in general, the better the medium for growth, the more resistant will be the cells or spores. The presence

of an adequate supply of accessory growth factors usually favors the production of heat-resistant cells or spores. This probably is the reason why vegetable infusions and liver extract increase heat resistance. According to Curran, spores formed and aged in soil or oats are more resistant than those in broth or agar. Carbohydrates, amino acids, and organic acid radicals have an effect, but it is difficult to predict. A small amount of glucose in a medium may lead to increased heat resistance, but more sugar may result in the formation of enough acid to cause decreased heat resistance. O. B. Williams, however, found that the initial pH of a medium had little influence on the resistance of *Bacillus subtilis* spores formed in it. Some salts seem to have an effect; phosphate and magnesium ions, for instance, are said to increase the heat resistance of bacterial spores produced in a medium containing them. Prolonged exposure to metabolic products reduces the heat resistance of cells and spores.

b. Temperature of incubation. The temperature of growth of cells and of production of spores influences their heat resistance. In general, resistance increases as the incubation temperature is raised toward the optimum for the organism, and for many organisms increases further as the temperature approaches the maximum for growth. *Escherichia coli,* for example, is considerably more heat-resistant when grown at 38.5 C, which is near its optimal temperature, than at 28 C. Spores of *Bacillus subtilis,* grown at different temperatures in 1 percent peptone water, were heated by Williams, with the results shown in Table 6–3.

TABLE 6–3. Effect of temperature of formation of spores of *Bacillus subtilis* on their heat resistance

Temperature of incubation, C	Time to kill at 100 C, min
21–23	11
37 (optimum)	16
41	18

c. Phase of growth or age. The heat resistance of vegetative cells varies with the stage of growth and of spores with their age. Bacterial cells show their greatest resistance during the late lag phase, but almost as great resistance during their maximum stationary phase, followed by a decline in resistance (see Chapter 5). The cells are least resistant during their phase of logarithmic growth. Very young (immature) spores are less resistant than mature ones. Some spores increase in resistance during the first weeks of storage, but later begin to decrease in resistance.

d. Desiccation. Dried spores of some bacteria are harder to kill by heat than those kept moist, but this apparently does not hold for all bacterial spores.

4. Composition of the substrate in which cells or spores are heated. The material in which the spores or cells are heated is so important that it must be stated if a thermal death time is to have meaning.

a. Moisture. Moist heat is a much more effective killing agent than dry heat, and as a corollary dry materials require more heat for sterilization than do moist ones. In the bacteriological laboratory about 20 to 30 min at 120 C (248 F) in the moist heat of an autoclave will effect sterilization of ordinary materials, but 3 to 4 hr at 160 to 180 C (320 to 356 F) is necessary when the dry heat of an oven is employed. Spores of *Bacillus subtilis* are killed in less than 10 min in steam at 120 C, but in anhydrous glycerol 170 C for 30 min is required.

b. Hydrogen-ion concentration (pH). In general, cells or spores are most heat-resistant in a substrate that is at or near neutrality. An increase in acidity or alkalinity hastens killing by heat, but a change toward the acid side is more effective than a corresponding increase in alkalinity. O. B. Williams heated spores of *B. subtilis* at 100 C in 1:15 molar phosphate solutions, adjusted to various pH values, with results shown in Table 6–4. Other examples will be encountered in the discussion of the heat processing of canned foods.

TABLE 6–4. Effect of pH on heat resistance of spores of *Bacillus subtilis*

pH	Time of survival, min
4.4	2
5.6	7
6.8	11
7.6	11
8.4	9

Cameron has divided canned foods into the **acid foods,** the pH values of which are below 4.5, and the **low-acid foods,** with a pH above 4.5. Acid foods include the common fruits and certain vegetable products, and the low-acid foods are those like meat, seafood, milk, and most of the common vegetables. A further subdivision has been suggested by Cameron.

(1) **Low-acid foods,** with a pH above 5.3, including such foods as peas, corn, lima beans, meats, fish, poultry, and milk (although Cameron included only vegetables and fruits in his original grouping).

(2) **Medium-acid foods,** with a pH between 5.3 and 4.5, including such foods as spinach, asparagus, beets, and pumpkin.

(3) **Acid foods,** with a pH between 4.5 and 3.7, including such foods as tomatoes, pears, and pineapple.

(4) **High-acid foods,** with a pH of 3.7 and below, including such foods as berries and sauerkraut.

The heat process required in the canning of foods will increase with their pH, and, as will be pointed out in Chapter 22, the most likely type of spoilage of the canned food will vary with this grouping on their pH values.

The effect of the pH of the substrate is complicated by the fact that heating at high temperature causes a decrease in the pH of low- or medium-acid foods; and the higher the original pH, the greater will be the drop in pH caused by heating. Foods with an original pH of less than 5.5 to 5.8 change little in acidity as the result of heating. Foods artificially adjusted to more alkaline pH values give increasing protection to spores against heat as the pH increases toward 9.0.

c. Other constituents of the substrate. The only salt present in appreciable amounts in most foods is sodium chloride, which in low concentrations has a protective effect on some spores. Viljoen, heating spores of a thermophile in pea liquor, found that NaCl had a marked protective effect in concentrations of 0.5 to 3.0 percent, but that more salt added to the destructive power of the heat. Other workers have had similar results with other spores.

Sugar seems to protect some organisms or spores, not others. The optimal concentration for protection varies with the organism: it is high for osmophilic organisms and low for others, high for spores and low for nonosmophilic cells. Laboratory class experiments at the University of Wisconsin have indicated that the best concentration of sugar for protection corresponds to the sugar tolerance of the organism, and that when this optimum is exceeded, sugar increases the killing effect of the heat.

Solutes differ in their effect on bacteria. Glucose, for example, protects *Escherichia coli* and *Pseudomonas fluorescens* against heat better than sodium chloride at a_w levels near the minimum for growth. On the other hand, glucose affords practically no protection or is even harmful to *Staphylococcus aureus*, whereas sodium chloride is very protective.

Since the concentration of solutes may affect the heat process necessary for sterilization, canners sometimes further classify foods as high-soluble-solids foods, such as sirups or concentrates, and low-soluble-solids foods, such as fruits, vegetables, or meats.

Colloidal materials, especially proteins and fats, are protective against heat. This is well illustrated in Table 6–5 by the data of Brown and Peiser, who used *thermal death points*.

TABLE 6–5.   Effect of protective substances on heat resistance of bacteria

Substance	*S. lactis,* C	*E. coli,* C	*L. bulgaricus,* C
Cream	69–71	73	95
Whole milk	63–65	69	91
Skim milk	59–63	65	89
Whey	57–61	63	83
Broth	55–57	61	

It will be observed that as the content of protective substances (proteins and fat) decreased in the media, the temperature needed to kill the organism in 10 min decreased. (C. W. Brown and K. Peiser. 1916. Mich. Agr. Exp. Sta. Tech. Bull. 30.)

Antiseptic or germicidal substances in the substrate aid heat in the destruction of organisms. Thus hydrogen peroxide plus heat is used to reduce the bacterial content of sugar and is the basis of a process for milk. The combination of an antibiotic or of plant extract or of irradiation with ionizing rays with heat has been suggested for the treatment of canned foods.

HEAT RESISTANCE OF MICROORGANISMS AND THEIR SPORES

The heat resistance of microorganisms usually is expressed in terms of their **thermal death time,** which is defined as the time it takes at a certain temperature to kill a stated number of organisms (or spores) under specified conditions. This sometimes is referred to as the **absolute thermal death time** to distinguish it from the **majority thermal death time** for killing most of the cells or spores present, and the **thermal death rate,** expressed as the rate of killing. **Thermal death point,** now little used, is the temperature necessary to kill all of the organisms in 10 min.

The reports of different workers on the comparative heat resistance of various kinds of yeasts, molds, and bacteria and their spores do not entirely agree because of differences between the cultures used and the conditions during heating. Therefore only generalizations will be made, with the results of some of the workers cited as examples.

Heat resistance of yeasts and yeast spores

The resistance of yeasts and their spores to moist heat varies with the species and even the strain and, of course, with the substrate in which they are heated. In general the ascospores of yeasts need only 5 to 10 C more heat for their destruction than the vegetative cells from which they are formed. Most ascospores are killed by 60 C for 10 to 15 min; a few are more resistant, but none can survive even a brief heating at 100 C. Vegetative yeasts usually are killed by 50 to 58 C (122 to 136.4 F) for 10 to 15 min. Both yeasts and their spores are killed by the pasteurization treatments given milk (62.8 C, or 145 F, for 30 min, or 71.7 C, or 161 F, for 15 sec), and yeasts are readily killed in the baking of bread, where the temperature of the interior reaches about 97 C (206.6 F).

Heat resistance of molds and mold spores

Most molds and their spores are killed by moist heat at 60 C (140 F) in 5 to 10 min, but some species are considerably more heat-resistant. The asexual spores are more resistant than ordinary mycelium and require a temperature 5 to 10 C higher for their destruction in a given time. Many species of *Aspergillus* and some of *Penicillium* and *Mucor* are more resistant to heat than other molds; a very heat-resistant mold on fruits is *Byssochlamys fulva,* with resistant ascospores. The pasteurizing treatments given milk usually kill all molds and their spores, although spores of some aspergilli, not commonly found in milk, could survive such a heat process.

Sclerotia are especially difficult to kill by heat. Some can survive a heat-treatment of 90 to 100 C (194 to 212 F) for a brief period and have been known to cause spoilage in canned fruits. It was found that 1,000 min at 180 F (82.2 C) or 300 min at 185 F (85 C) was necessary to destroy sclerotia from a species of *Penicillium.*

Mold spores are fairly resistant to dry heat. Data from various workers indicate that dry heat at 120 C (248 F) for as long as 30 min will not kill some of the more resistant spores.

Heat resistance of bacteria and bacterial spores

The heat resistance of vegetative cells of bacteria varies widely with the species, from some of the delicate pathogens that are easily killed to thermophiles that may require several minutes at 80 to 90 C (176 to 194 F). A few general statements can be made about the heat resistance of vegetative cells of bacteria: (1) cocci usually are more resistant than rods, although there are many notable exceptions; (2) the higher the optimal and maximal temperatures for growth, the greater the resistance to heat is likely to be; (3) bacteria that clump considerably or form capsules are more difficult to kill than those that do not; (4) cells high in lipid content are harder to kill than other cells.

A few examples of thermal death times of bacterial cells are shown in Table 6–6.

TABLE 6–6.   Thermal death times of bacterial cells

Bacterium	Time, min	Temperature, C
Gonococcus	2–3	50
Salmonella typhosa	4.3	60
Staphylococcus aureus	18.8	60
Escherichia coli	20–30	57.3
Streptococcus thermophilus	15	70–75
Lactobacillus bulgaricus	30	71

It should be kept in mind that these thermal death times (and those to be given later for spores) are for various concentrations of cells (or spores), heated in different substrates, and might be higher or lower under other conditions.

The heat resistance of bacterial spores varies greatly with the species of bacterium and the conditions during sporulation. Resistance at 100 C (212 F) may vary from less than a minute to over 20 hr. In general, spores from bacteria with high optimal and maximal temperatures for growth are more heat-resistant than those from bacteria growing best at lower temperatures. Or a spore-former growing with another of higher heat resistance may have increased resistance, e.g., *Clostridium perfringens* growing with *C. sporogenes*. Examples of thermal death times of bacterial spores are given in Table 6–7.

TABLE 6–7.   Thermal death times of bacterial spores

Spores of	Time to kill at 100C, min
Bacillus anthracis	1.7
Bacillus subtilis	15–20
Clostridium botulinum	100–330
Clostridium calidotolerans	520
Flat sour bacteria	Over 1,030

Heat resistance of enzymes

Although most food and microbial enzymes are destroyed by 175 F (79.4 C), some, such as peroxidases, may withstand higher temperatures, especially if high-temperature–short-time heating is employed.

DETERMINATION OF HEAT RESISTANCE (THERMAL DEATH TIME)

The simple glass-tube method of Esty and Williams for the determination of thermal death time will be outlined, and only brief mention will be made of the more involved methods used by the canning laboratories. In the tube method a quantity of standardized spore (or cell) suspension in a buffer solution or a food liquor is sealed in small glass tubes, which are heated in a bath thermostatically controlled at a selected temperature. Tubes are removed periodically, cooled immediately, and either incubated directly or subcultured into an appropriate medium for incubation to test for the growth of survivors. A brief outline of the method follows.

Preparation of the spore (or cell) suspension

The organisms to be tested are grown in a culture medium, at a temperature and for a period of time that will produce resistant spores (or cells). They are washed off from solid media or centrifuged from liquid media, usually are washed, and are made into a suspension. Care must be taken to break up or remove clumps, otherwise results will be irregular. This is accomplished by shaking with glass beads or sand, or by filtration through cotton, gauze, filter paper, etc. Sometimes spore suspensions are reincubated for 24 hr to complete sporulation. Spore suspensions then are pasteurized to kill the vegetative cells which might have a protective effect. Next the number of spores (or cells) per unit volume of suspension is determined by a cultural or direct counting method. This stock suspension is diluted in the heating medium to the desired concentration of spores or cells for the test, and distributed in 1-ml portions in glass vials, which are sealed and refrigerated. The medium in which the cells or spores are to be heated may be a buffered phosphate solution or food or food liquor. The volume of suspension of organisms added should be small (1 to 2 percent) to avoid changing the composition of the heating substrate.

Heating to determine the thermal death time

The heating of the vials is done in an agitated and thermostatically controlled bath of oil for higher temperatures or water for temperatures below 100 C (212 F). Tubes are heated at one temperature for varying intervals of time. The following precautions are taken: (1) The vials are brought to a definite temperature before introduction into the bath, usually at 0 C (32 F), obtained in a bath of ice water. An arbitrary preheating time (e.g., 30 sec) is allowed before timing begins. Vials of resistant spores sometimes are preheated to 100 C (212 F) in boiling water before being introduced into the bath. (2) When an oil bath is used, vials are wiped dry before their introduction and are carefully sealed, for water will cause foaming of the oil. (3) If many iced vials are introduced at the same time, an allowance is made for the cooling effect on the bath. (4) The multiple-tube method should be used if accuracy is desired. At least five replicate vials should be used in exploratory runs, and twenty-five to thirty when tubes are being removed at short intervals. The fewer the number of replicates, the more likely will be the occurrence of "skips," that is, killing after a certain period, followed by survival after a longer heating period. (5) Cooling should be prompt and rapid; this usually is done in ice water.

Test for viability (survival)

If the substrate in which the heating is done is a good culture medium for the organism and if it can grow anaerobically, the vials can be incubated to test for the growth of survivors. Otherwise the contents of the vial are subcultured into a good culture medium, which is incubated under optimal conditions for the organism. If the number of survivors is desired, a quantitative count is made by the agar plate or another cultural method. The composition of the medium used to test for survival is very important, for it has been shown by Curran and Evans that spores or cells that have been exposed to drastic heat-treatments are much more exacting in their nutritive requirements than the spores or cells before heating.

Instead of vials, the American Can Company laboratories use special flat cans that can be evacuated, sealed, and then heated and cooled in small steam sterilizers. Workers in the laboratories of the National Canners Association used a unit with valves which allowed the withdrawal of samples periodically under aseptic conditions. This technique is useful in the determination of thermal death rates.

THERMAL-DEATH-TIME CURVES

Thermal death times are plotted as logarithms on semilogarithmic paper and temperatures as arithmetic values to give a thermal-death-time curve like that in Figure 6–2 for flat sour spores from the data of Bigelow and Esty in Table 6–1. The straight line indicates that the order of death by heat is logarithmic, or, in other words, that the death rate is constant. From the straight line obtained or by an extension of it, thermal death times for temperatures and times not listed can be estimated. The slope of the line, after certain corrections, is termed the z value, and is the interval in temperature, in degrees Fahrenheit, required for the line to pass through one logarithmic cycle on the semilogarithmic paper. In other words, z represents the degrees Fahrenheit required to reduce the thermal death time tenfold. F is the time in minutes required to destroy the organism in a specified medium at 250 F (121.1 C). The uncorrected z value from Figure 6–2 is 19 and the F value[1] is 16.4 min. These values will vary with the heat resistance and concentration of the test organism and with the medium in which it is heated. From the z and F values, process times can be calculated (see page 96). The symbol D is used for the decimal reduction time, that is, the time of heating at a temperature to cause 90 percent reduction in the count of viable spores (or cells). This would be the time for the survivor curve to traverse one log cycle.

[1] The symbol F_0 is used to express the F value when z equals 18; the z values for many of the more resistant spore-forming bacteria, and hence those important in canning, often approximate 18, and therefore this value often is used when food-processing times are to be calculated.

As shown in Figure 6–3, the larger the initial number of cells or spores, the longer it will take to reduce the numbers to one per milliliter. With the D value of 10 min, as in the figure, this would require 60 min if the initial number of spores were a million per milliliter, and only 20 min with 100 spores. After 80 min theoretically there would be 0.01 spores per milliliter, or one spore per 100 ml of heating medium. Survivor curves are not always the straight lines indicated in the figure. Convex curves have been attributed to the clumping of cells or spores, or to the ability of injured cells to repair a certain amount of damage.

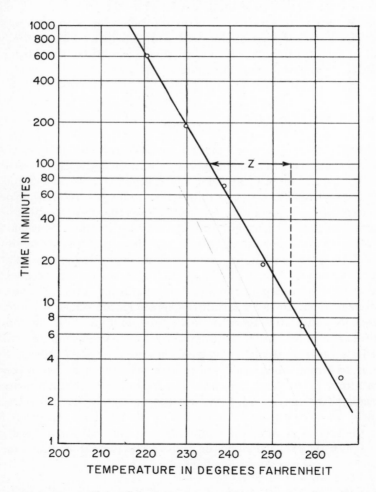

Figure 6–2. Thermal-death-time curve for spores of flat sour bacteria (*after data of Bigelow and Esty*); 115,000 spores per milliliter in corn at pH 6.1 ($z = 19$).

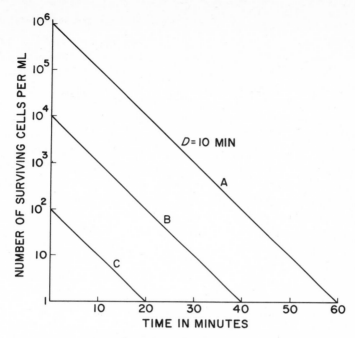

Figure 6–3. Hypothetical survivor curves; $D = 10$ minutes.

It has been assumed that concave curves are the result of unequal resistance among cells or spores. There may be a shoulder at the start of the curve, indicating slower inactivation of cells or spores, or there may be a "tailing off" when numbers have been reduced greatly, indicating a few cells or spores more resistant than the others.

As a safety measure, the canning industry has recommended a 12-D heat-treatment for *Clostridium botulinum* spores in low-acid foods, i.e., enough heat to reduce 10^{12} spores to one spore per milliliter.

Actually death of bacteria usually is not accomplished at a constant heating temperature, for it increases from the start of application of heat and decreases during cooling. Therefore a survivor curve at the temperature of the retort is not sufficient for the calculation of a thermal process. As will be noted later, lethal rates at different lethal temperatures during heating and cooling are considered.

HEAT PENETRATION

The rate of penetration of heat into a food must be known in order to calculate the thermal process necessary for its preservation. Since every part of the food in a can or other container must receive an adequate heat-treatment to prevent spoilage, that part that heats the most slowly is the critical one, and rates of change in the temperature of that part—usually near the center of containers of foods heating by conduction, and farther down when heating is by convection—are measured.

Heat penetration from an external source to the center of the can may take place by conduction, where heat passes from molecule to molecule, by convection, where heat is transferred by movement of liquids or gases, or, as is usually the case, by a combination of conduction and convection. Conduction is slow in foods and rapid in metals. The rate of heat transfer by convection depends upon the opportunity for currents in the liquid and the rate of flow of these currents.

When both conduction and convection are involved in the heating of foods, they may function simultaneously or successively. When solid particles of food are suspended in a liquid, the particles heat by conduction and the liquid heats by convection. Some foods change in consistency during heating, and a "broken" heating curve results. This is true of sugar sirups, brine-packed whole-grain corn, and certain thick soups and tomato juices.

The factors that determine the time required to bring the center of the container of food up to the sterilizing temperature are as follows:

1. The material of which the container is made. Glass has a slower rate of heat penetration than the tinned iron of a can.
2. The size and shape of the container. The larger a can is, the longer it will take to reach a given temperature at the center, because the distance to the center of the larger can is greater and it has less surface per volume or weight. Hence larger cans are heated longer proportionally, but not to as high a temperature at the center. Of course the shape of the can determines the radius: a long, slim, cylindrical can will heat faster than the same volume in a compact, cylindrical form.
3. Initial temperature of the food. Actually the temperature of the food in a can when it goes into the retort (steam sterilizer) makes practically no difference in the time required for the center of the can to reach the retort temperature, for a food at a low initial temperature heats faster than the same food at a higher initial temperature. However, the food with the higher starting temperature is in the lethal range for the microorganisms for a longer time, and its average temperature during heating is higher, than the food in the can with a lower initial temperature. A high initial temperature is important in the processing of canned foods that heat slowly, like cream-style corn, pumpkin, and meat.
4. Retort temperature. Replicate cans of food, placed in retorts of different temperatures, reach the respective temperatures at practically the same time; however, fastest heating would take place in the hottest retort, and the food would reach lethal temperatures most rapidly.

5. Consistency of can contents and size and shape of pieces. All of these are important in their effect on heat penetration. The size and kind of pieces of food and what happens to them during cooking warrant their division into three classes:

a. Pieces that retain their identity, that is, do not cook apart. Examples are peas, plums, beets, asparagus, and whole-grain corn. If the pieces are small and in brine, as with peas, heating is almost as in water. If the pieces are large, heating is delayed, because the heat must penetrate to the center of the pieces before the liquor can reach the retort temperature. Large beets or large stalks of asparagus would heat more slowly than small ones.

b. Pieces that cook apart and become mushy or viscous. These heat slowly, because heat penetration becomes mostly by conduction rather than convection. This takes place in cream-style corn, squash, pumpkin, and sweet potatoes.

c. Pieces that layer. Asparagus layers vertically; hence convection currents travel mostly up and down. Spinach layers horizontally, producing what is called the "baffleboard effect," that interferes with convection currents. Layering is greatly affected by the degree of fill of the can.

Consistency of the can contents is affected by some of the sauces added. Tomato sauce on baked beans slows down heat penetration more than plain sauce. Starch interferes increasingly as the concentration is raised toward 6 percent, but further increases in starch have little additional effect. Sodium chloride never is added in high enough concentrations to have an appreciable effect on rates of heating. The rate of heat penetration decreases with increasing concentrations of sugar, but this effect is counteracted somewhat by the marked decrease in the viscosity of sugar solutions, even heavy ones, with increase in temperature.

6. Rotation and agitation. Rotation or agitation of the container of food during heat processing will hasten heat penetration if the food is at all fluid, but it also may cause undesirable physical changes in some foods. It makes comparatively little difference in the process time of foods that allow free convection currents and have very small particles, like peas. On the other hand agitation is very helpful with foods that layer, like spinach, tomatoes, and peach halves. With older equipment it is practicable to roll cans no faster than 10 to 12 rpm, but newer methods of end-over-end rotation permit higher speeds. Rotation is used successfully with canned evaporated milk, and shaking is used with foods in the form of pastes or purees. A recent process for brined, whole-kernel corn employs heating in a continuous cooker of high-boiling liquid with the can contents mixed by being on the periphery of a rotating reel or by rotation on rollers.

The *cooling* operation involves the same principles of heat transfer as the heating process. Rapid, artificial cooling is recommended because it can be well controlled. Too slow cooling may cause overcooking of the food and may permit the growth of thermophiles.

DETERMINATION OF THERMAL PROCESSES

The following data are needed for the calculation of thermal processes for a canned food: (1) The thermal-death-time curve for the most heat-resistant organism likely to be present in the food. In low-acid foods this usually would be for the spores of a thermophile, e.g., the flat sour organism. (2) The heat penetration and cooling curves for the food in the size and type of container to be used. One of three methods is used to determine the heat process: (1) the graphical method, (2) the formula method, or (3) the nomogram method.

By each of these methods **equivalent processes** can be calculated, that is, heat processes that effect the same killing of microorganisms in the food. In Table 6–8 are equivalent processes for corn in No. 2

TABLE 6–8.   Equivalent processes for corn in no. 2 cans

Process temperature, F	Time, min
260	56
255	64
250	78
245	96
240	129

cans, inoculated with spores of a thermophile (*Nat. Canners Assoc. Bull.* 16-L, 1920).

The principle is similar for all three of the methods for the calculation of thermal processes, but since the graphical method is the simplest to explain, it will be outlined and the others only briefly mentioned.

General method for calculation of process time

To calculate the time in minutes (t) necessary to destroy a certain number of test organisms (or spores) in a given container of a food by heating at the temperature T, knowing the values for z and F, the following equations are employed (t/F = time to destroy organism at temperature T if $F = 1$; and F/t = lethal rate at T):

$$\frac{\log t - \log F}{\log 10 \, (= 1)} = \frac{250 - T}{z}$$

then $\quad \log \dfrac{t}{F} = \dfrac{250 - T}{z} \quad$ and $\quad \dfrac{t}{F} = \text{antilog} \dfrac{250 - T}{z}$

or $\quad t = F \, \text{antilog} \dfrac{250 - T}{z}$

The graphical method

Briefly, the graphical method, as described by Bigelow and coworkers, is as follows:

1. The thermal-death-time curve for the most resistant spoilage organism likely to be encountered is determined in the food being canned. Thermal death times from this curve are converted to **lethal rates** for the various heating temperatures. The lethal rate for a temperature is the reciprocal of the thermal death time: thus, if it took 400 min at 210 F to kill all the spores in a food, the lethal rate would be $\frac{1}{400} = 0.0025$.
2. The heat-penetration (and cooling) curve for the food and can size involved is determined.
3. Lethal rates for the different temperatures at the center of the can during the length of the heating and cooling process are plotted on the heat-penetration (and cooling) curve, as is illustrated in Figure 6–4. In this figure lethal rates are 0.01 unit per side of a square, and the times are 10 min per side of a square. An area equal to ten squares under the lethality curve is unity, which means that destruction of all of the spores (or cells) has been accomplished. If this area is less than ten squares the process is inadequate, and if more than ten squares it is greater than needed. The area beneath the curve is measured by means of a planimeter. In Figure 6–4 the heat-treatment for 56 min at 260 F was found to be adequate, and for 78 min at 250 F.

The formula method

The formula method applies data from the thermal-death-time and heat-penetration curves to an equation, by means of which the thermal

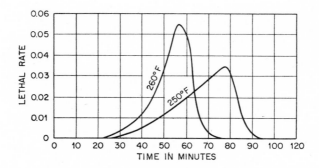

Figure 6–4. Equivalent lethality curves with retort at 260 and 250 F. Corn in No. 2 cans; 50,000 spores of a thermophile per milliliter. (*After Bigelow.*)

process is calculated mathematically. For details of this method the work of Ball should be consulted (see references at end of chapter).

The nomogram method

The nomogram method is the most rapid one for the estimation of thermal process times. It involves the application of the data on thermal death times and heat penetration to a graphic representation of these numerical relations and has the advantage over the previously described methods in that the "coming-up time" of the steam pressure sterilizers is considered. The original article by Olson and Stevens should be consulted for a description of this method.

Regardless of the method used for the calculation of thermal process times, they are verified by actual tests on canned food. An experimental pack is inoculated with a known concentration of spores of the resistant spoilage organism. These cans and uninoculated controls are processed for several time intervals near that calculated for the temperature chosen, are incubated to test for spoilage, and are subcultured to test for sterility. Usually a margin of safety is allowed beyond the minimal treatment for killing the spores being tested, when recommendations are made concerning a thermal process time. It should be kept in mind that the processes recommended will be successful only for the concentration of spores used and might not take care of gross contamination beyond that level.

HEAT-TREATMENTS EMPLOYED IN PROCESSING FOODS

The temperature and time used in heat-processing a food will depend upon what effect heat has on the food and what other preservative methods are to be employed. Some foods, like milk or peas, can be heated to only a limited extent without undesirable changes in appearance or loss in palatability, whereas others, like corn or pumpkin, can undergo a more rigorous heat-treatment without marked change. The greater the heat-treatment, the more organisms will be killed, up to the heating that will produce sterility of the product. If not all the organisms are killed, either the heating must destroy all potential spoilage organisms or the food must be handled thereafter so as to delay or prevent the growth of surviving spoilage organisms. In canning, an attempt is made to kill all organisms that could spoil the food during later handling; in pasteurization most of the spoilage organisms are killed, but others survive and must be inhibited by low temperatures or some other preservative method, if spoilage is to be prevented. The various degrees of heating used on foods might be classified as (1) pasteurization, (2) heating at about 100 C (212 F), and (3) heating above 100 C.

Pasteurization

Pasteurization is a heat-treatment that kills part but not all the microorganisms present and usually involves the application of temperatures below 100 C (212 F). The heating may be by means of steam, hot water, dry heat, or electric currents and the products are cooled promptly after the heat-treatment. Pasteurization is used (1) when more rigorous heat-treatments would harm the quality of the product, as with market milk; (2) when one aim is to kill pathogens, as with market milk; (3) when the main spoilage organisms are not very heat-resistant, like the yeasts in fruit juices; (4) when any surviving spoilage organisms will be taken care of by additional preservative methods to be employed, as is done in the chilling of market milk; and (5) when competing organisms are to be killed, allowing a desired fermentation, usually by added starter organisms, as in cheese making.

Preservative methods used to supplement pasteurization include (1) refrigeration, e.g., of milk; (2) keeping out microorganisms, usually by packaging the product in a sealed container; (3) maintenance of anaerobic conditions, as in evacuated, sealed containers; (4) addition of high concentrations of sugar, as in sweetened condensed milk; and (5) presence or addition of chemical preservatives, such as the organic acids on pickles.

Times and temperatures used in the pasteurizing process depend upon the method employed and the product treated. The high-temperature–short-time (HTST) method employs a comparatively high temperature for a short time; whereas the low-temperature–long-time, or holding (LTH), method uses a lower temperature for a longer time. Some examples follow of pasteurizing treatments given various types of foods. The minimal heat-treatment of market milk is at 62.8 C (145 F) for 30 min in the holding method and at 71.7 C (161 F) for at least 15 sec in the HTST method. One basis for the selection of this treatment is the thermal resistance of the rickettsia responsible for Q fever, *Coxiella burnetii*, an organism that may be transmitted by milk. The heat-treatment often is greater when milk is to be used for other purposes, but it sometimes is slighted in cheese making, in which event the cheese should be aged as is raw-milk cheese. Ice-cream mix is pasteurized at various temperatures for different times, usually receiving a greater heat-treatment than market milk. For example, ice-cream mix may be heated at 71.1 C (160 F) for 30 min, or at 82.2 C (180 F) for 16 to 20 sec.

Grape wines may be pasteurized for 1 min at 82 to 85 C (180 to 185 F) in bulk; whereas fruit wines sometimes are heated to 62.8 C (145 F) or over and bottled hot. Beer may be pasteurized at 60 C (140 F) or above, the time varying with the temperature. Dried fruits usually are pasteurized in the package at 65.6 to 85 C (150 to 185 F) for 30 to 90 min, the treatment varying with the kind of fruit and the size of the package. The pasteurizing treatment given fruit juices depends upon their acidity and whether they are in bulk or in the bottle or can. Recommended for bottled grape juice is 76.7 C (170 F) for 30 min, or flash treatment in bulk at 80 to 85 C (176 to 185 F), and for apple juice 60 C (140 F) if bottled, and 85 to 87.8 C (185 to 190 F) for 30 to 60 sec if in bulk. The average heat-treatment for carbonated juices would be 65.6 C (150 F) for 30 min. When vinegar is pasteurized in the bottle in a water bath, all of the vinegar is brought to at least 65.6 C (150 F). If flash-pasteurized, the vinegar is heated so as to be at 65.6 to 71.1 C (150 to 160 F) when the bottle is closed. When pasteurized in bulk, the vinegar is held at 60 to 65.6 C (140 to 150 F) for 30 min.

Heating at about 100 C (212 F)

Formerly, home canners processed all foods for varying lengths of time at 100 C or less. This treatment was sufficient to kill everything but bacterial spores in the food and often was sufficient to preserve even low- and medium-acid foods. Now, however, most home canners use pressure cookers for the less acid foods. Many acid foods can be processed successfully at 100 C or less, for example, sauerkraut or highly acid fruits. A temperature of approximately 100 C is obtained by boiling a liquid food, by immersion of the container of food in boiling water, or by exposure to flowing steam. Some very acid foods, e.g., sauerkraut, may be preheated to a temperature somewhat below 100 C, packaged hot, and not further heat-processed. Blanching of fresh vegetables before freezing or drying involves heating briefly at about 100 C.

During **baking,** the internal temperature of bread, cake, or other bakery products approaches but never reaches 100 C as long as moisture is present, although the oven is much hotter. The temperature of unsealed canned goods heated in the oven cannot exceed the boiling temperature of the liquid present. As will be indicated later, bacterial spores that survive the baking of bread (maximal temperature about 97 C) may cause ropiness. **Simmering** is incipient or gentle boiling, with the temperature about 100 C.

In the **roasting** of meat the internal temperature reaches only about 60 C (140 F) in rare beef, up to 80 C (176 F) in well-done beef, and 85 C (185 F) in a pork roast. **Frying** gets the outside of the food very hot, but the center ordinarily does not reach 100 C. **Cooking** is an indefinite term with little meaning. **Warming up** a food may mean anything from a small increase in temperature up to heating to 100 C.

Heating above 100 C (212 F)

Temperatures above 100 C (212 F) usually are obtained by means of steam under pressure in steam-pressure sterilizers or retorts. The temperature in the retorts increases with rising steam pressures. Thus with no pressure the temperature at sea level is 100 C (212 F); with 5 lb of pressure, 109 C (228 F); with 10 lb, 115.5 C (240 F); and with 15 lb, 121.5 C (251 F). When liquid foods are to be sterilized prior to their introduction into sterile cans, high steam pressures are used to apply a high temperature for a few seconds.

Heat-treatments used in the processing of canned foods will be discussed further in the following section.

CANNING

Canning is defined as the preservation of foods in sealed containers and usually implies heat-treatment as the principal factor in the prevention of spoilage. Most canning is in "tin cans," which are made of tin-coated steel, or in glass containers, but increasing use is being made of containers that are partially or wholly of aluminum, of plastics as pouches or solid containers, or of a composite of materials.

Spallanzani in 1765 preserved food by heating it in a sealed container. Other workers, as well, had employed heat processes in attempts to prevent the spoilage of food, but it remained for a Frenchman, Nicolas Appert, who has been called the "Father of Canning," to experiment on the heating of foods in sealed containers and to publish directions for preservation by canning. His work, mostly in the years 1795 to 1810, was prompted by the offer of a prize by the French government for a published method for preserving foods for the armed forces. Appert won the 12,000 francs in 1809 for a treatise that was published a year later under the name "The Book for All Households; or the Art of Preserving Animal and Vegetable Substances for Many Years." Appert gave exact directions for the preservation of a wide variety of foods in cork-stoppered, widemouthed glass bottles, which he heated for hours in boiling water. Nothing was known at the time about the relationship of microorganisms to the spoilage of foods; yet Appert worked out methods that were good enough to be followed for years thereafter by home and commercial canners.

Succeeding improvements in the canning process have been chiefly in methods of heat processing, in the construction of the container, and in the calculation of heat processes required. From the time of Appert until 1850, canners heat-processed food much as he did, although there was much secrecy about actual times used. About 1850 European workers began using baths of oil, salt brine, or calcium chloride solution to obtain temperatures above 100 C (212 F), and in 1860 Solomon introduced the use of the calcium chloride bath in this country, reducing process times from 5 or 6 hr to ½ hr and aiding in the canning of food during the American Civil War. Experiments on the use of steam under pressure for heat processing were conducted by Appert and his descendants, but the method was dangerous because of the lack of safety devices. Schriver, a Philadelphia canner, is credited with the development of the closed and controlled retort in this country and received a patent for it in 1874. Equipment for the rotation or agitation of containers during processing has been developed within the past quarter century and applied to various foods. Recently, high-temperature–short-time methods of heating, such as the Martin process, have been receiving increased attention.

The original container used by Appert was the cork-stoppered, wide-mouthed glass jar, and it was used in much of the early canning. The tin canister, abbreviated by Americans to "tin can," was patented by Peter Durand in England in 1810 and has been used increasingly since that time. The first cans were crudely made by hand with soldered seams and small openings in the tops. It took an expert to make 60 to 70 cans per day; modern machines can turn out 300 to 400 per minute. The modern sanitary, or open-top, can, with its double-crimped top and crimped side seam was developed during the last few years of the nineteenth century; since then there have been improvements in gaskets and sealing compounds.

Most modern cans are made of steel plate coated with tin. The trend is toward a thinner and more even coating of tin. **Enamels** are coated onto flat sheets of plate before the manufacture of cans to prevent or slow down discoloration or corrosion. Sanitary, or standard, enamel is used for cans for highly colored fruits and berries or for beets to prevent the fading of color caused by tin plate. C enamel contains zinc oxide, so that the white ZnS is formed instead of the dark FeS when low-acid, sulfur-bearing foods like corn are canned, and darkening of the interior of the can is thus avoided. This enamel is not usable for meat, for the fat would loosen and flake the enamel. Special enamels are employed for certain products, e.g., milk, meats, wine and beer, soups and entrees, and some fruit juices.

Glass containers are used for the canning of many foods and have been greatly improved since the days of Appert.

Aluminum containers are available, but do not as yet withstand strong mechanical stresses, and therefore are used mostly for products that do not require high vacuums or high-temperature processing, e.g., beer, frozen fruits, frozen juice concentrates, and cheese. Ends of fiber or metal containers sometimes are made of aluminum, as is the easy-opening device on beer cans; and some foods, e.g., for astronauts, are placed in collapsible aluminum tubes.

Flexible pouches or bags, made of plastic or of plastic laminated with foil, are being employed mostly for packaging frozen, dried, or unprocessed foods. They are also used for foods that can be packaged hot, although steam-pressure sterilization of foods in pouches has been accomplished. Pressurized canned foods will be discussed later.

The canning procedure

The general procedures in the canning of food will be mentioned only briefly here. A reference on food technology should be consulted for technical details. Raw food for canning should be freshly harvested, properly prepared, inspected, graded if desired, and thoroughly washed before introduction into the can. Many vegetable foods are **blanched** or scalded briefly by hot water or steam before packaging. The blanching washes the food further, "sets" the color, softens the tissues to aid packing, helps form the vacuum, and kills some microorganisms. A "brine," consisting of salt solution or salt plus sugar, is added to some canned vegetables; and sugar sirups may be added to fruits. The container is evacuated before sealing, usually by the heating of the "head space," or unfilled part of the container, but often by mechanical means.

The heat process

Methods for the calculation of heat processes for canned foods have been discussed earlier in this chapter. The canner aims for complete sterilization of most foods but does not always attain it. Instead of killing all microorganisms in the food, he may kill all that could spoil the food under normal conditions of storage and he may leave some that are unable to grow, making the can of food "commercially sterile," "practically sterile," or "bacterially inactive."

The heat processes necessary for the preservation of canned foods depend upon the factors that influence the heat resistance of the most resistant spoilage organism and those that affect heat penetration. The

Research Laboratory of the National Canners Association publishes bulletins that recommend minimal heat processes for various foods in different sizes of glass or metal containers. (Low-acid foods are treated in Bulls. 26-L and 30-L, and acid foods in Bull. 17-L). Examples of process times recommended for No. 2 (pint) cans of various foods are shown in Table 6–9.

With higher retort temperatures the times would be shorter, and the processes would vary with the varieties of food canned, the sauces used, the can size and shape, the initial temperature of the food, and other factors. The process time for food consisting of a mixture of discrete particles and a finely divided component in water or brine can be shortened by means of the "Strata-Cook" process. The components, e.g., creamed, whole-grain corn and brine, are stratified in the container and kept separated during heating. This method takes advantage of the fact that small differences in the water content of a food heating by conduction do not have much effect on the rate of heat penetration and that a thin layer of such food will heat faster than a thicker one. Starch that does not coagulate also aids in heat penetration. Table 6–9 shows that foods with loose liquor require less heat than solidly packed ones. Acid foods require less heat than those nearer neutrality. Some

TABLE 6–9.   Process times recommended for foods in no. 2 cans*

| Food | Temperature | | Time, min |
	Initial, F	Retort, F	
Asparagus spears (all green, tips down)	120	240	25
Green beans (whole or cut)	120	240	20
Lima beans (succulent)	140	240	35
Baked beans, heavy sauce, pork or none	140	240	95
Beets in brine	140	240	30
Carrots in brine	140	240	30
Corn, cream-style	140	240	105
Corn, whole-kernel, in brine	140	240	50
Peas in brine	140	240	35
Pumpkin	140	240	85
Spinach, whole-leaf	140	252	50
Oysters (cove, Atlantic, or Gulf Coast)	130	240	27

* From National Canners Association. 1962. Bull. 26-L, 9th ed. (When further data and information are available, the bulletin will be revised, and previous editions should be discarded.)

very acid foods, such as sauerkraut, may be packed hot and require no further heating. Some foods, e.g., globe artichokes, that can be damaged by high temperatures, are acidified and then processed at lower temperatures.

The heating ordinarily is done in retorts, with or without steam pressure as the food demands. High-temperature–short-time heat processes, now used for some fluid foods, require special equipment for sterilizing the food in bulk, sterilizing the containers and lids, and filling and sealing the sterile containers under aseptic conditions. The Dole process is an example of the HCF, or heat-cool-fill, method. In the Martin HTST Sterile Canning System, mixed liquid and solid pieces are heated directly by contact with high-temperature steam prior to aseptic canning. When a particular heat-resistant spoilage organism is feared, a high-temperature–short-time treatment may be given a liquid food prior to canning, followed by a milder heat-treatment of the food in the can. Thus tomato juice may be presterilized at 250 to 270 F (121 to 132 C) to kill spores of *Bacillus coagulans* before canning, and then the sealed cans of juice are given a milder heating. In the SC, or sterilizing and closing, process, sterilization of the food is accomplished before the can is sealed. In the PFC, or pressure-filler-cooker, method, the food is sterilized by high-pressure steam and filled into the can; then the can is sealed and the heat processing is continued as long as is necessary before cooling. In dehydrocanning, e.g., of apple slices, the food is dried to about half of its original weight before canning.

Other ways of heating cans are by means of a direct gas flame, by steam injection, by heating in a fluidized bed of granular solids, and by the hydrostatic sterilizer which consists of a vertical tank with conveyors that carry cans down through a water leg, up into live steam, and then up and out through a second water leg. In the "Flash 18" method, canning is done in a high-pressure (18 psi) chamber. The product is given an HTST treatment to bring it to processing temperature, and the cans are filled, closed, and partially cooled in the chamber. Dielectric heating will be discussed in Chapter 10.

Heat is also being combined with other preservative agencies, e.g., antibiotics, irradiation, or chemicals (for example, hydrogen peroxide). To date, most of this work has been experimental.

Pressurized packaged foods

Pressurized packaged liquids or pastes, sometimes termed "aerosols," are packed under pressure of a propellent gas, usually carbon dioxide, nitrogen, nitrous oxide, or an approved fluorocarbon, so as to dispense the food as a foam, spray, or liquid. Many foods are now being so packaged, such as whipped cream and other toppings, beverage concentrates, salad dressings, condiments, oils, jellies, flavoring substances, etc. The pressurized foods are subject to microbial spoilage unless adequate preservative methods are employed. Acid foods may be heated, canned, and then gassed, but the gassing process may contaminate the food. Aseptic canning is a possibility for low- and medium-acid foods. Process requirements for pasteurized foods, e.g., whipped cream or other toppings, are similar with or without gas pressure. The gas used as a propellant may have an influence on the kinds of organisms likely to grow. Nitrogen, for example, would not inhibit aerobes if a little oxygen were present, but carbon dioxide would be inhibitory under the same condition. Carbon dioxide under pressure inhibits many microorganisms, including aerobic bacteria and molds, but does not hold back lactic acid bacteria, such as *Bacillus coagulans* and *Streptococcus faecalis,* or yeasts. Nitrous oxide represses some fungi.

The cooling process

Following the application of heat the containers of food are cooled as rapidly as is practicable. The cans may be cooled in the retort or in tanks by immersion in cold water or by a spray of water. Glass containers and large cans are cooled more gradually to avoid undue strain or even breakage. This tempering process involves the use of warm water (or spray), the temperature of which is lowered as cooling progresses. Final cooling of containers usually is by means of air currents.

The leakage of cooling water through imperfections in the container or its seal and the resultant spoilage will be discussed later (Chapter 22).

Canning in the home

For many years home canners have used processes similar to those of Appert, with the temperature of heating not exceeding 100 C (212 F). This has been accomplished by one of three methods: (1) by means of a bath of boiling water, in which the containers were immersed; (2) by means of a steamer, in which the containers were exposed to flowing steam; or (3) by means of oven heat. When jars are heated in the oven, the temperature cannot exceed the boiling point of the food unless the containers are sealed and pressure develops, and this results in the risk of the explosion of the jars. These processes are sufficient for acid foods like fruits, but are not considered adequate for low-acid foods, such as peas, corn, or meats. Modern methods of home canning employ the pressure cooker for the heat processing of low-acid foods according to directions that are available in state and Federal bulletins on home canning. Low- or medium-acid home-canned foods, because they are less likely to be contaminated with very heat-resistant spores than the commercially canned foods, *may* be preserved successfully with less heat than used by the commercial canner, and even with a temperature of 100 C (212 F), but the lower heat-treatment adds to the risk of spoilage and of botulism (Chapter 26).

In the **cold-pack** method of home canning, the food is not heated before being placed into the jar or other container, in contrast to the **hot-pack** method, where the food is precooked, is hot when transferred to the container, and is immediately heat-processed. It is obvious that the cold-packed food will require more of a heat-treatment in the container than hot-packed food. The cold-pack method is not recommended for vegetables and meats.

Canning compounds, usually boric or salicylic acid or potassium pyrosulfite, have been used by home canners in an effort to reduce the heat-treatment necessary for the preservation of foods. There is doubt about their efficacy and their healthfulness.

REFERENCES

American Can Company. 1947. The canned food reference manual. 3rd ed. American Can Company, New York.
ANONYMOUS. 1964. HTST and aseptic techniques gain. Food Eng. 36(11):79.
BALL, C. O., and F. C. W. OLSON. 1957. Sterilization in food technology. McGraw-Hill Book Company, New York.
BRIGHTON, K. W., D. W. RIESTER, and O. G. BRAUN. 1963. Technical problems presented by new containers and materials. Food Technol. 17(9):22–31.
CURRAN, H. R. 1935. The influence of some environmental factors upon the thermal resistance of bacterial spores. J. Infect. Dis. 56:196–202.
Food Engineering Staff. 1962. Advances in processing methods. Food Eng. 34(2):37–52.
GELBER, P. 1963. Numerous refinements in aseptic processing. Food Processing 24(4):61–66.
GOLDBLITH, S. A., M. A. JOSLYN, and J. T. R. NICKERSON (*Eds.*) 1961. An introduction to the thermal processing of foods. Avi Publishing Co., Inc., Westport, Conn. (This volume contains most of the classic papers on canning referred to in this chapter. Therefore these papers are not listed separately among the References.)
HALVORSON, H. O. 1959. Symposium on initiation of bacterial growth: IV, Dormancy, germination, and outgrowth. Bacteriol. Revs. 23:267–272. (See also Symposium on spores, Spores II, and Spores III in Chapter 3 References.)
HANSEN, N.-H., and H. RIEMANN. 1963. Factors affecting the heat resistance of nonsporing organisms. J. Appl. Bacteriol. 26:314–333.

HAYS, G. L., and D. W. RIESTER. 1958. Microbiological aspects of pressure packaged foods. Soap Chem. Spec. 34(9):113, 115–119, 121.

HERSON, A. C., and E. D. HULLAND. 1964. Canned foods: an introduction to their microbiology. Chemical Publishing Company, Inc., New York.

LECHOWICH, R. V., and Z. J. ORDAL. 1962. The influence of the sporulation medium on heat resistance, chemical composition, and germination of *Bacillus megaterium* spores. Canad. J. Microbiol. 8:287–295.

LEVINSON, H. S., and MILDRED T. HYATT. 1964. Effect of sporulation medium on heat resistance, chemical composition, and germination of *Bacillus megaterium* spores. J. Bacteriol. 87:876–886.

MAYER, P. C., K. ROBE, and J. B. KLIS. 1961. "Canning" without cans. Food Processing 22(11):36–39.

NAIR, J. H. 1964. Hydrostatic sterilizers. Food Eng. 35(12):37–42.

National Canners Association. 1962. Processes for low-acid canned foods in metal containers. 9th ed. Nat. Canners Ass. Bull. 26-L. 1963. Processes for low-acid canned goods in glass containers. 3rd ed. Nat. Canners Ass. Bull. 30-L.

OLSON, F. C. W., and H. P. STEVENS. 1939. Nomograms for graphic calculation of thermal processes for non-acid canned foods exhibiting straight-line semi-logarithmic heating curves. Food Res. 4:1–20.

PFLUG, I. J. 1960. Thermal resistance of microorganisms to dry heat: design of apparatus, operational problems and preliminary results. Food Technol. 14:483–487.

PFLUG, I. J., J. H. BOCK, and F. E. LONG. 1963. Sterilization of food in flexible packages. Food Technol. 17:1167–1172.

PIGOTT, G. M. 1963. Fluidized-bed heat processing of canned foods. Food Processing 24(11):79–82.

SACHSEL, G. F. 1963. Fluidized-bed cooking. Food Processing 24(11):77–78.

SCHOTT, D. 1964. Latest developments in the sterilization of canned foods. Food in Canada 24(1):28–29.

STUMBO, C. R. 1964. Heat processing. Food Technol. 18:1373–1375.

WALTERS, A. H. 1964. Microbial resistance. Food Eng. 36(11):57.

WILLIAMS, O. B. 1929. The heat resistance of bacterial spores. J. Infect. Dis. 44:421–465.

XEZONES, HELEN, and I. J. HUTCHINGS. 1965. Thermal resistance of *Clostridium botulinum* (62A) spores as affected by fundamental food constituents: I, Effect of pH. Food Technol. 19:1003–1005.

ZIEMBA, J. V. 1964. "Flash 18"; sequel to aseptic canning. Food Eng. 36(3):122–126.

CHAPTER SEVEN

PRESERVATION BY USE OF LOW TEMPERATURES

Low temperatures are used to retard chemical reactions and action of food enzymes and to slow down or stop growth and activity of microorganisms in food. The lower the temperature, the slower will be chemical reactions, enzyme action, and microbial growth; and a low enough temperature will prevent the growth of any microorganism. Temperatures that are just above freezing maintain foods near their original condition without special pretreatments. Storage time is limited, however; artificial refrigeration is expensive; and the temperature of storage is critical for some foods. Freezing and storage in the frozen condition are still more costly and may result in undesirable changes in some foods.

Any raw plant or animal food may be assumed to contain a variety of bacteria, yeasts, and molds which need only conditions for growth to bring about undesirable changes in the food. Each microorganism present has an optimal or best temperature for growth and a minimal temperature below which it cannot multiply. As the temperature drops from this optimal temperature toward the minimal, the rate of growth of the organism decreases and is slowest at the minimal temperature. Cooler temperatures will prevent growth, but slow metabolic activity may continue. Therefore the cooling down of a food from ordinary temperatures has a different effect on the various organisms present. A drop of 10° might stop the growth of some organisms and slow down the growth of others, but to an extent that would vary with the kind of organism. A further decrease of 10° in temperature would stop growth of more organisms and make still slower the growth of the others. The farther the temperature is lowered toward 0 C (32 F), the fewer will be the organisms that can grow and the slower will be their multiplication. It will be recalled that some bacteria, yeasts, and molds can grow slowly at temperatures several degrees below the freezing point of water. Therefore, even a temperature of 0 C (32 F) or slightly below will not prevent microbial spoilage of most raw foods indefinitely, if the moisture has not been tied up by the freezing process. Freezing not only ties up most of the moisture present but also increases the concentration of dissolved substances in the unfrozen moisture, and hence reduces available water.

Most raw foods contain a number of enzymes that continue their action during storage. The lower the storage temperature, the slower will be the enzymatic action; but the action of some enzymes continues at an appreciable rate at temperatures well below freezing. For this reason vegetables usually are scalded or blanched before freezing.

GROWTH OF MICROORGANISMS AT LOW TEMPERATURES

A few examples, reported by various workers, of growth of microorganisms at low temperatures will be given. Of the molds, *Cladosporium* and *Sporotrichum* have been found growing on foods at −6.7 C (20 F), and *Penicillium* and *Monilia* at −4 C (24.8 F). Growth of yeasts has taken place at −2 to −4 C (28.4 to 24.8 F). Growth of bacteria at −4 to −7.5 C (24.8 to 18.5 F) has been reported. Many of the low-temperature bacteria are species of *Pseudomonas, Achromobacter, Alcaligenes, Micrococcus,* or *Flavobacterium.* These bacteria grow fairly well at chilling temperatures, as do some yeasts and many of the molds.

Bacteria have been reported growing at temperatures as low as −5 C (23 F) on meats, −10 C (14 F) on cured meats, −11 C (12.2 F) on fish, −12.2 C (10 F) on vegetables (peas), and −10 C (14 F) in ice cream; yeasts at −5 C (23 F) on meats, and −17.8 C (0 F) on oysters; and molds at −7.8 C (18 F) on meats and vegetables, and at −6.7 C (20 F) on berries.

LETHAL EFFECT OF FREEZING AND SUBFREEZING TEMPERATURES

Freezing usually effects a considerable reduction in the numbers of viable organisms in a food, but does not sterilize the food. The percentage of microorganisms killed during the freezing will vary with factors to be discussed, but is at least 50 to 80 percent if a quick-freezing process is employed, the lethal effect depending upon the substrate and the method and speed of freezing. Subzero temperatures cause metabolic injury to some bacteria, thereby increasing nutritional requirements. The lethal effect, believed to be caused by denaturation and flocculation of cell proteins due to the increased concentration of solutes in the unfrozen water (but perhaps partially because of physical damage by ice crystals), depends on the following factors:

1. *The kind of microorganism and its state.* Resistance to freezing temperatures varies with the kind of organism, its phase of growth, and whether it is in the vegetative or spore state. It would be expected that a bacterium in the logarithmic phase of growth would be more easily killed than in other phases and that spores might be more resistant than the corresponding vegetative cells. When freezing is accomplished in less than 15 sec at

−70 C (−94 F) it is found that while all of the *Bacillus megaterium* spores and cells of *Staphylococcus aureus* survive, only 70 percent of *Escherichia coli* cells and 20 percent of *Pseudomonas aeruginosa* cells survive. A large percentage of yeast cells are killed, but the survivors persist for a long time at very low temperatures. Mold spores are fairly resistant to freezing.

2. *The temperature during freezing and storage.* Haines has reported that bacteria die most rapidly in the range −1 to −5 C (30.2 to 23 F). The quicker the freezing process, the more rapidly the temperature of the food would pass through this critical zone and the less would be the killing. Haines also found that storage of frozen foods at temperatures within this critical range resulted in a more rapid reduction of numbers of bacteria than storage at lower temperatures. Very low temperatures of freezing and storage do not seem to be more lethal than moderately low temperatures. Very rapid freezing causes little reduction in numbers of bacteria. Slow freezing permits some growth of bacteria before the food becomes frozen. Although a greater percentage of organisms is killed by slow freezing than by fast, the greater total number of organisms present at freezing with the slow method may result in a greater number of surviving organisms than was in the original unfrozen food.

3. *The time of storage in the frozen condition.* The numbers of viable microorganisms decrease with lengthened time of storage. This decrease is gradual, and some organisms will survive after several years. Death supposedly is due to starvation.

4. *The kind of food.* The composition of the food influences the rate of death of organisms during freezing and storage. Sugar, salt, proteins, colloids, fat, and other substances may be protective, while high moisture and low pH may hasten killing.

5. *Alternate freezing and thawing.* Alternate freezing and thawing is reported to hasten the killing of microorganisms, but apparently does not do so in all instances.

TEMPERATURES EMPLOYED IN LOW-TEMPERATURE STORAGE

Many of the terms used in connection with low-temperature storage are applied rather loosely: for example, the term "cold storage," as commonly used, might refer to the use of temperatures above or below freezing, although the application of mechanical refrigeration is implied. For this reason the more exact British terminology is used here in the division of low-temperature storage into three types: (1) common, or cellar, storage, (2) chilling, and (3) cold storage, or freezing.

Common, or cellar, storage

The temperature in **common, or cellar, storage** usually is not much below that of the outside air, and seldom is lower than 15 C (60 F).

Root crops, potatoes, cabbage, celery, apples, and similar foods can be stored for limited periods. The deterioration of such fruits and vegetables, by their own enzymes and by microorganisms, is not prevented but is slower than at atmospheric temperatures. Too low a humidity in the storage cellar results in losses of moisture from the stored food, and too high a humidity favors spoilage by microorganisms. In many foreign countries no refrigeration is available in most homes, and common storage of all foods is the rule.

Chilling

Chilling storage is at temperatures not far above freezing, and usually involves cooling by ice or by mechanical refrigeration. It may be used as the main preservative method for foods or for temporary preservation until some other preservative process is applied. Most perishable foods, including eggs, dairy products, meats, seafood, vegetables, and fruits, may be held in chilling storage for a limited time with little change from their original condition. Enzymatic and microbial changes in the foods are not prevented but are slowed down considerably.

Factors to be considered in connection with chilling storage include the temperature of chilling, the relative humidity, air velocity and composition of the atmosphere in the storeroom, and the possible use of ultraviolet rays or other radiations.

TEMPERATURE. The lower the temperature of storage, the greater will be the cost. Therefore, although most foods will keep best at a temperature just above their freezing point, they are not necessarily stored at this low temperature. Instead the chilling temperature is selected on the basis of the kind of food and the time and conditions of storage. Certain foods have an optimal storage temperature or range of temperatures well above the freezing point and may be damaged by lower temperatures. A well-known example is the banana, which should not be kept in the refrigerator; it keeps best at about 13.3 to 16.7 C (56 to 62 F). Some varieties of apples undergo "low-temperature breakdown" at temperatures near freezing, and sweet potatoes keep best at 10 to 12.8 C (50 to 55 F). As will be pointed out, the minimal chilling temperature required for storage of a food will depend upon the relative humidity and composition of the storage atmosphere or upon special treatments like ultraviolet irradiation.

The temperature of an ordinary icebox varies from 4.4 to 12.8 C (40 to 55 F), depending upon the amount of ice, its rate of melting, the amount of food, the kind of icebox, etc. The temperature of a mechanical refrigerator is mechanically controlled, but varies in different parts, usually between 0 and 10 C (32 and 50 F). It formerly was recommended that the temperature of food in refrigerators be kept below 10 C (50 F), because a temperature this low was believed to prevent the growth of pathogens and prevent or greatly slow down the growth of the more important spoilage organisms. Now 5.6 C (42 F) or below is recommended to deter the growth of psychrophiles and prevent the growth of pathogens, for some have been found able to grow at 7.78 C (46 F), e.g., *Staphylococcus aureus*. It should be noted, however, that *Clostridium botulinum* type E can grow slowly and elaborate toxin at 3.3 C (38 F).

RELATIVE HUMIDITY. The optimal relative humidity of the atmosphere in chilling storage varies with the food stored and with environmental factors such as temperature, composition of the atmosphere, and ray treatments. Too low a relative humidity results in loss of moisture and hence of weight, in the wilting and softening of vegetables, and in the shrinkage of fruits. Too high a relative humidity favors the growth of spoilage microorganisms. The highest humidity, near saturation, is required for most bacterial growth on the surface of foods; less moisture is needed by yeasts, about 90 to 92 percent, and still less by molds, which can grow in a relative humidity of 85 to 90 percent. Changes in humidity, as well as in temperature, during storage may cause "sweating," or precipitation of moisture, on the food. A moist surface favors microbial spoilage, such as the slime on the moist surface of sausage.

Examples of how the optimal relative humidity and the temperature for chilling storage vary with the food stored are given in Table 7–1.

TABLE 7–1.   Optimal relative humidities and storage temperatures for raw foods*

Product	Temperature, F	Relative humidity, percent
Apricots	31–32	85–90
Bananas	53–60	85–90
Beans (snap), peppers	45	85–90
Cabbage, lettuce, carrots	32	90–95
Lemons	55–58	85–90
Melons (cantaloupe)	40–50	80–85
Nuts	32–36	65–70
Onions	32	70–75
Tomatoes (ripe)	40–50	85–90

* From U.S. Dep. Agr. Handbook 66.

The effect of storage temperature on the maximal relative humidity permissible is illustrated by some German recommendations (Table 7–2) for the storage of meat:

TABLE 7–2. Temperatures and relative humidities for storage of meat

Storage temperature, C	Recommended relative humidity, percent
4	75
2	88
0	92

Ozone or carbon dioxide in the storage atmosphere permits an increase in the relative humidity. Eggs have been reported to keep as well in a relative humidity of 90 percent in the presence of 1.5 parts per million of ozone as in a relative humidity of 85 percent in the absence of ozone. Both higher storage temperatures and humidities can be used for apples if controlled amounts of carbon dioxide are present in the atmosphere.

VENTILATION. Ventilation or control of air velocities of the storage room is important in maintaining a uniform relative humidity throughout the room, in removing odors, and in preventing the development of stale odors and flavors. The rate of air circulation will, of course, affect the rate of drying of foods. If adequate ventilation is not provided, food in local areas of high humidity may undergo microbial decomposition.

COMPOSITION OF STORAGE ATMOSPHERE. The amounts and proportions of gases in the storage atmosphere influence preservation by chilling. Usually no attempt is made to control the composition of the atmosphere, although stored plant foods continue to respire, using oxygen and giving off carbon dioxide. In recent years, however, increased attention has been given to "gas storage" of foods, where the composition of the atmosphere has been controlled by the introduction of carbon dioxide, ozone (experimentally) or other gas, or the removal of carbon dioxide. Gas storage ordinarily is combined with chilling storage; and it has been found that in the presence of optimal concentrations of carbon dioxide or ozone (1) a food will remain unspoiled for a longer period; or (2) a higher relative humidity can be maintained without harm to the keeping quality of certain foods; or (3) a higher storage temperature can be used without shortening the keeping time of the food than is possible with ordinary chilling storage. It is especially advantageous to be able to maintain a high relative humidity without added risk of microbial spoilage, because many foods keep their original qualities better if they lose little moisture.

The optimal concentration of carbon dioxide in the atmosphere varies with the food stored, from the approximately 2.5 percent reported best for eggs and 10 percent for chilled beef up to the 100 percent for bacon. For some foods, e.g., apples, the concentration of oxygen as well as of carbon dioxide is significant, and a definite ratio of these gases is sought. Respiring plant cells may evolve too much carbon dioxide into the storage room for some foods, and then part of it must be removed.

Ozone in a concentration of several parts per million has been tried as an aid in the chilling preservation of foods. Ozone is a strong oxidizing agent and therefore cannot be used with foods harmed by oxidation, such as butter or similar fatty foods. Ozone is irritating to the mucous membranes of workers.

Storage in the inert gas nitrogen has been tried experimentally but to date has not been used to any extent with chilling storage.

IRRADIATION. The combination of ultraviolet irradiation with chilling storage aids in the preservation of some foods and may permit the use of a higher humidity or storage temperature than is practicable with chilling alone. Ultraviolet lamps have been installed in rooms for the storage of meat and cheese. The use of these and other rays will be discussed in Chapter 10.

Freezing (cold storage)

The storage of foods in the frozen condition has been an important preservative method for centuries where outdoor freezing temperatures were available. With the development of mechanical refrigeration and of the quick-freezing processes, the frozen-food industry has expanded rapidly. Even in the home, the freezing of foods has become extensive, now that home deep-freezers are readily available. Under the usual conditions of storage of frozen foods microbial growth is prevented entirely and the action of food enzymes is greatly retarded. The lower the storage temperature the slower will be any chemical or enzymatic reactions, but most of them will continue slowly at any temperature now used in storage. Therefore, it is a common practice to inactivate enzymes of vegetables by scalding or blanching prior to freezing when practicable.

SELECTION AND PREPARATION OF FOODS FOR FREEZING. The quality of the food to be frozen is of prime importance, for the frozen food can be no better than the food was before it was frozen. Fruits and vegetables are selected on the basis of their suitability for freezing and their maturity, and are washed, trimmed, cut, or otherwise pretreated as desired. Most vegetables are scalded or blanched, and fruits may be packed in a sirup. Meats and seafood are selected for quality and are handled so as to minimize enzymatic and microbial changes in them. Most foods are packaged before freezing, but some foods in small pieces, e.g., strawberries, may be frozen before packaging.

The scalding or blanching of vegetables ordinarily is done with hot water or steam, the extent of the treatment varying with the food. This brief heat-treatment is supposed to accomplish the following: (1) inactivation of most of the plant enzymes which otherwise might cause toughness, change in color, mustiness, loss in flavor, softening, and loss in nutritive value; (2) reduction (as large as 99 percent) in the numbers of microorganisms on the food; (3) enhancement of the green color of vegetables such as peas, broccoli, and spinach; (4) the wilting of leafy vegetables such as spinach, making them pack better; (5) the displacement of air entrapped in the tissues.

FREEZING OF FOODS. The rate of freezing of foods depends upon a number of factors, such as the method employed, the temperature, circulation of air or refrigerant, size and shape of package, kind of food, etc. **Sharp freezing** usually refers to freezing in air with only natural air circulation or at best with electric fans. The temperature is usually —10 F (—23.3 C) or lower, but may vary from 5 to —20 F (—15 to —29 C), and freezing may take from 3 to 72 hr. This sometimes is termed **slow freezing** to contrast it to **quick freezing**, in which the food is frozen in a relatively short time. Quick freezing is defined variously, but, in general, implies a freezing time of 30 min or less, and usually the freezing of small packages or units of food. Quick freezing is accomplished by one of three general methods: (1) by direct immersion of the food or the packaged food in a refrigerant, as is done in the freezing of fish in brine or of berries in special sirups; (2) by indirect contact with the refrigerant, where the food or package is in contact with the passage through which the refrigerant at 0 to —50 F (—17.8 to —45.6 C) flows; or (3) by air-blast freezing, where frigid air at 0 to —30 F (—17.8 to —34.4 C) is blown across the materials being frozen.

A recent method for the overseas shipment of frozen, packaged foods involves nitrogen freezing of the cartoned foods in a special aluminum case to about —150 F and ordinary storage on the ship. The original low temperature plus the insulation guarantee that the food will remain in the frozen condition for the desired period. Certain fruits and vegetables, fish, shrimp, and mushrooms now are being frozen by means of liquid nitrogen at —320 F. For **dehydrofreezing**, fruits and vegetables have about half of their moisture removed before freezing.

The advantages claimed for quick freezing over slow freezing are that (1) smaller ice crystals are formed; hence there is less mechanical destruction of intact cells of the food; (2) there is a shorter period of solidification and therefore less time for diffusion of soluble materials and for separation of ice; (3) there is more prompt prevention of microbial growth; (4) there is more rapid slowing of enzyme action. Quick-frozen foods, therefore, are supposed to thaw to a condition more like that of the original food than slow-frozen foods. This seems to be true for some foods, e.g., vegetables, but not necessarily for all foods. Research on fish, for example, has indicated little advantage for quick freezing over slow freezing. The choice of the freezing process to be employed is likely to be based on economic reasons rather than on the effect on the quality of the food as used.

CHANGES DURING PREPARATION FOR FREEZING. The rate and kind of deterioration of foods prior to freezing will depend upon the condition of the food at harvesting or slaughter and the methods of handling thereafter. The changes in the food, due to chemical reaction, action of enzymes of the food, or action of microorganisms, will be discussed later. The higher the temperatures during handling, the more rapid will be the chemical and enzymatic changes. The temperatures at which the food is held and other environmental conditions will determine the kinds of microorganisms to grow and the changes to be produced. The condition of the food at the time of freezing will determine the potential quality of the frozen food.

Fruits and vegetables continue to mature after harvesting. Respiration continues, with intake of oxygen, evolution of carbon dioxide, and loss of sweetness and flavors. Oxidation, chemical and enzymatic, produces changes in flavor and color, with acceleration of these processes if cut surfaces are exposed. Hydrolytic enzymes continue to function except when foods, e.g., vegetables, are blanched. If conditions are favorable, microorganisms grow on surfaces of the foods and cause physical and chemical changes. Molds or bacteria may cause rotting, or microorganisms may cause sliminess, off-flavors, and off-colors.

After killing, meats and fish continue to support the change of glycogen to lactic acid; pigments change color because of oxidation-reduction reactions; fats become oxidized and hydrolyzed; hydrolytic enzymes break down tissues with a desirable tenderizing effect in beef, but with undesirable changes in the flavor and consistency of most meats or of fish; and bacteria, yeasts, and molds begin to grow on the surfaces to produce sliminess, discoloration, or undesirable flavors, and in time may affect the interior flesh. The food processor attempts to keep all of the above undesirable changes at a minimum until the food can be frozen.

Prior to freezing, eggs may undergo contamination from various sources, including that during the breaking process, and the inclusion of eggs containing large numbers of bacteria may not only increase the bacterial content of the frozen product but reduce its quality (see Chapter 18).

CHANGES DURING FREEZING.   The quick-freezing process rapidly slows down chemical and enzymatic reactions in the foods and stops microbial growth. A similar effect is produced by sharp or slow freezing, but with less rapidity. The physical effects of freezing are of great importance. There is an expansion in volume of the frozen food, and ice crystals form and grow in size. These crystals usually are larger with slow freezing, and more ice accumulates between tissue cells than with quick freezing and may crush cells. Water is drawn from the cells to form such ice, with a resultant increase in the concentration of solutes in the unfrozen liquor, which in this way has a constantly dropping freezing point until a stable condition is reached. It is claimed that the ice crystals rupture tissue cells or even microorganisms, but some workers minimize the importance of such an effect. The increased concentration of solutes in the cells hastens the salting out, dehydration, and denaturation of proteins, and causes irreversible changes in colloidal systems, such as the syneresis of hydrophilic colloids. Further, it is thought to be responsible for the killing of microorganisms. As was stated previously, bacteria, and probably other microorganisms, die most rapidly in the range −1 to −5 C (30.2 to 23 F). Therefore more cells should be killed by a slow-freezing process than by a quick one, but at the same time there would be more undesirable physical change with slow freezing.

CHANGES DURING STORAGE.   During storage of the food in the frozen condition chemical and enzymatic reactions proceed slowly. Meat, poultry, and fish proteins may become irreversibly dehydrated, the red myoglobin of meat may be oxidized—especially at surfaces—to brown metmyoglobin, and fats of meat and fish may become oxidized and hydrolyzed. The unfrozen, concentrated solution of sugars, salts, etc., may ooze from packages of fruits or concentrates during storage as a viscous material called the **metacryotic liquid**. Fluctuation in the storage temperature results in growth in the size of ice crystals and in physical damage to the food. Desiccation of the food is likely to take place at its surface during storage. When ice crystals evaporate from an area at the surface a defect called **freezerburn** is produced on fruits, vegetables, meat, poultry, and fish. The spot usually appears dry, grainy, and brownish; and in this area the chemical changes mentioned above take place, and the tissues become dry and tough.

The Association of Food and Drug Officials of the United States has recommended that all frozen foods be held at 0 F or lower, although brief and temporary deviation up to 10 F is permitted for defrost, loading or unloading, and other temporary conditions. Microbial growth is not apt to take place at these temperatures.

At freezing temperatures vegetative cells of microorganisms that are unable to multiply will, in time, die. There is a slow but continuous decrease in numbers of viable microorganisms as storage continues, with some species dying more rapidly than others but with representatives of most species surviving for months or even years. Some bacteria can increase in numbers during storage at —4 C (25 F) and above in frozen peas, green beans, cauliflower, spinach, and meat. Molds can grow slowly in frozen foods at —8.9 to —6.67 C (16 to 20 F) or even lower, and yeasts at —8.9 C (16 F) and above.

CHANGES DURING THAWING. Most of the changes that seem to appear during thawing are the result of freezing and storage but do not become evident earlier. When the ice crystals melt, the liquid either is absorbed back into the tissue cells or leaks out from the food. Slow, well-controlled thawing usually results in better return of moisture to the cells than rapid thawing and results in a food more like the original food that was frozen. The pink or reddish liquid that comes from meat on thawing is called **drip,** or **bleeding,** and the liquid oozing from fruits or vegetables on thawing is termed **leakage.** The wilting or flabbiness of vegetables and the mushiness of fruits on thawing are chiefly the result of physical damage during freezing. During thawing the rate of action of enzymes in the food will increase, but the time for action will be comparatively short if the food is utilized promptly.

If thawing is reasonably rapid and the food is used promptly there should be little trouble with growth of microorganisms because the temperatures will be too low for any appreciable amount of growth. Only when the thawing is very slow or when the thawed food is allowed to stand at room temperature is there opportunity for any considerable amount of growth and activity of microorganisms. The kinds of organisms growing depend upon the temperature of thawing and the time the food was allowed to stand after thawing.

The microbiology of most of the various kinds of frozen foods will be discussed in more detail when the preservation of specific foods is treated in following chapters.

DISPOSAL OF THAWED FOODS. Sometimes power failures will lead to partial or complete thawing of foods in freezers. Thawed fruits may be refrozen. Flesh foods and vegetables may be refrozen if the packages still contain some ice. Refrozen foods will contain large ice crystals, and may show leakage of liquid (syneresis) and mushiness. If thawed, flesh foods may be used if their temperature is below 38 F (3.3 C), but they should be cooked thoroughly. In the cases of doubt, the food should be discarded.

PRECOOKED FROZEN FOODS. Precooked frozen foods include such a variety of types of foods and food products that they can be most conveniently discussed together. Most such foods are meat, fish, or poultry products, for example soups, creamed products, stews, pies, fried fish or poultry, chow mein, barbecued meat, meat loaf, chicken à la king, etc. Some bakery products, fruits, and vegetables may be cooked and then frozen, however. The precooking process usually is enough to kill any pathogens in the raw material and greatly reduce the total number of microorganisms present. Therefore, the suggested standard of not over 100,000 colonies per gram by a plate count seems fairly lenient. Most samples of precooked frozen foods examined by various workers would have met this standard. The cooking process would not destroy any preformed staphylococcus toxin in the food. Enterococci survive freezing and persist longer during storage than coliform bacteria and therefore are recommended as "indicator bacteria" for possible fecal contamination.

It is especially important to prevent contamination of the food after the precooking, for any pathogenic or spoilage organism then introduced would find competition from other organisms greatly reduced and the cooked food probably a better culture medium than the original raw material if opportunity for growth were given. Therefore it also is important that cooling and freezing be done promptly after cooking, so as to give no opportunity for such growth.

If these precooked frozen foods are kept at warm temperatures too long after thawing, there may be growth and toxin production by staphylococci or *Clostridium botulinum,* if present, although no such occurrence has ever been reported. The final cooking or "warming up" of these products in the home or restaurant is not always enough of a heat-treatment to greatly reduce numbers of organisms present or guarantee that any pathogens present will be killed or toxins destroyed.

REFERENCES

American Society of Refrigeration Engineers. 1951. The refrigeration data book. American Society of Refrigeration Engineers, New York.

ANONYMOUS. 1961. Relative stabilities of frozen foods. Food Processing 22(1):41–43.

ANONYMOUS. 1961. Super cold for ultra-quick freezing. Food Processing 22(6):48–50.

ANONYMOUS. 1964. Nitrogen freezing arrives. Food Eng. 36(11):73–74.
ARPAI, J. 1964. The recovery of bacteria from freezing. Zeit. Allg. Mikrobiol. 4:105–113.
DOEBBLER, G. F., and A. P. RINFRET. 1963. Survival of microorganisms after ultrarapid freezing and thawing. J. Bacteriol. 85:485.
ELLIOTT, R. P., and H. D. MICHENER. 1960. Review of microbiology of frozen foods. In Conference on frozen food quality. Agricultural Research Service. U.S. Dep. Agr. ARS–74–21.
ELLIOTT, R. P., and H. D. MICHENER. 1965. Factors affecting the growth of psychrophilic micro-organisms in foods; a review. Agricultural Research Service. U.S. Dep. Agr. Tech. Bull. 1320.
FISHER, F. E. (Chairman). 1962. Report of the committee on frozen food sanitation. J. Dairy & Food Technol. 25:55–57.
GAVER, K. M. 1951. Processing: 2, Heat removal, chap. 13. In M. B. Jacobs (Ed.) The chemistry and technology of food and food products, 2nd ed. Interscience Publishers (Division of John Wiley & Sons, Inc.), New York.
GORTNER, W. A., F. S. ERDMAN, and NANCY K. MASTERMAN. 1948. Principles of food freezing. John Wiley & Sons, Inc., New York.
GUNDERSON, M. F. 1961. Mold problem in frozen foods, p. 299–310. In Proceedings low temperature microbiology symposium (Campbell Soup Co.)
HAINES, R. B. 1935–1936. The freezing and death of bacteria. Rep. Food Invest. Board [Britain]. 1934:47; 1935:31–34.
HAINES, R. B. 1938. The effect of freezing on bacteria. Roy. Soc. (London), Proc., B 124:451–463.
HUCKER, G. J., and ELIZABETH R. DAVID. 1957a. The effect of alternate freezing and thawing on the total flora of frozen chicken pies. Food Technol. 11:354–356.
HUCKER, G. J., and ELIZABETH R. DAVID. 1957b. The effect of alternate freezing and thawing on the total flora of frozen vegetables. Food Technol. 11:381–383.
MICHENER, H. D., and R. P. ELLIOTT. 1964. Minimum growth temperatures for food-poisoning, fecal-indicator, and psychrophilic microorganisms. Advances Food Res. 13:349–396.
PENNINGTON, M. E., and D. K. TRESSLER. 1951. Food preservation by temperature control, chap. 34. In M. B. Jacobs (Ed.) The chemistry and technology of food and food products. 2nd ed. Interscience Publishers (Division of John Wiley & Sons, Inc.), New York.
PORTER, J. R. 1946. Bacterial chemistry and physiology. John Wiley & Sons, Inc., New York.
ROGERS, J. L. 1958. Quick frozen foods. Food Trade Press, London.
ROSE, D. H., R. C. WRIGHT, and T. M. WHITEMAN. 1949. The commercial storage of fruits, vegetables and florists' stocks. U.S. Dep. Agr. Circ. 278.
STRAKA, R. P., and J. L. STOKES. 1959. Metabolic injury to bacteria at low temperatures. J. Bacteriol. 78:181–185.
TRESSLER, D. K., and C. F. EVERS. 1957. The freezing preservation of foods: Vol. I, Freezing of fresh foods; Vol. II, Freezing of precooked and prepared foods. 3rd ed. Avi Publishing Co., Inc., Westport, Conn.
WRIGHT, R. C., D. H. ROSE, and T. M. WHITEMAN. 1954. The commercial storage of fruits, vegetables, ornamentals and nursery stock. U.S. Dep. Agr. Handbook 66.

CHAPTER EIGHT

PRESERVATION BY DRYING

Preservation of foods by drying has been practiced for centuries. Some foods, e.g., grains, are sufficiently dry as harvested, or with a little drying, to remain unspoiled for long periods under proper storage conditions. Most foods, however, contain enough moisture to permit action by their own enzymes and by microorganisms, so that in order to preserve them by dryness the removal or binding (e.g., by solutes) of moisture is necessary.

Drying usually is accomplished by the removal of water, but any method that reduces the amount of available moisture in a food is a form of drying. Thus, for example, dried fish may be heavily salted so that moisture is drawn from the flesh and bound by the solute and hence is unavailable to microorganisms. Sugar may be added, as in sweetened condensed milk, to reduce the amount of available moisture.

Moisture may be removed from foods by any of a number of methods, from the ancient practice of drying by means of the sun's rays to the modern artificial ones. Many of the terms used in connection with the drying of foods are rather inexact. A **sun-dried** food has had moisture removed by exposure to the sun's rays without any artificially produced heat and without controlled temperatures, relative humidities, or air velocities. A **dehydrated** or **desiccated** food has been dried by artificially produced heat under controlled conditions of temperature, relative humidity, and air flow. **Condensed** usually implies that moisture has been removed from a liquid food, and **evaporated** may have a similar meaning or may be used synonymously with the term dehydrated.

METHODS OF DRYING

Methods of drying will be mentioned only briefly; references on food technology should be consulted for details. The drying of individual foods will be discussed in the chapters on the preservation of these foods.

Sun drying

Sun drying is limited to climates with a hot sun and a dry atmosphere and to certain fruits, such as raisins, prunes, figs, apricots, nectarines, pears, and peaches. The fruits are spread out on trays and may be turned during drying. Other special treatments will be discussed later.

Drying by mechanical driers

Most methods of artificial drying involve the passage of heated air with controlled relative humidity over the food to be dried or the passage of the food through such air. A number of devices are used for controlled air circulation and for the reuse of air in some processes. The simplest drier is the **evaporator** or **kiln,** sometimes used in the farm home, where the natural draft from the rising of heated air brings about the drying of the food. Forced-draft drying systems employ currents of heated air that move across the food, usually in tunnels. An alternative method is to move the food on conveyor belts or on trays in carts through the heated air.

Liquid foods, such as milk, juices, and soups, may be **evaporated** by the use of comparatively low temperatures and a vacuum in a vacuum pan or similar device; **drum-dried** by passage over a heated drum, with or without vacuum; or **spray-dried** by spraying the liquid into a current of dry, heated air.

Freeze drying

Freeze drying, or the sublimation of water from a frozen food by means of a vacuum plus heat applied at the drying shelf, is being used for a number of foods, including meats, poultry, seafood, fruits, and vegetables. Frozen thin layers of foods of low sugar content may be dried without vacuum by sublimation of moisture during passage of dry carrier gas.

Drying during smoking

As is indicated in Chapter 9, most of the preservative effect of the smoking of foods is due to the drying of the food during the process. Indeed, some workers maintain that drying is the main preservative factor, especially drying at the surface of the food.

Other methods

Electronic heating has been suggested for the removal of still more moisture from a food already fairly well dried. Foam-mat drying, in which liquid food is whipped to a foam, dried with warm air, and crushed to a powder, is receiving attention, as is pressure-gun puffing of partially dried foods to give a porous structure that facilitates further drying. Tower drying in dehumidified air at 86 F (30 C) or lower has been successful with tomato concentrate, milk, and potatoes.

FACTORS IN THE CONTROL OF DRYING

A consideration of the proper control of dehydration includes the following factors: (1) The temperature employed, which will vary with the food and the method of drying. (2) The relative humidity of the air. This, too, is varied with the food and the method of drying and also with the stage of drying. It usually is higher at the start of drying than later. (3) The velocity of the air. (4) The time of drying. Improper control of these factors may cause **casehardening** due to more rapid evaporation of moisture from the surface than diffusion from the interior, with a resulting hard, horny, impenetrable surface film that hinders further drying.

TREATMENTS OF FOODS BEFORE DRYING

Many of the pretreatments of foods to be dried are important in their effect on the microbial population as will be indicated. These pretreatments may include (1) selection and sorting for size, maturity, and soundness; (2) washing, especially of fruits and vegetables; (3) peeling of fruits and vegetables by hand, machine, lye bath, or abrasion; (4) subdivision into halves, slices, shreds, or cubes; (5) alkali dipping, which is used primarily for fruits such as raisins, grapes, and prunes (for sun drying) and employs hot 0.1 to 1.5 percent lye or sodium carbonate; (6) blanching or scalding of vegetables and some fruits (apricots, peaches); (7) sulfuring of light-colored fruits and of certain vegetables. Fruits are sulfured by exposure to sulfur dioxide gas produced by the burning of sulfur so that a level of 1,000 to 3,000 parts per million, depending upon the fruit, will be absorbed. Vegetables may be sulfured after blanching in a similar manner or by dipping into or spraying with sulfite solution. Sulfuring helps to maintain an attractive light color, conserve vitamin C and perhaps A, and repel insects; it also kills many of the microorganisms present.

PROCEDURES AFTER DRYING

The procedures after drying vary with the kind of dried food.

Sweating

"Sweating" is storage, usually in bins or boxes, for equalization of moisture or readdition of moisture to a desired level. It is used primarily with some dried fruits and some nuts (almonds, English walnuts).

Packaging

Most foods are packaged soon after drying for protection against moisture, contamination with microorganisms, and infestation with insects, although some dried foods (e.g., fruits and nuts) may be held as long as a year before packaging.

Pasteurization

Pasteurization is limited, for the most part, to dried fruits, and kills any pathogens that might be present, as well as destroys spoilage organisms. The fruit usually is pasteurized in the package, and the treatment, varying with the fruit, is from 30 to 70 min at 70 to 100 percent relative humidity at 150 to 185 F (65.6 to 85 C).

MICROBIOLOGY OF DRIED FOODS

Prior to reception at the processing plant

The microbiology of foods prior to their reception at the processing plant is likely to be similar whether the foods are to be dried, chilled, frozen, canned, or otherwise processed. Fruits and vegetables have soil and water organisms on them when harvested, plus their own natural surface flora, and spoiled parts contain the microorganisms causing the spoilage. Growth of some of these organisms may take place before the foods reach the processing plant if environmental conditions permit. Thus piled vegetables may heat and support the surface growth of slime-forming, flavor-harming, or even rot-producing organisms. Meats and poultry are contaminated by soil, intestinal contents, handlers, and equipment. Fish are contaminated by water, and their own slime and intestinal contents, as well as by handlers and equipment; and growth may take place before the fish reach the processing plant. Eggs are dirtied by the hen, the nests, and the handler and, unless they are well and promptly handled, may support some microbial growth. Milk is subject to contamination from the time of its secretion by the cow to its reception at the processing plant and may support the growth of some low-temperature bacteria.

In the plant prior to drying

Growth of microorganisms that has begun on foods before they have reached the drying plant may continue in the plant up to the time of drying. Also equipment and workers may contaminate the food. As will be noted, some of the pretreatments reduce numbers of organisms and others may increase them, but the foods may be contaminated after any of these treatments.

The grading, selection, and sorting of foods, especially those like fruits, vegetables, eggs, and milk, will influence kinds and numbers of microorganisms present. The elimination of spoiled fruits and vegetables or of spoiled parts will reduce numbers of organisms in the product to be dried. The rejection of cracked, dirty, or spoiled eggs serves a similar purpose, as does the rejection of milk that does not conform to bacteriological standards of quality.

The washing of fruits and vegetables removes soil and other adhering materials and serves in this way to remove microorganisms. There also is the possibility of the addition of organisms if the water is of poor quality; and the moisture on the surface may encourage microbial growth if opportunity is given for it. The washing of eggs may prove more harmful than helpful unless they are used promptly, for the moisture aids the penetration of the shell by bacteria.

The peeling of fruits or vegetables, especially with steam or lye, should reduce numbers of microorganisms, since the majority of organisms usually are on the outer surface. Slicing or cutting should not increase numbers of organisms, but will do so if equipment is not adequately cleansed and sanitized.

Dipping in alkali, as applied to certain fruits before sun drying, may reduce the microbial population.

The blanching or scalding of vegetables reduces bacterial numbers greatly, as much as 99 percent in some instances. Following blanching, numbers of bacteria may build up because of contamination from equipment and opportunities for growth.

Sulfuring of fruits and vegetables also causes a great reduction in numbers of microorganisms and serves to inhibit growth in the dried product.

During the drying process

Heat applied during a drying process causes a reduction in total numbers of microorganisms, but the effectiveness varies with the kinds and numbers of organisms originally present and the drying process employed. Usually all yeasts and most bacteria are destroyed; but spores of bacteria and molds commonly survive, as do vegetative cells of a few species of heat-resistant bacteria. As noted later, improper conditions during drying may even permit the growth of microorganisms.

More microorganisms are killed by freezing than by dehydration during the freeze-drying process, when up to 90 percent of bacterial cells and 50 percent of spores of *Clostridium botulinum* may be killed in fruits, vegetables, and meats (68 percent in chicken). Any method that involves abrupt and considerable changes in temperature, either an increase during drying by heat or a decrease during freeze drying, is likely to cause "metabolic damage" to some organisms, making them more demanding nutritionally.

After drying

If the drying process and storage conditions are adequate, there will be no growth of microorganisms in the dried food. Heath reports that bacteria do not grow below 18 percent available moisture; yeasts require 20 percent or more and molds require 13 to 16 percent (see Chapter 11). During storage there is a slow decrease in numbers of organisms, more rapid at first and slower thereafter. The microorganisms that are resistant to drying will survive best; therefore the percentages of such organisms will increase. Especially resistant to storage under dry conditions are the spores of bacteria and molds, some of the micrococci, and microbacteria. There may be some opportunity for contamination of the dried food during packaging and other handling subsequent to drying.

Special treatments given some dry foods will influence microbial numbers. The sweating of dry fruits to equalize moisture may permit some microbial growth. Pasteurization of dry fruits will reduce numbers of microorganisms. Some products are repackaged for retail sale, figs in the Near East, for example, and are subject to contamination then.

The microbial content and the temperature of water used to rehydrate dried foods will affect the keeping quality of the rehydrated product. Bacteria in freeze-dried chicken meat are further reduced in numbers by rehydration with water at 50 C (122 F), and almost eliminated when the water is at 85 to 100 C (185 to 212 F). Growth of bacteria in the rehydrated meat will occur at favorable temperatures, but there is good shelf life (keeping time) at 4 C (39.2 F). *Staphylococcus aureus* has been found to survive freeze drying and rehydration at 60 C (140 F); therefore rehydration at 100 C (212 F) is recommended.

Microbiology of specific dried foods

DRIED FRUITS. The numbers of microorganisms on most fresh fruits range from comparatively few to many, depending upon pretreatments, and on most dried fruits they vary from a few hundred per gram of fruit to thousands; and in whole fruits they are mostly on the outer surfaces. Spores of bacteria and molds are likely to be most numerous. When part of the fruit has supported growth and sporulation of mold before or after drying, mold spores may be present in large numbers.

DRIED VEGETABLES. Microbial counts on dried vegetables range from negligible numbers to millions per gram. The numbers on the vegetable just before drying may be high because of contamination and growth after blanching, and the percentage killed by the dehydrating process usually is less than with the more acid fruits. If drying trays are improperly loaded, souring of such vegetables as onions or potatoes by lactic acid bacteria with marked increase in numbers of bacteria may take place during the drying process. The risk is greater with onions because they are not blanched.

Chiefly bacteria are found on dried vegetables. A number of investigators have listed the genera of bacteria found (summarized by Vaughn) to include *Escherichia, Aerobacter, Achromobacter, Bacillus, Clostridium, Micrococcus, Pseudomonas,* and *Streptococcus.* Vaughn found *Lactobacillus* and *Leuconostoc* species predominant in many samples of dehydrated vegetables.

DRIED EGGS. Dried eggs may contain from a few hundred microorganisms, mostly bacteria, per gram up to over a hundred million, depending upon the eggs broken and the methods employed. Since the contents of fresh eggs of good quality are normally free of microorganisms or include only a few, dried eggs should be low in microbial numbers. However, the inclusion of poorly washed eggs, those that have been permitted to sweat, dirty and cracked eggs, and those already invaded by microorganisms may add large numbers of organisms; also contamination and growth may take place during breaking and other handling prior to drying. The drying process may reduce numbers ten- to a hundredfold but still permit large numbers to survive. A variety of kinds of organisms has been found in dried eggs, including micrococci, streptococci, coliforms, spore-formers, and molds. When egg white has been pretreated by a fermentation process, the counts on the dried product may be high. Egg yolk is a better culture medium than the white and is likely to have higher counts at breaking and to support growth better before drying.

DRIED MILK. The number of microorganisms in dry milk may vary from a few hundred per gram to millions, depending upon the milk being dried and the drying process. Roller or drum drying kills more organisms than the spray process. The American Dry Milk Institute has set bacterial standards (see Appendix) for the various grades of whole and nonfat dry milk, ranging from 30,000 to 100,000 viable bacteria per gram. The predominant kinds of organisms in dry milk are thermoduric streptococci (like *Streptococcus thermophilus* and *S. durans*), micrococci, spore-formers, and microbacteria (*Microbacterium lacticum*). The United States Department of Agriculture has set standards for numbers of bacteria in nonfat dry milk to be purchased by the government, as estimated by direct microscopic clump counts. At present this count should not be over 200 million per gram of dry milk, but this limit is 75 million for Extra Grade. Most of the bacteria so counted are not alive but are indicative of previous growth in the original milk or during the processing.

REFERENCES

ANONYMOUS. 1962. Tumbling freeze drying. Food Processing 23(10):67.

CRUESS, W. V. 1958. Commercial fruit and vegetable products, chaps. 17, 18, 19. 4th ed. McGraw-Hill Book Company, New York.

FRY, R. M., and R. I. N. GREAVES. 1951. The survival of bacteria during and after drying. J. Hyg. 49:220–246.

GOLDBLITH, S. A. 1963. Microbiological considerations in freeze-dehydrated foods. *In* S. A. Goldblith (*Ed.*) Exploration in future food-processing techniques. The M.I.T. Press, Cambridge, Mass.

GOLDBLITH, S. A., M. KAREL, and G. LUSK. 1963. The role of food science and technology in the freeze dehydration of foods. Food Technol. 17:139–144, 258–264.

Great Britain Ministry of Agriculture, Fisheries and Food. 1961. The accelerated freeze-drying (AFD) method of food preservation. Her Majesty's Stationery Office, London.

HARPER, J. C., and A. L. TAPPEL. 1957. Freeze-drying of food products. Advances Food Res. 7:172–234.

HEATH, B. 1947. Dehydrated foods and microorganisms. Australian Food Manufacture 16(12), July 5.

HUNZIGER, O. F. 1949. Condensed milk and milk powder. 7th ed. [Published by author] La Grange, Ill.

LAZAR, M. E., E. J. BARTA, and G. S. SMITH. 1963. Dry-blanch-dry method for drying fruit (DBD). Food Technol. 17:1200–1202.

MAY, K. N., and L. E. KELLY. 1965. Fate of bacteria in chicken meat during freeze-dehydration, rehydration, and storage. Appl. Microbiol. 13:340–344.

MRAK, E. M., and G. MACKINNEY. 1951. The dehydration of foods, chap. 33. In M. B. Jacobs (Ed.) The chemistry and technology of food and food products. 2nd ed. Interscience Publishers (Division of John Wiley & Sons, Inc.), New York.

National Agricultural Library. 1963. Freeze-drying of foods; a list of selected references. U.S. Department of Agriculture, Washington, D.C.

ROCKWELL, W. C., E. LOWE, A. I. MORGAN, JR., R. P. GRAHAM, and L. F. GINNETTE. 1962. How foam-mat dryer is made. Food Eng. 34(8):86–88.

VAN ARSDEL, W. B. 1963. Food dehydration: Vol. I, Principles. Avi Publishing Co., Inc., Westport, Conn.

VAN ARSDEL, W. B., and M. J. COPLEY (Eds.) 1964. Food dehydration: Vol. II, Processes and products. Avi Publishing Co., Inc., Westport, Conn.

VAUGHN, R. H. 1951. The microbiology of dehydrated vegetables. Food Res. 16:429–438.

VON LOESECKE, H. W. 1955. Drying and dehydration of foods. 2nd ed. Reinhold Publishing Corporation, New York.

WOODWARD, H. T. 1963. Freeze-drying without vacuum. Food Eng. 35(6):96–97.

ZIEMBA, J. V. 1962. Now—drying without heat. Food Eng. 34(7):84–85.

CHAPTER NINE

PRESERVATION BY PRESERVATIVES

Preservatives have been defined by Jacobs as "chemical agents which serve to retard, hinder, or mask undesirable changes in food." These changes may be caused by microorganisms, by the enzymes of food, or by purely chemical reactions. The inhibition of the growth and activity of microorganisms is one of the main purposes of the use of preservatives and therefore will be the chief subject of discussion here. Preservatives may inhibit microorganisms by interfering with their cell membranes, their enzyme activity, or their genetic mechanism. Preservatives also may be used as antioxidants to hinder the oxidation of unsaturated fats, as neutralizers of acidity, as stabilizers to prevent physical changes, as firming agents, and as coatings or wrappers to keep out microorganisms, prevent loss of water, or hinder undesirable microbial, enzymatic, and chemical reactions. The ideal preservative is constantly being sought but has not been found: one that would be harmless to the consumer and efficient in its preservative action, and would not cover up inferiority of the food or add undesirable color, odor, or taste.

In addition to the chemicals added to foods or put on them or around them to aid in their preservation, there are many chemicals that get on or into foods during production, processing, or packaging. Residues of pesticides, herbicides, and fungicides on fruits and vegetables, of detergents used in washing foods, and of detergents and sanitizers used on utensils and equipment are likely to carry over into foods.

The present discussion will be limited to antimicrobial preservatives added to foods or developed in them. Factors that influence the effectiveness of chemical agents in killing microorganisms or inhibiting their growth and activity are similar to those considered in Chapter 6 on effectiveness of heating: (1) concentration of the chemical; (2) kind, number, age, and previous history of the organism; (3) temperature; (4) time; and (5) the chemical and physical characteristics of the substrate in which the organism is found (moisture content, pH, kinds and amounts of solutes, surface tension, and colloids and other protective substances). A chemical agent may be bactericidal at a certain concentration, only inhibitory at a lower level, and ineffective at still greater dilutions.

Selection of a preservative

Ideally a preservative should kill rather than inhibit microorganisms. It should be effective against those likely to grow in the food, especially against food-poisoning organisms. It should not be inactivated by the food or any substance in the food or by products of microbial metabolism. If germicidal, it should decompose to innocuous products, or it should be destroyed by cooking. It should not encourage the development of resistant strains of microorganisms. Preferably the preservative, for example an antibiotic, should not be in common use for therapy or for animal feeds. When used with heat, the preservative should be heat-stable and should guarantee security against *Clostridium botulinum*.

Tests for added preservatives

Tests for added preservatives involve determination of their effect on growth of selected test organisms, e.g., of a yeast for acid foods and of bacteria for nonacid foods.

ADDED PRESERVATIVES

The Federal Food, Drug, and Cosmetic Act, as amended by the Food Additives Amendment of 1958, defines a **chemical preservative** as "any chemical which, when added to food, tends to prevent or retard deterioration thereof; but does not include common salt, sugars, vinegars, spices, or oils extracted from spices, or substances added by . . . wood smoke." A **food additive** is "any substance the intended use of which results or may reasonably be expected to result, directly or indirectly, in its becoming a component or otherwise affecting the characteristics of any food . . . , if such substance is not generally recognized, among experts qualified by scientific training and experience to evaluate its safety, as having been adequately shown through scientific procedures (or, in the case of a substance used in food prior to January 1, 1958, through either scientific procedures or experience based on common use in food) to be safe under the conditions of its extended use."

The foregoing definitions have led to the grouping of antimicrobial preservatives added to foods as follows: (1) Those added preservatives not defined as such by law: natural organic acids (lactic, malic, citric, etc.) and their salts, vinegars (acetic is a natural acid), sodium chloride, sugars, spices and their oils, wood smoke, carbon dioxide, and nitrogen. (2) Substances generally recognized as safe (**GRAS**) for addition to foods: propionic acid and sodium and calcium propionates, caprylic acid, sorbic acid and potassium, sodium, and calcium sorbates, benzoic acid and benzoates and derivatives of benzoic acid such as methylparaben and propylparaben, sodium diacetate, sulfur dioxide and sulfites, potassium and sodium bisulfite and metabisulfite, and sodium nitrite. (Limitations on the use of some of these will be mentioned later.) (3) Chemicals considered to be food additives, which would include all not listed in the first two categories. They can be used only when proved safe for man or animals, and they then fall into group 4. (4) Chemicals proved safe and approved by the Food and Drug Administration.

Preservatives added to inhibit or kill microorganisms may be classified on various other bases, such as their chemical composition, mode of action, specificity, effectiveness, and legality. Some, e.g., sugar, are effective because of their physical action, others because of their chemical action, as with sodium benzoate, and others because of a combination of these effects, as with sodium chloride. Some preservatives are incorporated into foods and usually are antiseptic rather than germicidal, while others are used only to treat outer surfaces and may kill organisms as well as inhibit them. Some are employed to treat wrappers or containers for foods, while others are used as gases or vapors about the food. Some have been incorporated in ice used to chill foods like fish. Preservatives may be fairly specific against microorganisms; for example, they may be effective against molds or yeasts and less so against bacteria, or vice versa, and may act against definite groups or species of bacteria or other organisms.

Chemical compounds and elements used as preservatives may be divided for convenience into inorganic and organic substances.

Inorganic preservatives

The inorganic preservatives are chiefly inorganic acids and their salts, although alkalies and alkaline salts, metals, halogens, peroxides, and gases also fall into this group.

INORGANIC ACIDS AND THEIR SALTS. Most commonly used are sodium chloride, hypochlorites, nitrates and nitrites, sulfites and sulfurous acid, and boric acid and borates. Acids injure cell membranes and encourage decarboxylation of amino acids.

Sodium chloride is used in brines and curing solutions or is applied directly to the food. Enough may be added to slow down or prevent the growth of microorganisms or only enough to permit an acid fermentation to take place. Salt has been reported to have the following effects: (1) it causes high osmotic pressure and hence plasmolysis of cells, the percentage of salt necessary to inhibit growth or harm the cell varying with the microorganism; (2) it dehydrates foods by drawing out and tying up moisture, as it dehydrates microbial cells; (3) it ionizes to yield the chlorine ion which is harmful to organisms; (4) it reduces the solubility of oxygen in the moisture; (5) it sensitizes the cell against carbon dioxide; (6) it interferes with the action of proteolytic enzymes. The effectiveness of NaCl varies directly with its concentration and the temperature.

Hypochlorites, usually of calcium or sodium, yield hypochlorous acid, a powerful oxidizing agent, and are effective germicidal agents; but their effectiveness is reduced by the presence of organic matter in any considerable amount. The hypochlorites are used in the treatment of water used in food plants for drinking, processing, and cooling, and on plant equipment. They have been incorporated in ice for icing fish in transit and in water for washing the exterior of fruits and vegetables. Microorganisms are harmed by oxidation or by direct chlorination of cell proteins.

Nitrates and nitrites are used in the curing of meats, primarily to fix a desirable red color, but nitrites have some bacteriostatic effect in acid solutions and have been recommended for the preservation of fish. Nitrates raise the oxidation-reduction potential and therefore are more favorable to aerobic than to anaerobic organisms; conversely, nitrites reduce the O-R potential. A nitrite is bacteriostatic because of its effect on sulfur metabolism, its action on α-amino groups of amino acids at low pH, its reaction with monophenols, e.g., tyrosine, and the reaction of its nitric oxide with heme pigments. The undissociated molecule is especially effective.

Sulfur dioxide (sulfurous acid), sulfites, and metabisulfites cannot be used in foods high in thiamine, such as meats, and are used for only a limited number of foods, such as molasses, fruits and fruit juices, and wines. In this country "sulfuring," or treatment with sulfur dioxide, is applied chiefly to dried fruits, where the main purpose is the conservation of color and not the inhibition of microorganisms, although molds are affected more readily than yeasts or bacteria. Potassium pyrosulfite (metabisulfite) has been used as a source of SO_2 in canning powders, and liquid sulfur dioxide is added to the musts for wine manufacture to inhibit competing microorganisms and hence favor the wine yeasts. Sulfur dioxide is effective because of free hydrogen ions released in solution. It affects germination of bacterial spores, and inhibits NAD-dependent (NAD: nicotinamide-adenine-dinucleotide) steps in the metabolism of carbohydrates.

Phosphoric acid is used in some of the soft drinks, such as the colas.

Boric acid and borates still are used in some countries as preservatives for foods, but their use is forbidden in the United States. Powdered boric acid has been dusted onto foods, for example meats, but it is a very weak antiseptic and is not considered healthful. Borax (sodium tetraborate) has been used to wash vegetables and whole fruits, such as oranges.

Fluorides are considered harmful to health, and their use as preservatives is forbidden. The tolerance for fluorides in food is 2 ppm, according to Food and Drug Administration regulations.

ALKALIES AND ALKALINE SALTS. These are employed chiefly as cleansing agents or detergents, which will be discussed later. Sodium hydroxide solution serves as an antiseptic, as well as a detergent, when milking-machine parts and other equipment are soaked in it. Sodium carbonate, sodium metasilicate, trisodium phosphate, and the polyphosphates are examples of alkaline salts. Alkaline agents are effective antiseptics because of the liberation of free hydroxyl ions in solutions of them. Alkaline agents injure cell membranes and encourage deamination of amino acids.

METALS. Heavy metals exert an antiseptic or germicidal effect on microorganisms that sometimes is termed an "oligodynamic" action, the result, probably, of injury to the cell membranes, denaturation of proteins, and direct combination with the proteins. Silver is the only metal that has been recommended for use in foods. In the "catadyn" process, silver ions are released in water from a silver electrode by a weak current. Depending upon the concentration of silver, microorganisms may be only inhibited or may be destroyed. Some workers report that yeasts are readily affected and others claim the reverse. *Aerobacter, Proteus,* and *Pseudomonas* species and intestinal pathogens have been reported most sensitive, lactics fairly sensitive, and most yeasts and molds fairly resistant. The catadyn process has been recommended for the treatment of drinking water, vinegar, and clarified fruit juices. Silver-lined flasks have been used in Germany for the storage of clear liquids in the belief that preservation would be furthered.

HALOGENS. Water for washing foods or equipment, for cooling, for addition to some products (e.g., washing of butter), and for drinking may be chlorinated by the direct addition of chlorine; or hypochlorites or chloramines may be used. Iodine-impregnated wrappers have been employed to lengthen the keeping time of fruits. Iodophors, which are combinations of iodine with nonionic wetting agents and acid, are being used in the sanitization of dairy utensils. Halogens kill organisms by oxidation, by injury to cell membranes, or by direct combination with cell proteins.

PEROXIDES. The oxidizing agent, hydrogen peroxide, has been used as a preservative, usually in conjunction with heat. One method for the pasteurization of milk for cheese involves the addition of H_2O_2 and the use of a comparatively low heating temperature. Excess peroxide is decomposed by means of catalase. Thermophiles are destroyed in the processing of sugar by a combination of heat and H_2O_2. Other peroxides are used in foods but not for the prevention of microbial growth.

GASES. Gas storage of foods has been mentioned in Chapter 7 in connection with preservation by chilling and will be discussed further in the chapters on the preservation of specific foods. Most often used is carbon dioxide in combination with chilling. Oxygen or air under pressure has been combined with chilling, as in the Hofius process for milk. Nitrogen is used as an inert gas over foods that should not be exposed to air. Ozone has been used in gas storage; and ethylene oxide (an organic compound) formerly was employed in the treatment of spices and, with activated hydrocarbons, in the preservation of fruits. Now propylene oxide is taking its place; it is permitted for treatment of dried prunes and glacé fruit. The two oxides, however, may combine with the chlorine of inorganic chlorides in foods, to form the very toxic chlorohydrins, which persist under most food processing conditions.

Organic preservatives

The present-day search for antiseptics, germicides, and preservatives is directed mainly toward the synthesis of organic compounds for these purposes. In recent years large numbers of such compounds have been introduced, so many that no attempt will be made to present even a partial list. Most of these, if used in connection with foods, are employed to impregnate food wrappers, are applied to surfaces to remove or kill microorganisms, or are in the form of a vapor or gas about the food. The mixture of most of them with the food would not be permitted. Most of the substances natural to foods or developed in them during fermentation may be added if the amount added is stated clearly on the label of the container. These substances include certain organic acids—lactic, acetic, propionic, and citric, for example—constituents of wood smoke if added by smoking, salt, sugars, spices, and ethyl alcohol. Compounds that may be present in foods, but may not be added as such, include oxalic, benzoic (in some states), and other organic acids, some of the alcohols, and certain aldehydes and ketones. Formaldehyde and cresols may be introduced in small amounts as wood smoke but may be added in no other manner.

ORGANIC ACIDS AND THEIR SALTS. Lactic, acetic, propionic, and citric acids or their salts may be added to or developed in foods. Their development in foods during fermentation will be discussed in a following section. Citric acid is used in sirups, drinks, jams, and jellies as a substitute for fruit flavors and for preservation. Lactic and acetic acids are added to brines for pickles of various kinds, green olives, etc. Acetic acid and sodium or calcium propionate are used for their bacteriostatic and mycostatic properties in bread and cakes and on cheese, dried fish, and other foods. The fatty acids damage cell membranes.

Sodium benzoate is a "chemical preservative" that may be added legally to foods in the United States, although some states forbid its addition. Not more than 0.1 percent is added, and the addition must be stated plainly on the package. Sodium benzoate is relatively ineffective at pH values near neutrality, and the effectiveness increases with increase in acidity, an indication that the undissociated acid is the effective agent. The pH at which sodium benzoate is effective is in itself enough to inhibit the growth of most bacteria; but some (not all) yeasts and molds are inhibited at pH levels that would otherwise permit their growth. Sodium benzoate has been used in catchup (but is omitted by present standards), oleomargarine, fruit juices, jellies, and other acid foods, and benzoic acid has been incorporated in ice for preserving fish. Methyl and propyl esters of parahydroxybenzoic acid, called methyl and propyl paraben (up to 0.1 percent), and the sodium and calcium salts are GRAS. They are reported to be effective against bacteria, yeasts, and molds, and to delay surface growth on sausages and prevent greening.

Salicylic acid and salicylates have been used as food preservatives in some of the European countries, but their use is forbidden here. They are comparable to benzoic acid and benzoates in effectiveness, but are supposedly more harmful to the consumer. Aspirin (acetylsalicylic acid), usually as the sodium salt, has been used by home canners for acid foods, but its addition is not recommended. Benzoic and salicylic acids interfere with sundry enzymatic processes in the cell.

Derivatives of acetic acid, such as monochloracetic acid, peracetic acid, dehydroacetic acid, and sodium diacetate, have been recommended as preservatives, but not all are approved by the Food and Drug Administration. Dehydroacetic acid has been used to impregnate wrappers for cheese to inhibit growth of molds and as a temporary preservative for squash. Sodium diacetate is permitted in bread (not over 0.32 part for each 100 parts by weight of flour).

Sorbic acid, a six-carbon α,β-unsaturated fatty acid, and sorbates are being used as fungistatic agents for foods, for packed or sliced hard cheeses, and on the surfaces and in the wrapping materials of cut cheese. They also have been recommended for the preservation of sweet pickles and for the control of the lactic acid fermentation of cucumbers and olives. They inhibit molds, yeasts, and some highly aerobic bacteria, but have little effect against lactic acid bacteria and anaerobes. It has been suggested that sorbic acid interferes with oxidative assimilation of carbon.

Organic acids exert their antiseptic effect by increasing the hydrogen-ion concentration in foods, but probably more important is the concentration of undissociated acid present, the effect of which varies with the kind of acid. Undissociated benzoic acid, for example, is thought to be especially effective.

FORMALDEHYDE. The addition of formaldehyde to foods is not permitted, except as a minor constituent of wood smoke, but this compound is effective against molds, bacteria, and viruses and can be used where its poisonous nature and irritating properties are not objectionable. Thus it is useful in the treatment of walls, shelves, floors, etc., to eliminate molds and their spores. Paraformaldehyde can be used to control bacterial and fungal growth in maple-tree tapholes (2 ppm tolerance of formaldehyde in the maple sirup). Formaldehyde probably combines with free amino groups of the proteins of cell protoplasm, injures nuclei, and coagulates proteins.

SUGARS. Sugars, such as glucose or sucrose, owe their effectiveness as preservatives to their ability to tie up moisture, thereby making it unavailable to organisms, and to their osmotic effect. Examples of foods preserved by means of high sugar concentrations are sweetened condensed milk, fruits in sirups, jellies, and candies.

ALCOHOLS. Ethanol, a coagulant and denaturizer of cell proteins is most germicidal in concentrations between 70 and 95 percent. Flavoring extracts, vanilla and lemon extracts, for example, are preserved by their content of alcohol. The alcohol content of beer, ale, and unfortified wine is not great enough to prevent their spoilage by microorganisms but limits the types able to grow. Liqueurs and distilled liquors usually contain enough alcohol to ensure freedom from microbial attack. Methanol is poisonous and should not be added to foods; the traces added to foods by smoking are not enough to be harmful. Glycerol is antiseptic in high concentrations because of its dehydrating effect but is unimportant in food preservation. Propylene glycol has been used as a mold inhibitor and as a spray to kill air-borne microorganisms.

ANTIBIOTICS.   In the future there probably will be more use of antibiotics as preservatives in foods. Most of the better-known antibiotics have been tested on raw foods, chiefly proteinaceous ones like meats, fish, and poultry, in an endeavor to lengthen the storage time at chilling temperatures. Aureomycin (chlortetracycline) has been found superior to other antibiotics tested, but Terramycin (oxytetracycline) is almost as good for lengthening the time of preservation of foods. Some success also is claimed with Chloromycetin (chloramphenicol). These three antibiotics inhibit protein synthesis in the cell. Streptomycin, neomycin, polymyxin, nisin, subtilin, bacitracin, and others were not as satisfactory, and penicillin was of little use. Nisin has been employed in Europe to suppress anaerobes in cheese and cheese products.

Experimentally, antibiotics have been combined with heat in attempts to reduce the thermal treatment necessary for the preservation of low- and medium-acid canned foods. Most tests have been with the peptides, subtilin and nisin, and the macrolide, tylosin. It has been suggested that a "botulinum cook," i.e., enough of a heat-treatment to inactivate all spores of *Clostridium botulinum,* be given canned foods, combined with the addition of enough antibiotic to inhibit germination and outgrowth of surviving spores of the more heat-resistant thermophilic spoilage bacteria and putrefactive anaerobes. Subtilin supposedly has no effect on the heat resistance of bacterial spores but inhibits heat-damaged cells during outgrowth; whereas nisin apparently interferes with spore germination and with lysis of the spore coat. Tylosin may inhibit cell growth.

Although food bacteriologists realize the advantages of the preservation of raw foods by means of a nontoxic antibiotic or the use of one in combination with reduced amounts of heat in the processing of canned foods, they raise certain questions about the use of antibiotics as preservatives. They agree that antibiotics never should be substituted for good hygiene. The effect of an antibiotic on microorganisms is known to vary with the species or even with the strain of the organism; hence the antibiotic may be effective against some spoilage organisms but not others, or against part of the population in a culture but not all organisms. Organisms are known to become adapted to increasing concentrations of an antibiotic so that new, resistant strains may develop. There also is the possibility that other organisms, not now significant in food spoilage but resistant to the antibiotic, might assume new importance in food spoilage. Then, too, there may be effects of the antibiotic on the consumer, such as his sensitization to it, changes in his intestinal flora, and the development of strains of pathogens in his body resistant to that antibiotic, although these effects probably would be minimized by the very low levels of antibiotics employed in foods as compared with amounts employed for therapy. It has been recommended that antibiotics selected for use in food preservation be other than those being used in the treatment of human diseases. Canners feel that, when used in the processing of canned foods, the antibiotic plus the heat-treatment must destroy all spores of *Clostridium botulinum* and allow a margin of safety, and preferably the treatment should destroy all spoilage organisms and their spores. If spores survive, the antibiotic must remain in sufficient concentration in the food to prevent germination or outgrowth or vegetative growth of cells. This means that the antibiotic must persist in bacteriostatic or sporostatic concentration throughout the storage life of the canned food.

Attempts have been made to test the bactericidal effect of edible extracts of plants, e.g., of carrot, green bean, tomato, and celery plants, in combination with a milder heat-treatment than usual, to destroy various bacteria and bacterial spores. The use of plant extracts in this manner would avoid most of the problems just discussed.

Until recently the use of antibiotics as preservatives for foods has not been permitted in the United States. Now, however, the Food and Drug Administration has approved the use of a chlortetracycline and oxytetracycline dip for preserving poultry, setting up a 7-ppm tolerance in the uncooked, dressed fowls. This quantity of antibiotic has been shown to double or triple the storage life of the poultry. Apparently approval was granted because evidence had been given to prove (1) that use of the material gives added protection to the consumer; (2) that basic sanitation procedures are not replaced because of the method; and (3) that the antibiotic is destroyed during cooking of the poultry, leaving no harmful end products. Now it also is permissible to use these tetracyclines at 5 ppm on fresh fish, shucked scallops, and unpeeled shrimp. The antibiotic may be applied as a dip or as an ice.

WOOD SMOKE. The smoking of foods usually has two main purposes: to add desired flavors and to aid in preservation. Other desirable effects may result, however, like improvement in the color of the inside of meat and in the finish, or "gloss," of the outside, and a tenderizing action on meats. The smoking process aids in preservation by impregnation of the food near the surface with chemical preservatives from the smoke, by combined action of the heat and these preservatives during smoking, and by the drying effect, especially at the surface. Commonly, smoke is obtained from the burning wood, preferably a hardwood like hickory, but it may be generated from burning corn cobs or other materials. Other woods used are oak, maple, beech, birch, walnut, and mahogany. Sawdust is added to the fire to give a heavy smudge. Temperature and humidity are controlled at levels favorable to the product being smoked, and the duration of smoking depends upon the kind of food.

Smoking temperatures for meats vary from 43 to 71 C (109.4 to 159.8 F) and the smoking period from a few hours to several days.

Wood smoke contains a large number of volatile compounds that differ in their bacteriostatic and bactericidal effect. Formaldehyde is considered the most effective of these compounds, with phenols and cresols next in importance. Other compounds in the smoke are: aliphatic acids from formic through caproic, primary and secondary alcohols, ketones, acetaldehyde and other aldehydes, waxes, resins, guaiacol and its methyl and propyl isomers, catechol, methyl catechol, and pyrogallol and its methyl ester. These compounds sometimes are grouped under the term "pyroligneous acid." As would be expected, wood smoke is more effective against vegetative cells than against bacterial spores, and the rate of germicidal action of the smoke increases with its concentration and the temperature and varies with the kind of wood employed. The residual effect of the smoke in the food has been reported to be greater against bacteria than against molds. The concentration of mycostatic materials from wood smoke necessary to prevent mold growth increases with a rise in the humidity of the atmosphere of storage.

The application of "liquid smoke," a solution of chemicals similar to those in wood smoke, to the outside of foods has little or no preservative effect, although it contributes to flavor.

SPICES AND OTHER CONDIMENTS. Spices and other condiments do not have any marked bacteriostatic effect in the concentrations customarily used but may aid other agents in the prevention of the growth of organisms in food. Different lots of spice vary in their effectiveness, depending upon the source, the freshness, and whether they have been stored whole or ground up. The inhibitory effect of spices differs with the kind of spice and the microorganism being tested. Mustard flour and the volatile oil of mustard, for example, are very effective against *Saccharomyces cerevisiae,* but are not as potent against most bacteria as are cinnamon and cloves. The essential oils of spices are more inhibitory than the corresponding ground spices.

Cinnamon and cloves, containing cinnamic aldehyde and eugenol, respectively, usually are more bacteriostatic than are other spices. Ground peppercorn and allspice are less inhibitory than cinnamon and cloves, and mustard, mace, nutmeg, and ginger still less. Thyme, bay leaves, marjoram, savory, rosemary, black pepper, and others have only weak inhibitory power against most organisms and may even stimulate some, for example, yeasts and molds. Fairly heavy concentrations of the more effective spices permit mycelial growth of some of the molds but inhibit the formation of asexual spores. Of the oils tested, the volatile oil of mustard is most effective against yeasts; oils of cinnamon and cloves are fairly effective and oils of thyme and bay leaves less effective.

Unless spices have been treated to reduce their microbial content, they may add high numbers and undesirable kinds of microorganisms to foods of which they are ingredients.

Other plant materials used in seasoning foods, such as horse-radish, garlic, and onion, may be bacteriostatic or germicidal. Extracts of these plants, as well as of cabbage and turnip, have been shown to be inhibitory to *Bacillus subtilis* and *Escherichia coli*. Acrolein is supposedly the active principle in onions and garlic, and butyl thiocyanate the one in horse-radish. These volatile compounds are lost from the condiment on exposure to the air, with a corresponding loss in bacteriostatic properties.

Other groupings of chemical agents

OXIDIZING AGENTS. Most of the oxidizing agents employed in food preservation have been mentioned under other headings. These include peroxides, bromine, chlorine, iodine, hypochlorites, chloramines, nitrates, and ozone. Oxidizing agents used in bleaching flour, oxides of nitrogen, chlorine, nitrosyl chloride, nitrogen trichloride, and benzoyl peroxide may be bacteriostatic or even bactericidal, but that is not the reason for their addition.

CLEANSING AND SANITIZING AGENTS. The use of chemical agents for cleansing and sanitizing equipment will be discussed in Chapter 28. Most of these chemicals are antiseptic and some are germicidal in the concentrations employed, but their main function in food preservation is to reduce the contamination of foods with microorganisms from the equipment contacted by these foods. There should not be enough residual chemical carry-over into the food to be of any significance in its preservation. Some of these chemicals are employed to remove dirt and remove or kill microorganisms on the outer surfaces of some foods, e.g., fruits, vegetables, and eggs, but none of these detergents or sanitizers would be incorporated in foods as preservatives.

FUNGISTATIC AND FUNGICIDAL AGENTS. It has been mentioned that chemical preservatives can be grouped on the basis of their specificity for certain microorganisms. Considerable attention has been given to compounds that are effective against molds and related fungi. Mycostats previously mentioned include propionic acid and propionates, caprylic acid, acetic acid, dehydroacetic acid, monochloroacetic acid, sorbic acid and sorbates, and propylene glycol. Of these only the propionates and sorbates have been incorporated in foods to any great extent. Chemical preservatives against molds have been applied in liquid form to the outside surfaces of foods or used as a vapor about them. Examples of mycostatic chemicals which have been tried in vapor or gaseous form about foods are propylene glycol, carbon dioxide, methyl bromide, and various derivatives of phenol. Antifungal antibiotics include griseofulvin, pimaricin, fulcine, actidione, rimocidin, and nyastatin. Antimycotics permitted in food-packaging material (under specified conditions) are: sodium and calcium propionates, sodium benzoate, sorbic acid and sorbates, and methyl and propyl parabens. Caprylic acid may be used in cheese wraps. Also recommended for impregnating wrappers for foods have been iodine, sulfites, o-phenylphenol or sodium o-phenylphenate, biphenyl, dimethylol-urea, and many other substances. A large number of compounds, too many to list, have been recommended for the treatment of surfaces of fruits, vegetables, eggs, cheese, meats, and fish. These range from simple, inorganic compounds such as nitrites and sulfur dioxide (sulfurous acid), through organic acids (crotonic, sorbic, levulinic, etc.) and esters of organic acids (such as esters of vanillic or parabenzoic acid), to complex phenolic compounds. Biphenyl and sodium o-phenylphenate plus ammonia, for example, have been applied to citrus fruits to reduce fungal spoilage. Most of these compounds cannot be used without the approval of the Food and Drug Administration.

Diethyl pyrocarbonate (DEPC), which is especially effective against yeasts, is permitted in bottled wines (not over 200 ppm). This compound hydrolyzes to ethanol and carbon dioxide.

Tolerances for some of the chemical preservatives added to foods are given in Table 9–1.

BOILER-WATER ADDITIVES. Boiler-water additives that have been approved by the Food and Drug Administration must be used for steam that comes in contact with foods. These are listed in a reference at the end of this chapter.

TABLE 9–1. Allowable tolerances for some chemical preservatives in foods

Chemical	Maximal tolerance	Foods
Benzoic acid, benzoate	0.1%	Oleomargarine
Dehydroacetic acid	65 ppm	Squash
Propylene oxide	700 ppm residue	Dried prunes, glacé fruit
Sodium nitrate	500 ppm	Meat-curing preparations, smoked, cured salmon, sablefish
Sodium nitrite	10 ppm	Smoked, cured tuna
	200 ppm	Meat-curing preparations, smoked, cured salmon, sablefish
Ethyl formate	250 ppm HCOOH	Raisins, dried currants
Sorbic acid, sorbate	0.2%	Hard cheese (sliced, packaged)
Na or Ca propionate	0.3%	Cheese foods, spreads

DEVELOPED PRESERVATIVES

Food fermentations may serve either or both of two purposes: (1) to produce new and desired flavors and physical characteristics and hence a different food product, and (2) to aid in the preservation of the food. The relative importance of these two aims varies with the food and is difficult to evaluate. Certainly the first fermented milks and sauerkraut were empirical discoveries and served primarily to keep milk or cabbage over long periods of storage; but a taste for the fermented products developed, and they now are made as much for their palatability as for their keeping quality.

The preservatives produced in foods by microbial action are, for the most part, acids (chiefly lactic) and alcohol. The preservative effect of these substances nearly always is supplemented by one or more additional preservative agents, such as low temperature, heat, anaerobic conditions, sodium chloride, sugar, or added acid.

Developed acidity plays a part in the preservation of sauerkraut, pickles, green olives, fermented milks, cheese, certain sausages, and in various fermented foods of plant origin. Development of the full amount of acidity from the sugar available may be permitted in the pickle and green-olive fermentations, or the fermentation may be stopped by chilling or canning before the maximum acidity is attained in other fermentations, such as that for fermented milks or sauerkraut. The approximate acidity developed in some of these products, expressed as lactic acid, is sauerkraut, 1.7 percent; salt-stock or dill pickles and green olives, 0.9 percent; and fermented milks, 0.6 to 0.85 percent. The acidity of cheese usually is expressed in terms of hydrogen-ion concentration; most freshly made cheeses have a pH of about 5.0 to 5.2 and become more alkaline during curing.

The alcohol content of beer, ale, and fermented fruit juices has a preservative effect but was not produced primarily for that purpose. Liqueurs and distilled liquors, for the most part, contain enough alcohol to prevent the growth of most microorganisms.

The microbiology of the fermentations for the production of sauerkraut, cucumber pickles, and green olives will be discussed in Chapter 15 and for the production of cheese in Chapter 20. The other food fermentations will be described in Chapter 24.

REFERENCES

ANONYMOUS. 1960. Additives in the White List. Food in Canada 20(8):30–39.
ANONYMOUS. 1960. FDA names 182 safe additives. Food Eng. 32(1):81–82.

ANONYMOUS. 1962. Boiler water additives approved for steam that contacts food. Food Processing 23(9):29–30.

BELL, T. A., J. L. ETCHELLS, and A. F. BORG. 1959. Influence of sorbic acid on the growth of certain species of bacteria, yeasts, and filamentous fungi. J. Bacteriol. 77:573–580.

BOSUND, ONGMAR. 1962. The action of benzoic and salicylic acids on the metabolism of microorganisms. Advances Food Res. 11:331–353.

CAMPBELL, L. L., Jr., and R. T. O'BRIEN. 1955. Antibiotics in food preservation. Food Technol. 9:461–465.

CLIFTON, C. E. 1957. Introduction to bacterial physiology. John Wiley & Sons, Inc., New York.

FABIAN, F. W. 1951. Food preservation by use of microorganisms, chap. 36. In M. B. Jacobs (Ed.) The chemistry and technology of food and food products. 2nd ed. Interscience Publishers (Division of John Wiley & Sons, Inc.), New York.

FARBER, L. 1959. Antibiotics in food preservation. Annu. Rev. Microbiol. 13:125–140.

Food Additives Amendment of 1958. Public Law 85-929, 85th Congress, H.R. 13254, Sept. 6, 1958.

GOLDBERG, H. S. 1964. Nonmedical uses of antibiotics. Advances Appl. Microbiol. 6:91–117.

GOULD, G. A. 1964. Gas sterilization of packaged, dried ingredients. Food Processing 25(9):96–97, 104–106.

HALL, L. A. 1964. Chemicals; twenty-five years of progress. Food Technol. 18:1377–1380.

INGRAM, M., ELLA M. BARNES, and J. M. SHEWAN. 1956. Problems in the use of antibiotics for preserving meat and fish. Food Sci. Abstr. 28:121–136.

MAHONEY, J. F. 1958. Food additives amendment enacted. Food Technol. 12:637–640.

McCONNELL, J. E. W., and C. P. Collier. 1962. Gases sterilize containers. Food Eng. 34(12):96–98.

McCULLOCH, E. C. 1945. Disinfection and sterilization. 2nd ed. Lea & Febiger, Philadelphia.

MOLIN, N., and A. ERICHSEN (Eds.) 1965. Microbial inhibitors in food. IV International Symposium on Food Microbiology. Almqvist and Wiksell, Stockholm, Sweden.

NEWELL, G. W. 1963. Food additives of tomorrow. Food Processing 24(1):70–74.

PORTER, J. R. 1946. Bacterial chemistry and physiology. John Wiley & Sons, Inc., New York.

REDDISH, G. F. 1954. Antiseptics, disinfectants and chemical and physical sterilization. Lea & Febiger, Philadelphia.

ROUNDY, Z. D. 1958. Treatment of milk for cheese with hydrogen peroxide. J. Dairy Sci. 41:1460–1465.

VAUGHN, R. H., H. NG, G. F. STEWART, C. W. NAGEL, and K. L. SIMPSON. 1960. Antibiotics in poultry meat preservation. Development in vitro of bacterial resistance to chlortetracycline. Appl. Microbiol. 8:27–30.

WHEATON, E., and G. L. HAYS. 1964. Antibiotics and the control of spoilage in canned foods. Food Technol. 18:549–551.

WOODBINE, M. (Ed.) 1962. Antibiotics in agriculture. Butterworth & Co. (Publishers), Ltd., London.

CHAPTER TEN

PRESERVATION BY RADIATIONS AND BY PRESSURE

In their search for new, improved methods of food preservation, investigators have paid special attention to the possible utilization of radiations of various frequencies, ranging from the low-frequency electric current to the high-frequency gamma rays (see Figure 10–1). Much of this work, as well as that on the use of high pressures in the preservation of foods, has not advanced beyond the experimental stage.

RADIATIONS

The electromagnetic spectrum, showing the approximate location of various types of radiations, is diagramed in Figure 10–1. The sonic and ultrasonic waves are not considered radiations by some authorities. There are two types of radiations, **corpuscular** and **electromagnetic**. Of the corpuscular radiations, which are streams of atomic or subatomic particles that transfer energy when they strike, the beta or cathode rays are being used experimentally in treating foods. Electromagnetic radiations, which disturb the internal structure of matter and thus dissipate their energy, include radio waves, microwaves, light waves, X-rays and gamma rays. The beta and gamma rays (and X-rays) are ionizing, that is, they cause ionization of molecules, chiefly water, of the absorbing materials and destroy microorganisms without raising the temperature appreciably. Therefore such sterilization sometimes is termed "cold sterilization."

HEATING RADIATIONS

Of the heating radiations, electrical, radio, micro- and infrared waves have been tried in food preservation, although infrared rays are used chiefly to heat foods, as in some vending machines.

Electric currents

Direct currents have not proved useful in the treatment of foods because of the extensive electrolytic changes produced, but with alternat-

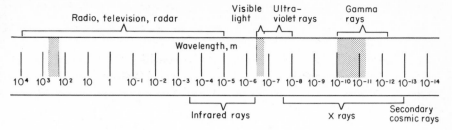

Figure 10–1. Chart of the electromagnetic spectrum, showing location of various types of radiation. (*Courtesy of Westinghouse Electric Corporation, Bloomfield, New Jersey.*)

ing currents these changes are of minor importance compared with the heating effect. Heating by electric current has the advantage that it is rapid and uniform throughout a liquid. The "Electropure" process for the pasteurization of milk is an example of the successful application of heat by means of a low-frequency electric current, as is a similar treatment of fruit juices. The continued passage of an alternating electric current through a liquid may result in the production of small amounts of ozone and chlorine, which may aid the heat in killing microorganisms.

Alternating currents of moderately high frequency have been reported to favor microbial growth at a controlled favorable temperature. Very high (radio) frequencies result in the death of microorganisms, the effect being due primarily to the heat produced, although killing effects without the development of much heat have been claimed but not proved. Experimentally, from several million to 3 billion cycles (reversals of current) per second have been tried in radio-frequency heating, but between 10 and 30 million cycles per second are most often recommended for foods.

RADIO-FREQUENCY HEATING. When a nonconducting material is placed between two metal electrodes and a rapidly alternating, high-voltage (radio-frequency) current is applied, **dielectric heating** (molecular motion) takes place uniformly and internally by friction losses as polar molecules try to align themselves with alternating currents at the electrodes. In most foods only part of the heating is dielectric, for much of the food conducts electricity and therefore **induction heating** (electron motion) results because of changes in particle momentum resulting from the very frequent reversals of potential per second. This continuous increase in translational motion results in an equivalent increase in kinetic particle energy throughout the mass of food and hence in a rise in temperature. The combined nonconductor-plus-conductor heating has been termed **dithermal processing** of the food (and also has been called "electronic heating"). The heating is rapid and uniform, in contrast to heating by steam, where the heat must penetrate slowly to the center of the food from the overheated outer surfaces.

Among the uses of dithermal heating in the food industries are dehydration, cooking, baking, melting, roasting, deinfestation, defrosting, mold inhibition or destruction, and blanching. Use of the method for killing microorganisms in foods and especially for sterilization of food is, for the most part, in the experimental stage. Experimentally, small portions of a variety of foods have been successfully treated to destroy all spoilage organisms. Attempts also have been made to sterilize canned foods by dithermal heating.

Although dithermal heating is being used successfully for processes that require only moderate heating, such as deinfestation, warming of foods, and defrosting, a number of difficulties are encountered in attempts to destroy the microorganisms in a food. Such difficulties include (1) undesirable side reactions, such as oxidation (unless in inert gas) and denaturation of proteins; (2) uneven heating of different parts of the food, between the lean and fat parts of meat, for example, or between the sharp edges of a loaf of Boston brown bread and the rest of the loaf; (3) the drying effect at high temperatures unless the foods are under pressure; (4) the greater resistance of enzymes than of most organisms; and (5) changes in the color and structure of foods, e.g., the change in the color of meat from red to purple, the bleaching of strawberries and carrots, and the structural breakdown of sliced fruits and some vegetables. Attempts to use dithermal heating in canning have demonstrated special problems such as (1) uneven heating or arcing and burning across corners; (2) induced heat across the ends of cans, which are good conductors; (3) the development of high pressures in cans, resulting in distortion or blowing of the can; and (4) the hindrance of dielectric heating because of electrolytes, e.g., $NaCl$, in the food.

Another disadvantage of dithermal heating at present is the greater cost as compared to that for conventional methods of heating.

MICROWAVE HEATING. Microwave heating may be used in freeze drying, electronic ovens, etc. Microwaves, which are produced by special oscillator tubes to give frequencies between those of radio or broadcast bands and infrared, are used mostly at frequencies of about 2,450 megacycles per second.

Sound waves

Sound waves, although not true radiations, act similarly in producing heat. Most attempts to apply sonic waves to food preservation have involved the use of ultrasonic frequencies, that is, above the audible range (over 20,000 cycles per second). Most of the killing effect of sonic vibrations on microorganisms is supposed to be the result of the heat produced, although disruption of cells and especially of their membranes may take place. The latter effect is supposedly caused by cavitation, which results from large differences in pressure within a liquid during the treatment. Physical damage to foods, destruction of vitamins, and lack of damage to enzymes have been listed as disadvantages of sonic treatments. As yet, sound waves have not found practical use in food preservation, although milk has been pasteurized by this method and cheese ripening has been hastened, both experimentally.

ULTRAVIOLET RAYS

Of the various electromagnetic radiations, only ultraviolet rays have had widespread, practical use in the food industries up to the present time. Waves whose length is from 136 to 3900 A (angstrom; one A is equal to 0.1 millimicron) are termed ultraviolet rays and are germicidal, but those between 2500 and 2800 A are especially effective, and a length of 2650 to 2660 A is most germicidal. Radiations in this range are absorbed strongly by the purine and pyrimidine groups of nucleic acids in the microbial cells, and mutation or death results.

GERMICIDAL LAMPS. The rays from commonly used quartz-mercury-vapor lamps are mostly at 2537 A, but spread to include both the rays in the visible range and those in the erythemic range, which have an irritating effect on skin and mucous membranes. The lamps available vary in size, shape, and power. Some of the older types of lamps gave off appreciable amounts of ozone, but the newer, cold-cathode type releases only negligible amounts, so that there is less than 0.1 ppm in the air under normal conditions. More ozone is released by a new lamp than by one that has been used for several days. Lamps weaken continuously as used and have a longer effective life when run steadily than when used intermittently.

FACTORS INFLUENCING EFFECTIVENESS. It should be emphasized that only direct rays are effective, unless they come from special reflectors, and even then their effectiveness is reduced. The factors that influence the effectiveness of ultraviolet rays are as follows: (1) *Time.* The longer the time of exposure at a given concentration, the more effective will be the treatment. (2) *Intensity.* The intensity of the rays upon reaching an object will depend upon the power of the lamp, the distance from lamp to object, and the kind and amount of interfering material in the path of the rays. Obviously the intensity will increase with the power of the lamp. Within the short distances common in industrial uses, the intensities of the rays vary inversely about as the distance from the lamp. A lamp is about 100 times as effective in killing microorganisms at 5 in. as at 8 ft from the irradiated object. Most tests are reported for a distance of 12 in. Dust in the air or on the lamp reduces effectiveness, as does too much atmospheric humidity. Over 80 percent relative humidity definitely reduces penetration through the air, but humidities below 60 percent have little effect. (3) *Penetration.* The nature of the object or material being irradiated has an important influence on the effectiveness of the process. Penetration is reduced even by clear water, which also exerts a protective effect for microorganisms. Dissolved mineral salts, especially of iron, and cloudiness greatly reduce the effectiveness of the rays. Even a thin layer of fatty or greasy material cuts off the rays. There is no penetration through opaque material. Therefore the rays take effect only on the outer surface of most irradiated foods and only on that part of the surface directly exposed to the lamp and do not penetrate to microorganisms inside the food. The lamps do serve, however, to reduce the numbers of viable microorganisms in the air surrounding foods.

EFFECTS ON MAN AND ANIMALS.   Gazing at an ultraviolet lamp produces irritation of the eyes in man within a few seconds, and longer exposure of the skin results in erythema, or reddening. The effect on animals usually is not as marked, although the eyes, especially of chicks, may be irritated.

ACTION ON MICROORGANISMS.   As has been stated, the intensity of the rays when they reach the organism, the time they act, and the location of the organism determine the germicidal effect. Each kind of microorganism has a characteristic resistance to ultraviolet irradiation that varies with the phase of growth and whether it is in the vegetative or spore state. It takes as much as five times as much exposure to kill vegetative cells of some bacteria as of others, but, in general, the killing exposure does not differ widely among different species. The location of the organism during the test has a marked influence. For example, 97 to 99 percent of *Escherichia coli* organisms in air were killed in 10 sec at 24 in. with a 15-watt lamp; but 20 sec at 11 in. was necessary if the bacteria were on the surface of an agar plate. Capsulation or clumping of bacteria increases their resistance. Bacterial spores usually take from two to five times as much exposure as the corresponding vegetative cells. Some types of pigmentation also have a protective effect.

Some yeasts are killed as easily as bacteria, but in general they are from two to five times as resistant. The resistance of molds is reported to be from ten to fifty times that of bacteria. Pigmented molds are more resistant than the nonpigmented, and spores more resistant than mycelium. Pigmented spores are more difficult to kill, in general, than nonpigmented ones. The killing effect of ultraviolet rays usually is explained by the "target theory," to be described later in the discussion of gamma and cathode rays.

APPLICATIONS IN THE FOOD INDUSTRIES.　　The use of ultraviolet irradiation in the food industries will be discussed in connection with the preservation of specific foods. Examples of the successful use of these rays include treatment of water as for beverages; aging of meats; treatment of knives for slicing bread; treatment of bread and cakes; packaging of sliced bacon; sanitizing of eating utensils; prevention of growth of film yeasts on pickle, vinegar, or sauerkraut vats; killing of spores on sugar crystals and in sirup; storage and packaging of cheese; prevention of mold growth on walls and shelves; and treatment of air for or in storage and processing rooms.

IONIZING RADIATIONS

Kinds of ionizing radiations

Radiations classified as ionizing include X- or gamma rays, cathode or beta rays, protons, neutrons, and alpha particles. Neutrons cause radioactivity, and protons and alpha particles have little penetration; therefore these rays are not practical for use in food preservation and will not be discussed.

X-rays are penetrating electromagnetic waves which are produced by the bombardment with cathode rays of a heavy-metal target within an evacuated tube.

Gamma rays are like X-rays but are emitted from by-products of atomic fission or from imitations of such by-products. Cobalt 60 and caesium 137 have been used as sources of these rays in most experimental work thus far.

Beta rays are streams of electrons (beta particles) emitted from a radioactive material. **Electrons** are small, negatively charged particles of uniform mass that form part of the atom. They are deflected by magnetic and electric fields. Their penetration depends on the speed with which they hit the target—the higher the charge of the electron, the deeper will be its penetration.

Cathode rays are the streams of electrons (beta particles) from the cathode of an evacuated tube. In practice these electrons are accelerated by artificial means.

X-rays, then, are indistinguishable from gamma rays; and beta and cathode rays both consist of moving electrons.

Definitions of terms

Before the utilization of ionizing radiations can be discussed, a few terms must be defined.

A **roentgen** (r) is the quantity of gamma or X-radiation which produces one electrostatic unit of electric charge of either sign in one cubic centimeter of air under standard conditions.

A **roentgen-equivalent-physical** (rep) is the quantity of ionizing energy which produces, per gram of tissue, an amount of ionization equivalent to a roentgen. A **megarep** is a million rep. One r or one rep is equivalent to the absorption of eighty-three to ninety ergs per gram of tissue.

The **rad** now is employed chiefly as the unit of radiation dosage, a rad being equivalent to the absorption of one hundred ergs per gram of irradiated material. A **megarad** (Mrad) is a million rads, and a **kilorad** (Krad) is a thousand rads.

An **electron-volt** (ev) is the energy gained by an electron in moving through a potential difference of one volt. A **mev** is a million electron-volts.

A mev, then, is a measure of the intensity of the irradiation, and a rep is a measure of the absorbed energy that is effective within the food.

X-rays

X-rays, gamma rays, and cathode rays are equally effective in sterilization for equal quantities of energy absorbed, but X-rays and gamma rays have good penetration, while cathode rays have comparatively poor penetration. The greatest drawback at present to the use of X-rays in food preservation is the low efficiency and consequent high cost of their production, for only about 3 to 5 percent of the electron energy applied is used in the production of X-rays. For this reason most recent research has been on the application of gamma rays and cathode rays. Since X-rays and gamma rays are indistinguishable, the following section on the application and effects of gamma rays is equally applicable to X-rays.

Gamma rays and cathode rays

Since these two types of rays are equally effective in sterilization for equal quantities of energy absorbed and apparently produce similar changes in the food being treated, they will be discussed together and compared where possible.

SOURCES.   Chief sources of gamma rays are (1) radioactive fission products of uranium, (2) the coolant circulated in nuclear reactors, and (3) other fuel elements used to operate a nuclear reactor. Cathode rays usually are taken from the cathode of an evacuated tube and accelerated by special electrical devices. The greater this acceleration (i.e., the more mev), the deeper will be the penetration into the food.

PENETRATION.   Gamma rays have good penetration, but their effectiveness decreases exponentially with depth. They have been reported to be effective at about 5.5 in. in most foods, but this depth will depend upon the time of exposure. Cathode rays, on the other hand, have poor penetration, being effective at only about ¼ in. per mev when "crossfiring" is employed, that is, irradiation from opposite sides. The absorption dose level in a material is not a uniformly decreasing fraction with depth but rather builds up to a maximum at a depth equal to about one-third of the total penetration, then decreases to zero.

EFFICIENCY.   Because cathode rays are directional, they can be made to hit the food and therefore be used with greater efficiency than gamma rays, which are constantly being emitted in all directions from the radioactive source. Various estimates of the maximal efficiency of utilization of cathode rays range between 40 and 80 percent, depending upon the shape of the irradiated material, but only a maximum of 10 to 25 percent utilization efficiency is estimated for gamma rays. Radioactive sources of gamma rays decay steadily and hence weaken with time.

SAFETY.   The use of cathode rays presents fewer health problems than the use of gamma rays since cathode rays are directional and less penetrating, can be turned off for repair or maintenance work, and present no hazard of radioactive materials after a fire, explosion, or other catastrophe. Gamma rays are emitted in all directions, are penetrating, are continuously emitted, and come from radioactive sources. Gamma rays require more shielding to protect workers. Tests thus far on animals and human volunteers have not indicated any ill effects from eating irradiated foods, but tests on toxicity of the foods still are in progress.

EFFECTS ON MICROORGANISMS. The bactericidal efficacy of a given dose of irradiation depends upon:

1. The kind and species of organism. The importance of this factor is illustrated by the data in Table 10–1.
2. The numbers of organisms (or spores) originally present. The more organisms there are, the less effective will be a given dose.
3. The composition of the food. Some constituents, e.g., proteins, catalase, and reducing substances (such as nitrites, sulfites, and sulfhydryl compounds), may be protective. Compounds that combine with the SH groups would be sensitizing. Products of ionization may be harmful to the organisms.
4. The presence or absence of oxygen. The effect of free oxygen varies with the organism, ranging from no effect to sensitization of the organism. Undesirable "side reactions," to be discussed later, are likely to be intensified in the presence of oxygen and be less in a vacuum or an atmosphere of nitrogen.
5. The physical state of the food during irradiation. Both moisture content and temperature affect different organisms in different ways.
6. The condition of the organisms. Their age, temperature of growth and sporulation, and their state—vegetative or spore—may affect the sensitivity of the organisms. These factors have been discussed previously in connection with other methods of processing foods.

The type of irradiation and, within limits, the pH of the food seem to have little influence on the dose needed to inactivate the organisms.

It has been stated by some workers that the resistance of a given species of microorganisms to ionizing radiations parallels, in general, its resistance to conventional heat processing, although there are notable exceptions. Spores of *Clostridium botulinum,* for example, have been found to be more resistant to gamma rays than spores of a flat sour bacterium (No. 1518) and a thermophilic anaerobe (T.A. No. 3814), although the latter two are the more heat-resistant. Table 10–1 summarizes reports from various sources on the approximate dosages of radiations necessary to kill various types of microorganisms. These figures will vary with the conditions listed in the preceding paragraph. It is to be noted, however, (1) that man is much more sensitive to radiations than are microorganisms; (2) that bacterial spores are considerably more resistant than vegetative cells; (3) that Gram-negative bacteria are, in general, less resistant than Gram-positive ones; and (4) that yeasts and molds vary considerably, but some are more resistant than most bacteria. Note *Candida krusei* in Table 10–1, which is as resistant as many bacterial spores. Many workers report having sterilized foods with 2.5

TABLE 10–1. Approximate killing doses of ionizing radiations in kilorads*
(From numerous sources)

Organism	Approx. lethal dose	Organism	Approx. lethal dose
Man	0.56–0.75	Bacteria (cells of saprophytes)	
Insects	22–93	Gram–negative	
Viruses	1,000–4,000	*Escherichia coli*	100–230
Yeasts (fermentative)	400–900	*Pseudomonas aeruginosa*	160–230
Saccharomyces cerevisiae	500	*Pseudomonas fluorescens*	120–230
Torula cremoris	470	*Aerobacter aerogenes*	140–180
Yeasts (film)	370–1,800	Gram–positive	
Hansenula sp.	470	*Lactobacillus* spp.	23–38
Candida krusei	1,160	*Streptococcus faecalis*	170–880
Molds (with spores)	130–1,100	*Leuconostoc dextranicum*	90
Penicillium spp.	140–250	*Sarcina lutea*	370
Aspergillus spp.	140–370	Bacterial spores	310–3,700
Rhizopus sp.	1,000	*Bacillus subtilis*	1,200–1,800
Fusarium sp.	250	*Bacillus coagulans*	1,000
Bacteria (cells of pathogens)		*Clostridium botulinum* (A)	1,900–3,700
Mycobacterium tuberculosis	140	*Clostridium botulinum* (E)	1,500–1,800
Staphylococcus aureus	140–700	*Clostridium perfringens*	310
Corynebacterium diphtheriae	420	Putrefactive anaerobe 3679	2,300–5,000
Salmonella spp.	372–475	*Bacillus stearothermophilus*	1,000–1,700

* Size of lethal dose depends on factors listed on page 154.

megarep of radiation, but as much as 4 megarep has been found necessary at times. For unknown reasons certain microorganisms may be much more resistant than anticipated. Thus, for example, a radiation-resistant micrococcus, resembling *Micrococcus roseus,* has been found in irradiated meat. This coccus survived 6 megarep on agar slopes, being more resistant than any bacterial spores tested. *Microbacterium* species in meat have been found to be especially resistant. There is, then, the possibility that resistant strains may build up on irradiated foods.

Preliminary gamma irradiation of bacterial spores has been found to make them more sensitive to heat, but preliminary heat shocking did not affect the lethal action of subsequent gamma irradiation. Previous ultrasonic treatment of organisms sensitizes them to radiations.

It is supposed that irradiated microorganisms are destroyed by passage of an ionizing particle or quantum of energy through, or in close proximity to, a sensitive portion of the cell, causing a direct "hit" on this target, ionization in this sensitive region, and death of the organism (this is called the "target theory"). It is assumed, also, that much of the germicidal effect results from ionization of the surroundings, especially of water, to yield free radicals, some of which may be oxidizing or reducing and therefore helpful in the destruction of the organisms. Irradiation also may cause mutations in organisms present.

EFFECTS ON FOODS. Radiation doses heavy enough to effect sterilization have been found to produce undesirable "side reactions," or secondary changes, in many kinds of foods, causing undesirable colors, odors, tastes, or even physical properties. Greatest promise of successful sterilization at the present time is with fresh and smoked pork products, baked beans, sweet potatoes, Brussels sprouts, green beans, prunes, raisins, chicken, organ meats, and some kinds of fish. Less promising are beef, lamb, milk and milk products, eggs, some seafoods, and certain fruits and vegetables. Special methods are being investigated for minimizing these side effects: (1) removal of oxygen; (2) irradiation of foods in the frozen state, or at least at reduced temperatures; (3) addition of free radical acceptors, e.g., ascorbic acid to counteract oxidation; (4) reduction of the moisture content; and (5) concurrent irradiation and vacuum distillation, as with milk.

Some of the changes produced in foods by sterilizing doses of radiation include: (1) in meats, a rise in pH, destruction of glutathione, and increase in carbonyl compounds, hydrogen sulfide, and methyl mercaptan; (2) in fats and lipids, destruction of natural antioxidants, oxidation followed by partial polymerization, and increase in carbonyl compounds; (3) in vitamins, reduction in most foods of levels of thiamine, pyridoxine, and vitamins B_{12}, C, D, E, and K; riboflavin and niacin are fairly stable. On the other hand, destruction of many of the food enzymes requires five to ten times the dosage of rays needed to kill all of the microorganisms, so that enzyme action may continue after all microorganisms have been destroyed, unless a special blanching treatment has preceded irradiation. The lower the dosage of irradiation, of course, the less will be the undesirable effects on the food.

The chief effect on the healthfulness of the foods is the destruction of vitamins. There is no indication of production of radioactivity with electron beams below 11 mev or with gamma rays from cobalt 60.

APPLICATIONS. Ionizing radiations are being applied successfully to the killing of insects and their eggs in foods and the destruction of trichinae (see Chapter 26) and liver fluke larvae in meats. Experiments also have shown that packaging materials can be sterilized by irradiation; that partial or complete sterilization of the surfaces of a number of foods, such as fruits, vegetables, baked goods, meats, etc., where spoilage usually begins at the surface, can be accomplished without deterioration of the food; and that less irradiation than that needed to sterilize foods can be used in combination with other preservative methods, such as refrigeration, heating, drying, and addition of chemicals, to extend the storage life of the foods. It also can be used for the selective inactivation of specific non-spore-forming pathogens ("radicidation"), e.g., Salmonella in egg products.

The Natick (Massachusetts) Laboratories of the United States Army and other laboratories are working out practical applications of the ray treatments of foods.

Application has been made to the Food and Drug Administration for the acceptance of various irradiated foods. Irradiation of fresh canned bacon with a dose of 4.5 or more megarads of gamma or X-rays has been approved, and it is believed that approval of commercial sterilization ("radappertization") of other foods by ionizing rays will come soon. The use of such rays for "pasteurizing" treatments ("radurization") to lengthen the storage life of foods to be preserved also by chilling, freezing, drying, heating, or added antibiotics or other chemicals probably is in the not too distant future. Low dosages (20,000 to 500,000 rad) have been shown to give four- to tenfold extension of shelf life of various meats, fish, fruits, vegetables, and baked goods. Of course economic considerations will determine how and if the ionizing radiations will be utilized, after methods for their use have been improved.

MECHANICAL PRESSURE

High, mechanically produced air pressures will kill microorganisms and inactivate enzymes, and sublethal pressures will affect growth and metabolism of organisms. Pressures are expressed as pounds per square inch (psi) or atmospheres, 1 atmosphere equaling 14.7 psi. A pressure of 6,000 atmospheres for 45 min has been found to kill many of the nonsporulating bacteria, but 20,000 atmospheres did not kill bacterial spores. Sudden release of high pressures has an added germicidal effect. Hite and coworkers at the West Virginia Experiment Station tried high pressures in the preservation of fruits and fruit juices and used pressures up to 100,000 psi (about 6,800 atmospheres). A pressure of 60,000 to 80,000 psi preserved apple juice for 5 years, but inconsistent results were obtained with berries. It was concluded that results were uncertain and mechanical difficulties great with this method of preservation.

More success has been attained by the use of gases under pressure. In the Hofius method for preserving milk, the normal gases are removed by bubbling oxygen through the milk. Then the milk is stored at 8 C (46.4 F) or lower under an oxygen pressure of 8 atmospheres or higher. This method has been reported to lengthen the keeping time of milk to 4 weeks or longer. The process has been used to some extent in Europe but not in this country. Carbon dioxide under a pressure of about 114 psi (7.7 atmospheres) at 59 F (15 C) has been used to preserve grape juice until it could be sterilized by filtration. Charged soft drinks contain carbon dioxide under pressure and keep better for that reason.

REFERENCES

ANDERSON, A. W., K. E. RASH, and P. R. ELLIKER. 1961. Taxonomy of a recently isolated radiation-resistant micrococcus. Bacteriol. Proc. (Soc. Amer. Bacteriol.) 1961:56 (Abstr.)

ANONYMOUS. 1963. European research (radiations). Food Eng. 35(8):45.

ASSELBERGS, E. A. 1961. New developments in infra-red radiation. Food in Canada 21(10):36–38.

BELLAMY, W. D. 1959. Preservation of foods and drugs by ionizing radiations. Advances Appl. Microbiol. 1:49–73.

BRIDGES, B. A., and T. HORNE. 1959. The influence of environmental factors on the microbicidal effect of ionising radiations. J. Appl. Bacteriol. 22:96–115.

CHANDLER, VELMA L., and COWORKERS. 1956. Relative resistance of microorganisms to cathode rays. Appl. Microbiol. 4:143–152.

COPSON, D. A. 1962. Microwave heating. Avi Publishing Co., Inc., Westport, Conn.

DEAN, E. E., and D. L. HOWIE. 1963. Safety of food sterilization by ionizing radiations. U.S. Army Natick Labs. Activities Rep. 15(4th quarter):174–183.

DESROSIER, N. W. 1963. The technology of food preservation. Rev. ed. Avi Publishing Co., Inc., Westport, Conn.

DESROSIER, N. W., and H. M. ROSENSTOCK. 1960. Radiation technology in food, agriculture, and biology. Avi Publishing Co., Inc., Westport, Conn.

GOLDBLITH, S. A. (Ed.) 1963. Exploration in future food-processing techniques. The M.I.T. Press, Cambridge, Mass.

GOLDBLITH, S. A. 1964. Radiation. Food Technol. 18:1384–1391.

GOLDBLITH, S. A., and B. E. PROCTOR. 1956. Radiation preservation of milk and milk products: I, Background and problems. J. Dairy Sci. 39:374–378.

GORESLINE, H. E., M. INGRAM, P. MACUCH, G. MOCQUOT, D. A. A. MOSSEL, C. F. NIVEN, JR., and F. S. THATCHER. 1964. Tentative classification of food irradiation processes with microbiological objectives. Nature. [London] 204:237–238.

HANNAN, R. S. 1956. Science and technology of food preservation by ionizing radiations. Chemical Publishing Company, New York.

HITE, B. H., N. J. GIDDINGS, and C. E. WEAKLY, JR. 1914. The effect of pressure on certain microorganisms encountered in preserving fruits and vegetables. West Va. Exp. Sta. Bull. 146.

HOLLAENDER, A. (Ed.) 1954–1956. Radiation biology. McGraw-Hill Book Company, New York. 3 vols.

HUBER, W., and J. L. HEID. 1956. Ionizing radiations in food products manufacture. Western Canner and Packer, Aug., p. 25–36.

KETCHUM, H. W., J. W. OSBURN, JR., and J. DEITCH. 1965. Current status and commercial prospects for radiation preservation of food. Business and Defense Services Administration, U.S. Dep. Commerce. U.S. Government Printing Office, Washington, D.C.

LEA, D. E. 1955. Action of radiation on living cells. 2nd ed. Cambridge University Press, London.

National Research Council of the National Academy of Sciences. 1965. Radiation preservation of foods. Publ. 1273.

159

PROCTOR, B. E., and S. A. GOLDBLITH. 1951. Electromagnetic radiation fundamentals and their applications in food technology. Advances Food Res. 3:119–196.

SHERMAN, V. W. 1946. Electronic heat in the food industries. Food Ind. 18:506–509, 628–630.

THORNLEY, MARGARET J. 1963. Radiation resistance among bacteria. J. Appl. Bacteriol. 26:334–345.

U.S. Army Quartermaster Corps. 1957. Radiation preservation of food. U.S. Government Printing Office, Washington, D.C.

WERTHEIM, R. A. P. 1949. Applications of ultrasonics. Food Sci. Abstr. 21:69–83.

WHITNEY, R. McL. 1955. How food ultrasonics are shaping up. Food Eng. 27(5):80–82, 159–160.

WIERBICKI, E. 1963. Radiation processing of foods: present status. U.S. Army Natick Labs. Activities Rep. 15(4th quarter):160–167.

CHAPTER ELEVEN

GENERAL PRINCIPLES UNDERLYING SPOILAGE

FITNESS OR UNFITNESS OF FOOD FOR CONSUMPTION

When is a food fit to eat? According to Thom and Hunter: "A product is fit for food if a discriminating consumer, knowing the story of its production and seeing the material itself, will eat it, and, conversely the same product is spoiled when such an examiner refuses it as food." According to this definition the fitness of the food will depend upon the person judging it, for what one person will eat another will not. Some of the British, for example, like their game meat "high," with a strong flavor developed by "hanging," or aging, the meat, while most Americans would call this meat spoiled. The buried fish, "titmuck," eaten by the Eskimos, is a malodorous, semiliquid product that most of us (as well as the Eskimos' huskies) would consider inedible. A starving person might eat food that he would not consume under normal conditions.

Despite differences between individuals in their judgment of fitness of food, they would agree on certain criteria for assurance of fitness: (1) The desired stage of development or maturity. Fruits should be at a certain but differing stage of ripeness; sweet corn should be young enough to be tender and milky; poultry preferably is from birds that are fairly young. (2) Freedom from pollution at any stage in production or handling. Vegetables should not be consumed raw if they had been fertilized with sewage; oysters from waters contaminated with sewage should be rejected; food handled by dirty or diseased vendors should be spurned; food contaminated by flies or rodents should be suspect. (3) Freedom from objectionable change resulting from microbial attack or action of enzymes of the food. Sometimes it is difficult to draw a line between spoilage by microorganisms and harmless growth, or the same type of change may be considered undesirable in one food and desirable in another. Thus the housewife says that sour milk has "spoiled," but the cultured buttermilk made by the same lactic acid fermentation is good. Putrefaction in meat means definite spoilage, but putrefactive changes in Limburger cheese are normal to the ripening process. Some changes termed spoilage may be only changes in appearance or physical characteristics, as in wilted lettuce or flabby carrots, although the product probably has undergone no microbial spoilage and there has been little loss in nutritive value. Yet each one of us has his own idea about whether a food is spoiled or not and usually can come to a decision about its edibility without much difficulty.

CAUSES OF SPOILAGE

Decay or decomposition of undesirable nature usually is implied when the term "spoiled" is applied to food, while food unfit to eat for sanitary reasons usually is not called spoiled. Spoilage may be due to one or more of the following:

1. Growth and activity of microorganisms (or higher forms occasionally). Often a succession of organisms is involved.
2. Insects.
3. Action of the enzymes of the plant or animal food.
4. Purely chemical reactions, that is, those not catalyzed by enzymes of the tissues or of microorganisms.
5. Physical changes, such as those caused by freezing, burning, drying, pressure, etc.

The discussion to follow will be devoted chiefly to spoilage caused by microorganisms.

CLASSIFICATION OF FOODS ON EASE OF SPOILAGE

On the basis of ease of spoilage, foods can be placed into three groups:

1. Stable or nonperishable foods. These foods, which do not spoil unless handled carelessly, include products such as sugar, flour, and dry beans.
2. Semiperishable foods. If these foods are properly handled and stored they will remain unspoiled for a fairly long period; examples are potatoes, some varieties of apples, waxed rutabagas, and nutmeats.
3. Perishable foods. This group includes most of our important daily foods, that spoil readily unless special preservative methods are used. Meats, fish, poultry, most fruits and vegetables, eggs, and milk belong in this classification.

Most foods fall into one of the above three groups, but some are near enough the border line to be difficult to place.

FACTORS AFFECTING KINDS AND NUMBERS OF
MICROORGANISMS IN FOOD

The kind of spoilage of foods by microorganisms and enzymes will depend upon the *kinds* and *numbers* of these agents present and upon the environment about them. Most raw foods contain a variety of bacteria, yeasts, and molds and may contain plant or animal enzymes as the case may be. Because of the particular environmental conditions, only a small proportion of the kinds of microorganisms present will be able to grow rapidly and cause spoilage—usually a single kind of organism but sometimes two or three types—and these may not have been predominant in the original food. If spoilage by the first organism or organisms is allowed to proceed, one or more other kinds of organisms are likely to produce secondary spoilage, or even a further succession of organisms and changes may be involved.

The kinds and numbers of microorganisms that will be present on or in food will be influenced by the kind and extent of contamination, previous opportunities for the growth of certain kinds, and pretreatments which the food has received.

Contamination may increase numbers of microorganisms in the food and may even introduce new kinds. Thus wash water may incorporate surface-taint bacteria in butter; plant equipment may add spoilage organisms to foods during processing; washing machines may add them to eggs; and dirty boats may add them to fish. The increased "load" of microorganisms, especially of those which cause spoilage, makes preservation more difficult, i.e., spoilage is more likely and more rapid, and perhaps takes a different form from that which would have appeared without the contamination.

Growth of microorganisms in or on the food obviously will increase numbers or the "load" of microorganisms, and presumably in most foods will bring about the greatest increase in the organisms most likely to be concerned with spoilage. The heavier "load" will add to the difficulty of preventing spoilage of the food and may influence the kind of spoilage to be anticipated.

Pretreatments of foods may remove or destroy some kinds of microorganisms, add organisms, or change the proportions of those present or inactivate part or all of the food enzymes, and thus limit the number of spoilage agents and hence the possible types of spoilage. Washing, for example, may remove organisms from the surface or may add some from the wash water. If washing is by means of an antiseptic or germicidal solution, numbers of organisms may be greatly reduced and some kinds eliminated. Treatment with rays, ozone, sulfur dioxide, or germicidal vapors will reduce numbers and be selective of kinds. High temperatures will kill more and more organisms and leave fewer and fewer kinds as the heat-treatment is increased. Storage under various conditions may either increase or decrease kinds and numbers. Any of the above, as well as other treatments not mentioned, will influence the numbers, kinds, proportions, and health of the microorganisms.

FACTORS AFFECTING THE GROWTH OF MICROORGANISMS IN FOOD

Associative growth

Associations of microorganisms with each other are involved in spoilage or fermentations of most kinds of food. Competition between the different kinds of bacteria, yeasts, and molds in a food ordinarily determines which one will outgrow the others and cause its characteristic type of spoilage. If conditions are favorable for all, bacteria usually grow faster than yeasts, and yeasts faster than molds. Therefore, yeasts outgrow bacteria only when they are predominant in the first place or when conditions are such as to slow down the bacteria. Molds can predominate only when conditions are better for them than for yeasts or bacteria. The different kinds of bacteria present compete amongst themselves, with one kind usually outstripping the others. Likewise, if yeasts are favored, one kind usually will outgrow others; and among the molds one kind will find conditions more favorable than will other kinds. Microorganisms are not always antagonistic, or **antibiotic,** to each other, however, and may sometimes be **symbiotic,** that is mutually helpful, or they may grow simultaneously without seeming to aid or hinder each other. Two kinds of microorganisms may be **synergistic,** that is, when growing together they may be able to bring about changes, such as fermentations, that neither could produce alone. *Pseudomonas syncyanea* growing alone in milk produces only a light-brownish tinge; *Streptococcus lactis* causes no color change in milk; but when the two organisms grow together a bright-blue color develops.

A most important effect of a microorganism upon another is the **metabiotic** one, when one organism makes conditions favorable for growth of the second. Both organisms may be growing at the same time, but more commonly one succeeds the other. Most natural fermentations or decompositions of raw foods illustrate metabiosis. Raw milk at room temperature normally first supports an acid fermentation by *Streptococcus lactis* and coliform bacteria until the bacteria are inhibited by the acid they have produced. Next the acid-tolerant lactobacilli increase the acidity further until they are stopped. Then film yeasts and molds grow over the top, finally reducing the acidity so that proteolytic bacteria can become active. Metabiosis in the sauerkraut fermentation is discussed in Chapter 15. The normal succession of organisms is first, miscellaneous bacteria, chiefly coliform; second, *Leuconostoc mesenteroides;* third, *Lactobacillus plantarum;* and last, *Lactobacillus brevis.* Other examples of metabiosis will be cited in the discussion of the spoilage of various foods in Part 3.

Effect of environmental conditions

The environment determines which of the different kinds of microorganisms present in a food will outgrow the others and cause its characteristic type of change or spoilage. The factors that make up this environment are interrelated, and their combined effect determines the organisms to grow and the effects to be produced. Chief of these factors are the physical and chemical properties of the food, the availability of oxygen, and the temperature.

PHYSICAL STATE AND STRUCTURE OF THE FOOD. The physical state of the food, its colloidal nature, whether it has been frozen, heated, moistened, or dried, together with its biological structure, may have an important influence on whether a food will spoil or not and the type of spoilage.

The *water* in food, its location and availability, is one of the most important factors influencing microbial growth. Water may be considered both as a chemical compound necessary for growth and as part of the physical structure of the food.

The moisture requirements of molds, yeasts, and bacteria have been discussed in Chapters 1, 2, and 3, respectively. It has been emphasized that all microorganisms require moisture for growth and that all grow best in the presence of a plentiful supply. This moisture must be *available* to the organisms, that is, not tied up in any way, such as by solutes or by a hydrophilic colloid such as agar. Solutes such as salt or sugar dissolved in the water cause an osmotic pressure that tends to draw water from the cells if the concentration of dissolved materials is greater outside the cells than inside. It should be recalled that when the relative humidity of the air about the food corresponds to the available moisture or the water activity (a_w) of the food, the food and air about it will be in equilibrium in regard to moisture. If the relative humidity of the air is correspondingly greater than the a_w of the food, the latter will take up moisture. If the relative humidity of the air is correspondingly lower than the a_w of the food, the food will lose moisture at its surface. This loss of moisture at the surface will cause the diffusion of water from the inner parts of the food toward the surface, tending to make the moisture content uniform (and lower) throughout the food. Spoilage of most solid pieces of food usually is initiated at the surface and may take place there almost exclusively. Therefore, a lack of available moisture at the surface may be an important preservative factor, or, conversely, a plenitude of available moisture at the surface is likely to favor microbial spoilage and the spread of spoilage organisms, especially of motile ones.

A consideration of the moisture requirements of microorganisms leads to some general conclusions:

1. Each organism has its own characteristic optimal a_w and its own range of a_w for growth for a given set of environmental conditions. Factors affecting the moisture requirements of organisms are (a) the nutritive properties of the substrate, (b) its pH, (c) its content of inhibitory substances, (d) availability of free oxygen, and (e) temperature. The range of a_w permitting growth is narrowed if any of these environmental factors is not optimal, and narrowed still more if two or more conditions are not favorable.
2. An unfavorable a_w will result not only in a reduction in rate of growth, but also in a lowered maximal yield of cells.
3. The more unfavorable the a_w of the substrate, the greater will be the delay (lag) in initiation of growth or germination of spores. This often is as important in food preservation as reduction in the rate of growth of the organism.
4. In general, bacteria require more moisture than yeasts, and yeasts more than molds as is shown in Table 11–1, after Mossel and Ingram, showing

TABLE 11–1.   Lowest a_w values permitting growth of spoilage organisms*

Group of microorganisms	Minimal a_w value
Normal bacteria	0.91
Normal yeasts	0.88
Normal molds	0.80
Halophilic bacteria	0.75
Xerophilic fungi	0.65
Osmophilic yeasts	0.60

* After Mossel and Ingram, 1955.

lower limits of a_w for bacteria, yeasts, and molds. There are notable exceptions to this generalization, however, for some molds have a higher minimal a_w for growth (and spore germination) than many yeasts and some bacteria.
5. Microorganisms that can grow in high concentrations of solutes, e.g., sugar and salt, obviously have a low minimal a_w. It is to be noted that most halophiles are bacteria and most osmophiles are yeasts.

A dry food like bread is most likely to be spoiled by molds; sirups and honey with their fairly high sugar content and hence lowered a_w favor the growth of osmophilic yeasts; and moist, neutral foods, such as milk, meats, fish, and eggs, ordinarily are spoiled by bacteria. However, environmental factors other than moisture should be kept in mind in predicting the type of microorganism apt to cause spoilage. Grape juice, for example, may favor yeasts because of its fairly high sugar content and low pH, but will support the growth of bacteria if incubation temperatures are too high or too low for fermentative yeasts. Refrigerated foods may mold in the presence of air but undergo bacterial spoilage in its absence. Honey, although its sugar content is too high for most yeasts but not for some of the molds, rarely is spoiled by molds because of fungistatic substances present.

An a_w as low as 0.70 makes unlikely any spoilage by microorganisms of a food held at room temperature. This is approximately the level of available moisture in dry milk at 8 percent total moisture, dried whole egg at 10 to 11 percent, flour at 13 to 15 percent, nonfat dry milk at 15 percent, dehydrated fat-free meat at 15 percent, seeds of leguminous crops at 15 percent, dehydrated vegetables at 14 to 20 percent, dehydrated fruits at 18 to 25 percent, and starch at 18 percent (Mossel and Ingram).

It is possible for microorganisms growing in food to change the level of available moisture by release of metabolic water or by changing the substrate so as to free water. In the production of ropiness in bread, for example, it is supposed that *Bacillus subtilis* causes the release of moisture as a result of the decomposition of starch and in this way makes conditions more favorable for its own growth. Destruction of moisture-holding tissues, as in fruits by molds, may make water available to yeasts or bacteria.

Freezing not only prevents microbial growth if the temperature is sufficiently low, but also is likely to damage tissues, so that juices released on thawing favor microbial growth. Freezing also increases the concentration of solutes in the unfrozen portion as the temperature is lowered, slowing down and finally stopping the growth of organisms able to grow at temperatures below 0 C (32 F). Freezing also effects the removal of water from hydrophilic colloids that is not wholly reabsorbed on thawing.

Heat processing may change not only the chemical composition of the food but also its structure by softening tissues, releasing or tying up moisture, destroying or forming colloidal suspensions, gels, or emulsions, and changing the penetrability of the food to moisture and oxygen. Protein may become denatured and therefore more available to some organisms than it was in the native state. Starch or protein may become gelated, releasing moisture and becoming more easily decomposed. For the reasons indicated, cooked food usually is more easily decomposed than the original fresh food.

Changes in the colloidal constituents of foods may be caused by agencies other than the freezing or heating processes, by sound waves, for instance, but the results are similar. Emulsions of fat and water are more likely to spoil, and the spoilage will spread more rapidly when water is the continuous phase and fat the discontinuous one, as in French dressing, as compared to butter, where the reverse is true.

The effect of the *biological structure* of food on the protection of foods against spoilage has been mentioned in Chapter 5. It has been noted that the inner parts of whole, healthy tissues of living plants and animals are either sterile or at least low in microbial content. Therefore, unless opportunity has been given for their penetration, spoilage organisms within may be few or lacking. Often there is a protective covering about the food, such as the shell on eggs, the skin on poultry, the shell on nuts, and the rind or skin on fruits and vegetables, or we may have surrounded the food with an artificial coating, e.g., plastic or wax. This physical protection to the food may not only aid in its preservation but may also determine the kind, rate, and course of spoilage. Layers of fat over meat may protect that part of the flesh, or scales may protect the outer part of the fish. On the other hand, an increase in exposed surface, brought about by peeling, skinning, chopping, or comminution, may serve not only to distribute spoilage organisms but also to release juices containing food materials for the invaders. The disintegration of tissues by freezing may accomplish a similar result.

In meat the growth of spoilage bacteria takes place mostly in the fluid between the small meat fibers, and it is only after rigor mortis that much of this food material is released from the fibers to become available to spoilage organisms.

CHEMICAL PROPERTIES OF THE FOOD. The chemical composition of a food determines how satisfactory it will be as a culture medium for microorganisms. Each organism has its own characteristic ability to utilize certain substances as food for energy and certain substances for growth, and it grows best within a certain range of available moisture content and of hydrogen-ion concentration (pH values). Moisture requirements have been discussed above.

Nutrients in the food, their kinds and proportions, are all-important in determining what spoilage organism is most likely to grow. Consideration must be given to (1) foods for energy, (2) foods for growth, and (3) accessory food substances or vitamins, which may be necessary for energy or growth.

1. Foods for energy. The carbohydrates, especially the sugars, are most commonly used as foods for energy, but other carbon compounds may serve, such as esters, alcohols, peptides, amino acids, and organic acids and their salts. Complex carbohydrates, e.g., cellulose, can be utilized by comparatively few kinds of organisms, and starch can be hydrolyzed by only a limited number of kinds. Microorganisms differ even in their ability to use some of the simpler soluble sugars. Many organisms cannot use the disaccharide lactose, or milk sugar, and therefore do not grow well in milk. Maltose is not attacked by some yeasts. It will be recalled that bacteria often are identified and classified on the basis of their ability or inability to utilize various sugars and alcohols. Most organisms, if they utilize sugars at all, can use glucose.

The ability of microorganisms to hydrolyze pectin, which is characteristic of some kinds of bacteria and many molds, is important, of course, in the softening or rotting of fruits and vegetables or fermented products from them.

A limited number of kinds of microorganisms can obtain their energy from fats, but will do so only if a more readily usable energy food, such as sugar, is absent. First, the fat must be hydrolyzed with the aid of lipase to glycerol and fatty acids, which then may serve as foods for energy for the hydrolyzing organism or others. In general, aerobic microorganisms are more commonly involved in the decomposition of fats than anaerobic ones, and the lipolytic organisms usually are also proteolytic. Direct oxidation of fats containing unsaturated fatty acids usually is chemical.

Split products of proteins, peptides and amino acids, for example, serve as foods for energy for many proteolytic organisms when a better energy source is lacking, and as foods for energy for other organisms that are not proteolytic. Meats, for example, may be low in carbohydrate and therefore decomposed by proteolytic species, e.g., *Pseudomonas* spp., with following growth of weakly or nonproteolytic species that can utilize the products of protein hydrolysis. Organisms differ in their ability to use individual amino acids for energy and in their action on them.

Not only the kind of energy food is important but also its concentration in solution and hence its osmotic effect and the amount of available moisture. For a given percentage of sugar in solution, the osmotic pressure will vary with the weight of the sugar molecule. Therefore, a 10 percent solution of glucose would have about twice the osmotic pressure of a 10 percent solution of sucrose or maltose or would tie up twice as much moisture. In general, molds can grow in the highest concentrations of sugars and yeasts in fairly high concentrations, but most bacteria grow best in fairly low concentrations. There are, of course, notable exceptions to this generalization: osmophilic yeasts grow in as high concentrations of sugar as molds, and some bacteria can grow in fairly high concentrations of sugar.

Of course, an adequate supply of foods for growth will favor utilization of the foods for energy. More carbohydrate will be used if a good nitrogen food is present in sufficent quantity than if the nitrogen is poor in kind or amount. Organisms requiring special accessory growth substances might be prevented from growing if one or more of these "vitamins" were lacking, and thus the whole course of decomposition might be altered.

2. Foods for growth. Microorganisms differ in their ability to use various nitrogenous compounds as a source of nitrogen for growth. Many organisms are unable to hydrolyze proteins and hence cannot get nitrogen from them without help from a proteolytic organism. One protein may be a better source of nitrogenous food than another because of different products formed during hydrolysis, especially peptides and amino acids. Peptides, amino acids, urea, ammonia, and other simpler nitrogenous compounds may be available to some organisms but not to others, or usable under some environmental conditions but not under others. Some of the lactic acid bacteria grow best with polypeptides as nitrogen foods, cannot attack casein, and do not grow well with only a limited number of kinds of amino acids present. The presence of fermentable carbohydrate in a substrate usually results in an acid fermentation and suppression of proteolytic bacteria, and hence in what is called a "sparing" action on the nitrogen compounds. Also, the production of obnoxious nitrogenous products is prevented or inhibited.

Many kinds of molds are proteolytic, but comparatively few genera and species of bacteria and very few yeasts are actively proteolytic. In general, proteolytic bacteria grow best at pH values near neutrality and are inhibited by acidity, although there are exceptions, such as proteolysis by the acid-proteolytic bacteria that are hydrolyzing protein while producing acid.

Carbon for growth may come partly from carbon dioxide, but more often from organic compounds.

The minerals required by microorganisms are nearly always present at the low levels required, but occasionally an essential mineral may be tied up so as to be unavailable, or even may be lacking or present in insufficient amounts. An example is milk drawn into a glass container, such milk containing insufficient iron for pigmentation of the spores of *Penicillium roqueforti*. Once milk has come in contact with steel equipment enough iron will be present.

3. Accessory food substances or vitamins. Some microorganisms are unable to manufacture some or all of the vitamins needed and must have them furnished. Most natural plant and animal foodstuffs contain an array of these vitamins, but some may be low in amount or lacking. Thus meats are high in B vitamins and fruits are low, but fruits are high in ascorbic acid. Egg white contains biotin but also contains avidin which ties it up, making it unavailable to microorganisms and eliminating those which must have biotin supplied as possible spoilage organisms. The processing of foods often reduces the vitamin content: thiamine, pantothenic acid, the folic acid group, and ascorbic acid (in air) are heat-labile; and drying causes a loss in vitamins such as thiamine, carotene, and ascorbic acid. Even storage of foods for long periods, especially if the storage temperature

is elevated, may result in a decrease in the level of some of the accessory growth factors.

Microorganisms growing in a food may either supply other organisms with vitamins or outcompete them in obtaining essential vitamins.

Hydrogen-ion concentration, or *pH*, and *buffering power* of the food are important for their influence on the kinds of microorganisms most apt to grow and hence the changes most likely to be produced. The effect of pH on the growth of molds, yeasts, and bacteria has been discussed briefly in Chapters 1, 2, and 3. It will be recalled that most molds can grow over a wider range of pH values than most yeasts and bacteria and that many molds grow at acidities too great for yeasts and bacteria. Most fermentative yeasts are favored by a pH of about 4.0 to 4.5, as in fruit juices, and film yeasts grow well on acid foods, such as sauerkraut and pickles. On the other hand, the majority of yeasts do not grow well in alkaline substrates and must be adapted to such media. Most bacteria are favored by a pH near neutrality, although some, such as the acid-formers, are favored by moderate acidity and others, e.g., the actively proteolytic bacteria, can grow in media with a high (alkaline) pH, as is found in the white of a stored egg.

The buffers in a food, that is, the compounds that resist changes in pH, are of importance not only for their buffering capacity but also for their ability to be especially effective within a certain pH range. Buffers permit an acid (or alkaline) fermentation to go on longer with a greater yield of products and organisms than would otherwise be possible. Vegetable juices have low buffering power, permitting an appreciable decrease in pH with the production of only small amounts of acid by lactic acid bacteria during the early part of sauerkraut and pickle fermentations, enabling the lactics to suppress the undesirable pectin-hydrolyzing and proteolytic competing organisms. Low buffering power makes for a more rapidly appearing succession of microorganisms during a fermentation than high buffering power. Milk, on the other hand, is fairly high in buffers and therefore permits considerable growth and acid production by lactic acid streptococci in the manufacture of fermented milks before growth is stopped.

Inhibitory substances, originally present in the food, added purposely or accidentally, or developed there by growth of microorganisms or by processing methods, may inhibit growth of all microorganisms or, more often, may deter certain specific kinds. Examples of inhibitors naturally present are the lactenins and anticoliform factor in freshly drawn milk, lysozyme in egg white, and benzoic acid in cranberries. Chapter 9 lists a number of chemical preservatives that may be added to foods, for example, propionates or sorbic acid added to bread to prevent ropiness and spoilage by molds, sodium benzoate added to catchup or oleomargarine, sulfur dioxide added to acid fruits and some cane sirups, constituents of wood smoke added to meat or fish during smoking, etc. Present by accident may be residues of detergents or sanitizers used in the treatment of equipment, insecticides, herbicides, etc. A microorganism growing in a food may produce one or more substances inhibitory to other organisms, products such as acids, alcohols, peroxides, or even antibiotics. Propionic acid produced by the propionibacteria in Swiss cheese is inhibitory to molds; alcohol formed in quantity by wine yeasts inhibits competitors; and nisin produced by certain strains of *Streptococcus lactis* may be useful in inhibiting lactate-fermenting, gas-forming clostridia in curing cheese and undesirable in slowing down some of the essential lactic acid streptococci during the manufacturing process. There also is the possibility of the destruction of inhibitory compounds in foods by microorganisms. Certain molds and bacteria are able to destroy some of the phenol compounds that are added to meat or fish by smoking or benzoic acid added to foods, sulfur dioxide is destroyed by yeasts resistant to it, and lactobacilli can inactivate nisin. Heating of foods may result in the formation of inhibitory substances: heating of lipids may hasten autooxidation and make them inhibitory; and browning of concentrated sugar sirups may result in the production of furfural and hydroxymethyl furfural that are inhibitory to fermenting organisms. Long storage at warm temperatures may produce similar results.

OXYGEN TENSION AND OXIDATION-REDUCTION POTENTIAL. The oxygen tension or partial pressure of oxygen about a food and the oxidation-reduction (O-R) potential or reducing and oxidizing power of the food itself influence the type of organisms to grow and hence the changes produced in the food. The O-R potential of the food is determined by (1) the characteristic O-R potential of the original food; (2) the **poising capacity,** that is, the resistance to change in potential, of the food; (3) the oxygen tension of the atmosphere about the food; and (4) the access which the atmosphere has to the food. Air has a high oxygen tension, but the head space in an "evacuated" can of food would have a low oxygen tension.

From the standpoint of ability to use free oxygen, microorganisms have been classified as aerobic when they require free oxygen, anaerobic when they grow best in the absence of free oxygen, and facultative when they grow well either aerobically or anaerobically. Molds are aerobic; most yeasts grow best aerobically; and bacteria of different kinds may be aerobic, anaerobic, or facultative. From the standpoint of O-R potential, a high (oxidizing) potential favors aerobes but will permit the growth of facultative organisms, and a low (reducing) potential favors anaerobic or facultative organisms. However, some organisms that are considered aerobic can grow, but not well, at surprisingly low O-R potentials. Growth of an organism may alter the O-R potential of a food enough to restrain other organisms. Anaerobes, for example, may lower the O-R potential to a level inhibitory to aerobes.

Most fresh plant or animal foods have a low and well-poised O-R potential in their interiors, the plants because of reducing substances such as ascorbic acid and reducing sugars, and animal tissues because of —SH and other reducing groups. As long as the plant or animal cells respire and remain active they tend to poise the O-R system at a low level, resisting the effect of oxygen diffusing from the outside. Therefore, a piece of fresh meat or a fresh whole fruit would have aerobic conditions only at and near the surface. The meat could support aerobic growth of slime-forming or souring bacteria at the surface at the same time that anaerobic putrefaction was proceeding in the interior. This situation may be altered by the application of processing procedures. Heating may reduce the poising power of the food by destruction or alteration of reducing and oxidizing substances and also allow more rapid diffusion of oxygen inwards, either because of the destruction of poising substances or because of changes in the physical structure of the food. Processing also may remove oxidizing or reducing substances; thus clear fruit juices have lost reducing substances by their removal during extraction and filtration and therefore have become more favorable to the growth of yeasts than was the original juice containing the pulp.

In the presence of limited amounts of oxygen the same aerobic or facultative organisms may produce incompletely oxidized products, such as organic acids, from carbohydrates, when with plenty of oxygen available complete oxidation to carbon dioxide and water might result. Protein decomposition under anaerobic conditions may result in putrefaction, whereas under aerobic conditions the products are likely to be less obnoxious.

TEMPERATURE.   Any nonsterile food is likely to spoil in time if moist enough and unfrozen. There is likelihood of spoilage at any temperature between —5 C (23 F) and 70 C (158 F). Since microorganisms differ so widely in their optimal, minimal, and maximal temperatures for growth, it is obvious that the temperature at which a food is held will have a great influence on the kind, rate, and amount of microbially induced change that will take place. Even a small change in temperature may favor an entirely different kind of organism and result in a different type of spoilage. Molds and yeasts, for the most part, do not grow well above 35 to 37 C (95 to 98.6 F) and therefore would not be important in foods held at high temperatures. On the other hand, molds and yeasts grow well at ordinary room temperatures and many of them grow fairly well at low temperatures, some even at freezing or slightly below. Although most bacteria grow best at ordinary temperatures, some (thermophiles) grow well at high temperatures and others (psychrophiles) at chilling temperatures. Therefore, molds often grow on refrigerated foods, and thermophilic bacteria grow in the hot pea blanchers. Raw milk held at different temperatures supports the initial growth of different bacteria. At temperatures near freezing, cold-tolerant bacteria, such as species of *Pseudomonas* and *Achromobacter*, are favored; at room temperatures *Streptococcus lactis* and coliform bacteria usually predominate; at 40 to 45 C (104 to 113 F) thermoduric lactics, e.g., *S. thermophilus* and *S. faecalis* grow first; and at 55 to 60 C (131 to 140 F) thermophilic bacteria like *Lactobacillus thermophilus* will grow.

It should be kept in mind that the temperature at which a raw food is stored may affect its self-decomposition and therefore its susceptibility to microbial spoilage. As noted in Chapter 7, wrong storage temperatures for fruits weaken them and may make them more likely to spoil.

Temperatures commonly used in handling and storing foods, especially in the market and in the home, are very different in different countries. In this country the refrigeration of most perishable foods is the rule, and keeping such foods at atmospheric temperatures for very long is the exception, but the reverse is true in many foreign lands. Therefore, the most commonly occurring type of spoilage of a food might be entirely different in different countries. In the United States, concern with spoilage of most perishable foods is mostly with changes at chilling temperatures, and psychrophilic strains of *Pseudomonas, Achromobacter, Flavobacterium, Alcaligenes,* and other genera and certain yeasts and molds would be important. Where foods are not refrigerated customarily, the prevailing atmospheric temperatures of the area would be significant, and differences during the seasons of the year would have to be considered. During seasons or at times when the climate is temperate, ordinary mesophilic bacteria, yeasts, and molds would assume importance. During hot weather, 80 to 110 F (26.7 to 43 C) and above, organisms favored by these temperatures would be important, such as coliform bacteria and species of *Bacillus, Clostridium, Streptococcus, Lactobacillus,* and other genera. Tropical temperatures would be unfavorable to most yeasts and molds. Under exceptional conditions, foods, especially canned varieties, sometimes are held at temperatures favoring thermophiles, as was often true during World War II, when canned foods were stored under tarpaulins in the tropics.

The combination of all of the factors just discussed, the kinds, numbers, and proportions of microorganisms present, and their environment, as controlled by physical and chemical properties of the food, oxygen tension and oxidation-reduction potential, and temperature, will determine the kinds of microorganisms most likely to grow in a food and hence the changes to be produced, and all should be considered in making predictions. Examples will be cited throughout the chapters to follow on the spoilage of specific foods. Sometimes a change in only one of the factors mentioned will be enough to limit the change to be expected, but more often several factors exert a combined effect. Thus a combination of low moisture, refrigerator temperature, high acidity, and high sugar would make mold growth more likely than growth of yeasts or bacteria. But increasing the moisture content and the temperature would change conditions to favor yeasts, and also decreasing the acidity and sugar content would encourage bacteria.

Chemical changes caused by microorganisms are discussed briefly in the next chapter.

REFERENCES

CLIFTON, C. E. 1957. Introduction to bacterial physiology. McGraw-Hill Book Company, New York.

DESROSIER, N. W. 1963. The technology of food preservation. Rev. ed. Avi Publishing Co., Inc., Westport, Conn.

GORESLINE, H. E. 1955. Food spoilage and deterioration, chap. 13. In F. C. Blank (Ed.) Handbook of food and agriculture. Reinhold Publishing Corporation, New York.

HALVORSON, H. O. 1951. Food spoilage and food poisoning, chap. 11. In M. B. Jacobs (Ed.) The chemistry and technology of food and food products. 2nd ed. Interscience Publishers (Division of John Wiley & Sons, Inc.), New York.

HALVORSON, H. O. 1953. Principles of food microbiology. J. Milk Food Technol. 16:73–76.

MOSSEL, D. A. A., and M. INGRAM. 1955. The physiology of the microbial spoilage of foods. J. Appl. Bacteriol. 18:233–268.

PORTER, J. R. 1946. Bacterial chemistry and physiology. John Wiley & Sons, Inc., New York.

SARLES, W. B., W. C. FRAZIER, J. B. WILSON, and S. G. KNIGHT. 1956. Microbiology: general and applied, chap. 25. 2nd ed. Harper & Row, Publishers, Incorporated, New York.

SCOTT, W. J. 1957. Water relations of food spoilage microorganisms. Advances Food Res. 7:83–127.

THOM, C., and A. C. HUNTER. 1924. Hygienic fundamentals of food handling. The Williams & Wilkins Company, Baltimore.

WEISER, H. H. 1962. Practical food microbiology and technology. Avi Publishing Co., Inc., Westport, Conn.

CHAPTER TWELVE

CHEMICAL CHANGES CAUSED BY MICROORGANISMS

Because of the great variety of organic compounds in foods and the numerous kinds of microorganisms that can decompose them, many different chemical changes are possible, and many kinds of products can result. The following discussion will be concerned only with the important types of decomposition of main constituents of foods and the chief products produced.

CHANGES IN NITROGENOUS ORGANIC COMPOUNDS

Most of the nitrogen in foods is in the form of proteins which must be hydrolyzed by enzymes of the microorganisms or of the food to polypeptides, simpler peptides, or amino acids before they can serve as nitrogenous food for most organisms. Proteinases catalyze the hydrolysis of proteins to peptides, which may give a bitter taste to foods. Peptidases catalyze the hydrolysis of polypeptides to simpler peptides and finally to amino acids. The latter give flavors, desirable or undesirable, to some foods; for example, amino acids contribute to the flavor of ripened cheeses.

For the most part these hydrolyses do not result in particularly objectionable products. Anaerobic decomposition of proteins, peptides, or amino acids, however, may result in the production of obnoxious odors and is then called **putrefaction.** It results in foul-smelling, sulfur-containing products, such as hydrogen, methyl, and ethyl sulfides and mercaptans, plus ammonia, amines (e.g., histamine, tyramine, piperidine, putrescine, and cadaverine), indole, skatole, and fatty acids.

When microorganisms act upon amino acids, they may deaminate them, decarboxylate them, or both, resulting in the products listed in Table 12–1. *Escherichia coli,* for example, produces glyoxylic acid, acetic acid, and ammonia from glycine; *Pseudomonas* also produces methylamine and carbon dioxide; and clostridia give acetic acid, ammonia, and methane. From alanine these three organisms produce: (1) an α-keto acid, ammonia, and carbon dioxide; (2) acetic acid, ammonia, and carbon dioxide; and (3) propionic acid, acetic acid, ammonia, and carbon dioxide, respectively. From serine, *E. coli* produces pyruvic acid and ammonia, and species of *Clostridium* give propionic acid, formic acid, and ammonia. As stated previously, the sulfur in sulfur-bearing amino acids may be reduced to foul-smelling sulfides or mercaptans.

Other nitrogenous compounds decomposed include: (1) amides, imides, and urea from which ammonia is the principal product; (2) guanidine and creatine which yield urea and ammonia; and (3) amines, purines, and pyrimidines which may yield ammonia, carbon dioxide, and organic acids (chiefly lactic or acetic).

CHANGES IN NONNITROGENOUS ORGANIC COMPOUNDS

The main nonnitrogenous foods for microorganisms, mostly used to obtain energy, but possibly serving as sources of carbon, include carbohydrates, organic acids, aldehydes and ketones, alcohols, glycosides, cyclic compounds, and lipids.

CARBOHYDRATES. Carbohydrates, if available, usually are preferred by microorganisms to other energy-yielding foods. Complex di-, tri- or polysaccharides usually are hydrolyzed to simple sugars before utilization. A monosaccharide, such as glucose, aerobically would be oxidized to carbon dioxide and water, and anaerobically would undergo decomposition involving any of six main types of fermentation: (1) an alcoholic fermentation, as by yeasts, with ethanol and carbon dioxide as the principal products; (2) a simple lactic fermentation, as by homofermentative lactic acid bacteria, with lactic acid as the main product; (3) a mixed lactic fermentation, as by heterofermentative lactic acid bacteria, with lactic and acetic acids, ethanol, glycerol, and carbon dioxide as chief products; (4) the coliform type of fermentation, as by coliform bacteria, with lactic, acetic, and formic acids, ethanol, carbon dioxide, hydrogen, and perhaps acetoin and butanediol as likely products; (5) the propionic fermentation, by propionibacteria, producing propionic, acetic, and succinic acids, and carbon dioxide; or (6) the butyric-butyl-isopropyl fermentations, by anaerobic bacteria, yielding butyric and acetic acids, carbon dioxide, hydrogen, and in some instances acetone, butylene glycol, butanol, and 2-propanol. A variety of other products is possible from sugars when different microorganisms are active, including higher fatty acids, other organic acids, aldehydes, and ketones.

ORGANIC ACIDS. Many of the organic acids, usually occurring in foods as salts, are oxidized by organisms to carbonates, causing the medium to become more alkaline. Aerobically the organic acids may be oxidized completely to carbon dioxide and water, as is done by film yeasts. Acids may be oxidized to other, simpler acids, or to other products similar to those from sugars. Saturated fatty acids or lower ketonic derivatives are degraded to acetic acid, two carbons at a time, aided by coenzyme A. Unsaturated or hydroxy fatty acids may be degraded partially in a similar manner, but must be converted to a saturated acid (or ketonic derivative) for complete beta oxidation.

OTHER COMPOUNDS. Alcohols usually are oxidized to the corresponding organic acid, e.g., ethanol to acetic acid. Glycerol may be dissimilated to products similar to those from glucose. Glycosides, after hydrolysis to release the sugar, will have the sugar dissimilated characteristically. Acetaldehyde may be oxidized to acetic acid or reduced to ethanol. Cyclic compounds are not attacked readily.

LIPIDS. Fats are hydrolyzed by microbial lipase to glycerol and fatty acids, which are then dissimilated as outlined previously. Microorganisms may be concerned in the oxidation of fats, but autooxidation is more common (see Chapter 21). Phospholipids may be degraded to their constituent phosphate, glycerol, fatty acids, and nitrogenous base (e.g., choline). Lipoproteins are made up of proteins, cholesterol esters, and phospholipids.

PECTIC SUBSTANCES. Protopectin, the water-insoluble parent pectic substance in plants, is converted to pectin, a water-soluble polymer of galacturonic acid which contains methyl ester linkages and varying degrees of neutralization by various cations. It gels with sugar and acid. Pectinesterase causes hydrolysis of the methyl ester linkage of pectin to yield pectic acid and methanol. Polygalacturonases destroy the linkage between galacturonic acid units of pectin or pectic acid to yield smaller chains and ultimately free D-galacturonic acid, which may be degraded to simple fatty acids.

TABLE 12–1. Products from the microbial decomposition of amino acids

Chemical process	Products
Oxidative deamination	Keto acid + NH_3
Hydrolytic deamination	Hydroxy acid + NH_3
Reductive deamination	Saturated fatty acid + NH_3
Desaturation deamination (at α and β positions)	Unsaturated fatty acid + NH_3
Mutual O-R between pairs of amino acids	Keto acid + fatty acid + NH_3
Decarboxylation	Amine + CO_2
Hydrolytic deamination + decarboxylation	Primary alcohol + NH_3 + CO_2
Reductive deamination + decarboxylation	Hydrocarbon + NH_3 + CO_2
Oxidative deamination + decarboxylation	Fatty acid + NH_3 + CO_2

REFERENCES

ALLEN, L. A. 1964. The biochemistry of industrial micro-organisms. Chem. Ind. May 23, p. 877–880.

BARKER, H. A. 1956. Bacterial fermentations. John Wiley & Sons, Inc., New York.

CLIFTON, C. E. 1957. Introduction to bacterial physiology. McGraw-Hill Book Company, New York.

DIENDOERFER, F. H., R. I. MATELES, and A. H. HUMPHREY. 1963. 1961 fermentation process review. Appl. Microbiol. 11:273–303.

GUNSALUS, I. C., and R. Y. STANIER (Eds.) 1961. The bacteria: Vol. 2, Metabolism. Academic Press Inc., New York.

MEYER, LILLIAN H. 1960. Food enzymes. Avi Publishing Co., Inc., Westport, Conn.

OGINSKY, EVELYN L., and W. W. UMBREIT. 1959. An introduction to bacterial physiology. 2nd ed. W. H. Freeman and Company, San Francisco.

PORTER, J. R. 1946. Bacterial chemistry and physiology. John Wiley & Sons, Inc., New York.

PRESCOTT, S. C., and C. C. DUNN. 1959. Industrial microbiology. 3rd ed. McGraw-Hill Book Company, New York.

RAINBOW, C., and A. H. ROSE (Eds.) 1963. Biochemistry of industrial micro-organisms. Academic Press Inc., New York.

SCHULTZ, H. W. (Ed.) 1960. Food enzymes. Avi Publishing Co., Inc., Westport, Conn.

THIMANN, K. V. 1963. The life of bacteria. 2nd ed. The Macmillan Company, New York.

PART THREE

CONTAMINATION, PRESERVATION, AND SPOILAGE OF DIFFERENT KINDS OF FOODS

The fruit or vegetable is harvested, milk is drawn, eggs are gathered, fish and other products are obtained from natural waters, and animals are collected and slaughtered, all carrying contaminating microorganisms from natural sources (Chapter 4). In most instances, with the start of human handling further contamination begins, and it continues while the product is being handled and processed. The processor attempts to cleanse and "sanitize" equipment coming in contact with food to reduce contamination from that source, and to employ packaging materials that will not add significant contamination. As has been stated the term **sanitize** is used rather than "sterilize" because although an attempt is made to sterilize the equipment, that is, free it of all living organisms, sterility seldom is attained.

In Part 3, methods of preservation discussed in Chapters 5 to 10 are applied to the main kinds of foods, and the chief types of spoilage of these foods are considered.

CHAPTER THIRTEEN

CONTAMINATION, PRESERVATION, AND SPOILAGE OF CEREALS AND CEREAL PRODUCTS

Cereal products discussed in this chapter include the grains themselves, meals, flours, alimentary pastes, and breads, cakes, and other bakery products.

CONTAMINATION

The exteriors of harvested grains contain some of the natural flora that they had while growing, plus contamination from soil and other sources. Freshly harvested grains contain loads of a few thousand to millions of bacteria per gram and from none to several hundred thousand mold spores and, perhaps, spores of smuts and rusts. Bacteria are mostly in the families *Pseudomonadaceae, Micrococcaceae, Lactobacillaceae,* and *Bacillaceae.* If the grains are stored under moist conditions, molds may grow and produce numerous spores. Scouring and washing of the grains removes part of the microorganisms, but most of the microorganisms are removed with the outer portions of the grain during milling. The milling processes, especially bleaching, reduce numbers of organisms, but there then is a possibility of contamination during other procedures, such as blending and conditioning.

Bacteria in wheat flour include spores of *Bacillus,* coliform bacteria, and a few representatives of the genera *Achromobacter, Flavobacterium, Sarcina, Micrococcus, Alcaligenes,* and *Serratia.* Mold spores are chiefly those of aspergilli and penicillia, with also some of *Alternaria, Cladosporium,* and other genera. Numbers of bacteria vary widely from a few hundred per gram to millions. Most samples of white wheat flour from the retail market contain a few hundred to a few thousand bacteria per gram, and average about twenty to thirty bacillus spores per gram and fifty to one hundred mold spores. Patent flours usually give lower counts than straight or clear, and numbers decrease with storage of the flour. Higher counts usually are obtained on prepared flours, and still higher (8,000 to 12,000 per gram on the average) on graham and whole-wheat flours which contain also the outer parts of the wheat kernel and are not bleached. Corn meal often contains from 5,000 to 70,000 bacteria per gram and 1,000 to 400,000 molds. Malts, because of incubation in a moist condition, contain high numbers of bacteria, usually in the millions per gram.

The surface of a freshly baked loaf of bread is practically free of viable microorganisms but is subject to contamination by mold spores from the air during cooling and before wrapping. During slicing contamination may take place from microorganisms in the air, on the knives, or on the wrapper. Cakes are similarly subject to contamination. Spores of bacteria able to cause ropiness in bread will survive the baking process.

PRESERVATION

Most cereals and cereal products have such a low moisture content that little difficulty is encountered in the prevention of the growth of microorganisms as long as the foods are kept dry. Such materials are stored in bulk or in containers so as to keep out vermin, especially insects and rodents, resist fire, and avoid rapid changes in temperature and hence increase in moisture. A storage temperature of about 40 to 45 F (4.4 to 7.2 C) is recommended for the dry products. Many bakery products, e.g., breads, rolls, cakes, pastries, pies, and canned mixes, contain enough moisture to be subject to spoilage unless special preservative methods are employed or turnover is rapid.

ASEPSIS

As in other food industries, the adequate cleansing and sanitization of equipment is essential for reasons of both sanitation and preservation. Improperly sanitized equipment may be a source of rope bacteria and the acid-forming bacteria that cause sourness of doughs. Bread, cakes, and other baked goods that may be subject to spoilage by molds should be protected against contamination by mold spores. Protection of bread is especially important. The bread leaves the oven free of live mold spores and should be cooled promptly in an atmosphere free of them, sliced with sporefree knives, and wrapped without delay.

USE OF HEAT

Bakery products may be sold unbaked, partially baked, or fully baked. The complete baking process ordinarily destroys all bacterial cells, yeasts, and mold spores, but not spores of the rope-forming or other bacteria; yet it has been reported that mold spores in proofer cloths in bakeries can build up enough heat resistance to survive baking. Unbaked or partially baked products usually are kept on the retailer's shelf for only a short period or are kept cool during longer storage. Some special breads, e.g., Boston brown bread or nut bread, have been successfully canned.

USE OF LOW TEMPERATURES

Although ordinary room temperatures are used by most housewives for the storage of baked goods, keeping times could be lengthened and risk of food poisoning lessened if really warm temperatures, such as those of hot kitchens or summer weather, were avoided and the foods were stored in a cool place, or even in the refrigerator.

The freezing-storage of bakery goods is on the increase. Unbaked or partially baked products, waffles, cheese cake, ice-cream pie, fish, poultry, and meat pies now often are frozen. Bread and rolls can be stored successfully for months in the frozen condition.

USE OF CHEMICAL PRESERVATIVES

Sodium and calcium propionates (0.1 to 0.32 percent of the weight of the flour) are used routinely by many bakers, mostly in bread but sometimes in other products, to delay or prevent mold growth and ropiness. Sorbic acid has been recommended as a substitute for propionate. Acidification of the dough, usually with acetic acid, has been employed to combat rope.

USE OF IRRADIATION

In bakeries, ultraviolet rays have been used to destroy or reduce numbers of mold spores in dough and proof rooms, on the knives of slicing machines, in the room where the bread is packaged, and on the surface of bread, cakes, and other bakery products. The application of radio-frequency radiations to loaves of bread to reduce the likelihood of mold spoilage has been reported, and ionizing radiations, gamma and cathode rays, have been applied experimentally for the preservation of baked goods.

SPOILAGE

Cereal grains, and meals and flours made from them, should not be subject to microbial spoilage if they are prepared and stored properly because their moisture content is too low to support even the growth of molds. If, however, these products are moistened above the minimum for microbial growth, that growth will follow. A little moistening will permit only mold growth, but more moisture will allow the growth of yeasts and bacteria.

CEREAL GRAINS AND MEALS

Since cereal grains and meals ordinarily are not processed so as to greatly reduce their natural flora of microorganisms, they are apt to contain molds, yeasts, and bacteria which are ready to grow if enough moisture is added. In addition to starch, which is unavailable to many organisms, these grains contain some sugar and available nitrogen compounds, minerals, and accessory growth substances; and the amylases will release more sugar if the grains are moistened, and proteinases will yield more available nitrogenous foods. A little added moisture will result in growth of molds at the surface where air is available. A wet mash of the grains or mash of the meals will undergo an acid fermentation, chiefly by the lactic acid and coliform bacteria normally present on plant surfaces. This may be followed by an alcoholic fermentation by yeasts as soon as the acidity has increased enough to favor them. Finally molds and perhaps film yeasts will grow on the top surface, although acetic acid bacteria, if present, may oxidize the alcohol to acetic acid and inhibit the molds.

FLOURS

The dry cleaning and washing of grains and the milling and sifting of flour reduce the content of microorganisms, but the important kinds still are represented in whole-grain flours, such as whole-wheat or buckwheat, and the spoilage would be similar to that described for cereal grains and meals.

White wheat flour, however, usually is bleached by an oxidizing agent, such as an oxide of nitrogen, chlorine, nitrosyl chloride, or benzoyl peroxide, and this process serves to reduce microbial numbers and kinds.

A moisture content of flour of less than 13 percent has been reported to prevent the growth of all microorganisms. Other workers claim that 15 percent permits good mold growth and over 17 percent the growth of both molds and bacteria. Therefore, slight moistening of white flour brings about spoilage by molds. Because of the variations in microbial content of different lots of flour, the type of spoilage in a flour paste is difficult to predict. If acid-forming bacteria are present, an acid fermentation begins, followed by aloholic fermentation by yeasts if they are there and then acetic acid by *Acetobacter* species. This succession of changes would be more likely in freshly milled flour than in flour that had been stored for a long period with a consequent reduction in kinds and numbers of microorganisms. In the absence of lactics and coliforms, micrococci have been found to acidify the paste, and in their absence species of *Bacillus* may grow, especially aerobacilli which produce lactic acid, gas, alcohol, acetoin, and small amounts of esters and other aromatic compounds. It is characteristic of most flour pastes to develop an odor of acetic acid and esters.

BREAD

The fermentations taking place in the doughs for various kinds of bread will be discussed in Chapter 24, where it will be noted that some changes caused by microorganisms are desirable and are even necessary in the making of certain kinds of bread. The acid fermentation by lactics and coliform bacteria that is normal in flour pastes or doughs may, however, be too extensive if too much time is permitted, with the result that the dough and bread made from it may be too "sour." Excessive growth of proteolytic bacteria during this period may destroy some of the gas-holding capacity so essential during the rising of the dough and produce what is called a "sticky" dough. Sticky doughs, however, are usually the result of overmixing or gluten breakdown by reducing agents, e.g., glutathione. There also is the possibility of the production by microorganisms of undesirable flavors other than the sourness.

The chief types of microbial spoilage of baked bread are moldiness and ropiness, usually termed "mold" and "rope."

Mold

Molds are the most common and hence the most important cause of the spoilage of bread (Figure 13-1) and, in fact, of most bakery products. The temperatures attained in the baking procedure usually

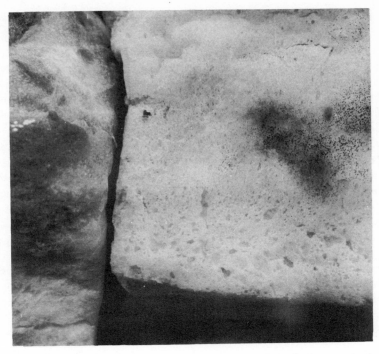

Figure 13–1. Moldy bread. (*Courtesy of Universal Foods Corporation.*)

are high enough to kill all mold spores in and on the loaf, so that molds must reach the outer surface or penetrate after baking. They can come from the air during cooling or thereafter, from handling, or from wrappers, and usually initiate growth in the crease of the loaf and between the slices of sliced bread.

Chief molds involved in the spoilage of bread are the so-called "bread mold," *Rhizopus nigricans,* with its white cottony mycelium and black dots of sporangia; the green-spored *Penicillium expansum* or *P. stoloniferum; Aspergillus niger* (Figure 13–2) with its greenish- or purplish-brown to black conidial heads and yellow pigment diffusing into the bread; and *Monilia* (*Neurospora*) *sitophila,* whose pink conidia give a pink or reddish color to its growth. Species of *Mucor* or *Geotrichum* or any of a large number of species of other genera of molds may develop. Mold spoilage is favored by (1) heavy contamination after baking, due, for example, to air heavily laden with mold spores, a long cooling time, considerable air circulation, or a contaminated slicing machine; (2) slicing, in that more air is introduced into the loaf; (3) wrapping, especially if the bread is warm when wrapped; and (4) storage in a warm, humid place. There is little growth of commercial importance on bread crust in a relative humidity below 90 percent. Bread with 6 percent of milk solids retains moisture somewhat better than milk-free bread, and hence there is less moisture between loaf and wrapper and hence less molding, but there is not enough effect to be of much practical importance. Molding often begins within a loaf of sliced bread, where more moisture is available than at the surface, especially in the crease.

Various methods are employed to prevent moldiness of bread: (1) Prevention of contamination of bread with mold spores in so far as is practicable. The air about the bread is kept low in spores by removal of possible breeding places for molds, such as returned bread, or walls and equipment. Spore-laden flour dust from other parts of the bakery has been incriminated in causing increased moldiness of bread. Filtration and washing of air to the room and irradiation of the room and more especially the air by means of ultraviolet rays cut down contamination. (2) Prompt and adequate cooling of the loaves before wrapping to reduce condensation of moisture beneath the wrapper. (3) Ultraviolet irradiation of the surface of the loaf and of slicing knives.

Figure 13–2. Growth of mold (*Aspergillus*) on bread crust. (*Courtesy of The Fleischmann Laboratories, Standard Brands Incorporated.*)

(4) Destruction of molds on the surface by electronic heating. (5) Keeping the bread cool to slow down mold growth, or freezing and storage in the frozen condition to prevent growth entirely. (6) Incorporation in the bread dough of some mycostatic chemical. Most commonly employed now is sodium or calcium propionate at the rate of 0.1 to 0.3 percent of the weight of the flour, a treatment that also is effective against rope. Sorbic acid, up to 0.3 percent, has been suggested as a substitute for propionate but is not permitted by present standards. An older remedy was the addition of vinegar or acetate to the dough or treatment of the exterior of the loaf with vinegar.

Rope

Ropiness of bread is fairly common in home-baked bread, especially during hot weather, but it is comparatively rare in commercially baked bread because of preventive measures now employed. Ropiness is caused by a mucoid variant of *Bacillus subtilis* or *B. licheniformis,* formerly called *B. mesentericus, B. panis,* and other species names. The spores of these species can withstand the temperature of the bread during baking, which does not exceed 100 C, and can germinate and grow in the loaf if conditions are favorable. The ropy condition apparently is the result of capsulation of the bacillus, together with hydrolysis of the flour proteins (gluten) by proteinases of the organism and of starch by amylase to give sugars that encourage rope formation. The area of ropiness is yellow to brown in color and is soft and sticky to the touch. In one stage the slimy material can be drawn out into long threads when the bread is broken and pulled apart (see Figure 13–3). The unpleasant odor is difficult to characterize, although it has been described as that of decomposed or overripe melons. First the odor is evident, then discoloration, and finally softening of the crumb, with stickiness and stringiness.

The production of ropiness is favored by the following factors: (1) Heavy contamination of the dough with spores of the causative bacillus, chiefly from the ingredients. Of these, the flour is likely to be the heaviest source of spores of the rope organism, and for that reason flours are tested for their content of spores of mesophilic bacteria. The yeast, dry milk, sugar, malt, and malted products also have been reported to be sources of spores of rope bacteria. (2) Contamination of the dough from equipment or of the bread by slicing knives. Spores of *B. subtilis* may build up on equipment, if trouble has been encountered previously. (3) Slow cooling of the bread after baking. This favors rapid germination of spores and multiplication of vegetative cells in the bread. (4) Lack of acidity in the bread. *B. subtilis* is favored by pH values near neutrality and is increasingly inhibited as the acidity is increased (pH drops) toward pH 5.0. (5) Storage of bread in a warm, humid atmosphere. This is more likely to happen in the home than elsewhere during the hot months of summer, August and September in particular. Bread containing spores may develop the defect at temperatures above 90 F (32.2 C), when the spores would not cause spoilage if the bread were kept at lower temperatures for the usual period before use.

Methods for preventing ropiness of bread are readily inferred from the conditions favoring it: (1) Use of ingredients of the dough that are low in numbers of spores of the rope bacterium. Some can be purchased with this low content specified. (2) Adequate cleansing and sanitizing of equipment coming in contact with the dough to prevent contamination from this source. Slicing knives also should be kept free from the bacterial spores. (3) Prompt cooling of the loaves after baking. (4) Use of a dough formula that will result in a pH of 5.0 to 5.15 in the bread. This has been accomplished by the addition of acetic, tartaric, citric, or lactic acid, or acid phosphate, but acidification must be used with caution or the rising of the dough will be adversely affected. (5) Addition of 0.1 to 0.3 percent of sodium or calcium propionate (or sorbic acid) on the basis of the weight of the flour. This probably is the best preventive procedure, and, as has been pointed out, also inhibits growth of molds. (6) Storage of the bread at a cool temperature. This, in itself, may be enough to prevent ropiness during the normal period of holding bread, if the original contamination of

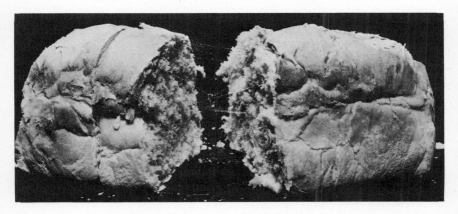

Figure 13–3. Ropy bread showing stringy capsular material stretching between the halves of the loaf; note also gummy crumbs. (*Courtesy of E. I. du Pont de Nemours and Company, Inc.*)

rope spores was not heavy. Freezing bread and storing it in the frozen condition will prevent ropiness entirely.

A number of methods have been suggested for testing flour or other ingredients for the presence of spores of rope bacteria. Most reliable is a bread-baking and incubation test. A plate count of spores of mesophilic *Bacillus* species has been employed where a flour suspension is pasteurized at 80 C for 10 min and plated quantitatively. A count of 20 or more spores per 100 g of flour is considered objectionable. Another method employs quantitative dilutions of the flour in nutrient broth followed by incubation at 37 C for 48 hr. A grayish-white surface pellicle on the broth is indicative of the growth of a *Bacillus* species. Other, less successful methods have been recommended.

Ropiness also can occur in doughnuts, brown bread, and cakes.

Red bread

Red, or "bloody," bread is striking in appearance but rare in occurrence. The red color results from the growth of pigmented bacteria, usually *Serratia marcescens,* an organism that often is brilliantly red on starchy foods. In ancient times the mysterious appearance of apparent drops of blood was considered miraculous. Necessary for the phenomenon is the accidental contamination of the bread with the red organisms and unusually moist conditions to favor their growth. Molds, such as *Monilia sitophila,* previously mentioned, may impart a pink to red color to bread. A red color in the crumb of dark bread has been caused by *Oidium* (*Geotrichum*) *aurantiacum.*

Chalky bread

Chalky bread, also uncommon in occurrence, is so named because of white, chalklike spots. The defect has been blamed on the growth of yeastlike fungi, *Endomycopsis fibuliger* and *Trichosporon variable.*

CAKES AND OTHER BAKERY PRODUCTS

Molds are the chief cause of the microbial spoilage of cakes and other bakery products. Methods of prevention are similar to those listed for bread. Ultraviolet irradiation of the surfaces of cakes before wrapping and packaging has been successful in reducing mold spoilage, and propionates can be employed for most kinds of cake, but not for some of the fruit pies. Sorbic acid is effective in preventing mold growth on cakes. As has been indicated, ropiness can develop in cakes or doughnuts; the cause and methods of prevention are similar to those for bread.

Staling of bakery products

The deterioration of bread, cakes, pies, and other bakery products called "staling" is due mostly to physical changes during holding and not to microorganisms. Freezing and storage in the frozen condition is an effective method of preventing the changes.

MACARONI AND TAPIOCA

Swelling of moist macaroni has been reported to be caused by gas production by bacteria resembling *Aerobacter cloacae*. During the drying of macaroni on paper a mold of the genus *Monilia* has been found responsible for purple streaks at the contact points with the paper. The appearance of these defects is uncommon, however, despite the long, slow drying process, and few reports of spoilage have been made.

Tapioca, prepared from the root starch of cassava, will spoil if moistened. Spoilage by an orange-pigmented, starch-hydrolyzing bacterium has been described.

REFERENCES

ANONYMOUS. 1937. Rope and mold. Standard Brands, Inc., New York.

CATHCART, W. H. 1951. Baking and bakery products, chap. 26. *In* M. B. Jacobs (*Ed.*) The chemistry and technology of food and food products. 2nd ed. Interscience Publishers (Division of John Wiley & Sons, Inc.), New York.

GLABE, E. F. 1942. Preventing spoilage by mold and bacteria. Food Ind. 14(2):46–48.

HEATH, B. 1947. Dehydrated foods and microorganisms. Australian Food Manuf. 16(12):July 5.

HESSELTINE, C. W., and R. R. GRAVES. 1963. Microbiological research on wheat and flour, p. 170–199. 2nd National Conference on Wheat Utilization Research, Peoria, Ill. Northern Utilization Res. Develop. Div., U.S. Dep. Agr.

JAMES, N., and A. R. LEJEUNE. 1952. Microflora and the heating of damp, stored wheat. Canad. J. Bot. 30:1–8.

JAMES, N., and K. N. SMITH. 1948. Studies on the microflora of flour. Canad. J. Res. 26C:479–484.

KENT-JONES, D. W., 1937. Flour spoilage. Analyst 62:649–653.

KNIGHT, R. A., and E. M. MENLOVE. 1961. Effect of the bread-baking process on destruction of certain mould spores. J. Sci. Food Agr. 12:653–656.

Matz, S. A. (*Ed.*) 1960. Baking technology and engineering. Avi Publishing Co., Inc., Westport, Conn.

Miller, F. W., Jr. 1942. The story of mold and rope. E. I. du Pont de Nemours & Co., Inc., Wilmington, Del.

Pyler, E. J. 1952. Baking science and technology. Seibel Publishing Co., Chicago. 2 vols.

Qasem, S. A., and C. M. Christensen. 1958. Influence of moisture content, temperature, and time on the deterioration of stored corn by fungi. Phytopathology 48:544–549.

Russ, J. J., W. R. Reeder, and D. W. Hatch. 1961. A rapid bacteriological method for predicting ropiness in bread. Cereal Sci. 6:89–91.

Spicher, G. 1959. Sur la microflora des cereales. Meunerie Franc. 148:23–30.

Tanner, F. W., 1944. Microbiology of foods. 2nd ed. The Garrard Press, Champaign, Ill.

Walter, A. H. 1964. Microbial resistance. *In* A. H. Woollen, London meeting unveils advances. Food Eng. 36(11):57.

CHAPTER FOURTEEN

CONTAMINATION, PRESERVATION, AND SPOILAGE OF SUGARS AND SUGAR PRODUCTS

Sugar products discussed in this chapter include sucrose (cane and beet sugar), molasses, sirups, maple sap and sugar, honey, and candy.

CONTAMINATION

SUCROSE

The raw juice expressed from sugar cane may become high in microbial content unless processing is prompt. The relevant microorganisms are those from the sugar cane and the soil contaminating it and therefore comprise slime producers, such as species of *Leuconostoc* and *Bacillus;* representatives of the genera *Micrococcus, Flavobacterium, Achromobacter,* and *Aerobacter;* a variety of yeasts, chiefly in the genera *Saccharomyces, Candida,* and *Pichia;* and a few molds. Numbers have been reported to range from 36,000 to 500,000,000 per milliliter at different stages. Much contamination may come from debris or fine particles on the sides or joints of troughs at the plant. If organisms grow to any extent, inversion of sucrose or even destruction of sugar may take place. Activity of the organisms continues from cutting of the cane through extraction to clarification of the juice, a process which kills yeasts and vegetative cells of bacteria. Bacterial spores are present from then on, through sedimentation, filtration, evaporation, crystallization, and centrifugation, but may be reduced in numbers by these processes, although spores of thermophiles may be added from equipment. Bagging of the raw sugar also may add some microorganisms. In raw sugars 400 to 68,000 organisms per gram have been found, and in molasses 100 to 190,000 per gram, although these numbers vary considerably. During the refining of the raw sugar, contamination may come from equipment, and organisms are added during bagging.

In the manufacture of beet sugar, cleaned beets are sliced into thin slices and the sugar is removed by a diffusion process at 60 to 85 C (140 to 185 F). Sources of contamination are flume waters and diffusion-battery waters. Thermophiles can grow in the latter up to 70 C (158 F). Contamination also may take place during refining and bagging of the sugar.

Granulated sugar now on the market is very low in microbial content for the most part, containing from a few to several hundred organisms per gram, mostly bacterial spores. Little information is available on the kinds of microorganisms in unspoiled sirups and molasses. The special flavor of Barbados molasses has been attributed to fermentation by *Zygosaccharomyces* yeasts and *Clostridium saccharolyticum*.

MAPLE SIRUP

Sap of the sugar maple in the vascular bundles is sterile or practically so, but becomes contaminated from outside sources in the tap holes and by the spout, plastic tubing, and buckets or other collection vessels. If a period of unusual warmth occurs before the sap is collected, considerable growth of yeasts and bacteria may take place in the sap.

Figure 14–1. Tapholes in maple tree and plastic tubing for transporting sap to plastic bag. (*Courtesy of C. O. Willets, Eastern Utilization Research and Development Division, U.S. Department of Agriculture.*)

MAPLE SAP SAMPLES COLLECTED SIMULTANEOUSLY FROM ONE NONSTERILE SPILE

IN SAP BUCKET   IN PLASTIC BAG

Figure 14–2.   Effect of sunlight on repression of microbial growth in maple sap. (*Courtesy of C. O. Willets, Eastern Utilization Research and Development Division, U.S. Department of Agriculture. Photo by M. C. Audsley.*)

Growth in the sap may be reduced by use of plastic tubing (Figure 14–1) and plastic bags (Figure 14–2) which permit the passage of sunlight. Usually the bucket is not sanitized after it is emptied, thus contaminating the next lot of sap, and the plastic tubing or bag may contribute its share of organisms unless adequately cleaned and sanitized before reuse. The concentration of the sap by boiling reduces its microbial content to low numbers.

As the season for collection progresses, microorganisms may grow sufficiently to block the tap holes. These are mostly psychrophilic, Gram-negative rods of the genera *Pseudomonas, Achromobacter,* and *Flavobacterium,* plus yeasts and molds.

HONEY

The chief sources of microorganisms in honey are the nectar of flowers and the honeybee. Yeasts have been shown to come from the nectar and from the intestinal contents of the bee, and also bacteria from the latter source.

CANDY

Candies from retail markets have been found to contain from 0 to 2,000,000 bacteria per piece, but most pieces harbored no more than a few hundred. Few coliform bacteria were found. The candies received most of their contamination from their ingredients, although some may be added to unwrapped pieces by air, dust, and handling.

PRESERVATION

As with cereals, sugars normally have a moisture content so low that microorganisms cannot grow. Only when moisture has been absorbed is there any chance for microbial spoilage. Storage conditions should be such that vermin are kept out and the sugar remains dry. The recommended storage temperature is similar to that for cereals.

During the manufacture of raw sugar and the subsequent refining process the numbers of microorganisms present, which may have been large during extraction from cane or sugar beet, are reduced by most subsequent processes such as clarification, evaporation, crystallization, centrifugation, and filtration. Special treatments to reduce numbers and kinds of organisms may be given during refining when the sugar is to be used for a special purpose, such as for soft drinks or canning. Care is taken to avoid build-up of organisms and their spores during processing, and numbers may be reduced by irradiation with ultraviolet rays or combined action of heat and hydrogen peroxide.

Because of their high sugar concentration most candies are not subject to microbial spoilage, although soft fillings of chocolate-covered candies may support the growth of microorganisms. The bursting of chocolates is prevented by means of a uniform and fairly heavy chocolate coating and use of a fondant or other filling that will not permit the growth of gas-formers.

Sirups and molasses usually have undergone enough heating to destroy most microorganisms, but should be stored at cool temperatures to prevent or slow down chemical changes and microbial growth. Some molasses may contain enough sulfur dioxide to inhibit microorganisms, but most sirups and molasses contain no added preservative and prevent microbial growth because of the high osmotic pressure of the sugar solution. The osmotic pressure increases with the extent of inversion (hydrolysis) of the sucrose. Mold growth on the surface is prevented by a complete fill of the container and is reduced by periodic mixing of the sirup or molasses.

The boiling process during evaporation of maple sap to maple sirup kills the important spoilage organisms. Such sirup, bottled hot and in a completely filled container, usually keeps well.

Honey distributed locally on a small scale usually is not pasteurized and therefore may be subject to crystallization and to possible spoilage in time by osmophilic yeasts. Commercially distributed honey usually is pasteurized at 160 to 170 F (71 to 77 C) for a few minutes. A recommended treatment is to heat fairly rapidly to at least 160 F (71 C), hold there for 5 min, and cool promptly to 90 to 100 F (32.2 to 38 C). Sodium benzoate has been suggested as a chemical preservative but has not been used much.

SPOILAGE

The spoilage of sugars or concentrated solutions of sugars is limited to that caused by osmophilic microorganisms, that is, those able to grow in such high concentrations. Among the osmophiles are species of *Leuconostoc* and *Bacillus,* certain yeasts, especially those of the genus *Zygosaccharomyces,* and certain molds. As the sugar concentrations decrease, increasing numbers of kinds of organisms can grow, so that sap from a maple tree would show types of spoilage that maple sirup could not.

SUCROSE

During the manufacture of sugar, the original cane or beet juice becomes more and more purified toward sucrose and the concentration of sugar in solution becomes greater and greater until finally crystalline sugar is attained plus molasses that is high in sugar. The purer the product becomes, the poorer it gets as a culture medium for microorganisms; and the more concentrated it becomes, the fewer kinds of organisms can grow in it.

Raw juice

The raw cane or beet juice is not high in sugar and contains a good supply of accessory foods for microorganisms, and therefore is readily deteriorated by numerous organisms present if sufficient time is allowed. Until the clarification process, gum and slime may be formed, e.g., dextran by *Leuconostoc mesenteroides* or *L. dextranicum* and levan by *Bacillus* species or, less commonly, by yeasts or molds. Gums cause clogging of pipes, strainers, and pumps, errors in polarization, and poor crystallization. Sucrose may be lost by inversion, fermentation to acid, or oxidation by bacteria, asporogenous yeasts and species of *Schizosaccharomyces* or *Zygosaccharomyces*, and molds such as species of *Aspergillus, Stemphylium, Sterigmatocystis, Cladosporium, Monilia,* etc.

Sugar in storage

Liquid sugar with sugar content as high as 67 to 72 brix will support the growth of yeasts (*Saccharomyces, Candida, Rhodotorula*) and molds which may enter from the air. Dilution by absorption of moisture at the surface may result in growth of microorganisms and hence deterioration of the product. This can be prevented by circulation of filtered sterile air across the top of the storage tank or exposure to ultraviolet lamps.

Molasses and sirups

Microbial spoilage of molasses is not common, although it is difficult to sterilize by heat because of the protective effect of the sugar. Chemical decomposition in stored tanks of mill molasses is common, however. The gassy or frothy "fermentation," which is attributed to the reaction of amino acids with glucose to release carbon dioxide gas, is favored by high temperatures (40 C, or 104 F, or above). Some workers claim that the carbon dioxide is released by the decomposition of calcium gluconate. At lower temperatures microorganisms may contribute somewhat to the gas production—organisms like asporogenous yeasts, *Zygosaccharomyces* species, and *Clostridium butyricum.*

Canned molasses or sirup may be subject to spoilage by osmophilic yeasts that survive the heat process. Gaseous chemical spoilage of the type described for molasses also may take place. Molasses or sirup exposed to air will mold, in time, on the surface, and this also may occur at the surface of a bottled or canned sirup if air is left there and contamination has taken place prior to sealing. Some kinds of molasses are acid enough to cause hydrogen swells (see Chapter 22) upon long storage.

MAPLE SAP AND SIRUP

As previously stated, sap from the sugar maple becomes contaminated when drawn; and although a moderate amount of growth may improve flavor and color, the sap often stands under conditions that favor excessive growth of microorganisms and hence spoilage. Five chief types of spoilage are recognized: (1) ropy or stringy sap, usually caused by *Aerobacter aerogenes,* although *Leuconostoc* species may be responsible; (2) cloudy, sometimes greenish, sap resulting from the growth of *Pseudomonas fluorescens,* with species of *Achrombacter* and *Flavobacterium* sometimes contributing to cloudiness; (3) red sap, colored by pigments of red bacteria, e.g., *Micrococcus roseus,* or of yeasts or yeast-like fungi; (4) sour sap, a catchall grouping for types of spoilage not showing a marked change in color, but having a sour odor and caused by any of a variety of kinds of bacteria or yeasts; and (5) moldy sap, spoiled by molds.

Maple sirup can be ropy because of *Aerobacter aerogenes,* yeasty as the result of growth of species of *Saccharomyces* or *Zygosaccharomyces* yeasts, pink from the pigment of *Micrococcus roseus,* or moldy at the surface, where species of *Aspergillus, Penicillium,* or other genera may grow. The sirup may become dark because of alkalinity produced by bacteria growing in the sap and inversion of sucrose.

Maple sugar keeps well unless moistened, at which time molds may grow.

HONEY

Honey is variable in composition, but must contain no more than 25 percent moisture. Because of its high sugar content, 70 to 80 percent, mostly glucose and levulose, and its acidity, pH 3.2 to 4.2, the chief cause of its spoilage is osmophilic yeasts: species of *Zygosaccharomyces,* such as *Z. mellis, richteri,* or *nussbaumeri,* or *Torula (Cryptococcus) mellis.* Most molds do not grow well on honey, although species of *Penicillium* and *Mucor* have been found to develop slowly.

Most honey yeasts do not grow in the laboratory in sugar concentrations as high as those usually found in honey. Therefore special theories for the initiation of growth of yeasts in honey have been advanced: (1) that honey, being hygroscopic, becomes diluted at the surface, where yeasts begin to multiply and soon become adapted to the high sugar concentrations; (2) that crystallization of glucose hydrate from honey leaves a lowered concentration of sugars in solution; or (3) that, on long standing, yeasts gradually become adapted to the high sugar concentrations. The critical moisture content for the initiation of yeast growth has been placed at 21 percent. The degree of inversion of sucrose to glucose and levulose by the bees and the content of available nitrogen also are listed as factors determining the likelihood of growth. The fermentation process usually is slow, lasting for months, and the chief products are carbon dioxide, alcohol, and nonvolatile acids which give an off-flavor to the honey. Darkening and crystallization usually accompany the fermentation.

CANDY

Most candies are not subject to microbial spoilage because of their comparatively high sugar and low moisture content. Exceptions are the chocolates with soft centers of fondant or of inverted sugar, which, under certain conditions, burst or "explode." Yeasts or species of *Clostridium* growing in these candies develop a gas pressure which may disrupt the entire candy, or more often will push out some of the sirup or fondant through a weak spot in the chocolate coating. Often this weak spot is on the poorly covered bottom of the chocolate, where a cylinder of fondant squeezes out. The defect is prevented by use of a filling that will not support growth of the gas-formers and by coating the candy with a uniformly thick and strong layer of chocolate. The defect is most common where chocolates are made on a small scale. Molds have been found growing under the coating of coconut bars.

REFERENCES

ALLEN, L. A., A. H. COOPER, ANNE CAIRNS, and MARY C. C. MAXWELL. 1946. Microbiology of beet-sugar manufacture. Soc. Appl. Bacteriol., Proc. 1946:5–9.

FRANK, H. A., and C. O. WILLITS. 1961. Maple sirup. XVIII. Bacterial growth in maple sap collected with plastic tubing. Food Technol. 15:374–378.

HAYWARD, F. W., and C. S. PEDERSON. 1946. Some factors causing dark-colored maple sirup. N.Y. State Agr. Exp. Sta. Bull. 718.

HUCKER, G. J., and R. F. BROOKS. 1942. Gas production in storage molasses. Food Res. 7:481–494.

HUCKER, G. J. and C. S. PEDERSON. 1942. A review of the microbiology of commercial sugar and related sweetening agents. Food Res. 7:459–480.

LOCHHEAD, A. G., and N. B. McMASTER. 1931. Yeast infection of normal honey and its relation to fermentation. Sci. Agr. 11:351–360.

MANSVELT, J. W., 1964. Microbiological spoilage in the confectionery industry. Confectionery Prod. 30(1):33–35, 37, 39.

MARVIN, G. E., 1928. Occurrence and characteristics of certain yeasts found in fermented honey. J. Econ. Entomol. 21:363–370.

MOROZ, R. 1963. Microbiology of the sugar industry, p. 373–449. In P. Honig (Ed.) Principles of sugar technology: Vol. III, Evaporation, centrifugation, microbiology, grading and classification of sugars and molasses. Elsevier, Amsterdam.

NAGHSKI, J. 1953. The organisms of maple sirup: their effect and control. 2nd Conf. Maple Products, Proc. p. 34–36.

OWENS, W. L. 1958. The deterioration of raw sugars in storage. Sugar J. 21(5):22–25.

PEDERSON, C. S., and G. J. HUCKER. 1948. The significance of bacteria in sugar mills. Int. Sugar J. 50:238–239.

SHENEMAN, J. M., and R. N. COSTILOW. 1959. Identification of microorganisms from maple tree tapholes. Food Res. 24:146–151.

UNDERWOOD, J. C., and C. O. WILLITS. 1963. Research modernizes the maple-sirup industry. Food Technol. 17:1380–1385.

WEINZIRL, J. 1922. The cause of explosion in chocolate candies. J. Bacteriol. 7:599–604.

CHAPTER FIFTEEN

CONTAMINATION, PRESERVATION, AND SPOILAGE OF VEGETABLES AND FRUITS

Vegetables and fruits may be fresh, dried, frozen, fermented, pasteurized, or canned. Spoilage of the canned products will be discussed in Chapter 22.

CONTAMINATION

As soon as fruits and vegetables are gathered into boxes, lugs, baskets, or trucks during harvesting, they are subject to contamination with spoilage organisms from each other and from the containers unless these have been adequately sanitized. During transportation to market or to processing plant, mechanical damage may increase susceptibility to decay and growth of microorganisms may take place. Precooling of the product and refrigeration during transportation will slow down growth.

Washing of the fruit or vegetable may involve a preliminary soaking, may be by agitation in water, or, preferably, may be by a spray treatment. Soaking and washing by agitation tend to distribute spoilage organisms from damaged to whole foods. Recirculated or reused water is likely to add organisms, and the washing process may moisten surfaces enough to permit growth of organisms during a holding period. Washing with detergent or germicidal solutions will reduce numbers of microorganisms on the foods.

Sorting of spoiled fruits or vegetables or trimming spoiled parts removes microorganisms, but additional handling may result in mechanical damage and therefore greater susceptibility to decay. When these products are sold in the retail market without processing, they are not ordinarily subjected to much further contamination, with the exceptions of storage in the market in contaminated bins or other containers, possible contact with decaying products, handling by sales people and customers, and perhaps spraying with water or packing with chipped ice (of green vegetables). This spraying gives a fresh appearance to the vegetables and delays decomposition, but also adds organisms, e.g., psychrophiles, from water or ice and gives a moist surface to encourage their growth on longer storage.

In the processing plant the fruits or vegetables are subjected to further contamination and chances for growth of microorganisms, or numbers and kinds of organisms may be reduced by some procedures. Adequate washing at the plant causes a reduction in numbers of microorganisms on the food, as do peeling by steam, hot water, or lye, and blanching (heating to inactivate enzymes, etc.). Sweating of products during handling increases numbers. Processes such as trimming, mechanical abrasion or peeling, cutting, pitting or coring, and various methods of disintegration may serve to add contaminants from the equipment involved. In fact every piece of equipment coming in contact with food can be a significant source of microorganisms unless it has been cleaned and sanitized adequately. Modern metal equipment with smooth surfaces and without cracks, dead ends, etc., is made to facilitate such treatments. Examples of possible sources of contamination of foods with microorganisms are trays, bins, tanks, pipes, flumes, tables, conveyor belts and aprons, fillers, blanchers, presses, screens, and filters. Wooden surfaces are difficult to clean and sanitize and therefore are especially likely to be sources of contamination, as are cloth surfaces, e.g., on conveyor belts. Neglected parts of any food-handling system can build up numbers of microorganisms to contaminate the food. Hot-water blanching, although it reduces total numbers of organisms on the food, may cause the build-up of spores of thermophilic bacteria causing the spoilage of canned foods, e.g., flat sour spores in peas.

Build-up of populations of microorganisms on equipment as the result of microbial growth in the exudates and residues from fruits and vegetables may influence greatly the amount of contamination of the foods and the growth of the contaminants. Not only is there the possibility of the addition of large numbers of organisms from this source, but there is also the likelihood that these will be organisms in their logarithmic phase of growth and therefore able to continue rapid growth. This effect is especially evident on vegetables following blanching. This heat-treatment reduces the bacterial content considerably, damages many of the surviving cells, and consequently lengthens their lag period. On the other hand, the actively growing contaminants from the equipment can attain large numbers if enough time is allowed before freezing, drying, or canning; such growth is usually the cause of very high bacterial counts.

Inclusion of decayed parts of fruits increases the numbers of microorganisms in fruit juices. Numbers in orange juice, for example, and numbers of coliforms are increased greatly by the inclusion of fruits with soft rots. Heating of grapes before extraction reduces numbers of organisms in the expressed juice, but pressing introduces contamination.

Average results from thirteen pea-freezing plants showing numbers of bacteria on the peas at different stages are shown in Table 15–1. The increase in numbers after blanching from 10,000 per gram to 736,000 per gram prior to freezing should be noted. Observe also the decreases in numbers as the result of washing, blanching, and freezing.

TABLE 15–1.   Average numbers of bacteria on peas during processing in thirteen pea-freezing plants*

Point of sampling	Bacteria on peas, nos/g
Platform	11,346,000
After washing	1,090,000
After blanching	10,000
End of flume	239,000
End of inspection belt	410,000
Entrance to freezer	736,000
After freezing	560,000

* Western Regional Laboratory. 1944. U.S. Dep. Agr., Albany, Calif.

The kinds of microorganisms from equipment will depend upon the product being processed, for that product will constitute the culture medium for the organisms. Thus pea residues would encourage bacteria that grow well in a pea medium and tomatoes those organisms that can develop in tomato juice. As the equipment is used throughout the day the organisms can continue to build up. At the end of the run, however, when equipment is cleansed and sanitized, total numbers of microorganisms thereon are greatly reduced, and if the operation is efficient only the resistant forms survive. Therefore spores of bacteria are likely to survive and, if conditions for growth are present while the equipment is idle, these spore-formers may increase in numbers, especially in poorly cleansed parts. The thermophilic spore-formers so troublesome to canners of vegetables build up in this manner and add to the difficulty of giving the foods an adequate heat process. The numbers of such organisms on poorly cleaned and sanitized equipment may be high at the start of a day's run and decrease as the day progresses, but the reverse usually is true. A layoff during the run permits a renewed increase in numbers. It is obvious that the numbers of microorganisms that enter foods from equipment are dependent upon the opportunities given these organisms for growth, and that these opportunities are the result of inadequacy of cleaning and sanitizing combined with favorable conditions of moisture and temperature for an appreciable time. Added ingredients like sugar or starch may add spoilage organisms, especially spores of thermophilic bacteria.

PRESERVATION OF VEGETABLES
AND VEGETABLE PRODUCTS

Microorganisms on the surfaces of freshly harvested fruits and vegetables include not only those of the normal surface flora but also those from soil and water and, perhaps, plant pathogens. Genera of bacteria usually present include *Achromobacter, Pseudomonas, Aerobacter, Alcaligenes, Bacillus, Bacterium, Chromobacterium, Flavobacterium, Lactobacillus, Leuconostoc, Micrococcus, Sarcina, Serratia, Staphylococcus, Streptococcus,* and others, plus perhaps genera containing plant pathogens, such as *Erwinia* and *Xanthomonas.* Any of a number of kinds of molds also may be there, and sometimes a few yeasts. If the surfaces are moist or the outer surface has been damaged, growth of some microorganisms may take place between harvesting and processing or consumption of the vegetables. Adequate control of temperature and humidity will reduce such growth.

ASEPSIS

While a limited amount of contamination of vegetables will take place between harvesting and processing or consumption, gross contamination can be avoided. Boxes, lugs, baskets, and other containers should be practically free of the growth of microorganisms and some will need cleansing and sanitation between uses. Examples are the lugs or other containers used for transporting shelled peas from the viner station to the canning factory. These containers may support a considerable amount of growth of bacteria on their moist interior and be a source of high numbers of organisms on the peas. Contact of vegetables undergoing spoilage with healthy vegetables will add contamination and may lead to losses. Contamination from equipment at the processing plant can be reduced by adequate cleansing and sanitizing. Especially feared is a build-up of heat-resistant spores of spoilage bacteria, for instance, the spores of flat sour bacteria, putrefactive anaerobes, or *Clostridium thermosaccharolyticum.*

REMOVAL OF MICROORGANISMS

Thorough washing of vegetables removes most of the casual contaminants on the surface but leaves much of the natural microbial surface flora. Unless the wash water is of good bacteriological quality it may add organisms, and subsequently growth may take place on the moist surface. Chlorinated water or borax solution sometimes is used for washing, and detergents may be added to facilitate the removal of dirt and microorganisms. Part of the mold growth on strawberries, for example, can be removed by washing with a nonionic detergent solution.

USE OF HEAT

Vegetables to be dried or frozen, and some to be canned, are scalded or blanched to inactivate their enzymes. At the same time the numbers of microorganisms are reduced appreciably.

The heat processing of canned vegetables has been discussed in Chapter 6, and examples of processes have been given. Canned vegetable juices are processed similarly.

USE OF LOW TEMPERATURES

As has been indicated, a few kinds of vegetables that are relatively stable, such as root crops, potatoes, cabbage, and celery, can be preserved for a limited time by common or cellar storage (see Chapter 7).

Chilling

Most vegetables to be preserved without special processing are cooled promptly and kept at chilling temperatures. The chilling is accomplished by use of cold water, ice, or mechanical refrigeration, or by vacuum cooling (moistening plus evacuation), as used for lettuce. Each kind of vegetable has its own optimal temperature and relative humidity for chilling storage, as illustrated in Chapter 7 and as recommended in special bulletins and manuals. The "freshening" of leafy vegetables (lettuce, spinach) by a water spray will cool the products if cold water is employed and will aid in their preservation.

Control of the composition of the atmosphere in the storage of vegetables has not been used as much as with fruits. The addition of carbon dioxide or ozone to the air has been recommended by some workers. Ultraviolet rays have not been successful because the rays do not hit all surfaces of the vegetables as they are packaged and handled.

Sweet potatoes were mentioned in Chapter 7 as an example of a vegetable requiring special conditions of chilling storage. Ordinary potatoes turn sweet at temperatures below 36 to 40 F (2.2 to 4.4 C) and are stored at higher temperatures if they are to be used for potato chips. Sweet potatoes and onions are subjected to special curing treatments prior to storage.

Freezing

The selection and preparation of vegetables for freezing, their blanching, their freezing, and the changes during these processes have been discussed in Chapter 7. On the surface of vegetables are the microorganisms of the natural flora, plus contaminants from soil and water (see Chapter 4). If the surfaces are moist, growth of some of these organisms will take place before the vegetable reaches the freezing plant. There, washing reduces the numbers of some organisms and adds some organisms, and scalding or blanching brings about a great reduction in numbers, as much as 90 to 99 percent in some instances (see Tables 15–1 and 15–2). But during the cooling and handling prior to freezing there is opportunity for recontamination from equipment and for growth of organisms, so that under poor conditions a million or more organisms per gram of vegetable may be present at freezing. The freezing process reduces the number of organisms by a percentage that varies with the kinds and numbers originally present, but, on the average, about half of them are killed. Table 15–2 illustrates the changes in numbers of organisms on snap beans as they pass through various operations in a freezing plant. During storage in the frozen condition there is a steady decrease in numbers of organisms, but there are at least some survivors of most kinds of organisms after the usual storage period. The kind of bacteria most likely to grow on thawing will depend on the temperature and the elapsed time. Micrococci are predominant on thawing vegetables like sweet corn and peas when the temperature of thawing is fairly low, although *Achromobacter* and *Aerobacter* species also are commonly present. Lactobacilli are also common on peas under such conditions. One species of *Micrococcus* may grow at first, followed by another spe-

TABLE 15–2.　Numbers of organisms per gram on snap beans after passing through unit operations in a freezing plant

Source of sample	Average plate count/g*
Before cutter	740,000
After cutter and shaker	573,000
After blancher	1,000
After shaker, before sorting belt	188,000
Final package	36,000

* Average of seventeen samples during 9 days. Plates incubated at 32 C. From paper by Hucker, Brooks, and Emery.

cies later. At higher temperatures, species of *Flavobacterium* also may multiply. As the small packages of quick-frozen vegetables usually are handled in the home, where the frozen food is placed directly into boiling water and cooked, there is no further opportunity for microbial growth. During freezing most vegetables wilt and become limp, and during storage frozen vegetables may undergo color changes.

When thawed vegetables are held at room temperature for any considerable period, there is a chance that food-poisoning bacteria may grow and produce toxin. Jones and Lochhead, for example, found enterotoxin-forming staphylococci in frozen corn. Sterilized corn inoculated with staphylococci of the food-poisoning type, then frozen, and then thawed and held at a room temperature of 20 C (68 F) for a total elapsed time of 1 day, permitted growth of these cocci and the production of enough enterotoxin to cause symptoms of food poisoning in kittens. Enterotoxin produced in this way would not be entirely destroyed by the customary cooking of the vegetable. It has been found that some strains of *Clostridium botulinum* can grow and produce toxin at 15 C (59 F), and most strains can grow and produce toxin at 20 C (68 F) within 2 or 3 days. The spores of this organism have been found in frozen vegetables and can be assumed to be present often. Fortunately the conditions for growth and toxin production would have to be unusual; power failure in freezers for several days during floods or hurricanes is an example of such conditions. The cooking of frozen vegetables will not kill all spores of *Clostridium botulinum,* and such cooked food should not be allowed to stand at room temperatures for any extended period. Boiling of the vegetable for 15 min will destroy any botulinum toxin that had been formed. Since *Aerobacter aerogenes* and enterococci are part of the surface flora of most vegetables, positive tests for coliforms or enterococci have no sanitary significance.

DRYING

As the methods of drying vegetables and vegetable products have improved, public acceptance has increased, so that now a number of dried food products have wide sale. Dried vegetables are used considerably during wartime but not as much during periods of peace, except for the long-used dry seeds of plants: beans, peas, corn, wheat, oats, rice, nuts, etc. Dried vegetables and vegetable products are used in dried soups, and dried spices and condiments are used as flavoring materials.

The drying of vegetables has been discussed in Chapter 8, including treatments before drying and the microbiology of the vegetables prior to reception at the processing plant. The removal of spoiled vegetables reduces numbers of microorganisms, and the washing removes more. Water may add organisms, and moist conditions may permit the growth of organisms until the blanching process. Contamination from equipment can add more microorganisms. Blanching causes a great reduction in numbers of organisms, as much as 99 percent in some instances. Growth of the microorganisms surviving blanching may take place up to the time of drying and add to the count of the dried product. Drying by heat destroys yeasts and most bacteria, but spores of bacteria and molds usually survive, as do the more heat-resistant vegetative cells. Microbial counts on dried vegetables, either after drying or as purchased in the retail market, usually are considerably higher than on dried fruits, because there are likely to be higher numbers before drying and a greater percentage survival afterwards. Most vegetables are less acid than fruits, and consequently the killing effect of the heat is less. Samples of dried vegetables from retail markets have been shown to contain microorganisms in the hundreds of thousands or even millions per gram, although these dried foods can be produced so as to contain a much smaller number of organisms.

When dried vegetables are sulfured to preserve a light color, their microbial content is reduced.

If the vegetables are dried adequately and stored properly, there will be no growth of microorganisms in them. During storage there is a slow decrease in the number of viable organisms, more rapid during the first few months and slower thereafter. The spores of bacteria and molds, some of the micrococci, and microbacteria are resistant to desiccation and will survive better than other microorganisms and will constitute an increasingly large percentage of the survivors as the storage time lengthens.

USE OF PRESERVATIVES

The addition of preservatives to vegetables is not common, although the surfaces of some vegetables may receive special treatment. Rutabagas and turnips sometimes are paraffined to lengthen their keeping time, and chlorinated water or borax solution may be used in the washing of some kinds of vegetables. Zinc carbonate has been reported to eliminate most mold growth on lettuce, beets, and spinach. A controlled atmosphere of carbon dioxide or ozone about chilled vegetables has been tried experimentally but has had little practical use.

ADDED PRESERVATIVE. Sodium chloride is the only added chemical preservative in common use. The amount added to vegetables may vary from the 2.25 to 2.5 percent in the making of sauerkraut up to saturation for cauliflower. The lower concentrations of salt permit an acid fermentation by bacteria to take place; as the percentage of salt is increased the rate of acid production becomes slower until a level of salt is reached that will permit no growth or production of acid.

Vegetables that are high in protein and low in carbohydrate, like green peas and lima beans, and some that soften readily, such as onions and cauliflower, are preserved by the addition of enough salt to prevent any fermentation: from 70 to 80° salometer (18.6 to 21.2 percent salt) up to saturation (26.5 percent salt, or 100° salometer).

It should be noted that, upon the addition of brine or salt to vegetables, water is drawn from them and serves to decrease the salt concentration in the liquid.

DEVELOPED PRESERVATIVES. At room temperature an acid fermentation is normal for shredded, chopped, or crushed vegetables containing sugar, but instead of a clean, acid flavor from the action of lactic acid bacteria, undesirable flavors and changes in body may result from growth of coliform bacteria, bacilli, anaerobes, proteolytic bacteria, and others. The addition of salt to such materials serves to reduce competition from undesirable organisms and hence to encourage the lactic fermentation. The salt also serves to draw the juice from the vegetables and bring about better distribution of the lactic acid bacteria. The amount of sugar in the vegetable affects the acidity that can be produced, while the amount of salt and the temperature determine the rate of acid production and the kinds of bacteria involved in it. In general, as the salt content is increased, the rate of acid formation becomes slower and the numbers of kinds of bacteria concerned become fewer. Some recipes call for a comparatively low salt content at the start, increasing amounts as the fermentation continues, and finally enough salt to prevent further growth of bacteria. This method is employed in the brining of vegetables like string beans and corn.

Sauerkraut

The Federal definition is as follows:

Sauerkraut is the clean, sound product, of characteristic flavor, obtained by full fermentation, chiefly lactic, of properly prepared and shredded cabbage in the presence of not less than two percent nor more than three percent of salt. It contains, upon completion of the fermentation, not less than one and one-half percent of acid, expressed as lactic acid. Sauerkraut which has been rebrined in the process of canning or repacking contains not less than one percent of acid, expressed as lactic acid.

GENERAL MAKING PROCEDURE. Closely filled, fully matured heads of a variety of cabbage preferred for kraut making are wilted for 1 or 2 days to bring the cabbage to a uniform temperature and to facilitate shredding. Spoiled spots and defective outer leaves are trimmed off, the heads are washed with pure water, and the core is drilled out and shredded to be added to the rest of the cabbage. The head is cut to shreds of desired size, usually fairly slim ones. Then 2.25 to 2.5 percent of salt by weight is mixed with the shredded cabbage before transfer to the vat or is added during the packing of the shreds into the vat. The first method is preferable because it results in more uniform salting, allows time for penetration of the salt into the cabbage, and makes packing easier. After the shreds have been packed into the vat, they are tamped down and finally weighted down, so that a layer of expressed, brined juice stands on the surface. A covering of some kind should protect the surface from contamination with dirt or insects. The temperature during the lactic acid fermentation should be about 70 to 75 F (21.1 to 23.9 C). If the temperature is below 60 F (15.6 C), the fermentation will be slow and incomplete, and if above 80 to 85 F (26.7 to 29.4 C), abnormal fermentations may result. During the fermentation, film yeasts or molds will grow on the surface of the liquor if the surface of the expressed juice is left uncovered. Formerly, these were skimmed off or their growth prevented by covering the surface with mineral oil (or paraffin in small fermentors) after the evolution of gas had ceased or by filling containers completely. At the present time, plastic bags filled with water are placed on top of the fermenting cabbage to seal the surface and serve as a weight. When the desired acidity has been attained, the fermentation is stopped by heat-treatment during canning or by low temperatures.

COMPOSITION OF CABBAGE. The composition of cabbage varies with the variety and the conditions during its growth. Especially significant in sauerkraut making is the sugar content because of its influence on the maximal acidity that can be produced by fermentation. Analyses have shown the sugar content to range from 2.9 to 6.4 percent in different lots of cabbage; the higher contents of sugar would permit the production of too much acidity, if steps were not taken to stop the fermentation. The sugars are about 85 percent glucose and fructose and 15 percent sucrose.

ADDITION OF SALT. Pederson and others have demonstrated that 2.25 to 2.5 percent of salt by weight should be added to the shredded cabbage to obtain kraut of the best quality and that this salt should be distributed evenly. The salt draws out the plant juices containing the sugars and other nutrients, helps control the flora of the fermentation (favoring the lactic acid bacteria), and has a dispersing effect on clumps of bacteria. Undesirable competitors of the lactics, e.g., proteolytic bacteria and aerobic and anaerobic spore-formers, are inhibited more by the salt than are the desired acid-formers.

WASHING OF THE HEADS. Keipper and coworkers demonstrated that, not only does washing of the cabbage heads reduce total numbers of microorganisms in the shredded cabbage, but it also increases the percentage of desirable lactics in the flora remaining. In their experiments the lactics increased from 24 percent to 35 percent and the percentages of undesirable organisms such as chromogens, coliform bacteria, yeasts, and molds decreased.

THE FERMENTATION. Anaerobic conditions develop rapidly in the salted, shredded cabbage and the surrounding juice, chiefly as the result of the removal of oxygen by the respiration of the plant cells, but with some help from the bacteria. The juice contains the natural flora of the cabbage plus contaminants from soil and water. At first, different kinds of bacteria begin to grow, but the acid-forming types soon predominate. Prominent among the bacteria that attain appreciable numbers early in the fermentation are the coliform bacteria, *Aerobacter cloacae* for example, which produce gas and volatile acids, as well as some lactic acid. *Flavobacterium rhenanus* also has been found early in the fermentation. These organisms must contribute some flavor. Soon, however, *Leuconostoc mesenteroides* bacteria begin to outgrow all other organisms and continue acid production up to 0.7 to 1.0 percent acid (as lactic acid). These streptococci, which appear in pairs or short chains, grow well at 65 to 70 F (18.3 to 21.1 C) and are not inhibited, but probably are even stimulated by 2.5 percent salt. They attack sugars to form lactic acid, acetic acid, ethanol, mannitol, dextran, esters, and carbon dioxide, which contribute to the flavor of good sauerkraut, as do the fatty acids produced by the lactic acid rods and cocci from the lipids of the cabbage. The volatile products inhibit yeasts, and the dextran and mannitol (having a bitter flavor) are available to the desired successor to the leuconostocs, *Lactobacillus plantarum*, but not to most competing organisms. *Streptococcus faecalis*, which may grow during the kraut fermentation, especially if the salt content is high (for example, 3.5 percent), produces chiefly lactic acid but may yield some diacetyl.

Next, non-gas-forming lactobacilli, chiefly of the species *Lactobacillus plantarum*, continue the production of acid and can raise the acidity to 1.5 to 2.0 percent. These bacteria produce chiefly lactic acid in their fermentation of the sugars. They also utilize the mannitol that had been produced by the *Leuconostoc* and thus remove the bitter flavor. *L. plantarum* completes the desired fermentation in the production of sauerkraut. Experience has shown that a final acidity of about 1.7 percent as lactic acid is most desirable in sauerkraut. The fermentation can be stopped at this stage by canning or refrigerating the sauerkraut.

If enough sugar and mannitol remain after *L. plantarum* has finished its work, gas-forming lactobacilli, chiefly of the species *L. brevis*, can grow and continue acid production up to 2.4 percent, an acidity that is attained rarely, however, because of lack of sugar and mannitol. The gas-forming lactobacilli, producing the same products as the leuconostocs, give an undesirable, sharply acid flavor to the sauerkraut.

Good sauerkraut should be light-colored and crisp, with an acidity of about 1.7 percent and a clean, acid flavor. Small amounts of diacetyl are present to add a pleasant aroma and taste. According to Pederson the average finished kraut has a pH of 3.4 to 3.6, a lactic acid content of 1.25 percent, about 0.3 percent of acetic acid, and 0.58 percent ethyl alcohol. When 3.5 percent salt is used instead of the usual 2.25 percent, or when the temperature is 32 to 37 C (89.6 to 98.6 F) instead of lower temperatures, *Pediococcus cerevisiae* may play a significant part in the fermentation, but the kraut is likely to be inferior to that produced under normal conditions of salt content and temperature. Low salt, e.g., 1.0 percent, favors the heterofermentative *Leuconostoc mesenteroides* and *Lactobacillus brevis*.

Experiments have shown that it is not necessary or even advantageous to add starters of lactic acid bacteria in the making of sauerkraut.

Sauerkraut may be canned by filling the cans at 73.9 C (165 F), exhausting, sealing, and cooling.

Pickles

Cucumber pickles may be prepared without fermentation, or with partial or complete fermentation. Unfermented, partially fermented, or fully fermented cucumbers may be pasteurized to improve their keeping quality. Usually brined, acidified (naturally or artificially) cucumbers are heated so that the interior of the cucumbers will be maintained at 165 F (73.9 C) for at least 15 min. Both heating and cooling should be fairly rapid. No attempt will be made here to describe different methods of pickle manufacture since these vary widely; and the discussion will be limited to pickles produced by fermentation. There are two chief types of fermented pickles—salt, or salt-stock pickles, and dill pickles. The salt pickles are prepared for use in making special products like sour, sweet-sour, and mixed pickles and relishes.

Salt or salt-stock pickles

In the preparation of fermented salt or salt-stock pickles immature cucumbers are washed, placed in barrels or tanks, and brined. Sometimes about 1 percent of glucose is added if the cucumbers are low in sugar, but some workers claim that the addition of sugar will favor the production of gassy pickles, or "bloaters."

ADDITION OF SALT. The rate of addition of salt and the total amount added are varied considerably by different makers. Two general methods of salting, the low-salt method and the high-salt method, are employed. In the low-salt method a comparatively low amount of salt is added and the concentration is gradually increased until enough is present to stop any growth of bacteria. For example, a 30°-salometer (nearly 8 percent NaCl) brine may be added to the cucumbers along with 9 lb of salt per 100 lb of cucumbers. Some makers start with lower than 8 percent salt but risk off-fermentation or spoilage; 6 percent salt has been found to be the minimal concentration to hold down undesirable spore-forming bacteria. In the high-salt method the first brine is 40° salometer (about 10.5 percent salt), and 9 lb of salt are added per 100 lb of cucumbers.

The cucumbers are "keyed down" under a surface layer of brine, and the fermentation begins. In both methods, salt is added at weekly intervals so as to increase the salometer reading by about 3° salometer up to 60° (about 15.9 percent salt). In the low-salt method the increase is about 2° per week up to 50° and 1° per week up to 60° salometer. In warm climates the salt content of the brine may be increased more rapidly than has just been indicated, and in cool climates a brine weaker than 30° salometer may be added initially.

THE FERMENTATION. The fermentation desired is lactic acid fermentation of the sugars. This process usually takes 6 to 9 weeks for completion, depending upon the salting method and temperature employed.

Any or all of a number of salt-tolerant species of bacteria may grow initially in the newly brined fresh cucumbers. In fact, there may be marked differences in the kinds of bacteria growing in different lots of cucumber brine, depending upon the numbers and kinds introduced by the cucumbers or dirt left on them and by the water of the brine, the initial concentration of sodium chloride and the rate of increase in that concentration, and the temperature of the brined cucumbers. In general, the lower the salt concentration, the greater will be the numbers of kinds of bacteria that will grow at the start, the more rapid will be the acid production, and the greater will be the total acidity produced. First to grow in most instances is a mixture of species of the genera *Pseudomonas, Flavobacterium,* and *Achromobacter,* types that are considered undesirable in that they would be classed as spoilage bacteria rather than acid-formers. Likewise, *Bacillus* species are likely to come from soil on the cucumbers and their growth would not be desired. In brines of low salt content, coliform bacteria (*Aerobacter*), *Leuconostoc mesenteroides, Streptococcus faecalis,* and *Pediococcus cerevisiae* may grow and form acid during the first few days of the fermentation, and in 15 percent brines unidentified gas-forming cocci may produce some acid. Later, *Lactobacillus brevis* may contribute to the acidity if the salt concentration is not too high. In most brines *L. plantarum* (Figure 15–1) is the most important bacterium, developing acidity in both low- and high-salt brines. It begins to attain appreciable numbers several days after the start of the fermentation. As the salt concentration is increased from about 10 percent toward 15 percent, *L. plantarum* becomes decreasingly active. The total titratable acidity,

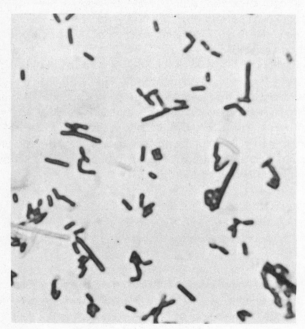

Figure 15–1. *Lactobacillus plantarum,* which normally is responsible for most of the acid production in the salt-pickle fermentation ($\times$1,600). (*Courtesy of J. L. Etchells and I. D. Jones, Glass Packer,* 30(5):358–360, 1951.)

Figure 15–2. Active gaseous fermentation of pickle brine by yeasts. Bloater pickles may result. (*Courtesy of J. L. Etchells and I. D. Jones, Glass Packer*, 30(5):358–360, 1951.)

as lactic acid, is about 0.6 to 0.8 percent on completion of the fermentation. Fermentations in which the heterofermentative lactics have been relatively effective generally yield pickles that are firm and have better density than those from homofermentative fermentations.

To complicate the fermentation further, yeasts may begin growth (Figure 15–2) after some acid has been formed by the bacteria. These yeasts are of two general types—the film, or oxidative, yeasts that grow on the surface of the brine and destroy lactic acid by oxidation, and the fermentative yeasts that grow down in the brine and ferment sugars to alcohol and carbon dioxide. Film yeasts of the genera *Debaryomyces*, *Endomycopsis*, and *Candida* have been found growing, with the first genus most widespread. Various methods for the reduction or elimination of the scum of yeasts in the brine during the fermentation include daily agitation of the surface or the addition of mineral oil, sorbic acid, oil of mustard, or other substances. Pickle vats often are located out in the sunlight, which inhibits surface growth on the brine. In the order of frequency of occurrence the following genera of fermentative yeasts have been found in the brines: *Torulopsis* (Figure 15–3), *Brettanomyces*, *Zygosaccharomyces* (a subgenus of *Saccharomyces*), *Hansenula*, *Torulaspora*, and *Kloeckera*. Gas produced by these yeasts bubbles from the brine and may be responsible for bloated pickles.

A fermentation as variable, complicated, and unpredictable as that of brined cucumbers is difficult to summarize. Most of the acid normally is produced by *Lactobacillus plantarum*, but also may be formed by *Leuconostoc mesenteroides*, *Lactobacillus brevis*, *Streptococcus faecalis*, *Pediococcus cerevisiae*, and possibly coliform bacteria. Acid is destroyed by film yeasts. Gas may be produced by *Leuconostoc mesenteroides*,

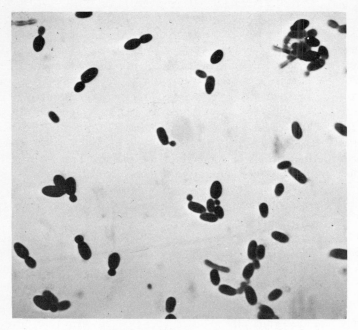

Figure 15–3. *Torulopsis caroliniana,* a fermentative yeast causing a gaseous fermentation in the brine during the salt pickle fermentation (×1,000). (*Courtesy of J. L. Etchells and I. D. Jones, Glass Packer,* 30(5):358–360, 1951.)

fermentative yeasts, and coliform bacteria. Pure cultures of pediococci and lactobacilli have been used experimentally in the production of pickles.

When the cucumbers are first brined they are chalky-white and opaque in cross section, but during the fermentation and cure the color changes from the original bright-green to an olive- or yellowish-green, and the flesh becomes increasingly translucent. Salt pickles in 15 percent salt brine can be kept for long periods if growth of film yeasts is prevented. They are too salty to be eaten and must be "freshened" by soaking before they are made into sour, sweet-sour, or mixed pickles, relishes, or other products.

Dill pickles

Dill pickles are so named because they are flavored by addition of dill herb in some form; usually spices are added as well, and to the kosher types, garlic or onion. Dill pickles may be unfermented or fermented or made from salt stock, but only the fermented types, "overnight" and "genuine" dill pickles, will be considered here.

The fermentation to produce dill pickles differs from that for salt stock in that a lower concentration of salt is employed than in salt stock and the brine usually is acidified with vinegar (acetic acid) at the start. The low salt content favors an increased rate and amount of acid production but adds to the risk of undesirable microbial changes. The flavoring materials, dill, spices, garlic, etc., do not markedly stimulate or inhibit the acid-forming bacteria, but they may be a source of considerable numbers of undesirable microorganisms and hence the cause of off-fermentations or spoilage of the pickles. Treated spices containing low numbers of microorganisms are now available.

OVERNIGHT DILL PICKLES. Overnight dill pickles are prepared by a slow acid fermentation at a low temperature in a comparatively weak acidified brine. For example, one formula calls for a 20°-salometer brine (5.3 percent salt), 10 lb of cured dill weed, 1 lb of mixed spices, and 1 qt of 100-grain vinegar per barrel. The brined cucumbers are then held at about 38 F (3.3 C), where they undergo a slow lactic acid fermentation until 0.3 to 0.6 percent of acid has developed. The pickles must be kept cold, and they have a comparatively short keeping time because of the low content of salt and acid. A higher acidity can be obtained by permitting fermentation to take place prior to storage, but there is increased danger of off-fermentation or spoilage, and the green color is not retained as well as in the usual process.

GENUINE DILL PICKLES. In the manufacture of genuine dill pickles a brine containing about 7.5 to 8.5 percent salt (28 to 32° salometer) is added to the cucumbers, usually in a 45-gal barrel, so that the concentration of salt in the finished pickles will be about 3.5 to 4.5 percent. Dill and pickle spices also are added. Most makers add vinegar, e.g., 1 qt of 100-grain vinegar per barrel, to keep down abnormal fermentations. A temperature between 60 and 85 F is usually employed, with lower temperatures preferred.

Because of the comparatively low concentration of salt, miscellaneous soil bacteria first begin growing, but acid production probably is begun by bacteria like *Leuconostoc mesenteroides*, *Streptococcus faecalis*, and *Pediococcus cerevisiae*, and is continued by *Lactobacillus plantarum*, with possible help from *L. brevis*. The final acidity ranges from 1.0 to 1.5 percent as lactic acid.

Green olives

Although olives are fruits, rather than vegetables, they will be discussed here because their fermentation resembles that of the fermented vegetables just mentioned.

In the preparation of green olives, the fruits are harvested when fully developed, but still green or straw-yellow; bruising is avoided to prevent defects in the product. In processing, they first are treated with and kept submerged in a 1.25 to 2.0 percent lye solution at about 60 to 70 F (15.6 to 21.1 C) until the lye has penetrated one-half to three-quarters of the way toward the pit, and then they are washed several times (with as little exposure to air as is practicable to avoid darkening) to remove the lye. This treatment removes most, but not all, of the bitterness caused by the glucoside, oleuropein. Next, the olives are barreled and covered with a salt brine, the concentration of which varies with the kind of olive. A 10 to 15 percent brine (40 to 50° salometer) is recommended for Manzanillo olives, resulting in a concentration of about 6 to 9 percent salt upon stabilization; the salt concentration is adjusted to and maintained at 7 to 8 percent salt throughout the fermentation. Sevillano olives, on the other hand, are started in brine with 5 to 6.25 percent salt (20 to 25° salometer) or less, resulting in a 2.5 to 4 percent brine, which is adjusted to 6 to 8 percent salt. Lost brine (leakage, evaporation, gas) is replaced promptly, and the olives are kept covered with brine at all times.

The lactic acid fermentation of the barreled olives may take as long as 6 to 10 months, the length of time depending upon the atmospheric temperatures. Vaughn, Douglas, and Gililland have divided the normal fermentation into three stages: (1) the first stage, lasting 7 to 14 days, during which the brine is becoming stabilized, foods for microorganisms are being leached from the olives, and potential spoilage organisms, e.g., *Aerobacter* and *Pseudomonas,* and perhaps *Clostridium, Bacillus,* and yeasts, may grow until growth of *Leuconostoc mesenteroides* has well begun; (2) the intermediate stage, lasting 2 or 3 weeks, during which *Leuconostoc* becomes predominant in growth and acid production and *Lactobacillus plantarum* and *L. brevis* begin to grow and produce acid; and (3) the final stage when these lactobacilli, especially *L. plantarum,* are predominant. Gas production by yeasts, coliforms, *Leuconostoc,* and the heterofermentative *L. brevis* and other lactobacilli takes place, especially during early stages of the fermentation. An average temperature of about 75 F (23.9 C) favors a rapid fermentation. The final acidity usually is about 0.7 to 1.0 percent acid as lactic and the final pH is 4.0 to 3.8 or lower. Sometimes, probably because of damage to the lactic acid bacteria by the lye treatment, the normal lactic fermentation is delayed or prevented. A starter of *Lactobacillus plantarum* or of actively fermenting brine has been recommended. Olives may be low in sugar or have too much leached out during lye treatments and washing and therefore require the addition of glucose in order for the desired lactic fermentation to take place.

Olives that have been pitted and stuffed with brined pimento are barreled, brined, and allowed to stand for a month or so until the pimento flesh has been fermented.

The fermented green olives are sorted and graded, and may be rebarreled and rebrined, or may be washed, packed into glass jars or other containers under vacuum, and rebrined with a brine of about 28° salometer (7 percent salt). Edible lactic acid may be added to the final brine. They may be pasteurized in the container at about 140 F (60 C) or brined at 175 to 180 F (79.4 to 82.2 C) to improve their keeping quality.

Ripe olives

Olives for the production of ripe olives are picked when green to straw-colored, shipped in boxes or bins to the factory, and held before processing in the factory in a 5 to 10 percent brine. A lactic acid fermentation usually takes place while the olives are being held in the brine. After grading and sorting of the olives they are given the following "pickling" process: a first lye treatment with 0.5 to 2.0 percent lye to barely penetrate the skin; an aeration treatment with stirring or administration of compressed air to darken the skin; and more lye treatments, followed by further aeration. Finally the lye is allowed to penetrate to the pit, dehydrolyzing all of the bitter glucoside, oleuropein; and then the olives are leached with water to remove all of the lye. Then comes stabilization of the ripe olives in a 2 to 3 percent brine for 2 or more days, during which period fermentation may take place, although it is not desired, for it may lead to color defects. Ripe olives usually are canned in weak brine and are processed in the glass or tin container at about 240 F (115.6 C) for 60 min.

Other fermented vegetables

Leafy vegetables such as spinach and chard, and "greens" such as beet, mustard, and turnip may be prepared in a manner similar to that for making sauerkraut from cabbage. The addition of 2.5 percent by weight of dry salt causes moisture to be drawn from the vegetable and permits a lactic acid fermentation similar to that for sauerkraut. Lettuce kraut prepared in this way has been compared favorably with sauerkraut. Sauerrüben are made from unpeeled, shredded young turnips to which is added about 2.2 percent salt by weight. The salted turnips, packed in half-gallon fruit jars, are held at 70 to 75 F (21.1 to 24 C) to undergo a lactic acid fermentation like that for sauerkraut.

Fermented green tomatoes may be prepared for incorporation in relishes and mixed pickles. The fermentation is similar to that for sauerkraut but slower, the same succession of bacteria being involved. The skins of the tomatoes may be punctured to hasten the process.

A number of Oriental fermented foods are made from vegetable products, and some of these are discussed in Chapter 24.

PRESERVATION BY IRRADIATION

In the future, vegetables and vegetable products may be preserved by means of radiations or radiations combined with other preservative methods (see Chapter 10), but such processing is still in the experimental stage. Experimental treatment with gamma rays to inactivate organisms causing decay, followed by storage, has resulted in discoloration, softening, or other deterioration of most vegetables tested.

PRESERVATION OF FRUITS AND FRUIT PRODUCTS

In general, principles similar to those for both vegetables and vegetable products are involved in the preservation of fruits and fruit products. The surfaces of healthy fruits include the natural flora plus contaminating microorganisms from soil and water and therefore have a surface flora much like that listed for vegetables, except that more yeasts are likely to be present. In addition, some fruits will contain plant pathogens or saprophytic spoilage organisms which may grow subsequent to harvesting. Such defective fruits should be sorted and spoiled portions may be trimmed out. A few microorganisms are present in the interior of occasional healthy fruits.

ASEPSIS

Fruits, like vegetables, may be subject to contamination between harvesting and processing from containers and from spoiling fruits and care should be taken to avoid such contamination as much as possible.

REMOVAL OF MICROORGANISMS

Thorough washing of fruits serves to remove not only dirt and hence casual contaminating microorganisms but also poisonous sprays. Washing may be by means of water, detergent solutions, or even bactericidal solutions, such as chlorinated water. Trimming also removes microorganisms. Clear fruit juices may be sterilized by filtration through bacteria-tight filters.

USE OF HEAT

Fruits seldom are blanched prior to other processing because blanching causes excessive physical damage.

The principles involved in the heat processing of canned fruits have been discussed in Chapter 6, and a few examples of the processes have been given. It will be noted that the fruits are in one of two groups on the basis of their pH (see page 85): the *acid* foods, such as tomatoes, pears, and pineapples, or the *high-acid* foods, such as berries. A steam-pressure sterilizer is not required for most fruits, for heating at about 100 C (212 F) is sufficient and can be accomplished by flowing steam or boiling water. Some high-acid fruits are preheated and placed in the container while hot, and require no further heat processing. In general, the more acid the fruit, the less heat is required for its preservation. Similar principles are involved in the canning of fruit juices.

USE OF LOW TEMPERATURES

A few fruits, such as apples, can be preserved for a limited time in common or cellar storage; but controlled lower temperatures usually are employed during most of the storage period of fruits.

Chilling

Each fruit has its own optimal temperature and relative humidity for chilling storage; even varieties of the same fruit may differ in their requirements. Bananas and apples were mentioned in Chapter 7 as fruits that should have special conditions of storage, and avocados and citrus fruits can be added to the list. Fruits have been treated with various chemicals before or during storage to aid in their preservation. Thus hypochlorites, sodium bicarbonate, borax, propionates, biphenyl, *o*-phenylphenols, and other chemicals have been recommended. Fruit also has been enclosed in wrappers treated with chemicals, e.g., sulfite paper on grapes, iodine paper on grapes and tomatoes, or borax paper on oranges. Waxed wraps, paraffin oil, paraffin, waxes, and mineral oil have been applied for mechanical protection.

There has been considerable research on the combination of the chilling storage of fruits with control of the atmosphere of the storage room. This control may consist merely of regulation of the concentrations of oxygen and carbon dioxide in the atmosphere or may involve the addition or removal of carbon dioxide or oxygen or the addition of ozone. The United States Department of Agriculture has recommended the use of high concentration of carbon dioxide in the air during the first few days after harvest, until the fruits or vegetables have cooled enough to slow down changes, and no further addition of CO_2 thereafter. Solid CO_2 (dry ice) can be used as the source of CO_2.

British workers have done much of the experimentation on the "gas storage" of fruits, especially of apples, controlling the carbon dioxide and oxygen of the air in the storage room by addition or removal of carbon dioxide or by ventilation. The optimal concentration of carbon dioxide and oxygen and proportion of these gases varies with the kind of fruit and even with the variety of fruit. Control of humidity is unnecessary, and a higher storage temperature may be used than for the usual chilling storage. The advantages of gas storage over ordinary chilling storage of apples are listed by Kidd and West as follows: (1) ripening proceeds at half the rate in air at a given temperature; (2) low-temperature breakdown is avoided because the temperature of about 40 F is above the limit for the disease; (3) firmness is almost unchanged over long periods; (4) changes in the ground color from green to yellow are markedly retarded; (5) the surface-eating Tortrix-moth larvae are killed; (6) life of the fruit after removal from gas storage is remarkably long. Too much CO_2 in the atmosphere causes brown heart of apples at higher temperatures and hastens low-temperature breakdown at lower temperatures. Too little oxygen favors alcohol formation and too much hastens ripening.

Although carbon dioxide storage has been employed chiefly with apples, it can be used successfully with pears, bananas, citrus fruits, plums, peaches, grapes, and other fruits.

Ozone in concentrations of 2 to 3 ppm in the atmosphere has been reported to double the storage time of loosely packed small, fresh fruits, such as strawberries, raspberries, currants, and grapes, and of delicate varieties of apples.

Ethylene in the atmosphere is used to hasten ripening or produce a desired color change, and is not considered preservative, although a combination of this gas and activated hydrocarbons has been suggested for the preservation of fruits.

Freezing

The surfaces of fruits contain the natural surface flora plus contaminants from soil and water. Any spoiled parts that are present will add molds or yeasts. During preparation of fruits for freezing, undesirable changes may take place, such as darkening, deterioration in flavor, and spoilage by microorganisms, especially molds. The washing of the fruit removes most of the soil microorganisms, and adequate selection and trimming get rid of many of the molds and yeasts involved in spoilage. With proper handling there should be little growth of microorganisms prior to freezing. The freezing process reduces the numbers of microorganisms but also usually causes some damage to the fruit tissues, resulting in flabbiness and release of some juice. During storage in the frozen condition the physical changes described in Chapter 7 occur as well as a slow but regular decrease in numbers of microorganisms. Yeasts (*Saccharomyces, Cryptococcus*) and molds (*Aspergillus, Penicillium, Mucor, Rhizopus, Botrytis, Fusarium, Alternaria,* etc.) have been reported to be the predominant organisms in frozen fruits, although small numbers of soil organisms, e.g., species of *Bacillus, Pseudomonas, Achromobacter,* etc., survive freezing. Yeasts are most likely to grow during slow thawing.

Numbers of viable microorganisms in frozen fruits are considerably lower than in frozen vegetables. Total counts are seldom attempted as a control measure, although the Howard method for estimating numbers of mold filaments has been employed by the Food and Drug Administration in the examination of frozen fruits. Large numbers of mold hyphae are indicative of the freezing of inferior fruit that included rotten parts.

The numbers of microorganisms in frozen fruit juices will depend upon the condition of the fruit, the washing process, the method of filtration, and the opportunities for contamination and growth prior to freezing. There may be from a few hundred to over a million organisms per milliliter present in the juice at the time of freezing. The inclusion of rotten parts of the fruit increases the numbers of organisms markedly. The washing process, especially the kind of solution used for washing, has a considerable influence on the numbers of organisms, for those on the surface of fruits are difficult to remove. Numbers can build up in the washing solution, on moist surfaces of the washed fruit, and in the juice itself before freezing. In the plant, too, there is opportunity for the addition of organisms from the equipment. The freezing process markedly reduces numbers, but added sugar or increased concentration of the juice has a protective effect against killing. The decrease in numbers of organisms during storage in the frozen condition is slow, but is faster than in most neutral foods. The kinds of organisms are chiefly those of soil, water, and rots, together with the natural surface flora of the fruit. Prominent usually are coliforms, enterococci, lactics, e.g., *Leuconostoc* and *Lactobacillus* species, *Achromobacter,* and yeasts.

Since coliform bacteria, mostly of the *Aerobacter aerogenes* type, form part of the natural flora of fruits, they are present in both fresh and frozen fruit juices. The use of decayed fruit for the juice increases the numbers of coliforms, but these organisms decrease during storage. Because coliforms normally are present, there are objections to the use of the presumptive test for coliforms to indicate sanitary quality of the juice. It has been suggested that tests be made for the intestinal coliform, *Escherichia coli,* or better still for the more persistent enterococcus, *Streptococcus faecalis.*

DRYING

The drying of fruits has been discussed in Chapter 8. It has been noted that the numbers of microorganisms in dried fruits are comparatively low, and that spores of bacteria and molds are likely to be most numerous. An occasional sample may contain high numbers of mold spores, indicating that growth and sporulation of molds has taken place on the fruit before or after dehydration. Alkali treatment, sulfuring, blanching, and pasteurization reduce numbers of microorganisms.

USE OF PRESERVATIVES

The use of chemical preservatives to lengthen the keeping time of fruits has been discussed in Chapter 9, where it was noted that chemicals have been applied to fruits chiefly as a dip or spray or impregnated in wrappers for the fruits. Among substances that have been applied to the outer surfaces of fruit are waxes, hypochlorites, biphenyl, and alkaline sodium orthophenylphenate. Wrappers for fruits have been impregnated with a variety of chemicals including iodine, sulfite, biphenyl, *o*-phenylphenol plus hexamine, and others. As a gas or fog about the fruit, carbon dioxide, ozone, and ethylene plus chlorinated hydrocarbons have been tried. Sulfur dioxide and sodium benzoate are preservatives that have been added directly to fruits or fruit products. Most of the chemical preservatives mentioned have been primarily antifungal in purpose.

Green olives are the only fruits which are preserved on a commercial scale with assistance from an acid fermentation. Locally, other fermented fruits sometimes are prepared, such as fermented green tomatoes and Rumanian preserved apples. In all of these products the lactic acid fermentation is of chief importance.

USE OF IONIZING RAYS

Experimentally, strawberries and dehydrated vegetables have been preserved successfully by means of ionizing rays.

SPOILAGE

The deterioration of raw vegetables and fruits may result from physical factors, action of their own enzymes, microbial action, or combinations of these agencies. Mechanical damage resulting from action of animals, birds, or insects, or from bruising, wounding, bursting, cutting, freezing, desiccation, or other mishandling may predispose toward increased enzymatic action or the entrance and growth of microorganisms. Previous damage by plant pathogens may make the part of the plant used as food unfit for consumption or may open the way for growth of saprophytes and spoilage by them. Contact with spoiling fruits and vegetables may bring about transfer of organisms, causing spoilage and increasing the wastage. Improper environmental conditions during harvesting, transit, storage, and marketing may favor spoilage. Most of the discussion to follow will be concerned with microbial spoilage, but it always should be kept in mind that the plant enzymes continue their activity in raw plant foods. If oxygen is available the plant cells will respire as long as they are alive; and hydrolytic enzymes can continue their action after death of the cell. As was stated in Chapter 11, the fitness of foods for consumption is judged partly on their maturity. If the desired stage of maturity is exceeded greatly, the food may be considered inedible or even spoiled. An example is an overripe banana, with its black skin and brown, mushy interior.

Diseases of vegetables and fruits may result from the growth of an organism that obtains its food from the host and usually damages it or from adverse environmental conditions that cause abnormalities in functions and structures of the vegetable or fruit. The diseases caused by pathogens and the decompositions caused by saprophytic organisms will be of chief interest in the following discussion, although clear distinction between these types of organisms is not possible; but diseases not caused by organisms should be mentioned because they may, at times, be confused with those caused by organisms in that they may be rather similar in appearance. Examples of nonpathogenic diseases are brown heart of apples and pears, blackheart of potatoes, black leaf speck of cabbage, and red heart of cabbage.

No attempt will be made to deal with changes caused by plant pathogens growing on the plants, or on parts of the plants used for food, prior to harvesting, but rather those microbial changes will be considered that may take place during harvesting, grading, packing, transportation, storage, and handling by wholesaler and retailer, although some of these changes may have begun before harvesting. Space will permit only a general treatment of the subject, and that only from the viewpiont of the food microbiologist rather than the plant pathologist. The reader should consult references listed at the end of the chapter for more detail, especially the series of bulletins from the Bureau of Plant Industry of the United States Department of Agriculture.

GENERAL TYPES OF MICROBIAL SPOILAGE

The most common or predominant type of spoilage varies not only with the kind of fruit or vegetable, but also to some extent with the variety. Microbial spoilage may be due to (1) plant pathogens acting on the stems, leaves, flowers, or roots of the plant, on the fruits or other special parts used as foods, e.g., roots or tubers, or on several of these locations; (2) saprophytic organisms, which may be secondary invaders after action of a plant pathogen, or may enter a healthy fruit or vegetable as in the case of various "rots," or grow on its surface, as when bacteria multiply on moist, piled vegetables. At times a saprophyte may succeed a pathogen, or a succession of saprophytes may be concerned in the spoilage. Thus, for example, coliform bacteria may grow as secondary invaders and be present in appreciable numbers in fruit and vegetable juices if rotten products have been included.

Although each fruit or vegetable has certain types of decomposition and kinds of microorganisms predominant in its spoilage, there are some general types of microbial spoilage that are found more often than the rest in vegetables and fruits. The most commonly occurring types of spoilage are as follows:

1. **Bacterial soft rot,** caused by *Erwinia carotovora* and related species which are fermenters of pectins. It results in a water-soaked appearance and a soft, mushy consistency and often in a bad odor.
2. **Gray mold rot,** caused by species of *Botrytis*, e.g., *B. cinerea*, a name derived from the gray mycelium of the mold. It is favored by high humidity and a warm temperature.
3. **Rhizopus soft rot,** caused by species of *Rhizopus*, e.g., *R. nigricans*. A rot results that often is soft and mushy. The cottony growth of the mold with small, black dots of sporangia often covers masses of the foods.

4. **Anthracnose,** usually caused by *Colletotrichum lindemuthianum*. The defect is a spotting of leaves and fruit or seed pods.
5. **Alternaria rot,** caused by species of *Alternaria*. Areas become greenish-brown early in the growth of the mold and later turn to brown or black spots.
6. **Blue mold rot,** caused by species of *Penicillium*. The bluish-green color that gives the rot its name results from the masses of spores of the mold.
7. **Downy mildew,** caused by species of *Phytophthora*, *Bremia*, and of other genera. The molds grow in white, woolly masses.
8. **Watery soft rot,** caused chiefly by *Sclerotinia sclerotiorum*, is found mostly in vegetables.
9. **Stem-end rots,** caused by species of molds of several genera, e.g., *Diplodia*, *Phomopsis*, *Fusarium*, and others, involve the stem ends of fruits.
10. **Black mold rot,** caused by *Aspergillus niger*. The rot gets its name from the dark-brown to black masses of spores of the mold, termed "smut" by the layman.
11. **Black rot,** often caused by species of *Alternaria*, but sometimes of *Ceratostomella*, *Physalospora*, and other genera.
12. **Pink mold rot,** caused by pink-spored *Trichothecium roseum*.
13. **Fusarium rots,** a variety of types of rots caused by species of *Fusarium*.
14. **Green mold rot,** caused usually by species of *Cladosporium*, but sometimes by other green-spored molds, e.g., *Trichoderma*.
15. **Brown rot,** caused chiefly by *Sclerotinia* species.
16. **Sliminess** or **souring,** caused by saprophytic bacteria in piled, wet, heating vegetables.

Fungal spoilage of vegetables often results in water-soaked, mushy areas, while fungal rots of fleshy fruits like apples and peaches frequently show brown or cream-colored areas in which mold mycelia are growing in the tissue below the skin and aerial hyphae and spores may appear later. Some types of fungal spoilage appear as "dry rots," where the infected area is dry and hard and often discolored. Rots of juicy fruits may result in leakage.

SPOILAGE OF SPECIFIC KINDS OF FRUITS AND VEGETABLES

In the discussion to follow of the spoilage of vegetables and fruits, these products will be handled in groups when possible and only the most important types of microbially caused spoilage will be listed. Whenever one of the sixteen most commonly occurring types of spoilage listed in the preceding section is mentioned, the name of the causal organism will not be repeated, nor will the defect be described unless it is appreciably different from the usual one. In general, the grouping of products will be similar to those employed in the government bulletins listed at the end of the chapter. It should be reemphasized that the discussion will be limited to microbial spoilages or "diseases" during storage, transit, and marketing and only to the most common ones, for many kinds of organisms may be causes of spoilage on occasion.

Lily family

Asparagus, onions, and garlic are representatives of this group. In asparagus, bacterial soft rot and fusarium rot are the most common market diseases, but gray mold rot, phytophthora rot (*Phytophthora* spp.), and watery soft rot may occur. A disagreeable odor is evident in later stages of bacterial soft rot. In the odorless fusarium rot, the mold growth initially is white and fluffy and later is pink. The affected tissues, water-soaked at first, become yellow and brown. Onions (Figure 15–4) and garlic are subject to a number of market diseases, the most important of which are bacterial soft rot, black mold rot, blue mold rot (garlic), gray mold rot, and smudge (anthracnose). The bacterial soft rot is very malodorous. Gray mold rot or "neck rot" of onions brings about a softening and water-soaking of the scales which have a grayish, then grayish-brown color. Later a dense, grayish mycelium develops over the infected part. Smudge may begin in the field and increase thereafter. Black blotches, or groups of small black or dark-green dots appear on the outer scales.

Pulse or legume family

Green, wax, and lima beans and peas of this group, which bear their seeds in pods, are subject to a large number of kinds of microbial spoilage, especially when they are marketed in the pods, as is usually the case. The most important market diseases are bacterial soft rot, gray mold rot, rhizopus soft rot (Figure 15–5), watery soft rot of both peas and beans, and, in addition, cottony leak of beans. There is, moreover, an array of microbial diseases in the field that affect the pods and perhaps the seeds, e.g., anthracnose, blights (Figure 15–6), wilts scab, rust, etc. Cottony leak or wilt, caused by *Pythium butleri*, gives the pods a white and cottony appearance.

Parsley family

Carrots and parsnips are grown for their taproot; celery is grown for its leafstalks and parsley for its leaves. Important market diseases

Figure 15–4. Spoilage of onion: (A, B, C) onion black mold rot; (D, E, F) onion smudge. (*Courtesy of U.S. Department of Agriculture.*)

of carrots are bacterial soft rot, black rot, fusarium rot, gray mold rot, rhizoctonia crown rot, rhizopus soft rot, and watery soft rot. Black rot is caused by the dark mold *Alternaria radicina*. In rhizoctonia crown rot, which begins in the field and continues during storage, the infected tissues of the root become brown and soft. Important market diseases of parsnips are bacterial soft rot, gray mold rot, and watery soft rot.

Most of the defects of celery occur in the field, but some continue to develop after harvesting. Examples are bacterial soft rot and watery soft rot. Gray mold rot gives trouble during transit and storage. Bacterial soft rot and watery soft rot are important market diseases of parsley. A number of field diseases can make parsley unmarketable.

Figure 15–5.   Spoilage of string beans. Bean rhizopus rot. (*Courtesy of U.S. Department of Agriculture.*)

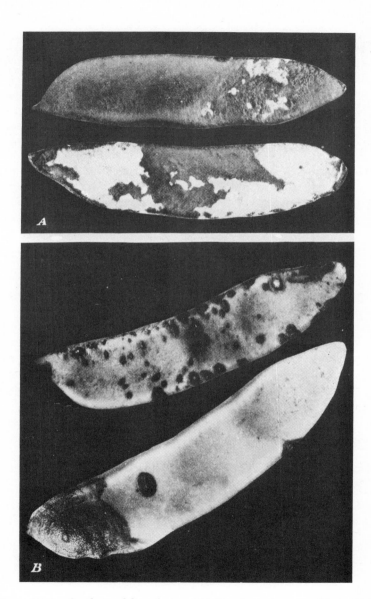

Figure 15–6. Spoilage of lima bean: (*A*) lima-bean downy mildew; (*B*) lima-bean pod blight. (*Courtesy of U.S. Department of Agriculture.*)

Beets

Both roots and tops are subject to spoilage, but the roots are less likely to be spoiled. The more common types of spoilage of roots are bacterial soft rot, black rot, blue mold rot, and fusarium rot. Internal black spot is caused by a boron deficiency in the soil. The tops may be decayed by bacterial soft rot or gray mold rot.

Endive and globe artichokes

The endive is most often spoiled by bacterial soft rot, although watery soft rot, downy mildew, and gray mold rot may occur. Gray mold rot is the main market disease of globe artichokes.

Lettuce

Leaves or leaf heads of lettuce are susceptible to many diseases during growth and are affected by diseases of roots and stems. The principal market diseases are bacterial soft rot, downy mildew, gray mold rot, and watery soft rot. When head lettuce is cooled by a water spray in the retail store it becomes contaminated with bacteria of the water and a reddish-brown color appears between the leaves and finally penetrates most of the head. This type of spoilage often develops during storage of the lettuce in the home refrigerator. Investigators disagree as to the cause of the spoilage, but species of *Achromobacter* have been blamed.

Rhubarb

Growing leafstalks of rhubarb are attacked by various plant pathogens. Bacterial soft rot and gray mold rot are the main market diseases.

Spinach

Like other leaf vegetables, growing spinach is deteriorated by plant pathogens. Chief market diseases are bacterial soft rot and downy mildew.

Sweet potatoes

Sweet potatoes in storage and on the market are subject to alternaria rot, black rot, blue mold rot, charcoal rot, dry rot, end rot, foot rot, gray mold rot, Java black rot, mucor rot, rhizopus soft rot, and a number of other rots. Black rot, showing brown to black spots, is caused by *Ceratostomella fimbriata;* charcoal rot, a spongy, dark-brown decay, by *Sclerotium bataticola;* dry rot by *Diaporthe batatatis;* end rots by a number of different organisms but mostly by species of *Fusarium;* foot rot by *Plenodomus destruens;* and Java black rot, in which the potato appears mummified, by *Diplodia tubericola.* Troubles with storage diseases are reduced if the sweet potatoes are cured for 10 to 14 days at 85 F (29.4 C) and 85 to 90 percent relative humidity before they are stored at 50 to 55 F (10 to 12.8 C) and 80 to 85 percent relative humidity.

Potatoes

Most important in the spoilage of market potatoes are the fusarium tuber rots, caused by species of *Fusarium.* Sunken, shriveled, wrinkled, or broken areas appear on the surface, often at stem and eyes. The rot usually is brown to black and becomes covered with whitish, dark, or brightly colored mold mycelium. The rot may be wet and jellylike, mushy and leaky, or dry and brittle. Other common rots include bacterial ring rot, alternaria tuber rot, bacterial soft rots, "leak," in which watery tissues and various colors are produced by *Pythium debaryanum,* soil rot, rhizopus soft rot, and sclerotium rot.

Crucifers

The crucifers include cabbage, Brussels sprouts, cauliflowers, broccoli, radishes, rutabagas, and turnips. A rot common to all of these is bacterial soft rot. Broccoli has few other market diseases. Cabbages, Brussels sprouts, cauliflower, rutabagas, and turnips all have bacterial soft rot, gray mold rot, black rot, and watery soft rot. Cabbages and Brussels sprouts also are spoiled by alternaria leaf spot, downy mildew, and rhizopus soft rot, and the root crops—radishes, rutabagas, and turnips—are subject to clubroot and rhizoctonia rot. Alternaria rot of cauliflower is called "brown rot."

Cucurbits

This gourd family includes the cucumber, muskmelon (cantaloupe), pumpkin, squash, and watermelon. All these may be spoiled by rhizopus soft rot and all may be subject to various stem rots. The cucumber can support the growth of a number of spoilage organisms, the more important of which may cause blue mold rot, bacterial soft rot, anthracnose, diplodia or stem-end rot, gray mold rot, cottony leak, pink mold rot, and fusarium rot. The muskmelon and similar melons are subject to blue mold rot, diplodia rot, bacterial soft rot, pink mold rot, fusarium rot (Figure 15–7), charcoal rot (rhizoctonia rot on honeydews), and phytophthera rot. Stem-end rots of watermelons due to species of

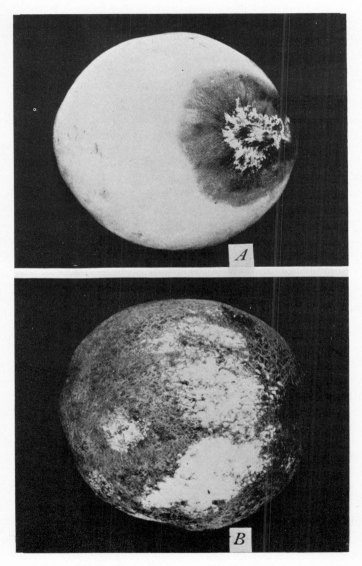

Figure 15–7. Spoilage of melons: (A) pink mold rot of honeydew melon; (B) fusarium rot of cantaloupe. (*Courtesy of U.S. Department of Agriculture.*)

Diplodia cause much of the wastage; other rots are bacterial soft rot, anthracnose, and phytophthera rot; pythium rot also may cause damage. Pumpkin and squash may show gray mold rot, fusarium rot, and anthracnose, in addition to those defects mentioned above.

Tomatoes

A large number of field diseases attack the tomato vines and still more attack both vines and fruits, and many diseases of the fruit carry over into market handling, e.g., alternaria rot, anthracnose, bacterial canker, bacterial spot, gray mold rot, green mold rot, late blight, phoma rot, sclerotium rot, soil rot, and mosaic. Affecting only the fruit are bacterial soft rot, basisporium rot, blue mold rot, buckeye rot, center rot, fusarium rot, isaria rot, melanconium rot, watery rot, and yeast spot. The government bulletin should be consulted for descriptions of these rots. Most important of the diseases are alternaria rot, anthracnose, phoma rot, and rhizopus rot.

Peppers and eggplants

For the most part, the diseases of peppers are similar to those attacking tomatoes. The chief market diseases are alternaria rot, anthracnose, and gray mold rot. Fruit rot, or phomopsis blight, is the most important market rot of eggplants. Caused by *Phomopsis vexans*, it is characterized by tan to light-brown lesions that darken with age. The decay is soft, spongy, and penetrating.

Small fruits

Blue mold rot, gray mold rot, and rhizopus rot are the chief market diseases of blackberries, currants, and dewberries. Russeted areas appear on gooseberries spoiled by powdery mildew, caused by *Sphaerotheca mors-uvae*. Grapes (Figure 15–8) are subject mainly to black mold rot, blue mold rot, gray mold rot, green mold rot (species of *Cladosporium* and *Alternaria*), and rhizopus rot. Strawberries are spoiled by gray mold rot (Figure 15–9), leather rot (*Phytophthera cactorum*), rhizoctonia rot, and, most important, rhizopus rot (Figure 15–10). *Cladosporium* species cause an olive to olive-green growth on raspberries.

Citrus fruits

Lemons, limes, oranges, and grapefruit have a number of market diseases in common (Figure 15–11), such as alternaria rot, anthracnose,

Figure 15–8. Spoilage of grapes: (A) gray mold rot on Zinfandel grapes; (B, C, D, E) various stages of black rot on Concord grapes. (*Courtesy of U.S. Department of Agriculture.*)

the blue mold rots, and the stem-end rots, but the relative importance of these diseases differs with the fruit. The most wastage of oranges and lemons results from the so-called "blue mold rots," which may be due to the green-spored *Penicillium digitatum* or the blue-green–spored "contact mold," *Penicillium italicum.* Alternaria rot, caused by *Alternaria citri,* is the worst market disease of grapefruit. The causes of stem-end rots include species of *Phomopsis* and *Diplodia.* Most of the rots are soft rots, as contrasted to the "leathery rot" of lemons by species of *Phytophthera* and the "firm, dry rot" of oranges by *Alternaria citri,* an organism which causes a soft rot in lemons and grapefruit. Other rots of lemons and limes are brown rot by *Phytophthera* spp., gray mold rot, cottony rot by *Sclerotinia sclerotiorum*, and sour rot by *Oöspora citri-aurantii*. Additional rots of oranges are gray mold rot, blossom-end rot by *Alternaria* and *Fusarium*, and trichoderma rot by *Trichoderma viride*, a green-spored mold.

Figure 15–9.   Spoilage of strawberries: (*A*) gray mold rot; (*B*, *C*) leather rot; (*D*, *E*) rhizoctonia brown rot; (*F*) tan rot; (*G*) beginning of gray mold rot on green strawberry. (*Courtesy of U.S. Department of Agriculture.*)

Figure 15–10. Spoilage of strawberries; rhizopus rot. (*Courtesy of U.S. Department of Agriculture.*)

Other subtropical fruits

Avocados are deteriorated by anthracnose and rhizopus rot and bananas by anthracnose. The main banana stalk is subject to stalk rot by *Fusarium* and *Gleosporium* species, and the stalks of individual bananas of a "hand" have finger rots by species of *Pestalozzia, Fusarium,* or *Gleosporium.* Microorganisms are reputed to enter the fig through the eye, carried there by insects. The spoilage types are grouped as (1) smut and mold; (2) soft rot, or endosepsis; and (3) souring. The first group includes alternaria spot by *Alternaria tenuis;* blue mold rot, which usually is from a secondary invasion; cladosporium rot, due to *Cladosporium herbarum;* gray mold rot, making the tissue soft and leaky; rhizopus rot, producing similar changes plus a browning of the flesh; and black mold rot, or "smut," resulting from the growth and sporulation of *Aspergillus niger.* The last rot first makes the skin dirty-white to slightly pink and the pulp firm to cheesy. A white mycelium grows within the fig and later fills the fig with black spores, or "smut." Secondary growth of other molds may follow. In soft rot, or endosepsis, caused

Figure 15–11. Spoilage of citrus fruits: (A) stem-end rot on orange; (B) blue and green mold rots on orange; (C) blue mold rot on lemon; (D) sulfur-spray injury on orange. (*Courtesy of U.S. Department of Agriculture.*)

by *Fusarium moniliforme,* pink or purple water-soaked areas are formed on the skin and an offensive odor comes from the eye end. The mold supposedly is spread among caprified figs by the fig wasp during pollination. "Souring" of figs results from action by yeasts, both fermentative and oxidative (film), so that pink liquid oozes through the eye and there are gas bubbles in the watery pulp. Dates also are "soured" by yeasts and are subject to spoilage by various molds.

Stone fruits

These fruits include peaches, apricots, plums, cherries, etc., which have a single large seed. In general, the chief causes of wastage are alternaria or green mold rot, especially of cherries, black mold rot, blue mold rot, brown rot, cladosporium rot, gray mold rot, and rhizopus rot. The same mold may produce different types of spoilage on different fruits. Thus, for example, rhizopus rot of peaches and apricots results in light-brown soft flesh with a cottony surface growth, while the rot of plums, prunes, and cherries produces a soft, leaky condition.

Pomes

It was mentioned earlier that although a few kinds of microorganisms cause most of the spoilage of fruits and vegetables, many kinds of organisms may happen to cause decay. The apple is a good example of this fact, for several workers have tried to list the kinds of molds causing spoilage. One list gives forty-two different genera of molds and many more species involved in the spoilage. Apples, too, are good examples of functional, nonparasitic diseases—for example, internal breakdown, bitter pit, scald, and brown heart. The chief market diseases caused by molds are blue mold rot, caused chiefly by *Penicillium expansum*; the black, firm alternaria rot; black rot, caused by *Physalospora malorum;* brown rot, bull's-eye rots, and core rots, caused by various molds; gray mold rot; pink mold rot; powdery mildew, with russeting caused by *Podosphaera leucotricha;* rhizopus rot; and so-called "miscellaneous rots," caused by other molds or often successions of them. Most rots cause a brown or cream-colored, soft, and watery area on the apple.

Pears are subject to black rot, blue mold rot, brown rot, gray mold rot, pink mold rot, powdery mildew, and rhizopus rot. Most rots are similar in appearance to those on apples. Quinces are spoiled by black rot, blue mold rot, and brown rot.

The preceding discussion of market diseases or spoilages of fruits and vegetables serves to illustrate the types of decomposition common to many kinds and the types most important to specific kinds. The composition of the fruit or vegetable influences the likely type of spoilage. Thus, bacterial soft rot is widespread for the most part among the vegetables which are not very acid, and among the fruits is limited to those that are not highly acid. Because most fruits and vegetables are somewhat acid and are fairly dry at the surface and are deficient in B vitamins, molds are the most common causes of spoilage. The composition, too, must determine the particular kinds of molds most likely to grow; thus some kinds of fruits or vegetables support a large variety of spoilage organisms and other kinds comparatively few.

The likelihood of the entrance of spoilage organisms also is important in influencing the possibility of spoilage and the kind that takes place. Damage by mechanical means, plant pathogens, or bad handling will favor entrance. The location of the plant part used also is important; thus underground parts like roots, tubers, or bulbs as in radishes, beets, carrots, and potatoes are in direct contact with moist soil and become infected from that source. Fruits like strawberries, cucumbers, peppers, and melons may be in direct contact with the surface of the soil. Leaves, stems, and flowers, as in lettuce, the greens, cabbage, asparagus, rhubarb, and broccoli, are especially exposed to contamination by plant pathogens or damage by birds and insects, as are most fruits, whether ordinarily classified as vegetables or "fruits."

The character of the spoilage will depend upon the product attacked and the attacking organism. When the food is soft and juicy, the rot is apt to be soft and mushy and some leakage may result. There are, however, some kinds of spoilage organisms that have a drying effect so that dry or leathery rots or discolored surface areas may result. In some instances most of the mycelial growth of the mold is subsurface and only a rotten spot shows, as in most rotting of apples. In other types of spoilage the growth of the mold mycelium on the outside is apparent and may be colored by spores.

The identification of a type of spoilage of a fruit or vegetable makes possible the application of available methods for the prevention of such decay.

SPOILAGE OF FRUIT AND VEGETABLE JUICES

Juices may be squeezed directly from fruits or vegetables, may be squeezed from macerated or crushed material so as to include a considerable amount of pulp, or may be extracted by water, as for prune juice. These juices may be used in their natural concentrations or may be concentrated by evaporation or freezing, and may be preserved by canning, freezing, or drying.

Juices squeezed or extracted from fruits are more or less acid, depending upon the product, the pH ranging from about 2.4 for lemon or cranberry juice up to 4.2 for tomato juice, and all contain sugars, the amounts varying from about 2 percent in lemon juice up to almost 17 percent in some samples of grape juice. Although molds can and do grow on the surface of such juices if the juices are exposed to air, the high moisture content favors the faster-growing yeasts and bacteria. Which of the latter will predominate in juices low in sugar and acid will depend more upon the temperature than the composition. The removal of solids from the juices by extraction and sieving raises the oxidation-reduction potential and favors the growth of yeasts. Most fruit juices are acid enough and have sufficient sugar to favor the growth of yeasts within the range of temperature that favors them, namely, from 60 to 95 F (15.6 to 35 C). The deficiency of B vitamins discourages some bacteria.

Therefore, the normal change to be expected in raw fruit juices at room temperatures is an alcoholic fermentation by yeasts, followed by the oxidation of alcohol and fruit acids by film yeasts or molds growing on the surface if it is exposed to air or the oxidation of the alcohol to acetic acid if acetic acid bacteria are present. The types of yeasts growing depend upon the kinds predominant in the juice and upon the temperature, but usually wild yeasts, such as the apiculate ones, producing only moderate amounts of alcohol and considerable amounts of volatile acid, will carry out the first fermentation. At temperatures near the extremes of the range indicated (60 to 95 F), the undesirable yeasts are more likely to grow than those producing desirable flavors. At temperatures above 90 to 95 F (32.2 to 35 C) lactobacilli would be likely to grow and form lactic and some volatile acids because these temperatures are too high for most yeasts. At temperatures below 60 F (15.6 C) wild yeasts may grow, but the more the temperature drops toward freezing, the more likely is the growth of bacteria and molds rather than yeasts. The acidity may be reduced by film yeasts and molds growing on the surface.

In addition to the usual alcoholic fermentation, fruit juices may undergo other changes caused by microorganisms: (1) the lactic acid fermentation of sugars, mostly by heterofermentative lactic acid bacteria such as *Lactobacillus pastorianus, L. brevis,* and *Leuconostoc mesenteroides* in apple or pear juice, and by homofermentative lactic acid bacteria such as *Lactobacillus arabinosus, L. leichmannii,* and *Microbacterium;* (2) the fermentation of organic acids of the juice by lactic acid bacteria, e.g., *Lactobacillus pastorianus,* malic acid to lactic and succinic acids, quinic acid to dehydroshikimic acid, and citric acid to lactic and acetic acids; (3) slime production by *Leuconostoc mesenteroides, Lactobacillus brevis,* and *L. plantarum* in apple juice and by *L. plantarum* and streptococci in grape juice. Some of these changes will be discussed under Wines in Chapter 24.

Vegetable juices contain sugars but are less acid than fruit juices, having pH values in the range of 5.0 to 5.8 for the most part. Vegetable juices also contain a plentiful supply of accessory growth factors for microorganisms and hence support good growth of the fastidious lactic acid bacteria. Acid fermentation of the raw juice by these and other acid-forming bacteria would be a likely cause of spoilage, although yeasts and molds can grow.

Concentrates of fruit and vegetable juices, because of their increased acidity and sugar concentration, favor the growth of yeasts and of acid- and sugar-tolerant *Leuconostoc* and *Lactobacillus* species. Such concentrates usually are canned and then heat-treated or frozen. Heat processing kills the important microorganisms that could cause spoilage, and freezing prevents the growth of such organisms.

SPOILAGE OF FERMENTED PRODUCTS

The spoilage of wines and other fermented fruit juices and of malt liquors and vinegar will be discussed in Chapter 24.

Sauerkraut

Sauerkraut may be of inferior quality because of an abnormal fermentation. An excessively high temperature may inhibit the growth of *Leuconostoc* and, consequently, the flavor production by that organism, and may permit the growth of *Pediococcus cerevisiae* and the development of undesirable flavors. An excessively low temperature may prevent adequate activity of the desired succession of lactic bacteria and encourage the growth of contaminants from the soil, e.g., *Aerobacter* and *Flavobacterium*. Too long a fermentation may favor the growth of the gas-forming *Lactobacillus brevis*, which yields a sharply acid flavor. Too much salt may encourage microorganisms other than the desired ones, *Pediococcus cerevisiae* and yeasts, for example. Abnormal fermentation of cabbage may result in a cheeselike odor caused by propionic, butyric, caproic, and valeric acids, along with isobutyric and isovaleric acids.

Soft kraut may result from a faulty fermentation and from exposure to air, or from excessive pressing or tamping.

Dark-brown or **black kraut** usually is due to oxidation during exposure to air and is caused by the combined action of plant enzymes and microorganisms. Destruction of acid by film yeasts and molds makes conditions favorable for proteolytic and pectolytic organisms to rot the kraut rapidly. Darkening is encouraged by uneven salting ("salt burn") and a high temperature. A brown color may result from iron in hoops and tannin from barrels.

Pink kraut often is caused by red, asporogenous yeasts in the presence of air and high salt, and is found especially when the salt has been distributed unevenly. The development of pink color is favored by high temperature, dirty vats, low acidity, and iron salts. A light-pink color has been attributed to the pigments in some varieties of cabbage.

Slimy or **ropy kraut** is caused by encapsulated varieties of *Lactobacillus plantarum*. The product is edible but unsalable. The sliminess may disappear on longer holding and usually disappears during the cooking of the kraut.

Sauerkraut, then, is especially subject to spoilage at its surface, where it is exposed to air. There film yeasts and molds destroy the acidity, permitting other microorganisms to grow and causing softening, darkening, and bad flavors.

Pickles

Fermented pickles are subject to a number of defects or "diseases," most of which are caused by microorganisms. Shriveling results from the physical effect of too strong salt, sugar, or vinegar solutions. **Hollow pickles** grow that way, according to most authorities, and get worse if the cucumbers are allowed to stand for a while after harvesting and before fermenting them. Other workers believe that improper conditions during fermentation, such as loose packing in the vat, insufficient weighting, too rapid a fermentation, and too strong or too weak a brine, cause hollow pickles.

Floaters, or **bloaters** (Figure 15–12), may be due to the original cucumbers being hollow or to gas being formed by yeasts (or *Lactobacillus brevis*) inside the cucumber. Floaters are favored by a thick skin that does not allow gas to diffuse out, by rapid gas production during fermentation, by high initial amounts of salt, by added acid, and by added sugar. Bloating is increased by factors that interfere with absorption of brine.

Slippery pickles occur when the cucumbers are exposed to the air, permitting the growth of encapsulated bacteria. Slipperiness also may be due to the broken scums of film yeasts that have grown on the surface of the brine and dropped onto the cucumbers. An early stage of softening gives a slippery surface to pickles.

Soft pickles (Figure 15–13) are made so by pectolytic enzymes, mostly from molds and from cucumber flowers, which enter the fermentation vat. These molds are mostly in the genera *Penicillium, Fusarium, Ascochyta, Cladosporium,* and *Alternaria.* The first step in the degradation of the pectin of the cucumber may be the removal of methoxyl groups to form pectic acid by action of pectinesterase from brine yeasts and the cucumber and accessory parts. Pectin also may be converted to intermediate uronides by the polymethylgalacturonase produced by molds. Further hydrolysis to galacturonic acid is catalyzed by polygalacturonases from molds and from cucumber flowers and other parts. The cucumber flowers, which usually support a fairly heavy fungal population, are believed to be a main source of the pectolytic enzymes. Pectolytic bacteria of the genera *Bacillus, Aeromonas,* and *Achromobacter* and of the coliform group play only a minor role, if any, in pickle softening. Terminal hydrolyzing enzymes may be involved in the final degradation of the pectic substances. Molds on the cucumbers also can cause softening, as can enzymes of the cucumber or weak or strong acids. Deesterification of pectic substances by yeasts may hasten softening by polygalacturonase from other sources. Softening is favored by

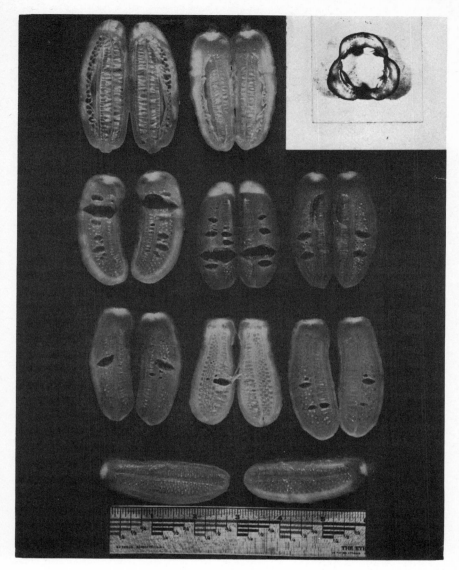

Figure 15–12. Bloaters (hollow cucumbers) formed by gaseous fermentation. Observe lens-shaped gas pockets, which may combine to form a single large cavity (see inset). (*Courtesy of J. L. Etchells and I. D. Jones, Glass Packer, 30(5): 358–360, 1951.*)

Figure 15–13. Soft pickles. These salt-stock pickles have become mushy as the result of action of pectolytic enzymes. (*Courtesy of J. L. Etchells and I. D. Jones, Glass Packer,* 30(5):358–360, 1951.)

(1) an insufficient amount of salt and hence an abnormal fermentation; (2) too high a temperature, also affecting the acid production; (3) low acidity, either because not enough was formed or because it was destroyed by film yeasts or molds; (4) presence of air favoring the growth of film yeasts, molds, or pectin-fermenting bacteria; and (5) inclusion of many blossoms (usually with very small cucumbers), on which molds have grown and produced polygalacturonase, which causes softening.

Black pickles may owe their color to the formation of hydrogen sulfide by bacteria (or less commonly by chemical reaction) and combination with iron in the water to yield black ferrous sulfide. Therefore, water for pickles should have a low iron and gypsum ($CaSO_4$) content. The defect is favored by low acidity of the brines. Another cause of black pickles is the growth of black-pigmented *Bacillus nigrificans,* a variety of *Bacillus subtilis,* which is favored by (1) presence of an available carbohydrate like glucose; (2) a low level of available nitrogen; and (3) a neutral or slightly alkaline brine.

Ropy pickle brine, due to various unidentified motile, Gram-negative, encapsulated rods, is favored by (1) low salt, (2) low acid, and (3) high temperature.

Green olives

Gassy spoilage of green olives usually is caused by coliform bacteria, especially *Aerobacter* species, but sometimes by *Bacillus* (aerobacillus) and *Clostridium* species. Evidences of gas are blisterlike spots under the skin, or fissures or gas pockets within the olives producing the defect termed "fisheye" and causing the olives to become "floaters." Gas-forming yeasts and heterofermentative lactics are less commonly responsible for gassy defects.

If conditions are not favorable for the normal lactic fermentation, an abnormal butyric fermentation by *Clostridium* species (e.g., *C. butyricum*) may result in bad odors and tastes. If the pH is above 4.2, a "sagy" off-odor and -taste, called "zapatera" spoilage, may develop, caused by species of *Clostridium* aided by propionic bacteria. The odor also has been described as from mildly cheesy (butyric) to foul and fecal (putrefactive).

Softening resulting from the destruction of pectic substances in the olive may have physical or chemical causes or may be caused by pectolytic coliform bacteria, by species of *Bacillus, Aeromonas,* or *Achromobacter,* by yeasts (*Rhodotorula*), or by molds (*Penicillum, Aspergillus,* etc.).

Raised white spots or pimples just under the epidermis of the olive, wrongly termed "yeast spots," are really colonies of *Lactobacillus plantarum* or *L. brevis.* Similar spots are sometimes found on fermented cucumber pickles.

Cloudiness of brine of glass-packed olives may be caused by microorganisms: (1) the lactics, if fermentation of residual sugar resumes; (2) salt-tolerant bacteria; or (3) film yeasts or molds, if air is available. Earlier, during the processing of the olives, surface growths of film yeasts may destroy lactic acid and produce undesirable odors and tastes that will carry over into the finished product.

Storage olives, i.e., those held for weeks to months in brines before processing, undergo a lactic acid fermentation, and then, if exposed to air may become subject to softening by pectinolytic action of an aerobic flora of film yeasts and molds developing on the surface. Most of the molds are in the genera *Fusarium* and *Penicillium.*

Ripe olives

Spoilage of ripe olives may take place while they are in the holding brine, with bleaching of the color and perhaps some softening. Coliform bacteria have been blamed for such defects, as well as for blistering or "fisheye" spoilage. Unless the temperature of the wash water is below 70 F (21 C) or is in the range 120 to 140 F (49 to 60 C) during the removal of lye from the olives, bacterial growth may take place with softening of the fruit. It has been demonstrated that pectolytic enzymes of *Bacillus subtilis* and *Bacillus pumilus* can cause the defect.

Summary on spoilage of fermented fruit and vegetable products

The most likely causes of the spoilage of fermented fruit and vegetable products, then, are (1) abnormal fermentations due to the wrong conditions during the fermentation; and (2) the oxidation of lactic and other acids in the fermented product by film yeasts and molds to permit the growth of microorganisms that cause undesirable appearance, texture, taste, or odor.

SPOILAGE OF CANNED FRUIT AND VEGETABLE PRODUCTS

The spoilage of jellies, jams, marmalades, and butters and of canned fruit and vegetable products will be discussed in Chapter 22.

REFERENCES

Fruits and vegetables

BIGELOW, W. D., and P. H. CATHCART. 1921. Relation of processing to the acidity of canned foods. Nat. Canners Ass. Bull. 17-L.

BROOKS, C., E. V. MILLER, C. O. BRATLEY, P. V. MOOK, and H. B. JOHNSON. 1932. Effect of solid and gaseous carbon dioxide upon transit diseases of certain fruits and vegetables. U.S. Dep. Agr. Tech. Bull. 318.

CRUESS, W. V. 1958. Commercial fruit and vegetable products. 4th ed. McGraw-Hill Book Company, New York.

FABIAN, F. W. 1943. Home food preservation. Avi Publishing Co., Inc., Westport, Conn.

PENTZER, W. T. 1951. Temperatures required by fruits and vegetables after harvest. Food Technol. 5:440–442.

Rose, D. H., R. C. Wright, and T. M. Whiteman. 1949. The commercial storage of fruits, vegetables and florists' stocks. U.S. Dep. Agr. Circ. 278.

Tomkins, R. G. 1951. The microbiological problems in the preservation of fresh fruits and vegetables. J. Sci. Food Agr. 2:381–386.

United States Department of Agriculture. 1932–1944. Market diseases of fruits and vegetables [series]. U.S. Dep. Agr. Misc. Publ:

No. 98: Potatoes. G. K. K. Link and G. B. Ramsey. 1932.

No. 121: Tomatoes, peppers, eggplants. G. B. Ramsey and G. K. K. Link. 1932.

No. 168: Apples, pears, quinces. D. H. Rose, C. Brooks, D. F. Fisher, and C. O. Bratley. 1933.

No. 228: Peaches, plums, cherries and other stone fruits. D. H. Rose, D. F. Fisher, C. Brooks, and C. O. Bratley. 1937.

No. 292: Crucifers and cucurbits. G. B. Ramsey, S. S. Wiant, and G. K. K. Link. 1938.

No. 340: Grapes and other small fruits. D. H. Rose, C. O. Bratley, and W. T. Pentzer 1939.

No. 440: Asparagus, onions, beans, peas, carrots, celery and related vegetables. G. B. Ramsey and S. S. Wiant, 1941.

No. 498: Citrus and other subtropical fruits. D. H. Rose, C. Brooks, C. O. Bratley, and J. R. Winston. 1843.

No. 541: Beets, endive, escarole, globe artichokes, lettuce, rhubarb, spinach, Swiss chard and sweetpotatoes. G. B. Ramsey and S. S. Wiant. 1944.

Von Schelhorn, Mathilde. 1951. Control of microorganisms causing spoilage in fruit and vegetable products. Advances Food Res. 3:429–482.

Fruits

Beraha, L., G. B. Ramsey, M. A. Smith, W. R. Wright, and F. Heiligman. 1961. Gamma radiation in the control of decay in strawberries, grapes, and apples. Food Technol. 15:94–98.

Cruess, W. V. 1938. Commercial fig products. Fruit Products J. 17:337–339, 343.

Esau, P., and W. V. Cruess. 1933. Yeasts causing "souring" of dried prunes and dates. Fruit Products J. 12:144.

Harvey, J. M., and W. T. Pentzer. 1960. Market diseases of grapes and other small fruits. Agricultural Marketing Service. U.S. Dep. Agr. Handbook 189.

Kidd, F., and C. West. 1937. Recent advances in the work on refrigerated gas-storage of fruit. J. Pomol. Hort. Sci. 14:299–316.

Klotz, L. J., and H. S. Fawcett. 1941. Color handbook of citrus diseases. University of California Press, Berkeley, Calif.

Luepschen, N. S., and M. A. Smith. 1962. Watermelon diseases on the Chicago market, 1960–1961. Plant Dis. Reporter 46:41–42.

Miller, M. W., and H. J. Phaff. 1962. Successive microbial populations in Calimyrna figs. Appl. Microbiol. 10:394–400.

Saravacos, G. D., L. P. Hatzipetrou, and E. Georgiadou. 1962. Lethal doses of gamma radiation of some fruit spoilage microorganisms. Food Irradiation 3:A6–A9.

VAUGHN, R. H., H. C. DOUGLAS, and J. R. GILILLAND. 1943. Production of Spanish-type green olives. Calif. Agr. Exp. Sta. Bull. 678.

Vegetables

ETCHELLS, J. L., and I. D. JONES. 1943*a*. Bacteriological changes in cucumber fermentation. Food Ind. 15(2):54–56.

ETCHELLS, J. L., and I. D. JONES. 1943*b*. Commercial brine preservation of vegetables. Fruit Products J. 22:242–246, 251, 253.

ETCHELLS, J. L., and I. D. JONES. 1951. Progress in pickle research. Glass Packer, 30(5):358–360.

FABIAN, F. W., and H. B. BLUM. 1943. Preserving vegetables by salting. Fruit Products J. 22:228–236.

FABIAN, F. W., and L. J. WICKERHAM. 1935. Experimental work on cucumber fermentation (dills), VIII. Mich. State Coll. Agr. Exp. Sta. Bull. 146.

FULDE, R. C., and F. W. FABIAN. 1953. The influence of gram-negative bacteria on the sauerkraut fermentation. Food Technol. 7:486–488.

HUCKER, G. J., R. F. BROOKS, and A. J. EMERY. 1952. The source of bacteria in processing and their significance in frozen vegetables. Food Technol. 6:147–155.

JONES, A. H., and A. G. LOCHHEAD. 1939. A study of micrococci surviving in frozen-pack vegetables and their enterotoxic properties. Food Res. 4:203–216.

KEIPPER, C. H., W. H. PETERSON, E. B. FRED, and W. E. VAUGHN. 1932. Sauerkraut from pretreated cabbage. Ind. Eng. Chem. 24:884–889.

LYNCH, L. J., R. S. MITCHELL, and D. J. CASIMIR. 1959. The chemistry and technology of the preservation of green peas. Advances Food Res. 9:61–151.

PEDERSON, C. S., 1930. Floral changes in the fermentation of sauerkraut. N. Y. State Agr. Exp. Sta. Tech. Bull. 168.

PEDERSON, C. S. 1931. Sauerkraut. N.Y. State Agr. Exp. Sta. Bull. 595.

PEDERSON, C. S., 1947. Significance of bacteria in frozen vegetables. Food Res. 12:429–438.

PEDERSON, C. S., and MARGARET N. ALBURY. 1953. Factors affecting the bacterial flora in fermenting vegetables. Food Res. 18:290–300.

PEDERSON, C. S., and MARGARET N. ALBURY. 1954. The influence of salt and temperature on the microflora of sauerkraut fermentation. Food Technol. 8:1–5.

PEDERSON, C. S., and L. WARD. 1949. Effect of salt on the bacteriological and chemical changes in fermenting cucumbers. N.Y. State Agr. Exp. Sta. Bull. 288.

SPLITTSTOESSER, D. F., W. P. WETTERGREEN, and C. S. PEDERSON. 1961. Control of microorganisms during preparation of vegetables for freezing: I, Green beans; II, Peas and corn. Food Technol. 15:329–331; 332–334.

SPLITTSTOESSER, D. F., and W. P. WETTERGREEN. 1964. The significance of coliforms in frozen vegetables. Food Technol. 18:392–394.

VAUGHN, R. H., 1954. Lactic acid fermentation of cucumbers, sauerkraut and olives, vol. II, chap. 11. *In* L. A. Underkofler and R. J. Hickey (*Eds.*) Industrial fermentations. Chemical Publishing Co., Inc., New York.

WHITE, ANNE, and HELEN R. WHITE. 1962. Some aspects of the microbiology of frozen peas. J. Appl. Bacteriol. 25:62–71.

Juices

BERRY, J. M., L. D. WITTER, and J. F. FOLINAZZO. 1956. Growth characteristics of spoilage organisms in orange juice and concentrate. Food Technol. 10:553–556.

BOWEN, J. F., and F. W. BEECH. 1964. The distribution of yeasts on cider apples. J. Appl. Bacteriol. 27:333–341.

CARR, J. G. 1958. Lactic acid bacteria as spoilage organisms of fruit juice products. J. Appl. Bacteriol. 21:267–271.

CARR, J. G. 1959. Some special characteristics of the cider lactobacilli. J. Appl. Bacteriol. 22:377–383.

FAVILLE, L. W., and E. C. HILL. 1952. Acid-tolerant bacteria in citrus juices. Food Res. 17:281–287.

HAYS, G. L., and D. W. RIESTER. 1952. The control of "off-odor" spoilage in frozen concentrated orange juice. Food Technol. 6:386–389.

LÜTHI, H. 1959. Microorganisms in noncitrus juices. Advances Food Res. 9:221–284.

MARSHALL, C. R., and V. T. WALKLEY. 1952. Some aspects of microbiology applied to commercial apple juice production: III, Isolation and identification of apple juice spoilage organisms; IV, Development characteristics and viability of spoilage organisms in apple juice. Food Res. 17:123–131; 197–203.

TRESSLER, D. K., and M. A. JOSLYN. 1961. Fruit and vegetable juice processing technology. Avi Publishing Co., Inc., Westport, Conn.

CHAPTER SIXTEEN

CONTAMINATION, PRESERVATION, AND SPOILAGE OF MEATS AND MEAT PRODUCTS

Meats may be fresh, cured, dried, or otherwise processed. Spoilage of the canned products will be discussed in Chapter 22.

CONTAMINATION

The healthy inner flesh of meats has been reported to contain few or no microorganisms, although they have been found in lymph nodes, bone marrow, and even flesh. The important contamination, however, comes from external sources during bleeding, handling, and processing. During bleeding, skinning, and cutting, the main sources of microorganisms are the exterior of the animal (hide, hoofs, and hair) and the intestinal tract. Recently approved "humane" methods of slaughter, mechanical, chemical, or electrical, have little effect on contamination, but each method is followed by sticking and bleeding, which can introduce contamination. As' with the older methods of use of a knife on hogs and poultry, any contaminating bacteria on the knife soon will be found in meat in various parts of the carcass, carried there by blood and lymph. The exterior of the animal harbors large numbers and many kinds of microorganisms from soil, water, feed, and manure, as well as its natural surface flora, and the intestinal contents contain the intestinal organisms (see Table 16–1). Knives, cloths, air, and hands and clothing of the workers can serve as intermediate sources of contaminants. During the handling of the meat thereafter, contamination can come from carts, boxes, or other containers, from other contaminated meat, from air, and from personnel. Especially undesirable is the addition of psychrophilic bacteria from any source, e.g., from other meats that have been in chilling storage. Special equipment such as grinders, sausage stuffers and casings, and ingredients in special products, e.g., fillers and spices, may add undesirable organisms in appreciable numbers, and sawdust on floors of processing rooms may contaminate meat with mold spores. Growth of microorganisms on surfaces contacting the meats and on the meats themselves increase their numbers. Accord-

TABLE 16–1. Average numbers of microorganisms contaminating beef
in packing-plant slaughter room[*]

Sample	Bacteria	Yeasts	Molds
Beef, dressed, on floor	$6,400-830,000/cm^2$		
Soil from animals (dry)	$110,000,000/g$	$50,000/g$	$120,000/g$
Animal feces (fresh)	$90,000,000/g$	$200,000/g$	$60,000/g$
Rumen content	$2,000,000,000/g$	$180,000/g$	$1,600/g$
Room air	$140/cm^2$ of plate		$2/cm^2$
Water, washing beef	$20-10,000/ml$		
Water, washing floor	$1,000-16,000/ml$		

*From W. A. Empey and W. J. Scott. 1939. Investigations on chilled beef: I,
Microbial contamination acquired in the meatworks. Council Sci. Ind. Res. [Australia]
Bull. 126.

ing to European workers, numbers of microorganisms contaminating
meats may be reduced by treatment of the surface with hot water.

Because of the varied sources, the kinds of microorganisms likely
to contaminate meats are many. Molds of many genera may reach the
surfaces of meats and grow there. Especially important are species of
the genera *Cladosporium, Sporotrichum, Oöspora (Geotrichum), Thamnidium, Mucor, Penicillium, Alternaria,* and *Monilia.* Yeasts, mostly
asporogenous ones, often are present. Bacteria of many genera are found,
among which some of the more important are *Pseudomonas, Achromobacter, Micrococcus, Streptococcus, Sarcina, Leuconostoc, Lactobacillus,
Proteus, Flavobacterium, Bacillus, Clostridium, Escherichia, Salmonella,*
and *Streptomyces.* Many of these bacteria can grow at chilling temperatures. There also is the possibility of the contamination of meat and
meat products with human pathogens, especially those of the intestinal
type.

In the retail market and in the home additional contamination usually takes place. In the market knives, saws, cleavers, slicers, grinders,
chopping blocks, scales, sawdust, and containers, as well as the market
operators, may be sources of organisms. In the home the refrigerator
containers used previously to store meats can serve as sources of spoilage
organisms.

PRESERVATION

The preservation of meats, as of most perishable foods, usually is
accomplished by a combination of preservative methods. The fact that
most meats are very good culture media—high in moisture, nearly neutral in pH, and high in nutrients—coupled with the facts that some

organisms may be in the lymph nodes, bones, and muscle, and contamination with spoilage organisms is almost unavoidable, makes the preservation of meats more difficult than that of most kinds of food. Unless cooling is prompt and rapid after slaughter, meat may undergo undesirable changes in appearance and flavor and may support the growth of microorganisms before being processed in some way for its preservation. Long storage at chilling temperatures may allow some increase in numbers of microorganisms.

ASEPSIS

Asepsis or keeping microorganisms away from meats as much as practicable during slaughtering and handling permits easier preservation by any method. Storage time under chilling conditions may be lengthened, aging for tenderizing becomes less of a risk, curing and smoking methods are more certain, and heating processes are more successful.

Asepsis begins with avoidance, as much as possible, of contamination from the exterior of the animal. Water spraying of the animal before slaughter has been recommended to remove as much gross dirt as possible from hair and hide, and a foot bath may be employed to remove dirt from the hoofs. Even so, the hide and hair of the animal are important sources of contamination of the surfaces of the carcass during skinning. The knife used to bleed animals after slaughter may contribute microorganisms to the still circulating blood stream and also introduce organisms while penetrating the hide. Organisms may be added to hide and lungs of hogs during scalding. There is not only contamination from the hide during skinning but also from knives and from workers and their clothes. During evisceration, contamination may come from the animal's intestine, the air, the water for washing and rinsing the carcass, cloths and brushes employed on the carcass, the various knives, saws, etc., used, and the hands and clothing of the workmen; and some organisms may come from walls touched by the carcass or from splash or mist from the floors. Meat in the chill room may be subject to contamination from air, walls, floors, and workers. Of special interest as a source of mold spores is the sawdust usually spread on the floor. Further contamination during cutting and trimming comes from knives, saws, conveyors, tables, air, water, and workmen.

The fact that the microorganisms added from the above-mentioned sources normally include practically all of the organisms involved in the spoilage of meats, many in appreciable numbers, emphasizes the importance of aseptic methods.

Once meat is contaminated with microorganisms, their removal is difficult. Gross soil may be washed from surfaces, but the wash water may add organisms. Moldy or otherwise spoiled surface areas of large pieces of meat, especially "hung," or aged, meat, may be trimmed off, but this should not be considered effective as a preservative method.

Films used to wrap meats keep out bacteria and affect the growth of those already there. These films differ considerably in their penetrability to water, oxygen, and carbon dioxide. Meats have been reported to have a shorter storage life in films less permeable to water. Fresh meats keep their red color better in an oxygen-permeable film without evacuation. With a gastight film, more carbon dioxide from bacteria would be retained; this would result in a poorer color, but would favor lactic acid bacteria over others. Cured meats preferably are packed in an oxygentight film with evacuation. Evacuation helps restrict the growth of aerobes, especially molds, reduces the rate of growth of staphylococci, and favors the growth of lactics, but apparently does not favor the growth of *Clostridium botulinum* any more than plain overwrapping does.

USE OF HEAT

The canning of meat is a very specialized technique in that the procedure varies considerably with the meat product to be preserved. Most meat products are low-acid foods that are good culture media for any surviving bacteria. Rates of heat penetration range from fairly rapid in meat soups to very slow in tightly packed meats or in pastes. Chemicals added to meats, such as spices, salt, or nitrates and nitrites in curing processes, also affect the heat processing, usually making it more effective. Nitrates in meat aid in the killing of spores of anaerobic bacteria by heat and inhibit germination of surviving spores.

Commercially canned meats may be divided into two groups on the basis of the heat processing used: (1) meats that are heat-processed in an attempt to make the can contents sterile or at least "commercially sterile," as for canned meats for shelf storage in retail stores; and (2) meats that are heated enough to kill part of the spoilage organisms but must be kept refrigerated to prevent spoilage. Canned hams and loaves of luncheon meats are so handled.

Although the National Canners Association publishes minimal heat processes for vegetables and fruits, it does not do so for meats, but recommends that a research laboratory connected with the canning industry be consulted for directions. Processes that have been used for meat products in 1-lb cans at 250 F (121 C) are 45 min for boiled beef, 60 min for beef stew, 55 min for veal or beef loaf and corned-beef hash.

Bulletins on the home canning of meats are available from Federal

TABLE 16–2. Recommended process times for canned meat in a pressure cooker at 240 F (115.5 C)*

Product	Container	Time, min
Beef	Quart jars	90
Chicken	Quart jars	75
Chicken, boned	Quart jars	90
Pork	Pint jars	75
Pork	Quart jars	90
Pork	No. 2 cans	65
Pork	No. 3 cans	90

* U.S. Dep. Agr. Tech. Bull. 930, 1946.

and state agencies, however, giving processing times and temperatures that allow a good margin of safety. Use of a pressure cooker is mandatory, and the meats usually are precooked to facilitate packing. In Table 16–2 are examples of recommendations of process times at 240 F (115.5 C).

Heat may be applied to meat products in other ways than canning. Treatment of meat surfaces with hot water to lengthen the keeping time has been suggested, although this may lessen nutrients and damage color. The cooking of wieners at the packing plant by steam or hot water reduces the numbers of microorganisms and aids in the preservation. Heat applied during the smoking of meats and meat products helps reduce microbial numbers. The precooking or tenderizing of hams reduces bacterial numbers somewhat but does not sterilize. Such products should be refrigerated, for they are perishable and they may support the growth of food-poisoning organisms if they are held at room temperatures. Similar considerations hold for cooked sausages like frankfurters and liver sausage, which also are spiced, but should be kept refrigerated. The cooking of meats for direct consumption greatly reduces the microbial content and hence lengthens the keeping time. Precooked frozen meats should contain few viable microorganisms.

USE OF LOW TEMPERATURES

More meat is preserved by the use of low temperatures than by any other method, and much more by chilling than by freezing.

Chilling

Modern packing-house methods involve chilling meat promptly and rapidly to temperatures near freezing, and chilling storage at only slightly above the freezing point. The more prompt and rapid this cooling, the less opportunity there will be for growth of mesophilic micro-organisms. The principles concerned in chilling storage, discussed in Chapter 7, apply to meats as well as other foods. Storage temperatures vary from 29.5 to 36 F (−1.4 to 2.2 C), with the lower temperatures favored by most storage men. The time limit for chilling storage of beef is about 30 days, depending upon the numbers of microorganisms present, the temperature, and the relative humidity; for pork, lamb, and mutton 1 to 2 weeks; and for veal a still shorter period. Uncooked sausage, like uncured pork sausage in bulk or in links, must be preserved by refrigeration. It was emphasized in Chapter 7 that the relative humidity usually is lowered with an increase in storage temperature. Storage time can be lengthened by storage of meats in an atmosphere containing added carbon dioxide or ozone, or the temperature and relative humidity can be raised without shortening storage time. Although considerable experimental work has been done on the gas storage of meats, the method has not been used extensively. Ships equipped for storage of meat in a controlled atmosphere of carbon dioxide have been employed successfully to carry meats from Australia to the British Isles. Increasing amounts of carbon dioxide in the atmosphere increasingly inhibit microorganisms but also hasten the formation of metmyoglobin and methemoglobin and hence the loss of "bloom," or natural color (see Figure 16–1). The storage life of meat has been doubled, according to reports, by such gas storage. Experts do not agree upon the optimal concentration of carbon dioxide, recommendations varying from 10 to 30 percent for most meats and up to 100 percent for bacon.

Storage time also can be increased by the presence of 2.5 to 3 parts per million of ozone in the atmosphere. Storage up to 60 days at 36 F (2.2 C) and 92 percent relative humidity without development of molds or slime has been reported. Ozone is an active oxidizing agent, however, that may give an oxidized or tallowy flavor to fats. It has been observed that while the levels of ozone cited will inhibit microorganisms, much higher concentrations are necessary to stop growth that already has begun.

The microorganisms that give trouble in the chilling storage of meats are the psychrophilic bacteria, chiefly of the genus *Pseudomonas*, although bacteria of the genera *Achromobacter, Micrococcus, Lactobacillus, Streptococcus, Leuconostoc, Pediococcus, Flavobacterium,* and *Proteus,* and yeasts and molds can grow in meats at low temperatures.

Freezing

Most meat sold in retail stores has not been frozen, but freezing often is used to preserve meats during shipment over long distances or for holding until times of shortage and, of course, considerable quantities of meat now are frozen in home freezers. Large pieces of meat, e.g., halves or quarters, are sharp-frozen, while hamburger and smaller, fancier cuts may be quick-frozen in wrapped packages. There is less drip from thawed mutton than from beef, and less from beef than from veal. The preservation of frozen meats is increasingly effective as the storage temperature drops from 10 F toward —20 F.

Meats for freezing are subject to the same risks of contamination by and growth of microorganisms as meats for any other purpose. The freezing process kills about half the bacteria, and numbers decrease slowly during storage. The low-temperature bacteria that grow on meat during chilling, species of *Pseudomonas, Achromobacter, Micrococcus, Lactobacillus, Flavobacterium,* and *Proteus,* can resume growth during the thawing of meat if this is done slowly. If directions are followed, packaged, quick-frozen meats are thawed too rapidly for appreciable growth of microorganisms. At temperatures above 15 C (59 F) there is a possibility of growth and toxin production by *Clostridium botulinum* types A and B in thawed meat if enough time is allowed; and at temperatures as low as 3.3 C (38 F) by type E. Salmonellae survive freezing and may remain viable for months at low temperatures of storage. It has been reported that bacteria and spores dried in meat by freezing are more resistant to salt, curing ingredients, and heat than their parent strains.

USE OF IRRADIATION

Irradiation with ultraviolet rays has been used in conjunction with chilling storage to lengthen the keeping time. It has been employed chiefly on large, hung pieces of meat in plant storage rooms but is used some in coolers in retail markets. The rays serve to reduce numbers of microorganisms in the air and to inhibit or kill them on the surfaces of the meat reached directly by the rays. To be affected, the microorganisms must be on the immediate surface, unprotected by fatty or opaque materials.

Irradiation also is used in the rapid aging of meats that are "hung" at higher than the usual chilling temperatures to reduce the growth of microorganisms, especially molds, on the surface. The aging, or hanging, process is for the purpose of tenderizing the meat by means of its own proteolytic enzymes and is used especially for obtaining tender steaks and other fancy cuts. Ordinary aging is for several weeks at 36 to 38 F (2.2 to 3.3 C) with the relative humidity between 80 and 90 percent and an air movement of 10 to 30 fpm, but with exposure to ultraviolet rays the time is reduced to 2 to 3 days at 60 to 65 F in a relative humidity of 85 to 90 percent. Some oxidation, favored by ultraviolet rays, and hydrolysis of fats may take place during aging.

Electronic irradiation of meats still is in the experimental stage. When sterilization is effected, undesirable changes in color and flavor may appear.

Irradiation with cathode (beta) rays or gamma rays has been used experimentally in the preservation of small cuts or packages of meat. In February, 1963, the Food and Drug Administration approved irradiation of fresh canned bacon with a dose of 4.5 megarads and later gave permission for use of X-rays. As was stated in Chapter 10, present interest is in irradiation of meats with ionizing rays at levels lower than those needed for sterilization, permitting a considerably lengthened storage life thereafter at chilling temperatures. This reduced dose of rays is made large enough to kill most of the important spoilage organisms on or near the surface of the meat, without noticeable harm to color, odor, or taste of the meat. Ham can be sterilized with rays, without marked changes.

PRESERVATION BY DRYING

Drying of meats for their preservation has been practiced for centuries. Jerky, or sun-dried strips of beef, was a standard food of American pioneers. Some types of sausage are preserved primarily by their dryness. In dried beef, made mostly from cured, smoked beef hams, growth of microorganisms may take place prior to processing and may develop in the "pickle" during curing, but numbers of organisms are reduced by the smoking and drying process. Organisms may contaminate the dried ham during storage and the slices during cutting and packing.

Meat products like the dry sausages, dry salamis, and dry cervelats, for example, are preserved chiefly by their low moisture content, for some varieties are not smoked. A dry outer surface on the casing of any sausage is protective.

Older methods of drying meats are usually combined with salting and smoking. During World War II pieces of freshly cooked beef and pork were dried by heat. Another method of drying pork involves a short nitrate-nitrite cure before drying and addition of lecithin as an antioxidant and stabilizer. Drying may be by vacuum, in trays, or by other methods. The final product keeps without refrigeration.

Freeze drying of meats is on the increase, with greater success with processed products such as meat patties, meat balls, and stew, than with fresh meats. The U.S. Armed Services, however, have been developing fresh-meat products, as have commercial plants here, in the

British Isles, and on the European Continent. The efficiency of the process is being improved enough to reduce costs to where production for retail sale has become practicable.

Meat for drying should be of good bacteriological quality, without previous development of appreciable numbers of microorganisms or of undesirable flavor.

USE OF PRESERVATIVES

The utilization of a controlled atmosphere containing added carbon dioxide or ozone in the chilling storage of meats has been discussed. The use of sulfur dioxide to give an unnaturally bright red color to meats is prohibited. Preservation by heavy salting is an old method that usually results in an inferior product. Ordinarily salting is combined with curing and smoking in order to be effective.

Curing

The curing of meats is limited to beef and pork, either ground meat or certain cuts like hams, butts, jowls, sides, loins, and bellies of hogs, and the hams, brisket, and leg muscles of beef. Originally, curing of meats was for the purpose of preserving by salting without refrigeration, but most cured meats of the present day have other ingredients added and are refrigerated, and many also are smoked, and hence dried to some extent. The curing agents permitted are sodium chloride, sugar, sodium nitrate, sodium nitrite, and vinegar, but only the first four are commonly used. The functions of the ingredients are as follows:

Sodium chloride, or common salt, is used primarily as a preservative and flavoring agent. The cover pickle, used for immersing the meat, may contain about 15 percent of salt, in contrast to the pumping pickle, injected into the meat, which has a higher concentration, approximating 24 percent. Its primary purpose is to lower the a_w.

Sugar adds flavor and also serves as an energy source for nitrate-reducing bacteria in the curing solution or pickle. Sucrose is used chiefly, but glucose can be substituted if a short cure is employed, or no sugar may be added.

Sodium nitrate is indirectly a color fixative and is mildly bacteriostatic in acid solution, especially against anaerobes. It also serves as a reservoir from which nitrite can be formed by bacterial reduction during the long cure.

Sodium nitrite is the source of nitric oxide, which is the real color fixative (Figure 16–1) and has some bacteriostatic effect in acid solution. The more spores of putrefactive clostridia there are in meats, the more sodium nitrite is needed to suppress them.

Most of the preservative effect of the curing agents, then, is attributed to the sodium chloride, with some bacteriostatic effect from the nitrite, and little effect from the nitrate. The salts, sugar, and meat protein combine to lower the a_w value of the cured meats, e.g., of hams to about 0.95 to 0.97. Other preservative factors are the low curing temperature and smoking.

The purplish-red color of meats (Figure 16–1) is due to blood hemoglobin and muscle myoglobin, and oxygenation of these compounds produces oxyhemoglobin and oxymyoglobin, which are bright red. Under acid and reducing conditions in the presence of nitrite, the red nitrosomyoglobin and nitrosohemoglobin are produced from myoglobin and hemoglobin (see Figure 16–1). The acid condition is produced by the meat itself, the reduced condition by the bacteria, and the nitric oxide for the reaction by reduction of the nitrite.

There are four methods for introducing the curing agents into meat: (1) the dry cure, in which dry ingredients are rubbed into the meat, as in curing belly bacon; (2) the pickle cure, in which the meats are immersed in a solution of the ingredients; (3) the injection cure, in which a concentrated solution of the ingredients is injected by needle into the arteries and veins of the meat via an artery or into the muscular tissue in various parts of the meat, as is done with pork hams; and (4) direct-addition method, in which the curing agents are added directly to finely ground meats, such as sausage, and aid in their preservation.

The curing temperature, especially when a pickling solution is employed, usually is about 2.2 to 3.3 C (36 to 38 F), and the time of the cure varies with the methods used and the meats to be cured. The older methods of curing in the pickle require several months, but the newer "quick cure," in which the pickling solution is pumped into the meat, greatly shortens that time.

Most meats are smoked after curing to aid their preservation; whereas others, like corned beef, are not smoked, but then must be refrigerated.

Some types of sausage, such as Thuringer, cervelat, Lebanon, bologna, the salamis, and the dry and semidry summer sausages, undergo an acid fermentation, preferably of a mixed lactic acid kind, during their curing. This not only has a preservative effect, preventing undesirable fermentations, but also adds a desired tangy flavor. Many proces-

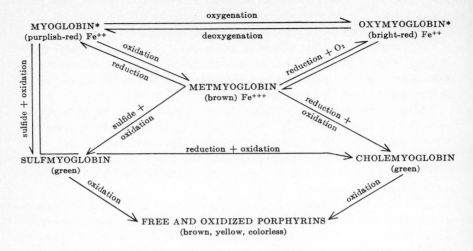

* Desirable pigments

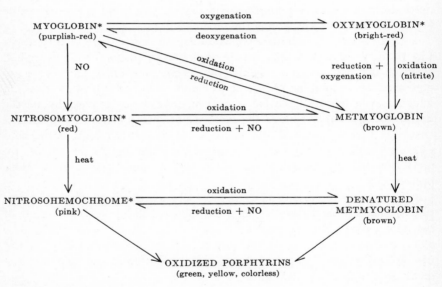

* Desirable pigments

Figure 16–1. Color changes in meats. Upper figure: Changes that may occur in raw meats. Lower figure: Changes that may occur during the curing of meats. Similar changes take place in the blood pigment, hemoglobin. (From American Meat Institute Foundation. 1960. The Science of Meat and Meat Products. W. H. Freeman and Company, San Francisco.)

TABLE 16–3.   Microorganisms reported in cured meats

Meat	*Microorganisms*
Sausages:	
Salami	Homofermentative lactobacilli
Big bologna	*Leuconostoc mesenteroides*, heterofermentative lactobacilli
Smoked links	*Leuconostoc mesenteroides*, heterofermentative lactobacilli
Frankfurters	Streptococci, pediococci, leuconostocs, lactobacilli, micrococci, spore-formers, yeasts
Fresh pork	Leuconostocs, microbacteria, lactobacilli
Bacon:	
Sliced, packaged	Mostly lactobacilli; also micrococci, enterococci
Wiltshire	Micrococci, lactobacilli
Vacuum-wrapped	Streptococci, leuconostocs, pediococci, lactobacilli
Ham:	
Raw	Lactobacilli, micrococci, microbacteria, enterococci, leuconostocs
Sliced, packaged	*Streptococcus faecium, Microbacterium* sp.
Pressed, spiced	Heterofermentative lactobacilli, leuconostocs
Canned	Enterococci, bacilli
Irradiated	Enterococci
Heated, irradiated	Bacilli, clostridia

sors carry over the mixed lactic flora from a previous lot of sausage, but Niven has recommended inoculation with pure cultures of *Pediococcus cerevisiae* (not over 0.5 percent of starter is permitted) to favor the desired lactic acid fermentation. A *Micrococcus* species also has been used. Kinds of microorganisms that have been found in cured meats are shown in Table 16–3.

Vinegar is added to the pickling solution in the preservation of foods like pickled pigs' feet, pickled spiced beef, and souses. Pigs' feet are cured in a solution of salt, sodium nitrate, and sodium nitrite, cooked, and then held in a brine of salt and vinegar. Then they are packed into jars or other containers and covered with a fresh salt-vinegar brine, and the jar is sealed. Unless the acidity is unduly low, the product will not spoil.

MICROBIOLOGY OF MEAT-CURING BRINES.   The microorganisms in curing brines and on immersed meats in them will vary with the initial condition of the meat and with the method of curing employed. The microbial content of the salt seems to have little significance except on salted meat, which after removal from brine or after dry salting sometimes develops red surface colonies of halophilic bacteria like those carried by the salt. In modern American short methods of curing meats, such as ham, bacteria in the brine apparently have little to do with changes that occur in the meat, for they do not reach high numbers and are killed mostly on the meat by the smoking that follows. Such brines contain principally lactic acid bacteria, except at the surface where micrococci and yeasts may develop. The lactics are chiefly lacto-bacilli and pediococci. In the old long cure, bacteria, especially micro-cocci, may function in reducing nitrate to nitrite, thus fixing the red color in the meat.

Foreign methods of curing bacon usually involve immersion in fairly concentrated brines and use of these brines for a long period. The brines appear to build up, besides micrococci, a special mixture of cocci and Gram-positive and Gram-negative rods that, for the most part, form tiny colonies on agar media. They are halotolerant to halophilic and reduce nitrates to nitrites. When hog bellies are treated with the dry curing mixture and compressed in boxes, growth of salt-tolerant, nitrate-reducing psychrophiles is permitted. Some beef-curing brines have been found to contain micrococci, lactobacilli, streptococci, *Achromobacter*, vibrios, and perhaps pediococci, plus other bacteria in small numbers. Species of *Micrococcus* are active in many of the pickling solutions and have been found especially in those of high salt concentration used in curing British and Canadian bacons.

Smoking

Use of wood smoke as a preservative has been discussed in Chapter 9, where it was pointed out that smoking has two main purposes, to add desired flavors and to aid in preservation. It was noted that the preservative substances added to the meat, together with the action of the heat during smoking, have a germicidal effect, and that the drying of the meat together with chemicals from the smoke inhibit microbial growth during storage.

Older methods of curing and smoking, where high salt concentrations were used in curing and greater drying and incorporation of preservative chemicals was accomplished in smoking, produced hams, dried beef, etc., that would keep without refrigeration. Many of the newer methods, however, yield a perishable product that must be refrigerated. Precooked or tenderized hams and sausages of high moisture content are examples.

Spices

Spices and condiments added to meat products like meat loaves and sausages are not in concentrations high enough to be preservative, but they may add their effect to those of the other preservative factors. Certainly products like bologna, Polish, or frankfurter and other sausages owe their keeping quality to the combined effect of spicing, curing, smoking (drying), cooking, and refrigerating.

Antibiotics

Although in this country the only permitted use of antibiotics in flesh foods now is in poultry and fish, experiments have indicated that antibiotics can be used successfully in meats to prolong storage life at chilling or higher temperatures. The antibiotics most often recommended have been chlortetracycline, oxytetracycline, and chloramphenicol. The antibiotics may be applied to meats in various ways: (1) the antibiotic may be fed the animal over a long period, (2) it may be fed more intensively for a short period before slaughter, (3) it may be infused into the carcass or into parts of it, or (4) it may be applied to the surface of pieces of meat or mixed with comminuted meat. The feeding of an antibiotic brings about a selection of microorganisms in the animal's intestinal tract, presumably reducing the numbers of spoilage bacteria there and therefore reducing the numbers that are likely to reach the meat from that source during slaughter and dressing. It has been suggested that injection of antibiotic before slaughter might be employed to prolong the keeping time of carcasses at atmospheric temperatures before they reach the refrigerator or to hold beef briefly at temperatures that will favor tenderization of special cuts, as well as lengthen the keeping time of meats held at chilling temperatures. Infusion of an antibiotic into the carcass immediately after slaughter or into special parts would serve similar purposes. Storage life of meats could be lengthened by means of an antibiotic dip, such as that now permitted for poultry, or by inclusion of antibiotic in ground meats.

SPOILAGE

Raw meat is subject to change by its own enzymes and by microbial action, and, in addition, its fat may be oxidized chemically. A moderate amount of autolysis is desired in the tenderizing of beef and game by "hanging" or aging, but is not encouraged in most other raw meats. Autolytic changes include some proteolytic action on muscle and connective tissues and slight hydrolysis of fats. The defect caused by excessive autolysis has been called "souring," an inexact term that, it will be noted, is applied to a variety of kinds of spoilage of food, and, in fact to almost any kind that gives a sour odor. "Souring" due to autolysis is difficult to separate or distinguish from defects caused by microbial action, especially from simple proteolysis. However, this preliminary hydrolysis of proteins by the meat enzymes undoubtedly helps microorganisms start growing in the meat by furnishing the simpler nitrogen compounds needed by many microorganisms that cannot attack complete native proteins.

GENERAL PRINCIPLES UNDERLYING MEAT SPOILAGE

It has been pointed out that during slaughter, dressing, and cutting, microorganisms come chiefly from the exterior of the animal and its intestinal tract, but that more are added from knives, cloths, air, workers, carts, boxes, and equipment in general. A great variety of kinds of organisms are added, so that it can be assumed that under ordinary conditions most kinds of potential spoilage organisms are present and will be able to grow if favorable conditions present themselves.

Invasion of tissues by microorganisms

Upon the death of the animal invasion of the tissues by contaminating microorganisms takes place. Factors that influence that invasion are:

1. The load in the gut of the animal. The greater the load the greater will be the invasion of tissues. For that reason starvation for 24 hr before slaughter has been recommended.
2. The physiological condition of the animal immediately before slaughter. If the animal is excited, feverish, or fatigued, bacteria are more likely to enter the tissues, bleeding is apt to be incomplete, thus encouraging the spread of bacteria, and chemical changes may take place more readily in the tissue, such as those due to better bacterial growth because of a higher pH, earlier release of juices from the meat fibers, and more rapid denaturation of proteins. Because glycogen is used up in fatigue, the pH will not drop from 7.2 to about 5.7, as it would normally.
3. The method of killing and bleeding. The better and more sanitary the bleeding, the better will be the keeping quality of the meat. Little has been reported on the effect of humane methods of slaughter on the keeping quality of meat, although it has been claimed that more greening was found in pork and bacon from electrically stunned animals than from those killed with carbon dioxide.
4. The rate of cooling. Rapid cooling will reduce the rate of invasion of the tissues by microorganisms.
   Microorganisms are spread in the meat through the blood and lymph vessels and connective-tissue interspaces, and, in ground meat, by grinding.

Growth of microorganisms in meat

Meat is an ideal culture medium for many organisms because it is high in moisture, rich in nitrogenous foods of various degrees of complexity, plentifully supplied with minerals and accessory growth factors, usually has some fermentable carbohydrate (glycogen), and is at a favorable pH for most microorganisms.

The factors that influence the growth of microorganisms and hence the kind of spoilage are those mentioned in Chapter 11. Briefly these factors are:

1. The kind and amount of contamination with microorganisms and the spread of these organisms in the meat. For example, meat with a contaminating flora that is high in percentage of low temperature *Pseudomonas* (or *Achromobacter*) species would spoil at chilling temperatures more rapidly than meat with a low percentage of these psychrophiles.
2. The physical properties of the meat. The amount of exposed surface of the flesh has considerable influence on the rate of spoilage because the greatest load of organisms usually is there and air is available for aerobic organisms. Fat may protect some surfaces, but is subject to spoilage itself, chiefly enzymatic and chemical. The grinding of meat greatly increases the surface and encourages microbial growth for this reason and because it releases moisture and distributes bacteria throughout the meat. Skin on meat serves to protect the meat inside, although microorganisms grow on it.
3. Chemical properties of the meat. It has been pointed out that meat in general is a fine culture medium for microorganisms. The moisture content is important in determining whether organisms can grow and what kinds can grow, especially at the surface, where drying may take place. Thus the surface may be so dry as to permit no growth, a little moist to allow mold growth, still moister to encourage yeasts, and very moist to favor bacterial growth. The relative humidity of the storage atmosphere is important in this regard.

    Food for microorganisms is plentiful, but the low content or absence of fermentable carbohydrate and the high protein content tend to favor the nonfermenting types of organisms, those that can utilize proteins and their decomposition products for nitrogen, carbon, and energy.

    The pH of raw meat may vary from about 5.7 to over 7.2, depending upon the amount of glycogen present at slaughter and subsequent changes in the meat. A higher pH value favors microbial growth; a lower one usually makes it slower and may be selective for certain organisms, such as yeasts.
4. Availability of oxygen. Aerobic conditions at the surface of meat are favorable to molds, yeasts, and aerobic bacteria. Within the solid pieces of meat, conditions are anaerobic and tend to remain that way because the oxidation-reduction potential is strongly poised at a low level, although oxygen will diffuse slowly into ground meat and slowly raise the oxidation-reduction potential unless the casing or packaging material is impervious to oxygen. True putrefaction is favored by anaerobic conditions.
5. Temperature. Meat should be stored at temperatures not far above freezing, where only low-temperature microorganisms can grow. Molds, yeasts, and

psychrophilic bacteria grow slowly and produce characteristic defects to be discussed later. True putrefaction is rare at these low temperatures but is likely at room temperature. As for most foods, the temperature is most important in selecting the kinds of organisms to grow and the types of spoilage to result. At chilling temperatures, for example, psychrophiles are favored and proteolysis is likely, caused by a dominating species of bacterium, followed by utilization of peptides and amino acids by secondary species. At ordinary atmospheric temperatures, mesophiles would grow, such as coliform bacteria and species of *Bacillus* and *Clostridium,* with the production of moderate amounts of acid from the limited amounts of carbohydrates present.

General types of spoilage of meats

The common types of spoilage of meats can be classified on the basis of whether they occur under aerobic or anaerobic conditions and whether they are caused by bacteria, yeasts, or molds.

SPOILAGE UNDER AEROBIC CONDITIONS.   Under aerobic conditions bacteria may cause:

1. Surface slime, which may be caused by species of *Pseudomonas, Achromobacter, Streptococcus, Leuconostoc, Bacillus,* and *Micrococcus.* Some species of *Lactobacillus* can produce slime. The temperature and the availability of moisture influence the kind of microorganisms causing surface slime. At chilling temperatures, high moisture will favor the *Pseudomonas-Achromobacter* group; with less moisture, as on frankfurters, micrococci and yeasts will be encouraged; and with still less moisture molds may grow. At higher temperatures, up to that of the room, micrococci and other mesophiles compete well with the pseudomonads and related bacteria. The numbers of microorganisms necessary before detection of off-odor or slime in meats and other proteinaceous foods are shown in Table 16–4. Numbers in the millions per square centimeter or gram are required.

TABLE 16–4.   Numbers of microorganisms at time of appearance of odor and slime in proteinaceous foods (from numerous sources)

Food	Nos. when odor evident	Nos. when slime evident
Poultry meat	2.5–$100 \times 10^6/cm^2$	10–$60 \times 10^6/cm^2$
Beef	1.2–$100 \times 10^6/cm^2$	3–$300 \times 10^6/cm^2$
Frankfurters	100–$130 \times 10^6/cm^2$	$130 \times 10^6/cm^2$
Processed meats		10–$100 \times 10^6/cm^2$
Wiltshire bacon		1.5–$100 \times 10^6/cm^2$
Fish	1–$130 \times 10^6/cm^2$	
Shell or liquid eggs	$10 \times 10^6/g$	

2. Changes in color of meat pigments (see Figure 16–1). The red color of meat, called its "bloom," may be changed to shades of green, brown, or gray as the result of the production of oxidizing compounds, e.g., peroxides, or of hydrogen sulfide, by bacteria. Species of *Lactobacillus* (mostly heterofermentative) and *Leuconostoc* are reported to cause the greening of sausage.

3. Changes in fats. The oxidation of unsaturated fats in meats takes place chemically in air and may be catalyzed by light and copper. Lipolytic bacteria may cause some lipolysis and also may accelerate the oxidation of the fats. Some fats, like butterfat, become tallowy on oxidation and rancid on hydrolysis; but most animal fats develop "oxidative rancidity" when oxidized, with off-odors due to aldehydes and acids. Hydrolysis adds the flavor of the released fatty acids. Rancidity of fats may be caused by lipolytic species of *Pseudomonas* and *Achromobacter* or by yeasts.

4. Phosphorescence. This rather uncommon defect is caused by phosphorescent or luminous bacteria, e.g., *Photobacterium* spp., growing on the surface of the meat.

5. Various surface colors due to pigmented bacteria. Thus "red spot" may be caused by *Serratia marcescens* or other bacteria with red pigments. *Pseudomonas syncyanea* can impart a blue color to the surface. Yellow discolorations are caused by bacteria with yellow pigments, usually species of *Micrococcus or Flavobacterium. Chromobacterium lividum* and other bacteria give greenish-blue to brownish-black spots on stored beef. The purple "stamping-ink" discoloration of surface fat is caused by yellow-pigmented cocci and rods. When the fat becomes rancid and peroxides appear, the yellow color changes to a greenish shade and later becomes purplish to blue.

6. Off-odors and -tastes. "Taints," or undesirable odors and tastes, that appear in meat as the result of the growth of bacteria on the surface often are evident before other signs of spoilage. "Souring" is the term applied to almost any defect that gives a sour odor that may be due to volatile acids, e.g., formic, acetic, butyric, and propionic, or even to growth of yeasts. "Cold-storage flavor" or taint is an indefinite term for a stale flavor. Actinomycetes may be responsible for a musty or earthy flavor.

Under aerobic conditions yeasts may grow on the surface of meats causing sliminess, lipolysis, off-odors and -tastes, and discolorations—white, cream, pink, or brown—due to pigments in the yeasts.

Aerobic growth of molds may cause:

1. Stickiness. Incipient growth of molds makes the surface of the meat sticky to the touch.

2. Whiskers. When meat is stored at temperatures near freezing, a limited amount of mycelial growth may take place without sporulation. Such white, fuzzy growth can be caused by a number of molds, including *Thamnidium chaetocladioides,* or *T. elegans, Mucor mucedo, M. lusitanicus,* or *M. racemosus, Rhizopus,* and others. Controlled growth of a special strain

of *Thamnidium* has been recommended for improvement in flavor during aging of beef.

3. Black spot. This usually is caused by *Cladosporium herbarum,* but other molds with dark pigments may be responsible.

4. White spot. *Sporotrichum carnis* is the most common cause of white spot, although any mold with wet, yeastlike colonies, e.g., *Geotrichum,* could cause white spot.

5. Green patches. These are caused for the most part by the green spores of species of *Penicillium* such as *P. expansum, P. asperulum,* and *P. oxalicum.*

6. Decomposition of fats. Many molds have lipases and hence cause hydrolysis of fats. Molds also aid in the oxidation of fats.

7. Off-odors and -tastes. Molds give a musty flavor to meat in the vicinity of their growth. Sometimes the defect is given a name indicating the cause, e.g., "thamnidium taint."

Spots of surface spoilage by yeasts and molds usually are localized to a great extent and can be trimmed off without harm to the rest of the meat. The time that has been allowed for diffusion of the products of decomposition into the meat and the rate of that diffusion will determine the depth to which the defect will appear. Extensive bacterial growth over the surface may bring fairly deep penetration. Then, too, facultative bacteria may grow inwards slowly.

SPOILAGE UNDER ANAEROBIC CONDITIONS. Facultative and anaerobic bacteria are able to grow within the meat under anaerobic conditions and cause spoilage. The terminology used in connection with this spoilage is inexact. Most used are the words "souring," "putrefaction," and "taint," but these terms apparently mean different things to different people.

1. Souring. The term implies a sour odor and perhaps taste. This could be caused by formic, acetic, butyric, propionic, and higher fatty acids, or other organic acids such as lactic or succinic. Souring can result from (*a*) action of the meat's own enzymes during aging or ripening; (*b*) anaerobic production of fatty acids or lactic acid by the bacterial action; or (*c*) proteolysis without putrefaction, caused by facultative or anaerobic bacteria and sometimes called "stinking sour fermentation." Acid and gas formation accompany the action of the "butyric" *Clostridium* species and the coliform bacteria on carbohydrates. Vacuum-packed meats, especially those in gastight wrappers, commonly support the growth of lactic acid bacteria.

2. Putrefaction. True putrefaction is the anaerobic decomposition of protein with the production of foul-smelling compounds like hydrogen sulfide, mercaptans, indole, skatole, ammonia, amines, etc. It usually is caused by species of *Clostridium,* but facultative bacteria may cause putrefaction or assist in its production, as evidenced by the long list of species with

the specific names "putrefaciens," "putrificus," "putida," etc., chiefly in the genera *Pseudomonas* and *Achromobacter*. Also, some species of *Proteus* are putrefactive. The confusion in the use of the term putrefaction arises from the fact that any type of spoilage with foul odors, whether from the anaerobic decomposition of protein or the breakdown of other compounds, even nonnitrogenous ones, may erroneously be termed putrefaction. Thus, for example, trimethylamine in fish, or isovaleric acid in butter, are described as "putrid" odors. Gas formation accompanies putrefaction by clostridia, the gases being hydrogen and carbon dioxide.

3. Taint. Taint is a still more inexact word applied to any off-taste or -odor. The term "bone taint" of meats refers to either souring or putrefaction next to the bones, especially in hams. Usually it means putrefaction.

Not only air but temperature has an important influence on the type of spoilage to be expected in meat. When meat is held at temperatures near 0 C (32 F), as recommended, microbial growth is limited to that of molds, yeasts, and bacteria able to grow at low temperatures. These include many of the types that produce sliminess, discoloration, and spots of growth on the surface and many that can cause souring, such as *Pseudomonas*, *Achromobacter*, *Lactobacillus*, *Leuconostoc*, *Streptococcus*, and *Flavobacterium* species. Most true putrefiers. like those in the genus *Clostridium*, require temperatures above those of the refrigerator.

SPOILAGE OF DIFFERENT KINDS OF MEATS

The processing of meats by curing, smoking, drying, or canning usually changes them and their microbial flora enough to encourage types of spoilage not undergone by fresh meats.

Spoilage of fresh meats

The spoilage of fresh meats has been covered in the preceding discussion of general types of spoilage. Little has been reported on the spoilage of fresh veal, pork, lamb, or mutton, although presumably the spoilage would be similar to that of beef. Perhaps pork spoils more readily than other meats because of its high content of B vitamins.

Lactic acid bacteria, chiefly of the genera *Lactobacillus*, *Leuconostoc*, *Streptococcus*, *Brevibacterium*, and *Pediococcus*, are present in most meats, fresh or cured, and can grow even at refrigerator temperatures. Ordinarily their limited growth does not detract from the quality of the meat; on the contrary, in certain types of sausage, such as salami, Lebanon, and Thuringer, the lactic fermentation is encouraged. However, the lactic acid bacteria may be responsible for three types of spoilage: (1) slime formation at the surface or within, especially in the presence of sucrose; (2) production of a green discoloration; or (3) souring, when excessive amounts of lactic and other acids have been produced.

FRESH BEEF. Fresh beef undergoes the changes in color mentioned: (1) changes in the hemoglobin and myoglobin, the red pigment in the blood and muscles, respectively, so as to cause loss of bloom and the production of reddish-brown methemoglobin and metmyoglobin and the green-gray-brown other oxidation pigments by action of oxygen and microorganisms; (2) white, green, yellow, greenish-blue to brown-black spots, and purple discolorations due to pigmented microorganisms; (3) phosphorescence; and (4) spots due to various bacteria, yeasts, and molds. Beef also is subject to sliminess on the surface due to bacteria or yeasts, stickiness due to molds, whiskers resulting from mycelial growth of molds, and souring and putrefaction by bacteria. Pseudomonads usually predominate in beef held at 10 C (50 F) or lower, but at 15 C (59 F) or above micrococci and pseudomonads grow in about equal numbers.

HAMBURGER. Hamburger held at room temperature usually putrefies, but at temperatures near freezing acquires a stale, sour odor. The sourness at low temperatures has been found to be caused chiefly by species of *Pseudomonas*, with help from lactic acid bacteria. *Achromobacter, Micrococcus,* and *Flavobacterium* species may grow in some samples. A large number of kinds of microorganisms have been found in hamburger held at higher temperatures, but no distinction has been made between mere presence and actual growth. Among the genera reported are *Bacillus, Clostridium, Escherichia, Aerobacter, Proteus, Pseudomonas, Achromobacter, Lactobacillus, Leuconostoc, Streptococcus, Micrococcus,* and *Sarcina* of the bacteria, and *Penicillium* and *Mucor* of the molds. A few yeasts also have been found. Ground beef packaged in polyvinylidene chloride and held refrigerated has been found to contain pseudomonads, lactic acid bacteria, micrococci, and microbacteria.

FRESH PORK SAUSAGE. Fresh sausage is made mostly of ground, fresh pork to which salt and spices have been added. It may be sold in bulk or in natural or artificial casings. Pork sausage is a perishable food that must be preserved by refrigeration and then can be kept only a relatively short time without spoilage. Souring, the most common type of spoilage at refrigerator temperatures of 0 to 11 C (32 to 51.8 F), has been attributed to growth and acid production by lactobacilli and leuconostocs, although *Microbacterium* and *Micrococcus* organisms may grow at higher storage temperatures. The encased pork sausages, and especially the "little-pig" type, are subject to slime formation on the outside of the casing on long storage, or to variously colored spots due to mold growth. Thus *Alternaria* has been found to cause small, dark spots on refrigerated links.

Spoilage of cured meats

Most of the cured meats are pork, although some cuts of beef may be cured. The inhibitory effect of nitrates against anaerobes has been mentioned previously. They hinder both growth and spore formation of anaerobes but favor aerobes. It is claimed that hydroxylamine and hydrogen peroxide produced from nitrates harm the anaerobes. Sodium nitrate is alleged to favor lactic acid bacteria in sausages like Thuringer or Essex that support a lactic fermentation. Curing salts make meats more favorable to growth of Gram-positive bacteria, yeasts, and molds than to the Gram-negative bacteria which usually spoil meats. They also reduce the thermal processing necessary to produce stable heated meat foods, such as pork luncheon meats. Some meats, e.g., bulk chipped beef, are preserved by their high content of sodium chloride.

The load of microorganisms on the piece of meat to be cured and any deterioration that has taken place will influence the success of the curing operation. Thus undesirable changes in the meat pigments will result in a discolored cured product, incipient spoilage will give an inferior appearance and flavor to the product, and large numbers of spoilage bacteria may interfere with the cure.

DRIED BEEF OR BEEF HAMS.   Beef hams are made spongy by species of *Bacillus*, sour by a variety of bacteria, red by *Halobacterium cutirubrum* or a red *Bacillus* species, and blue by *Pseudomonas syncyanea*, *Penicillium spinulosum* (purplish), and species of *Rhodotorula* yeasts.

Gas in jars of chipped dried beef has been attributed to a denitrifying, aerobic organism that resembles *Pseudomonas fluorescens*. The gases are oxides of nitrogen. *Bacillus* species have been known to produce carbon dioxide in the jars.

SAUSAGE.   In encased sausages, spoilage microorganisms may grow on the outside of the casing, between the casing and meat, or in the interior.

Growth of organisms can take place on the outside of the casings

Figure 16–2. Slimy spoilage of wieners. Note slimy growth of microorganisms on outside, developing into colonies here and there.

only if sufficient moisture is available. If moisture is available, micrococci and yeasts can form a slimy layer, as often occurs on frankfurters (Figure 16–2) that have become moist because of removal from refrigerator to warmer temperatures. With less moisture, molds (Figures 16–3, 16–4) may produce fuzziness and discoloration. Carbon dioxide, produced mostly by heterofermentative lactic acid bacteria, may swell packages of wieners or breakfast sausages when they are packaged in gastight, flexible film.

Growth between the casing and the meat is favored by an accumulation of moisture there during cooking, if the casing is penetrable to water. Or, when two casings are employed, the inner casing may be wetted before the outer casing is applied, entrapping water between the casings. The slime at the surface of the meat or between the casings is formed chiefly by acid-producing micrococci. The penetrability of the inner casing to soluble nutrients favors the bacterial growth.

Various kinds of bacteria have been reported able to grow within sausages on long chilling storage or at storage temperatures above 10.5 C (51 F). Acid-forming micrococci, such as *Micrococcus candidus,* may grow in liver sausage and bologna, and species of *Bacillus* have been found growing in liver sausage. Low-temperature leuconostocs and lactobacilli (Figure 16-5) also can grow and cause a souring that is not encouraged in most sausage but is favored in certain varieties, such as Lebanon, Thuringer, and Essex sausages. Fading of the red color of sausage to a chalky gray has been attributed to oxygen and light and may be hastened by bacteria. Various causes have been suggested for "chill rings," such as oxidation, the production of organic acids or reducing substances by bacteria, excessive water, and undercooking.

The greening of sausage may appear as a green ring not far from the casing, a green core, or a green surface. The cause of greening is probably the production of peroxides, e.g., hydrogen peroxide, by heterofermentative species of *Lactobacillus,* and *Leuconostoc* or other

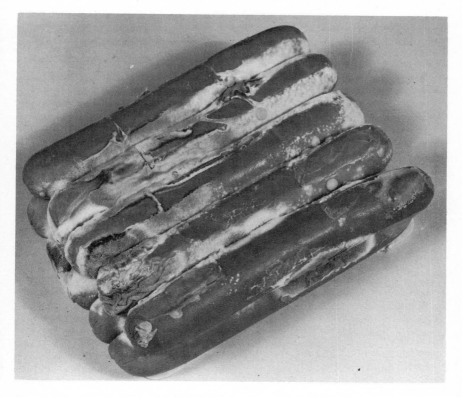

Figure 16–3. Moldy spoilage of wieners. Observe how mold growth is heaviest where the wieners touch. Some slimy bacterial colonies also are evident.

Figure 16–4. Moldy spoilage of smoked, highly spiced sausages. Mold growth has taken place over the exposed surfaces.

catalase-negative bacteria, according to Niven. Jensen states that hydrogen sulfide also may be involved. Greening is favored by a slightly acid pH and by the presence of small amounts of oxygen. The green ring below the surface of large sausages or green core in small sausages develops within 12 to 36 hr after the sausage has been processed, even under refrigeration; it is evident as soon as the sausage is cut and usually is not accompanied by surface slime. Bacterial growth and the production of heat-stable peroxide have taken place before smoking and cooking, and the peroxide continues to act to produce greening after the processing. Green cores in large sausage, e.g., big bologna, develop usually after 4 or more days of holding and within 1 to 12 hr after slicing, after large numbers of causative bacteria develop as a result of underprocessing and inadequate refrigeration. Greening of a cut surface indicates contamination with and growth of salt-tolerant, peroxide-forming bacteria (probably lactics) which can grow at low temperatures. Surface sliminess often accompanies the greening. The defect can be spread from sausage to sausage.

Production of nitric oxide gas in sausage by nitrate-reducing bacteria has been reported. Unless the casing or packaging material permits the passage of carbon dioxide, carbon dioxide may accumulate as the result of the action of heterofermentative lactics and cause swelling. This also can take place in packaged sliced, cured meats, in sandwich spreads, and in similar products in plastic casings or packages.

BACON. Since the parts of the hog used for bacon and the curing processes vary in different parts of the world, the types of spoilage and the organisms concerned also vary. The bellies employed in the American process usually are subject to little change and are reported to emerge from the smokehouse comparatively free from molds and yeasts and low in bacteria. Because of its salt tolerance and ability to grow at low temperatures, *Streptococcus faecalis* often is present. Molds are the chief spoilage organisms on the cured bacon, especially on the sliced, packaged bacon (Figure 16–6) when stored in the home

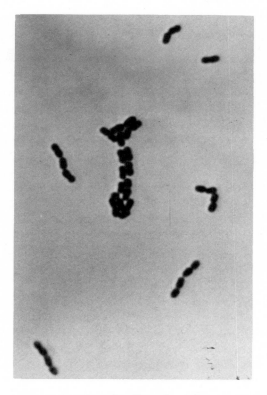

Figure 16–5. *Lactobacillus salimandus*, a meat lactic (×1,000). (*Courtesy of J. R. Allen.*)

Figure 16–6.   Moldy spoilage of sliced bacon.

refrigerator. Most trouble is encountered in late summer and early fall with species of *Aspergillus, Alternaria, Monilia, Oidium, Fusarium, Mucor, Rhizopus, Botrytis,* and *Penicillium.* Few microbiological problems are encountered with dry-salt bellies and Oxford-style bellies. Any rancidity that develops usually is due to chemical changes. Sliced bacon may be deteriorated by oxidizing and lipolytic bacteria on long storage, although chemical oxidation also may take place. Oxidizing and sulfide-forming bacteria also may be concerned in producing a poor color in the flesh part of bacon, although wrong concentrations of nitrite are more often responsible, and chromogenic bacteria may cause discolored areas. A yellowish-brown discoloration, showing the presence of tyrosine, has been blamed on proteolytic bacteria. Gumminess of pickle and bellies, now uncommon, results from the formation of gum by any of a large number of species of bacteria and yeasts.

An extensive study of the bacteriology of Wiltshire bacon has been made in Canada. In the manufacture of this bacon, the sides of the hog are cured in a very concentrated brine for a short period (6 to 8 days) at a low temperature (3.3 to 4.5 C, or 37.9 to 40.1 F), permitting the growth of only psychrophilic, salt-tolerant bacteria. Little growth takes place in the curing pickle, but marked increases in bacteria take place on the sides of meat, sometimes enough to give sliminess to the surface. Visible growth or slime usually appears when the count is over 71,500,000 per square centimeter. Micrococci are most common in the brine, but other organisms, unable to grow in the cold brine, may grow before brining or during the storage of the pickled sides after baling.

Unopened, packaged, sliced bacon is spoiled mostly by lactobacilli, but micrococci and fecal streptococci may grow, especially if the wrapper is somewhat permeable to oxygen. Opened bacon may be spoiled by molds. Vacuum-packed bacon supports the growth of coagulase-negative staphylococci at 37 C; at 20 C micrococci and lactobacilli grow. In canned bacon micrococci remove oxygen, then fecal streptococci grow, and finally lactobacilli predominate.

HAM. The term "souring," as used for the spoilage of hams, covers all important types of spoilage, from a comparatively nonodorous proteolysis to genuine putrefaction with its very obnoxious odors of mercaptans, hydrogen sulfide, amines, indole, etc., and may be caused by a large variety of psychrophilic, salt-tolerant bacteria. Jensen lists a number of genera, species of which may cause souring: *Achromobacter*, *Bacillus*, *Pseudomonas*, *Lactobacillus*, *Proteus*, *Serratia*, *Bacterium*, *Micrococcus*, *Clostridium*, and others, as well as some unnamed, hydrogen sulfide–producing streptobacilli that cause flesh-souring of ham. The types of souring are classified according to their location as sours of shank or tibial marrows, body or meat, aitchbone, stifle joint, body-bone or femur marrow, and butt. "Puffers," or gassy hams, are not encountered commercially but occur occasionally when inexpert curing is done.

When the long cure was used on hams, putrefaction by *Clostridium putrefaciens* was more common. This organism is able to grow at near-freezing temperatures and will begin growth even when the ideal rate of chilling of the ham is employed. Many of the bacteria causing souring cannot initiate growth under these conditions but must get started at higher temperatures to be able to grow at the low ones. Thus spoilage of southern country-style hams, chiefly loin or flesh sours, has been found to be proteolytic, putrefactive, or even gassy, as the result of localized growth of various species of *Clostridium*. Presumably these bacteria grow before or after the curing process and are not inhibited much by the strengths of curing solutions employed.

The present method for curing hams by the quick-curing method, in which the curing solution is pumped into the ham by way of the veins, has reduced greatly the incidence of souring. The reduction of bacterial contamination and growth by proper slaughter and bleeding of hogs, adequate refrigeration, sealing of the marrows by sawing in the right places, prompt handling, use of bacteriologically satisfactory pickling solution, and good over-all sanitation, all have helped reduce the amount of souring. The reader should turn to the "Microbiology of Meats" by L. B. Jensen for a more detailed discussion of the spoilage of ham.

Tenderized hams are really precooked and are given a mild cure. Such hams are perishable and should be protected from contamination and should be refrigerated during storage to prevent their deterioration by microorganisms. Improperly handled tenderized hams may be spoiled by any of the common meat-spoilage bacteria, among which *Escherichia coli*, *Proteus* spp., and food-poisoning staphylococci (*Staphylococcus aureus*) have been reported.

REFRIGERATED PACKAGED MEATS. Packaging films, permitting good penetration of oxygen and hence of carbon dioxide, favor the more aerobic bacteria, such as *Pseudomonas*, and their production of off-flavors, slime, and even putrefaction. This spoilage is much like that in the unwrapped meat. Films with poor gas penetration encourage lactic acid bacteria, especially when combined with vacuum packing. These bacteria in time cause souring, slime, and atypical flavors. Spoilage of canned meats is discussed in Chapter 22.

CURING SOLUTIONS OR PICKLES. Spoilage of the pickle or curing solution for ham and other cured meats is likely in the presence of available sugar and a pH well above 6.0. Spoilage of multiuse brines usually is putrefactive and is caused by *Vibrio*, *Achromobacter*, or *Spirillum*. Souring can be caused by *Lactobacillus* and *Micrococcus*, and slime by *Leuconostoc* or *Micrococcus lipolyticus*.

Turbid and ropy vinegar pickles about pigs' feet or sausages are caused chiefly by lactic acid bacteria from the meats, although yeasts may be responsible for cloudiness. Black spots on pickled pigs' feet may be caused by hydrogen sulfide–producing bacteria, and gas in vacuum-packed pickles may come from heterofermentative lactic acid bacteria or yeasts.

REFERENCES

ALLEN, J. R., and E. M. FOSTER. 1960. Spoilage of vacuum-packed, sliced processed meats during refrigerated storage. Food Res. 25:19–25.
American Meat Institute Foundation. 1960. The science of meat and meat products. W. H. Freeman and Company, San Francisco.
ANDERTON, J. I. 1963. Pathogenic organisms in relation to pasteurized cured

meats. Sci. and Technol. Surveys, No. 40. The British Food Manufacturing Industries Research Association, Leatherhead, Surrey, England.

AYRES, J. C. 1951. Some bacteriological aspects of spoilage of self-service meats. Iowa State Coll. J. Sci. 26:31–48.

AYRES, J. C. 1955. Microbiological implications in the handling, slaughtering, and dressing of meat animals. Advances Food Res. 6:110–161.

AYRES, J. C. 1960a. Temperature relationships and some other characteristics of the microbial flora developing on refrigerated beef. Food Res. 25:1–18.

AYRES, J. C. 1960b. The relationship of organisms of the genus *Pseudomonas* to the spoilage of meat, poultry and eggs. J. Appl. Bacteriol. 23:471–486.

BROOKS, F. T., and C. G. HANSFORD. 1923. Mould growths upon cold-store meat. Gt. Brit., Dep. Sci. Ind. Res., Food Invest. Spec. Rep. 17.

BROWN, W. L., C. VINTON, and C. E. GROSS. 1960. Radiation resistance of the natural bacterial flora of cured ham. Food Technol. 14:622–625.

CAVETT, J. J., 1962. The microbiology of vacuum packed sliced bacon. J. Appl. Bacteriol. 25:282–289.

COLEBY, B., M. INGRAM, D. N. RHODES, and H. J. SHEPHERD. 1962. Treatment of meats with ionising radiations: I, The irradiation preservation of pork sausages. J. Sci. Food Agr. 13:628–633.

DRAKE, SUZANNE D., J. B. EVANS, and C. F. NIVEN, JR. 1958. The microbial flora of packaged frankfurters and their radiation resistance. Food Res. 23:291–296.

DRAKE, SUZANNE D., J. B. EVANS, and C. F. NIVEN, JR. 1959. The identity of yeasts in the surface flora of packaged frankfurters. Food Res. 24:243–246.

DRAKE, SUZANNE D., J. B. EVANS, and C. F. NIVEN, JR. 1960. The effect of heat and irradiation on the microflora of canned hams. Food Res. 25:270–278.

DRAUDT, H. N. 1963. The meat smoking process: a review. Food Technol. 17:1557–1562.

DYETT, E. J. 1963. Microbiology of raw materials for the meat industry. Chem. Ind. 6:234–237.

EDDY, B. P. (*Ed.*) 1958. Microbiology of fish and meat curing brines. II Int. Symp. Food Microbiol., Proc. Her Majesty's Stationery Office, London.

EMPEY, W. A., and W. J. SCOTT. 1939. Investigations on chilled beef: I, Microbial contamination acquired in the meatworks. Council Sci. Ind. Res. [Australia] Bull. 126.

EVANS, J. B., and C. F. NIVEN, JR. 1955. Slime and mold problems with prepackaged processed meat products. Amer. Meat Inst. Found. Bull. 24.

Food Engineering Staff. 1962. Humane slaughter. Food Eng. 34(2):52.

HAINES, R. B. 1937. Microbiology in the preservation of animal tissues. Gt. Brit., Dep. Sci. Ind. Res., Food Invest. Spec. Rep. 45.

HALLECK, F. E., C. O. BALL, and ELIZABETH F. STIER. 1958. Factors affecting quality of prepackaged meat: IV, Microbiological studies. B. Effect of package characteristics and of atmospheric pressure in package upon bacterial flora of meat; C. Effects of initial bacteria count, kind of meat, storage time, storage temperature, antioxidants, and antibiotics on the rate of bacterial growth in packaged meat. Food Technol. 12:301–306; 654–659.

INGRAM, M. 1960. Bacterial multiplication in packed Wiltshire bacon. J. Appl. Bacteriol. 23:206–215.

INGRAM, M., 1962. Microbiological principles in prepacking meats. J. Appl. Bacteriol. 25:259–281.

JAYE, M., R. S. KITTAKA, and Z. J. ORDAL. 1962. The effect of temperature and packaging material on the storage life and bacterial flora of ground beef. Food Technol. 16(4):95–98.

JENSEN, L. B. 1954. Microbiology of meats. 3rd ed. The Garrard Press, Champaign, Ill.

KITCHELL, A. G. 1962. Micrococci and coagulase negative staphylococci in cured meats and meat products. J. Appl. Bacteriol. 25:416–431.

LEISTNER, L. 1960. Microbiology of ham curing. 12th Res. Conf. Amer. Meat Inst. Found., Proc., Circ. 61:17–23.

MILLER, W. A. 1961.The microbiology of some self-service packaged luncheon meats. J. Milk Food Technol. 24:374–377.

MILLER, W. A. 1964. The microbiology of self-service, prepackaged, fresh pork sausage. J. Milk Food Technol. 27:1–3.

NIINIVAARA, F. P., M. S. POHJA, and SAIMA E. KOMULAINEN. 1964. Some aspects about using bacterial pure cultures in the manufacture of fermented sausages. Food Technol. 18:147–153.

NIVEN, C. F., JR. 1951. Influence of microbes upon the color of meats. Amer. Meat Inst. Found. Circ. 2.

NIVEN, C. F., JR. 1956. Vinegar pickled meats. A discussion of bacterial and curing problems encountered in processing. Amer. Meat Inst. Found. Bull. 27.

NIVEN, C. F., JR. 1961. Microbiology of meats. Amer. Meat Inst. Found. Circ. 68.

NOTTINGHAM, P. M. 1960. Bone-taint in beef: II, Bacteria in ischiatic lymph nodes. J. Sci. Food Agr. 11:436–441.

ORDAL, Z. J. 1962. Anaerobic packaging of fresh meat. 14th Res. Conf., Amer. Meat Inst. Found., Proc., Circ. 70:39–45.

PHILLIPS, A. W., H. R. NEWCOMB, T. ROBINSON, F. BACH, W. L. CLARK, and A. R. WHITEHILL. 1961. Experimental preservation of fresh beef with antibiotics and radiation. Food Technol. 15:13–15.

RIEMANN, H. 1963. Safe heat processing of canned cured meats with regard to bacterial spores. Food Technol. 17:39–49.

SHANK, J. L., and B. R. LUNDQUIST. 1963. The effect of packaging conditions on the bacteriology, color, and flavor of table-ready meats. Food Technol. 17:1163–1166.

SHANK, J. L., J. H. SILLIKER, and P. A. GOESER. 1962. The development of a nonmicrobial off-condition in fresh meat. Appl. Microbiol. 10:240–246.

SHARPE, M. ELISABETH. 1962. Lactobacilli in meat products. Food Manufacture 37:582–589.

STEINKE, P. K. W., and E. M. FOSTER. 1961. Effect of temperature of storage on microbial changes in liver sausage and bologna. Food Res. 16:372–376.

STEINKRAUS, K. H., and J. C. AYRES. 1964. Biochemical and serological relationships of putrefactive anaerobic sporeforming rods isolated from pork. J. Food Sci. 29:100–104.

CHAPTER SEVENTEEN

CONTAMINATION, PRESERVATION, AND SPOILAGE OF FISH AND OTHER SEAFOODS

Seafoods discussed in this chapter include fresh, frozen, dried, pickled, and salted fish, as well as various shellfish. Fresh-water fish also are considered.

CONTAMINATION

The flora of living fish depends upon the microbial content of the waters in which they live. The slime that covers the outer surface of fish has been found to contain bacteria of the genera *Pseudomonas*, *Achromobacter*, *Micrococcus*, *Flavobacterium*, *Corynebacterium*, *Sarcina*, *Serratia*, *Vibrio*, and *Bacillus*. The bacteria on fish from northern waters are mostly psychrophiles, whereas fish from tropical waters carry more mesophiles. Fresh-water fish carry fresh-water bacteria, which include members of most genera found in salt water plus species of *Aeromonas*, *Lactobacillus*, *Brevibacterium*, *Alcaligenes*, and *Streptococcus*. In the intestines of fish from both sources are found bacteria of the genera *Achromobacter*, *Pseudomonas*, *Flavobacterium*, *Vibrio*, *Bacillus*, *Clostridium*, and *Escherichia*. Boats, boxes, bins, fish houses, and fishermen soon become heavily contaminated with these bacteria and transfer them to the fish during cleaning. The numbers of bacteria in slime and on the skin of newly caught ocean fish may be as low as 100 and as high as several million per square centimeter, and the intestinal fluid may contain from one thousand to 100 million per milliliter. Gill tissue may harbor a thousand to a million per gram. Washing reduces the surface count. Unopened fish, or fish "in the round," have been reported to keep better for a while than opened fish because contamination of the body cavity is avoided. Bacteria are supposed to spread through the fish flesh mostly via the gills.

Oysters and other shellfish that pass large amounts of water through their bodies pick up soil and water microorganisms in this way, including pathogens if they are present. *Achromobacter* and *Flavobacterium* predominate.

Shrimps, crabs, lobsters, and similar seafood have a bacteria-laden slime on their surfaces that probably resembles that of fish. Species of *Achromobacter, Bacillus, Micrococcus, Pseudomonas, Flavobacterium, Alcaligenes,* and *Proteus* have been found on shrimp.

Since the surface flora of fish and other sea animals seems to consist chiefly of water bacteria, the icing of these foods would not be expected to add much in the way of contamination. Ices containing antiseptic or germicidal chemicals have been used; chlortetracycline may now be used as a dip or in ice for fish in many parts of the world and for scallops and unpeeled shrimp in the United States.

PRESERVATION

Of all the flesh foods, fish is the most susceptible to autolysis, to oxidation and hydrolysis of fats, and to microbial spoilage. Its preservation, therefore, involves prompt treatment by preservative methods, and often these methods are rigorous compared to those used on meats. When fish are gathered far from the processing plant, preservative methods must be applied even on the fishing boat.

Rigor mortis is especially important in the preservation of fish, for it retards post-mortem autolysis and bacterial decomposition. Therefore any procedure that lengthens rigor mortis lengthens keeping time. It is longer if the fish have had less muscular activity before death and have not been handled roughly and bruised during catching and later processing, and is longer in some kinds of fish than in others. Reducing the holding temperature will lengthen the period.

Aseptic methods to reduce the contamination of seafood are difficult to apply, but some of the gross contamination prior to processing can be avoided by general cleansing and sanitization of boats, decks, holds, bins, or other containers and processing equipment in the plant and by use of ice of good bacteriological quality. The removal of soil from contaminating surfaces and from the fish by adequate cleansing methods, including effective detergent solutions, helps greatly to reduce the microbial "load" on the fish.

The removal of organisms is difficult, but the fact that most of the contamination is on the outer surface of the fish and other seafood permits the removal of many of the microorganisms by washing off slime and dirt.

USE OF HEAT

Canning has proved to be a successful method for preserving seafoods. Like meats they are low-acid foods and for the most part have a slow rate of heat penetration and hence are difficult to heat-process. Also some kinds of seafoods soften considerably or even fall apart when sterilization in the can is attempted.

Some seafoods, e.g., oysters, are packed into cans and are not heat-processed, but are preserved by refrigeration. A few kinds are pasteurized, crab meat, for example, and refrigerated until use. Most canned seafoods, however, are heat-processed so as to be sterile, or at least "commercially sterile." The process varies with the product being canned and the size and shape of the container. In general, the heat processes are more severe than those used for meats, but some special products are lightly processed. A few recommendations for minimal processes for marine products given by the National Canners Association are shown in Table 17–1.

The cooking of fish or other seafood for human consumption reduces their content of viable microorganisms and lengthens their keeping time.

USE OF LOW TEMPERATURES

It is only after death of the fish or other sea animal that autolysis gets underway with softening and production of off-flavor, and microbial

TABLE 17–1. Recommended heat processes for canned marine products*

Product	Can name	Initial temperature, F	Time, min 240 F	Time, min 250 F
Crab meat (brine), no liner	½ tuna	70	35	20
Mackerel (brine)	8Z tall	70	75	60
		130	70	50
Oysters, cove	No. 1	70	24	14
Salmon	½ flat	60	75	
Shrimp (wet)	No. 1	70	26	14
		90	25	13
Tuna (oil)	½ tuna	70	75	55

* From National Canners Association. 1955. Bull. 26-L. 8th ed. (These processes may be revised in a new edition.)

growth becomes uncontrolled; as has been stated these changes are de-layed by rigor mortis. Oysters in their shells, for example, will not decom-pose so long as they remain alive, and life is lengthened by chilling storage of shell oysters. Carp seined from Middle Western lakes have been kept alive and hence in good condition by shipment in tanks to the New York market. "Feedy" fish, that is, those stuffed with food, seem to decompose faster than normal fish.

Chilling

Because fish flesh autolyzes and the fats become oxidized at tem-peratures above freezing—rapidly at summer temperatures and more slowly as the temperature is dropped toward freezing—preservation by chilling temperatures is, at best, temporary. When fish or other seafood is obtained at some distance from the receiving plant, the necessity for chilling on the boat depends upon the kind of fish, whether it is dressed there or not, and the atmospheric temperature. In general, small fish are more perishable than large ones; and dressed fish autolyze more slowly than whole fish but are spoiled more readily by bacteria. When outside temperatures are warm and distances of transportation are great, it becomes necessary to chill the fish and related foods on the fishing boat by packing in crushed ice or by mechanical refrigeration in order to slow down autolysis and microbial growth until the products are marketed or are processed for longer preservation. The incorporation of preservatives in the ice used for chilling fish will be discussed later. The time allowable for holding in ice or in chilling storage will vary considerably with the kind of fish or other seafood, but will not be long in most instances. In general, chilling storage on shore is useful only when retail markets are near at hand and turnover is rapid. Other-wise, some other method of preservation is applied, such as freezing, salting, drying, smoking, or canning, or combinations of these methods.

Freezing

Most of the modern methods of freezing foods initially were devel-oped for freezing fish. In olden days, ice with added salt was employed. With the advent of mechanical refrigeration, sharp freezing was employed and the fish were "glazed," that is, a layer of ice was frozen around the outside. Whole fish, especially the larger ones, usually are sharp-frozen in air or in a salt brine. Quick freezing is applied to wrapped fillets or steaks, although whole smaller fish may be so frozen. As with meats, quick-frozen fish may thaw to more like their original condition than fish frozen more slowly. During storage the fats of frozen fish are subject to hydrolysis and oxidation. Fatty fish deteriorate more rapidly than lean ones, probably because of more hydrolysis.

Decapitated raw shrimp are frozen and glazed, and some cooked shrimp are frozen. Other seafoods preserved by freezing include scallops, clams, oysters, spiny lobster tails, and cooked crab and lobster meat. Most of these products are packaged before freezing.

As with meats, freezing kills part but not all of the microorganisms present, and growth will take place after thawing if time permits. Fish carry a flora of psychrophilic bacteria, most of which survive freezing and are ready to grow on thawing, e.g., *Pseudomonas, Achromobacter,* and *Flavobacterium* species. Spores of type E *Clostridium botulinum* will survive freezing and storage and may grow and produce toxin when temperatures reach 3.3 C (38 F) or above. Frozen raw seafoods contain few enterococci, coliforms, or staphylococci. Numbers of these organisms may be increased in the processing plant by cutting, breading, and battering operations. Precooking reduces only coliforms to any extent.

USE OF IRRADIATION

Preservation of fish by ultraviolet rays has been tried but not put into practice. Experiments have indicated that gamma or cathode irradiation of some kinds of fish may be successful.

PRESERVATION BY DRYING

The dry salting of fish or immersion in brine constitutes a method of drying, in that moisture is removed or tied up. Oxidation of fish oils is not retarded and may cause deterioration. Salting of fish is being done to a lesser extent in the United States than formerly but still is used widely throughout the world. Salt cod is prepared by a combination of salting and air drying. The flesh is then removed from bones and skin.

Sun drying of fish, either of small fish or of strips of flesh, is not practiced extensively in this country.

Part of the preservative effect of smoking is the result of the drying of the fish.

USE OF PRESERVATIVES

The salting or marination of fish by dry salt or in brine is effective not only because of the drying effect, mentioned in the preceding section, but also because of the effect of the sodium chloride as a chemical preservative. This method is used to a considerable extent in many countries of the world. The chemical and bacteriological qualities of the salt are important, for impurities such as calcium and magnesium salts may hinder the penetration of the sodium chloride, and halophilic or salt-tolerant bacteria that are introduced may cause discolorations of the fish.

Because of the great perishability of fish, investigators have tried numerous chemicals as preservatives, either applied directly to the fish or incorporated in the ice used in chilling them.

Preservatives used on fish

In the intensive search for chemical preservatives that could be applied directly to round fish or fillets or as dips, a large number of chemicals have been tried, ranging from those that most control agencies would approve to those whose use would be questionable. Sodium chloride, an acceptable preservative, has been discussed.

Fish may be dry-salted so as to contain 4 to 5 percent of salt. The salt contributes halophiles which may discolor the fish (e.g., a red color from *Serratia salinaria*). Species of *Micrococcus* usually grow on the fish, and there is a decrease in *Flavobacterium, Achromobacter, Pseudomonas*, and others. Curing of fish may be "mild," that is, with light salting, or may be in heavy brine or with solid salt, and may be followed by smoking. Benzoic acid and benzoates have been only moderately successful as preservatives. Sodium and potassium nitrites and nitrates have been reported to lengthen the keeping time and are permitted in some countries. Sorbic acid has been found to delay spoilage of smoked or salted fish. Boric acid has been used in Europe with some improvement in the keeping quality, but its use is illegal here. Other chemicals for which claims of success have been made, but whose use is contraindicated, include formaldehyde, hypochlorites, hydrogen peroxide, sulfur dioxide, undecylenic acid, capric acid, p-oxybenzoic acid, and chloroform.

Antibiotics also have been tried experimentally, usually in a dip or in ice. Of those tested, chlortetracycline and oxytetracycline seemed best, and now their use is permitted. Chloramphenicol is fairly effective, and penicillin, streptomycin, and subtilin are poor or useless. Some of the objections to the use of antibiotics as preservatives for food are listed on page 139.

Storage of fish in an atmosphere containing over 20 percent of carbon dioxide has been found to lengthen the keeping time, but this method has not come into commercial use.

Pickling of fish may mean salting or acidification with vinegar, wine, or sour cream. Herring is treated in various ways: salted, spiced, and acidified. Various combinations of these treatments, coupled with an airtight container, preserve the fish, although refrigeration also must be employed for some products.

Formerly, fish was smoked primarily for its preservation, and the smoking was heavy, but now that canning, chilling, and freezing are available to lengthen keeping time, much of the smoking of fish is primarily for flavor and hence is light. The smoke treatment and other preservative methods combined with it vary with the kind of fish, size of pieces, and keeping time desired. Fish to be smoked usually are eviscerated and decapitated, but may be in the round, split, or cut into pieces. Commonly, salting, light or heavy, precedes smoking and serves not only to flavor the fish but also to improve its keeping quality by reducing the moisture content. Drying may be aided by air currents. The smoking may be done at comparatively low temperatures, 80 to 100 F (26.7 to 37.8 C), or at high temperatures like 150 to 200 F, which result in partial cooking of the fish.

The principles of preservation involved in smoking fish are similar to those discussed in Chapter 9.

MICROBIOLOGY OF FISH BRINES. Numbers of bacteria in fish curing brines vary with the concentration of salt, temperature of the brine, the kind and amount of contamination from the fish introduced, and the duration of use of the brine, and range from 10,000 to 10,000,000 bacteria per milliliter. Salt concentrations usually are between 18 percent and saturation, but may be lower, especially after fish are introduced. The higher the temperature of the brine, the more salt is necessary to prevent its spoilage. Contamination comes from the fish, which ordinarily introduce chiefly species of *Pseudomonas, Achromobacter,* and *Flavobacterium,* from ice, which introduces these genera plus *Corynebacterium* and cocci, and from mechanically introduced sources, e.g., dust, which add cocci. On continued use of the brine, numbers of organisms increase, because of addition from successive lots of fish and because of growth of salt-tolerant bacteria such as the micrococci. As the brine ages there is a decrease in numbers of *Achromobacter* and an increase chiefly in corynebacteria in low-salt brines and in micrococci in high-salt brines.

Preservatives incorporated in ice

So-called "germicidal ices" are prepared by adding a chemical preservative to water prior to its freezing. These ices are **eutectic** when the added chemical is uniformly distributed throughout, as with sodium chloride, or **noneutectic,** when distribution is not uniform, as with sodium benzoate. Noneutectic ice is finely crushed for use on fish so as to get the chemical evenly spread in it.

Many investigators have sought the ideal chemical to be incorporated in ice for icing fish and have tested with some success a large number of chemicals, including hypochlorites, chloramines, benzoic acid and benzoates, colloidal silver, hydrogen peroxide, ozone, sodium nitrite, sulfonamides, antibiotics, propionates, levulinic acid, and many others. Both the American and Canadian governments, and those of other nations, now permit the incorporation of chlortetracycline (Aureomycin) in ice to be used by fishermen to preserve fish on trawlers and during transportation.

The purpose of the application of preservative chemicals to fish either directly or as dips or germicidal ices is to kill or inhibit microorganisms on the surfaces of the fish where at first they are most numerous and active.

Antioxidants

Fats and oils of many kinds of fish, especially the fatter ones, such as herring, mackerel, mullet, and salmon, are composed to a great extent of unsaturated fatty acids and hence are subject to oxidative changes producing oxidative rancidity and sometimes undesirable alterations in color. To counteract these undesirable changes, antioxidants may be applied as dips, coatings, glazes, or gases. Good results have been reported with nordihydroguaiaretic acid, ethyl gallate, ascorbic acid, and other compounds and with storage in carbon dioxide.

SPOILAGE

Like meat, fish and other seafood may be spoiled by autolysis, oxidation, or bacterial activity, or most commonly by combinations of these. Most fish flesh, however, is considered more perishable than meat because of more rapid autolysis by the fish enzymes and because of the less acid reaction of fish flesh that favors microbial growth. Also, many of the fish oils seem to be more susceptible to oxidative deterioration than most animal fats. The experts agree that the bacterial spoilage of fish does not begin until after rigor mortis, when juices are released from the flesh fibers. Therefore, the more this is delayed or protracted, the longer will be the keeping time of the fish. Rigor mortis is hastened by struggling of the fish, lack of oxygen, and a warm temperature, and is delayed by a low pH and adequate cooling of the fish. The pH of the fish flesh has an important influence on its perishability, not only because of its effect on rigor mortis, but also because of its influence on the growth of bacteria. The lower the pH of the fish flesh, the slower in general will be bacterial decomposition. The lowering of the pH of the fish flesh results from the conversion of muscle glycogen to lactic acid.

FACTORS INFLUENCING KIND AND RATE OF SPOILAGE

The kind and rate of spoilage of fish will vary with a number of factors:

1. The kind of fish. The various kinds of fish differ considerably in their perishability. Thus some flat fish spoil more readily than round fish because they pass through rigor mortis more rapidly, but a flat fish like the halibut keeps longer because of the low pH (5.5) of its flesh. Certain fatty fish deteriorate rapidly because of oxidation of the unsaturated fats of their oils. Fishes high in trimethylamine oxide soon yield appreciable amounts of the "stale-fishy" trimethylamine.
2. The condition of the fish when caught. Fish that are exhausted as the result of struggling, lack of oxygen, and excessive handling spoil more rapidly than those brought in with less ado, probably because of the exhaustion of glycogen and hence smaller drop in pH of the flesh. "Feedy" fish, that is, those full of food when caught, are more perishable than those with an empty intestinal tract.
3. The kind and extent of contamination of the fish flesh with bacteria. These may come from mud, water, handlers, and the exterior slime and intestinal content of the fish and are supposed to enter the gills of the fish, from which they pass through the vascular system and thus invade the flesh, or to penetrate the intestinal tract and thus enter the body cavity. Even then, growth probably is localized for the most part, but the products of bacterial decomposition penetrate the flesh fairly rapidly by diffusion. In general, the greater the load of bacteria on the fish, the more rapid will be the spoilage. This contamination may take place in the net (mud), in the fishing boat, on the docks, or, later, in the plants. Fish in the "round," that is, not gutted, have not had the flesh contaminated with intestinal organisms, but it may become odorous because of decay of food in the gut and diffusion of decomposition products into the flesh. This process is hastened by the digestive enzymes attacking and perforating the gut wall and the belly wall and viscera, which in themselves have a high rate of autolysis. Gutting the fish on the boat spreads intestinal and surface-slime bacteria over the flesh, but thorough washing will remove most of the organisms and adequate chilling will inhibit the growth of those left. Any damage to skin or mucous membranes will harm the keeping quality of the product.

4. Temperature. Chilling of the fish is the most commonly used method for preventing or delaying bacterial growth and hence spoilage until the fish is used or is otherwise processed. The cooling should be as rapid as possible to 32 to 30 F (0 to −1 C), and this low temperature should be maintained. Obviously, the warmer the temperature, the shorter will be the storage life of the fish. Prompt and rapid freezing of the fish is still more effective in its preservation.
5. Use of an antibiotic ice or dip.

EVIDENCES OF SPOILAGE

Since the change is gradual from a fresh condition to staleness and then to inedibility, it is difficult to determine or agree on the first appearance of spoilage. A practical test to determine the quality of fish has been sought for many years, but none has proved entirely satisfactory. A chemical test for trimethylamine is backed by most workers for use on salt-water fish, although some support other methods, such as an estimate of volatile acids or volatile bases, or a test for pH, hydrogen sulfide, ammonia, etc. Bacteriological tests are too slow to be useful.

Reay and Shewan describe the succession of external changes in a fish as it spoils and finally becomes "putrid." The bright, characteristic colors of the fish fade, and dirty, yellow, or brown discolorations appear. The slime on the skin of the fish increases, especially at the flaps and gills. The eyes gradually sink and shrink, the pupil becoming cloudy and the cornea opaque. The gills turn a light-pink and finally grayish-yellow color. Most marked is the softening of the flesh, so that it exudes juice when squeezed and becomes easily indented by the fingers. The flesh is easily stripped from along the backbone, where a reddish-brown discoloration develops toward the tail and is a result of the oxidation of hemoglobin.

Meanwhile a sequence of odors is evolved: first the normal, fresh, seaweedy odor, then a sickly sweet one, then a stale-fishy odor due to trimethylamine, followed by ammoniacal and final putrid odors, due to hydrogen sulfide, indole, and other malodorous compounds. Fatty fish also may show rancid odors. Cooking will bring out the odors more strongly.

BACTERIA CAUSING SPOILAGE

The bacteria most often involved in the spoilage of fish are part of the natural flora of the external slime of fishes and their intestinal contents. The predominant kinds of bacteria causing spoilage vary with the temperatures at which the fish are held, but at the chilling temperatures usually employed, species of *Pseudomonas* are most likely to predominate, with *Achromobacter* and *Flavobacterium* species next in order of importance. Appearing less often, and then at higher temperatures, are bacteria of the genera *Micrococcus* and *Bacillus*. Reports in the literature list other genera as having species involved in fish spoilage, such as *Escherichia, Proteus, Serratia, Sarcina,* and *Clostridium*. Most of these would grow only at ordinary atmospheric temperatures and probably would do little at chilling temperatures.

Normally pseudomonads increase in numbers on chilled fish during holding, achromobacters decrease, and flavobacteria increase temporarily and then decrease. The bacteria grow first on the surfaces and later penetrate the flesh. Fish have a high content of nonprotein nitrogen, and autolytic changes caused by their enzymes increase the supply of nitrogenous foods (e.g., amino acids and amines) and glucose for bacterial growth. From these compounds the bacteria make trimethylamine, ammonia, amines (e.g., putrescine and cadaverine), lower fatty acids, and aldehydes, and eventually hydrogen and other sulfides, mercaptans and indole, which products are indicative of putrefaction. A musty or muddy odor and taste of fish has been attributed to the growth of *Streptomyces* species in the mud at the bottom of the body of water and the absorption of the flavor by the fish.

As has been indicated, discolorations of the fish flesh may occur during spoilage; yellow to greenish-yellow colors caused by *Pseudomonas fluorescens*, yellow micrococci, and others; red or pink colors from growth of *Sarcina, Micrococcus,* or *Bacillus* species, or by molds or yeasts; and a chocolate-brown color by an asporogenous yeast. Pathogens parasitizing the fish may produce discolorations or lesions.

SPOILAGE OF SPECIAL KINDS OF FISH AND SEAFOODS

The previous discussion has been limited for the most part to the spoilage of fish preserved by chilling. Salt fish are spoiled by salt-tolerant or halophilic bacteria of the genera *Serratia, Micrococcus, Bacillus, Achromobacter, Pseudomonas,* and others, which often cause discolorations, a red color being common. Molds are the chief spoilage organisms on smoked fish. Marinated (sour pickled) fish should present no spoilage problems unless the acid content is low enough to permit growth of lactic acid bacteria, or the entrance of air permits mold growth. Frozen fish, too, should present no bacteriological problems after freezing, but, of course, their quality depends on what has happened to the fish prior to freezing. The spoilage of canned fish and other seafood will be discussed in Chapter 22. Japanese fish sausage is subject to souring caused by volatile-acid production by bacilli or to putrefaction, despite the addition of nitrite and permitted preservatives.

In general, shellfish are subject to types of microbial spoilage similar to those for fish. However, in chilled shrimp *Achromobacter* seems to increase the most and is chiefly responsible for spoilage, although there may be a temporary increase in pseudomonads and a decrease in *Flavobacterium, Micrococcus,* and *Bacillus.* Crab meat is deteriorated by *Pseudomonas* and *Achromobacter* at chilling temperatures, and mainly by *Proteus* at higher temperatures. Species of *Pseudomonas, Achromobacter, Flavobacterium,* and *Bacillus* have been incriminated in the spoilage of raw lobsters.

Oysters remain in good condition as long as they are kept alive in the shell at chilling temperature, but they decompose rapidly when they are dead, as in shucked oysters. The type of spoilage of the shucked oysters depends upon the temperature at which they are stored. Oysters are not only high in protein but also contain sugars which result from the hydrolysis of glycogen. At temperatures near freezing, *Pseudomonas* or *Achromobacter* species are the most important spoilage bacteria, but *Flavobacterium* and *Micrococcus* species also may grow. The spoilage is termed "souring" although the changes are chiefly proteolytic. At higher temperatures the "souring" may be the result of the fermentation of the sugars by coliform bacteria, streptococci, lactobacilli, and yeasts to produce acids and a sour odor. Early growth of *Serratia, Pseudomonas, Proteus,* and *Clostridium* may take place. An uncommon type of spoilage by an asporogenous yeast causes pink oysters.

REFERENCES

AKAMATSU, M. 1959. Bacteriological studies on the spoilage of fish sausage: I, Number of bacteria present in the meat of fish sausage on the market. Jap. Soc. Sci. Fish. Bull. 25:545–548.

American Public Health Association. 1962. Recommended procedures for the bacteriological examination of sea water and shellfish. 3rd ed. American Public Health Association, New York.

BERNARDE, M. A. 1961. Technology of crabmeat production—a bibliography. J. Milk Food Technol. 24:211–217.

BORGSTROM, G. (*Ed.*) 1961. Fish as food: Vol. I, Production, biochemistry, and microbiology. Academic Press Inc., New York.

BROOKE, R. O., and M. A. STEINBERG. 1964. Preservation of fresh unfrozen fishery products by low-level radiation: I, Introduction. Food Technol. 18:1056–1057.

COLWELL, RITA R., and J. LISTON. 1960. Microbiology of shellfish. Bacteriological study of the natural flora of Pacific oysters (*Crassostrea gigas*). Appl. Microbiol. 8:104–109.

EDDY, B. P. (*Ed.*) 1958. Microbiology of fish and meat curing brines. II Int. Symp. Food Microbiol., Proc. Her Majesty's Stationery Office, London.

EVELYN, T. P. T., and L. A. McDERMOTT. 1961. Bacteriological studies of fresh-water fish. I. Isolation of aerobic bacteria from several species of Ontario fish. Canad. J. Microbiol. 7:375–382.

HESS, E. 1950. Bacterial fish spoilage and its control. Food Technol. 4:477–480.

MacLEAN, D. P., and CAMILLE WELANDER. 1960. The preservation of fish with ionizing radiation: bacterial studies. Food Technol. 14:251–254.

MASUROVSKY, E. B., J. S. VOSS, and S. A. GOLDBLITH. 1963. Changes in the microflora of haddock fillets and shucked soft-shelled clams after irradiation with Co⁶⁰ gamma rays and storage at 0 C and 6 C. Appl. Microbiol. 11:229–234.

REAY, G. A., and J. M. SHEWAN. 1949. The spoilage of fish and its preservation by chilling. Advances Food Res. 2:343–398.

SHEWAN, J. M. 1962. The bacteriology of fresh and spoiling fish and some related chemical changes. Recent Advances Food Sci. 1:167–193.

SPENCER, R. 1959. The sanitation of fish boxes: I, The quantitative and qualitative bacteriology of commercial wooden fish boxes. J. Appl. Bacteriol. 22:73–84.

SPENCER, R., and R. B. HUGHES. 1963. Recent advances in fish processing technology. Food Manufacture 38:407–412.

STANSBY, M. E. 1962. Speculations on fishy odors and flavors. Food Technol. 16(4):28–32.

TANIKAWA, E. 1963. Fish sausage and ham industry in Japan. Advances Food Res. 12:367–424.

TARR, H. L. A. 1954. Microbiological deterioration of fish post mortem, its detection and control. Bacteriol. Rev. 18:1–15.

TOMIYASU, Y., and B. ZENTANI. 1957. Spoilage of fish and its preservation by chemical agents. Advances Food Res. 7:42–82.

U.S. Department of Health, Education, and Welfare, Public Health Service. 1957. Manual of recommended practices for the control of the shellfish industry. Pub. 33.

U.S. Department of Health, Education, and Welfare, Public Health Service. 1959. Sanitary control of the shellfish industry; 1959 manual of recommended practice.

CHAPTER EIGHTEEN

CONTAMINATION, PRESERVATION, AND SPOILAGE OF EGGS

Although only hens' (chickens') eggs are discussed here, it is assumed that the microbiology of the eggs of other domestic poultry will be similar.

CONTAMINATION

Although the majority of freshly laid eggs are sterile inside, the shells soon become contaminated: by fecal matter from the hen, by the lining of the nest, by wash water if the eggs are washed, by handling and, perhaps, by the material in which the eggs are packed. Molds and bacteria from these sources can grow through a moistened shell and into the egg. Since eggs usually are cooled promptly and stored at chilling temperatures, contamination with low-temperature bacteria such as those of the genera *Pseudomonas, Proteus,* and *Achromobacter* is especially undesirable. Also commonly present are species of *Alcaligenes, Flavobacterium, Bacillus, Micrococcus,* and *Streptococcus,* as well as coliform bacteria and molds. Most of the bacteria normally on egg-shells are Gram-positive cocci and rods that are not concerned in rotting, and there are few Gram-negative bacteria (some of which cause rots) unless larger numbers have been added by washing methods. *Salmonella* species may be on the shell or in the egg as laid, build up during processing, and be present in significant numbers in frozen or dried eggs.

PRESERVATION

Methods for the preservation of eggs have received considerable attention because of the perishability of this food. The protection of the shell egg against microbial attack by commonly used methods depends upon keeping the shell dry and the egg cool and avoiding destruction of the thin surface layer of proteinaceous material on the shell

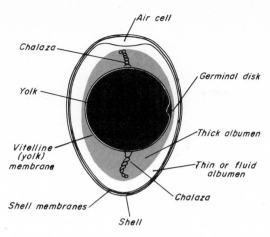

known as the cuticle or "bloom." The outer and inner membranes within the shell (see Figure 18–1) mechanically hold back all bacteria, but permit certain kinds to grow through. Changes in the membranes occur with aging and favor more rapid bacterial multiplication. The rates of physical and chemical changes within the eggs will depend upon the time and temperature of holding, the relative humidity, and the composition of the atmosphere about the eggs. The effect of special treatments of the shell on the keeping quality of the egg will be discussed below.

Egg white is reported to inhibit bacterial growth because of its content of the enzyme lysozyme, which dissolves the cell wall of Gram-positive bacteria and flocculates the cells; of avidin, which ties up biotin; of ovomucoid, which is antitryptic; and of conalbumin, which chelates iron and thus prevents the growth of bacteria at high pH. Wash water with a high iron content (5 to 10 ppm), however, may supply enough iron to favor the growth of spoilage bacteria. The low content of non-protein nitrogen, the combination of riboflavin with protein, and the high pH (about 9.6) that develops on storage combine to make egg white a poor culture medium.

ASEPSIS

Greater care is taken nowadays than formerly to reduce the contamination of the outside of the shell with hen feces and dirt from the nests. When eggs are broken for drying or freezing, care is taken to keep out those in which microbial growth has taken place, and to reduce contamination from equipment by adequately cleansing and sanitizing it.

REMOVAL OF MICROORGANISMS

Because dirty eggs command a lower price than clean ones, various methods have been tried for the removal of this soil. Dry cleaning, as by sandblasting, removes dirt and also the bloom (mucin). Washing with warm, plain water removes dirt, the bloom, and part of the microorganisms, but encourages the penetration of bacteria into the egg through the pores in the shell. Unless precautions are taken, the wash water will build up numbers of spoilage bacteria so that increased contamination will take place during washing. Experiments have shown that hand-washed eggs are more subject to rotting than unwashed ones, and that machine-washed eggs show more rots than hand-washed ones. The amount of rotting that results varies with the kind of washing machine and the kind of washing solution. Attempts to reduce contamination with rot bacteria by cleansing of machines and disinfection with 1 percent hypochlorite solution have not always been successful, but use of that disinfectant solution as the wash water has reduced the percentage of rots in the washed eggs. Lye, acids, formalin, hypochlorites, quaternary ammonium compounds, various detergents, and detergent-sanitizer combinations have been tried in washing solutions, with strong hypochlorite solution or a detergent-sanitizer solution receiving the strongest recommendations. These solutions are used warm or hot, at temperatures ranging from 90 to 140 F (32.2 to 60 C), depending upon the chemicals employed. A warm washing solution is essential so that the liquid will not be drawn into the cool egg. These chemical solutions not only remove microorganisms but also kill many of them.

USE OF HEAT

The heat coagulability of the egg white determines to a great extent the maximal heat-treatment that can be given shell eggs. Scott and coworkers in Australia have determined the maximal time at different temperatures for heating in water that will just avoid coagulation: for example, 800 sec at 57.5 C (135.5 F), 320 sec at 60 C (140 F), and 128 sec at 62.5 C (144.5 F). Temperatures of 65 C (149 F) or above resulted in some coagulation of the white when heat-treatments were employed that would control rotting. Rotting was well controlled, for instance, by heating at 60 C (140 F) for 320 sec. Heating in water was superior to heating in oil.

Treatments suggested by other workers include heating shell eggs in oil for 10 min at 60 C (140 F) or in water for 30 min at 54.4 C

(130 F), and heating of egg contents at 61.7 C (143 F) for 30 min. Immersion of shell eggs in boiling water for a few seconds has been suggested, or in hot oil (135 F, or 57.2 C) with or without a vacuum. Immersion of eggs in a hot detergent-sanitizer solution (at 110 to 130 F, or 43.3 to 54.4 C) has been recommended, the sanitizer being a quaternary ammonium compound. A new "thermostabilization" method of dipping into hot water reduces evaporation of moisture from the egg by a slight coagulation of the outermost part of the egg albumin.

As a safeguard against salmonellae it has been recommended in the Code of Federal Regulations (Amendments, May 1, 1965) that all liquid eggs, except whites, be pasteurized at 140 F (60 C) for not less than 3.5 min or be found free of salmonellae by laboratory test. Frozen whites are to be pasteurized or found free of salmonellae, and dried whites so dehydrated as to be free of these organisms. Pasteurization of liquid eggs was made compulsory on January 1, 1966, and pasteurization of liquid whites on June 1, 1966.

USE OF LOW TEMPERATURES

The most common method of preservation of eggs is by use of low temperatures—using chilling temperatures for shell eggs and freezing for egg "meats." Oiling, gas storage, or treatment with chemical preservatives may be combined with the chilling of shell eggs.

Chilling

In this country, most shell eggs are preserved by chilling. They are selected for storage on the basis of their general appearance and the result of "candling." To candle an egg, it is held and rotated in front of a light to examine it for defects such as cracks, rots, molds, blood, developing embryo, crusted or sided yolk, weak white, or large air cell. The eggs should be cooled as promptly as is practicable after production and held at a temperature and a relative humidity that will depend upon the anticipated time of storage. The lower the relative humidity is below 99.6 percent, the more rapidly the egg will lose moisture and hence weight, and the larger the air cell will become. The higher the relative humidity, the more likely will be microbial spoilage of the egg. The higher the temperature is above 29 F (−1.67 C), the more rapid will be penetration of microorganisms into the egg and growth there, as well as physical and chemical changes such as thinning of the white and weakening of the yolk membrane. For commercial storage for 6 months or longer a temperature of 29 to 31 F (−1.7 to

—0.55 C) and a relative humidity of 80 to 85 percent are recommended, although the present trend is toward 90 percent humidity. Overseas, temperatures of 0 to 1 C (32 to 33.4 F) are more common, with 80 to 85 percent relative humidity. Air circulation in the storage room is important, in order that the desired relative humidity will be maintained around the eggs; and a constant storage temperature is essential to avoid the condensation of moisture on the eggshells. Eggs for chilling (often called "cold storage") are gathered mostly in periods of abundance and stored for distribution during times of shortage.

Special treatments given eggs to improve their keeping quality during chilling storage may be for the purposes of keeping the shell surface dry, preventing loss of moisture, keeping air away, or inhibiting changes due to microorganisms or enzymes of the egg. Impregnation of the eggshell with a colorless and odorless mineral oil is a commonly employed method that keeps out moisture, slows down desiccation and air penetration, retains carbon dioxide, and retards physical and chemical changes within the egg. Early methods involved the use of hot oil (135 F, or 57.2 C) and a vacuum; and in one method the vacuum was replaced by carbon dioxide gas. At present, the egg is immersed for a few seconds in a thinner mineral oil at atmospheric temperature, or preferably at about 100 to 110 F (37.8 to 43.3 C) and then drained. Oiling is used primarily for eggs to be stored commercially for long periods. It appears to have little effect upon rates of microbial changes within the egg.

Gas storage, and the treatment of eggs, wrappers, or containers with chemicals will be discussed under Use of Preservatives later in this section.

Freezing

The chief bacteriological problems in the freezing of egg "meats" or "pulp" are in connection with the selection and the preparation of the egg contents for freezing. As with other frozen foods, the frozen egg is no better than the egg was before it was frozen. Many of the eggs broken for freezing or drying are defective in some way that lowers their grade as shell eggs but does not affect the quality of the contents, e.g., off-sized, dirty, cracked, and weak-shelled eggs. Dirty eggs should be washed before they are broken. The eggs first are selected by candling at about 40 F (4.4 C) and are broken in a dry room at about 55 to 60 F (12.8 to 15.5 C). The egg is broken on a dull knife blade and the contents are divided into white and yolk and placed in separate small cups if they are to be frozen separately, or are placed as a whole into a small cup. Not more than three eggs go into a cup, and if the appearance or odor is unsatisfactory the whole tray—knife, cups, and all—is sent to be cleansed and sanitized, a clean, sanitized tray replaces it, and the breaker washes and dries her hands before returning to work. The careful elimination of spoiled eggs has a very important influence on the kinds and numbers of bacteria in the frozen product, as does sanitation throughout the breaking process. All equipment that touches the shell eggs or their contents must be cleansed and sanitized daily or more often; and the breaking room and breakers must measure up to standards of sanitation. The egg meats, whole or separated, are filtered to remove pieces of shell and stringy material (chalazae), mixed or churned, standardized as to solids content, and frozen in 30- or 50-lb tin cans or other containers, usually by a sharp-freezing process. Yolks, when frozen separately and stored frozen, will form a jelly which does not become fluid on thawing. This difficulty is avoided by the incorporation of 5 percent or more of sugar, salt, or glycerol prior to freezing. The frozen eggs are stored at 0 to —5 F (—17.8 to —20.5 C).

Unless proper precautions are taken, frozen eggs may contain high numbers of bacteria, up to millions per gram. These large numbers may result from the use of heavily contaminated egg pulps, contamination from pieces of shell and from equipment in the breaking room, and growth before freezing. Bacteria that spoil eggs stored at low temperatures, especially *Pseudomonas* organisms, are likely to be numerous, as well as bacteria of the genera *Alcaligenes, Proteus,* and *Flavobacterium.* Gram-positive cocci and rods and coliform bacteria from the eggshell would occur in smaller numbers, and possibly anaerobes and miscellaneous bacteria as chance contaminants. Occasionally, *Salmonella* bacteria may be in eggs from infected hens, and rarely hemolytic bacteria may be introduced. The freezing process reduces numbers, but some of each kind of organism present are likely to survive. The numbers decrease slowly during storage in the frozen condition. If thawing is done at too high a temperature or the thawed eggs are held unduly long before use, numbers of bacteria will increase as a result of growth of the predominating low-temperature bacteria. Thawing at 55 F (12.8 C) for 48 hr has been recommended.

PRESERVATION BY DRYING

In the preparation of whole pulp, white, or yolk of eggs for drying, the principles involved are similar to those just discussed in the preparation of these materials for freezing. Egg white requires some additional treatment before drying to retain its whipping properties. The old Chinese method utilized a natural fermentation of the egg white, chiefly by coliform bacteria, in large casks for 35 to 60 hr. This resulted in high numbers of bacteria in the egg white both before and after drying. Removal of the glucose, the apparent purpose of the fermentation, has been accomplished also by inoculation with pure cultures of yeast, *Pseudomonas, Aerobacter aerogenes, Escherichia freundii,* and other organisms. Addition of sucrose or lactose before drying, and thinning of the white by trypsin also have been suggested. A new pretreatment employs an enzyme, glucose oxidase, obtained from fungi, to oxidize the glucose to gluconic acid. Hydrogen peroxide is added to furnish oxygen, which is released by activity of the enzyme catalase in the preparation.

Most of the drying of eggs in this country is by means of the spray dryer, where the liquid is sprayed into a current of dry, heated air. Another method is the roller or drum process in which the liquid egg is passed over a heated drum, with or without vacuum. Air drying is accomplished by means of open pans, as used by the Chinese, or by the belt system where the egg liquid is on a belt that passes through a heated (140 to 160 F, or 60 to 71.1 C) tunnel. Spray drying or pan drying, combined with tunnel drying, is used for egg white. Formerly, eggs were dried to a moisture content of about 5 percent, but it has been shown that the keeping quality of dried white or whole egg is improved as the moisture content is decreased toward 1 percent, and the trend is in that direction.

Dried egg may contain from a few hundred microorganisms per gram up to over a hundred million, depending upon the eggs broken and the methods of handling employed. The same reasons for high numbers would hold as were mentioned previously for large numbers in frozen eggs. The drying process may reduce the microbial content ten- to a hundredfold from the numbers in the liquid egg, but still permit large numbers to survive. A variety of kinds of organisms has been found in dried eggs, including micrococci, streptococci (enterococci), coliform bacteria, *Salmonella* species, spore-forming bacteria, and molds. Cocci and the Gram-positive rods are likely to be present as a result of contamination from the shell during breaking or from handlers or equipment rather than from spoiled eggs. *Salmonella* organisms are from infected hens. When the egg white has been pretreated by a fermentation process, the dry product may have a high content of microorganisms. Adequately dried egg products contain too little moisture for growth of microorganisms. In fact, during storage the numbers of organisms in dry egg will decrease, more rapidly at first and gradually later. Organisms resistant to desiccation, such as micrococci and spores of bacteria and molds, will make up an increasing percentage of the survivors as storage continues. The more the moisture content of the dry egg has been reduced from 5 percent, the more rapid will be the death of bacteria therein.

USE OF PRESERVATIVES

Preservatives may be used on the shells of eggs, in the atmosphere around them, or on wraps or containers for eggs. An enormous number of different substances have been applied to the surface of the shells of eggs or used as packing material about eggs to aid in their preservation. Some of these substances are primarily to keep the shell dry and reduce penetration of oxygen into the egg and passage of carbon dioxide and moisture out; waxing, oiling the shells, and otherwise sealing are examples. Other materials inhibit the growth of microorganisms, and some are germicidal. Materials used for the dry packing of eggs in the home include salt, lime, sand, sawdust, and ashes. Immersion in water glass, a solution of sodium silicate, long has been a successful home method of preservation. The solution is inhibitory because of its alkalinity. Other inhibitory chemicals that have been tried are borates, permanganates, benzoates, salicylates, formates, and a host of other compounds. The utilization of warm or hot solutions of germicides in the washing of eggs has been mentioned, such as solutions of hypochlorites, lye, acids, formalin, quaternary ammonium compounds, and detergent-sanitizer combinations. Sealing of the shell with a solution of dimethylol urea has been found effective in inhibiting mold growth.

Some attempts have been made to reduce spoilage of eggs by molds during storage by treatment of the flats and fillers of storage cases with a mycostatic or mold-inhibiting chemical. Sodium pentachlorophenate and related compounds have been used with some success.

Fumigation of eggs with gaseous ethylene oxide before storage has been reported to protect them against bacterial spoilage.

The only two gases that are added to the atmosphere about eggs to improve their keeping quality are carbon dioxide and ozone, although nitrogen has been used experimentally. Recommendations vary concerning the optimal concentration of carbon dioxide in the air for this purpose. A low concentration, 2.5 percent, for example, has been shown to slow down physical and chemical changes in the egg that accompany the rise in pH as carbon dioxide escapes from the egg but to have little effect on microbial growth, particularly of molds. Concentrations as high as 60 percent will markedly delay microbial spoilage, especially if the temperature is near freezing, even if the relative humidity approaches saturation, but the white becomes thinner and an unpleasant flavor develops. Thus the recommendations vary from 0.5 to 0.6 percent at -1 C (30.2 F), and from 2.5 to 5 percent at 0 C (32 F) up to a 15 percent maximum. There has been some disagreement about the effectiveness of ozone. It has been claimed that from 0.6 ppm of ozone (for clean eggs) to 1.5 ppm (for dirty eggs) at 31 F (-0.55 C) and

90 percent relative humidity will keep eggs fresh for 8 months and that 3.5 ppm will not injure them; but some workers claim that 0.5 to 1.5 ppm have little effect on microorganisms. Low concentrations of ozone are reported to improve the flavor of stored eggs because of the deodorizing effect of the gas.

USE OF IRRADIATION

Experiments have indicated that pathogens (e.g., *Salmonella*) in liquid, frozen, and dried eggs can be inactivated by means of ionizing radiation.

SPOILAGE

Some of the defects of eggs are obvious from their general appearance, others are shown by "candling" with transmitted light, and some show up only in the broken egg.

DEFECTS IN THE FRESH EGG

Fresh eggs may exhibit cracks, leaks, loss of bloom or gloss, or stained or dirty spots on the exterior, or "meat spots" (blood clots), general bloodiness, or translucent spots in the yolk when candled. Of these, any breaks in the shell or dirt on the egg will favor spoilage on storage.

CHANGES DURING STORAGE

The changes that take place in eggs while they are being held or stored may be divided into those due to nonmicrobial causes and those resulting from the growth of microorganisms.

Changes not caused by microorganisms

Untreated eggs lose moisture during storage and hence lose weight. The amount of shrinkage is shown to the candler by the size of the air space or air cell at the blunt end of the egg, a large cell indicating much shrinkage. Of more importance is the change in the physical state of the contents of the egg, as shown by candling or by breaking out the egg. The white of the egg becomes thinner and more watery as the egg ages and the yolk membrane becomes weaker. The poorer the egg, the more movement there is of the yolk and the nearer it approaches the shell when the egg is twirled during candling. When an old egg is broken onto a flat dish the thinness of the white is more evident, and the weakness of the yolk membrane permits the yolk to flatten out or even break. By contrast, a broken fresh egg would show a thick white and a yolk that stands up strongly in the form of a hemisphere. During storage, the alkalinity of the white of the egg increases from a normal pH of about 7.6 to about 9.5. Any marked growth of the chick embryos in fertilized eggs also serves to condemn the eggs.

Changes caused by microorganisms

To cause spoilage of an undamaged shell egg the causal organisms must do the following: (1) Contaminate the shell. (2) Penetrate the pores of the shell to the shell membranes. Usually the shell must be moist for this to occur. (3) Grow through the shell membranes to reach the white (or to reach the yolk if it touches the membrane). (4) Grow in the egg white, despite the previously mentioned unfavorable conditions there, to reach the yolk where they can grow readily and complete spoilage of the egg. Bacteria unable to grow in the white can reach the yolk and flourish there only when the yolk touches the inner cell membrane.

The time required for bacteria to penetrate the shell membranes varies with the organisms and the temperature, but may take as long as several weeks at refrigerator temperatures. The special set of environmental conditions, the selective egg-white medium, and the low storage temperature of about 0 C (32 F) combine to limit the number of kinds of bacteria and molds that can cause spoilage chiefly to those to be mentioned.

In general, more spoilage of eggs is caused by bacteria than by molds. The types of bacterial spoilage, or "rots," of eggs go by different names. Scott and his coworkers list five groups of rots that are found in Australian eggs for export. Among the three chief ones are the **green rots,** caused chiefly by *Pseudomonas fluorescens,* a bacterium that grows at 0 C (32 F); the rot is so named because of the bright-green color of the white during early stages of development. This stage is noted with difficulty in candling, but shows up clearly when the egg is broken. Later the yolk may disintegrate and blend with the white so as to mask the green color. Odor is lacking or is fruity or "sweetish." The contents of eggs so rotted fluoresce strongly under ultraviolet light. A second important group of rots are the **colorless rots,** which may be caused by *Pseudomonas, Achromobacter,* certain coliform bacteria, or other types of bacteria. These rots are detected readily by candling, for the yolk usually is involved, except in very early stages, and disintegrates or at least shows a white incrustation. The odor varies from a scarcely detectable one to fruity to "highly offensive." The third important group of rots are the **black rots,** where the eggs are almost opaque to the candling lamp because the yolks become blackened and then break down to give the whole-egg contents a muddy-brown color. The odor is putrid, with hydrogen sulfide evident, and gas pressure may develop in the egg. Species of *Proteus* most commonly cause these rots, although some species of *Pseudomonas* and *Aeromonas* can cause black rots. *Proteus melanovogenes* causes an especially black coloration in the yolk and a dark color in the white. The development of black rot and of red rot usually means that the egg has at some time been held at temperatures higher than those ordinarily used for storage. **Pink rots** occur less often, and **red rots** are still more infrequent. Pink rots are caused by strains of *Pseudomonas* and may, at times, be a later stage of some of the green rots. They resemble the colorless rots, except for a pinkish precipitate on the yolk and a pink color in the white. Red rots, caused by species of *Serratia,* are mild in odor and not offensive.

The Haines grouping of bacterial rots of eggs is similar for the most part, although he divides the black rots into two types: Type I (Figure 18–2), caused by species of *Proteus,* such as *Proteus melanovogenes,* to give gas, a hard, black, solid yolk and liquefied white that may be murky or greenish-brown; and Type II, caused primarily by species of *Pseudomonas,* where the white is liquefied and is fluorescent-green or greenish-brown in color and the yolk is a soft, greenish-black mass. Type I has a fecal odor and Type II a "cabbage-water" smell. Haines did not describe the colorless rots.

Florian and Trussell have listed rots by ten different species of the genera *Pseudomonas, Alcaligenes, Proteus, Flavobacterium,* and *Paracolobactrum.* These rots have been characterized as fluorescent, green and yellow, custard, black, red, rusty red, colorless, and mixed. These authors also list secondary invaders in the genera *Achromobacter, Aerobacter, Alcaligenes, Escherichia, Flavobacterium,* and *Paracolobactrum.* These bacteria can grow in the egg but not initiate penetration.

The spoilage of eggs by fungi goes through stages of mold growth that give the defects their names. Very early mold growth is termed **pin-spot** molding because of the small, compact colonies of molds appearing on the shell and usually just inside it. The color of these pin spots varies with the kind of mold: *Penicillium* species cause yellow or blue or green spots inside the shell, *Cladosporium* species give dark-green or black spots, and species of *Sporotrichum* produce pink spots. In stor-

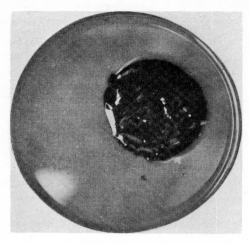

Figure 18–2. Black rot of egg, caused by *Proteus melanovogenes*. (*From R. B. Haines, Microbiology in the Preservation of the Hen's Egg, Gt. Brit., Dept. Sci. Ind. Res., Food Invest. Special Rept. 47, 1939. Reproduced by permission of Her Britannic Majesty's Stationery Office.*)

age atmospheres of high humidity a variety of molds may cause **superficial fungal spoilage,** first in the form of a fuzz or "whiskers" covering the shell and later more luxuriant growth. When the eggs are stored at near-freezing temperatures, as they usually are, the temperatures are high enough for slow mycelial growth of some molds, but too low for sporulation, while other molds may produce asexual spores. Molds causing moldiness of eggs include species of *Penicillium, Cladosporium, Sporotrichum, Mucor, Thamnidium, Botrytis, Alternaria,* and other genera. The final stage of spoilage by molds is **fungal rotting,** after the mycelium of the mold has grown through the pores or cracks in the egg. Jellying of the white may result, and colored rots may be produced, e.g., fungal red rot by *Sporotrichum* and a black color by *Cladosporium,* the cause of "black spot" of eggs, as well as of other foods. The hyphae of the mold may weaken the yolk membrane enough to cause its rupture, after which the growth of the mold is stimulated greatly by the food released from the yolk.

Off-flavors sometimes are developed in eggs with little other outward evidence of spoilage. Thus mustiness may be caused by any of a number of bacteria, such as *Achromobacter perolens, Pseudomonas graveolens,* and *P. mucidolens.* The growth of *Streptomyces* on straw or elsewhere near the egg may produce musty or earthy flavors that are absorbed by the egg. Molds growing in the shell also give musty odors and tastes. A hay odor is caused by *Aerobacter cloacae*, while fishy flavors are produced by certain strains of *Escherichia coli*. The "cabbage-water" flavor mentioned in connection with Type II of the black rots of Haines may appear before rotting is obvious. Off-flavors, such as the "cold-storage taste," may be absorbed from packing materials.

REFERENCES

ALFORD, L. R., N. E. HOLMES, W. J. SCOTT, and J. R. VICKERY. 1950. Studies in the preservation of shell eggs: I, The nature of wastage in Australian export eggs. Australian J. Appl. Sci. 1:208–214.

BOARD, R. G. 1964. The growth of gram-negative bacteria in the hen's egg. J. Appl. Bacteriol. 27:350–364.

BOARD, R. G., and J. C. AYRES. 1965. Influence of temperature on bacterial infection of the hen's egg. Appl. Microbiol. 13:358–364.

BOARD, R. G., J. C. AYRES, A. A. KRAFT, and R. H. FORSYTHE. 1964. The microbiological contamination of egg shells and egg packing materials. Poultry Sci. 43:584–595.

BROOKS, J. 1960. Mechanism of the multiplication of *Pseudomonas* in the hen's egg. J. Appl. Bacteriol. 23:499–509.

BROOKS, J., R. S. HANNAN, and BETTY C. HOBBS. 1958. The use of gamma radiation to destroy *Salmonella* in frozen whole egg. 2nd Int. Conf. Peaceful Uses At. Energy, Proc. 27:374–376.

FLORIAN, M. L. E., and P. C. TRUSSELL. 1957. Bacterial spoilage of shell eggs: IV, Identification of spoilage organisms. Food Technol. 11:56–60.

FROMM, D. 1960. Permeability of the egg shell as influenced by washing, ambient temperature changes and environmental temperature and humidity. Poultry Sci. 39:1490–1495.

GARIBALDI, J. A. 1960. Factors in egg white which control growth of bacteria. Food Res. 25:337–344.

GARIBALDI, J. A., and H. G. BAYNE. 1962. Iron and the bacterial spoilage of shell eggs. J. Food Sci. 27:57–59.

HAINES, R. B. 1939. Microbiology in the preservation of the hen's egg. Gt. Brit., Dep. Sci. Ind. Res., Food Invest. Spec. Rep. 47.

HARTUNG, T. E., and W. J. STADELMAN. 1962. Influence of metallic cations on the penetration of the egg shell membranes of *Pseudomonas fluorescens*. Poultry Sci. 41:1590–1596.

HARTUNG, T. E., and W. J. STADELMAN. 1963. Penetration of egg shell membranes by *Pseudomonas fluorescens* as influenced by shell porosity, age of egg and degree of bacterial challenge. Poultry Sci. 42:147–150.

KRAFT, A. A., E. H. MCNALLY, and A. W. BRANT. 1958. Shell quality and bacterial infection of shell eggs. Poultry Sci. 37:638–644.

LIFSHITZ, A., R. C. BAKER, and H. B. NAYLOR. 1964. The relative importance of chicken egg exterior structures in resisting bacterial penetration. J. Food Sci. 29:94–99.

MILLER, W. A. 1957. A comparison of various wash-water additives in preventing microbial deterioration in washed eggs that formerly were dirty. Poultry Sci. 36:579–584.

Murdock, C. R., E. L. Crossley, J. Robb, Muriel E. Smith, and Betty C. Hobbs. 1960. The pasteurization of liquid whole egg. Month. Bull. Med. Res. Counc. 19:134–152.

Scott, W. J., et al. 1950, 1951. Studies in the preservation of shell eggs. Australian J. Appl. Sci. 1:208–214, 215–223, 313–329, 514–538; 2:205–222.

Thatcher, F. S., and J. Montford. 1962. Egg-products as a source of salmonellae in processed foods. Canad. J. Public Health 53:61–69.

CHAPTER NINETEEN

CONTAMINATION, PRESERVATION, AND SPOILAGE OF POULTRY

The discussion of poultry is concerned mostly with chicken meat, but the principles should apply to meat of other fowl, such as turkey, goose, duck, squab, etc.

CONTAMINATION

Sources of contamination discussed under the heading of meats apply as well to poultry. Undrawn poultry is not subject to contamination from the alimentary tract of the fowl but develops off-flavors as the result of microbial growth in that tract. To the natural flora on the skin of the fowl is added heavier contamination from feathers and feet during plucking and washing, plus intestinal bacteria if the fowl is eviscerated. The lining of the body cavity is similarly contaminated. Table 19–1 shows microbial counts obtained in six poultry processing plants at different stages during processing. Microorganisms commonly found included bacteria of the genera *Pseudomonas, Achromobacter, Flavobacterium,* and *Micrococcus,* together with coliform bacteria, miscellaneous bacteria, and yeasts (*Trichosporon, Torulopsis, Candida,* and *Rhodotorula*).

The type of feed given the fowl before slaughter may have an influence on the development of visceral taints; there is less taint, for example, with a buttermilk ration. Although methods of management of the birds prior to dressing influence the kinds and numbers of bacteria on the skin at the moment of dressing, these ante-mortem conditions are not as important as the post-mortem conditions in delaying or preventing spoilage.

PRESERVATION

Most of the principles of preservation discussed for meats in Chapter 16 apply likewise to poultry, although the plucking and dressing of the fowls raise different problems. As in the slaughter of animals for meat, the method of killing and bleeding the fowl has an important effect on the quality of the product. Modern methods involve the severing of the jugular vein of the bird, suspended by its feet, and the drainage of blood. The method of picking or plucking has some influence on the keeping quality of the bird. Dry-plucked birds are more resistant to decomposition than semiscalded or scalded ones because the skin is less likely to be broken, but more pinfeathers are left. Most picking is by means of the semiscald method, in which the fowl is immersed in water at 125 to 135 F (51.7 to 57.2 C) for about 30 sec, rather than the older scalding method at 150 to 190 F (65.5 to 87.8 C) or over. Experiments have shown that the water in the semiscald method is not an important source of contaminating microorganisms if reasonable precautions are taken to change the water; in fact the process is one of mild pasteurization (See Table 19–1). Nor does the dressing (plucking) procedure normally add much contamination. Counts of numbers of bacteria on the skins of dressed fowls over a long period have shown less than 250,000 organisms per gram of skin and few in the adjoining flesh. Evisceration of the fowl adds contaminating bacteria from the alimentary tract. Microorganisms found on the skin, feet, and cut surfaces of poultry include *Pseudomonas, Achromobacter, Flavobacterium,*

TABLE 19–1. Numbers of microorganisms on poultry and in processing waters†

Point of sampling	Usual range of nos. ($\times 10^3$) *
Skin of live bird	0.6–8.1
Scald water (137–140 F)	5.9–17
Fresh chill water	50–210
Aerated chill water	34–240
Skin after rough pick	8.1–45
Skin after neck pick	3.3–32
Skin after pinning	10–84
Skin after singe	13–210
Skin after evisceration	11–93
Cavity after evisceration	1.4–12
In chill tank	50–600
After aeration	3.4–240
Final product	4–330

* Counts on waters are per milliliter, on surfaces per square centimeter.
† H. W. Walker and J. C. Ayres. 1956. Appl. Microbiol. 4:345–349.

Micrococcus, coliforms, *Alcaligenes, Proteus, Bacillus,* and others, including some yeasts and molds.

ASEPSIS

Comparatively little can be done to keep microorganisms away from the dressed poultry. The sanitation of the housing of the birds before killing has some influence on the numbers of microorganisms on the skin at dressing, but even under the best conditions enough spoilage organisms contaminate the skin to permit microbial deterioration if conditions of handling and storage are not good. Contamination of the lining of the body cavity of the bird can be prevented if the fowl is not eviscerated until sold in the retail market, but visceral taints may develop unless the birds are adequately refrigerated. The shackles holding feet and head of the fowl (Figure 19–1) may be a source of heavy contamination. Contamination can be reduced if equipment is adequately cleaned and sanitized at intervals.

REMOVAL OF MICROORGANISMS

Some microorganisms may be removed from the skin during the scalding process that precedes picking and some will be killed, but

Figure 19–1. Scrubbing poultry shackle hooks in a poultry plant. (*Courtesy of Klenzade Products, Beloit, Wis.*)

not enough to be very significant. During drawing or evisceration of the bird, microorganisms from the intestinal tract will contaminate the lining of the body cavity, but thorough washing with good water will remove many of the organisms. Chlorinated water is recommended.

USE OF HEAT

Dressed chickens and other fowls may be canned, whole or dissected, in their own juices or in jelly. Heat processes are analogous to those for canned meats (Chapter 16). The chicken or other fowl may be salted in a weak brine before being packed into the glass jars or cans.

USE OF LOW TEMPERATURES

Most poultry is preserved by either chilling or freezing.

Chilling

Poultry for either chilling or freezing storage should be cooled promptly and rapidly after dressing and drawing (if that is done) down to a temperature of about 35 F (1.7 C). Chilling can be done by cold air, which leaves the bird relatively dry, or by contact with ice or ice water or a spray of refrigerant, which leaves the skin wet. The moist methods not only may add organisms but also may favor their growth.

Chilling storage of poultry is for only a short period, usually less than a month; birds to be stored longer should be frozen. Packing the dressed birds in ice has been used for short periods of holding and where mechanical refrigeration is not available. Most large-scale chilling, however, is now by means of mechanical refrigeration. The closer the temperature of storage is to freezing, the longer the birds can be stored without undesirable amounts of change taking place. In tests on cut-up chicken, Ayres found that, as compared with room temperature, storage life was extended 2 days at 50 F (10 C), 6 days at 40 F (4.4 C), and 14 days at 32 F (0 C).

Freezing

Poultry can be kept in good condition for months if freezing is prompt and rapid and the storage temperature is low enough. Fairly rapid freezing is desirable, for this produces a light-appearing bird because fine ice crystals are formed within the fibers. Slow freezing, on the other hand, causes large crystals to accumulate outside the fibers and the flesh to appear darker. A bird that is frozen rapidly while fresh will have smaller crystals than one frozen after a delay. Regardless of the rate of freezing, intact birds do not drip upon thawing. Most chickens frozen commercially are packed ten to a box that is lined with moisture- and airproof paper, and for the most part are dressed but not drawn.

Poultry should be frozen fast enough to retain most of the natural bloom or external appearance of a freshly dressed fowl, and the storage temperature should be below 0 F (−17.8 C) and the relative humidity above 95 percent to reduce drying of the surface. Most poultry is sharp-frozen at about −20 F (−29 C) or less in circulating air or on a moving belt in a freezing tunnel. For "quick freezing," a smaller package is necessary, usually a whole bird, a dissected one, or a boned fowl, packed into a special watertight and almost airtight wrapper. Actually, a large chicken would not freeze fast enough to measure up to the definition of quick freezing in Chapter 7.

Although part of the bacteria are killed by the freezing process and numbers decrease slowly during storage, enough remain to cause spoilage when the bird is thawed. The growth of bacteria can take place during picking, dressing, drawing, chilling, and also during the freezing process, until the temperature of the bird drops below 32 F (0 C). Deterioration that has developed due to bacterial growth, to diffusion of taints from the viscera, and to activity of enzymes of the bird before freezing will carry over into the frozen product, and some of the enzymatic action will continue, unless the storage temperature is very low, although bacterial growth has been stopped. If thawing is incorrectly done, e.g., for too long a time or at too high a temperature, spoilage can begin.

Unless adequate sanitary precautions are taken, a marked increase in bacteria will take place during the removal of the bones of cooked fowl for subsequent canning or quick freezing. Low-temperature bacteria of the genera *Achromobacter, Proteus,* and *Alcaligenes,* as well as coliform bacteria, have been found in considerable numbers. Canning will destroy these, but quick freezing permits many to survive.

USE OF PRESERVATIVES

Not over 7 ppm of chlortetracycline or oxytetracycline is permitted in the flesh of uncooked, dressed birds, as has been mentioned in Chapter 9. To obtain this level of antibiotic, about 10 ppm are added to the chilling bath through which the birds are passed following evisceration.

Results indicate that these antibiotics lengthen the keeping time by at least several days, but that bacteria resistant to the antibiotic, e.g., pigmented pseudomonads and certain *Achromobacter* species, grow on the birds and build up in the plant. They 'may be spread by means of reused ice and water from chill tanks, grow on the surface of the tanks, or be redistributed during cutting and packaging operations. The feeding of these antibiotics to birds may lead to increased percentages of resistant microorganisms in the fecal matter and hence on the birds, although low levels of antibiotic may be deposited in the flesh. Low levels of antibiotic in the meat of treated birds are mostly destroyed by cooking. Soaking cut-up poultry in solutions of organic acids (acetic, adipic, succinic, etc.) at pH 2.5 has been reported to lengthen shelf life.

Turkey sometimes is cured in a solution of salt, sugar, and sodium nitrate for several weeks at about 38 F (3.3 C), washed, dried, and then smoked. Usually a light smoking process is used, more for flavor than for preservation. Recommended temperatures during smoking range from 110 to 140 F (43.3 to 60 C) and the time from a few hours to several days.

USE OF IRRADIATION

Irradiation of poultry with cathode or gamma rays can be a successful preservative method, for the rays apparently produce less objectionable change in appearance and flavor than in some other foods, but the method has not been approved by the Food and Drug Administration to date.

SPOILAGE

While the enzymes of the fowl contribute to the deterioration of the dressed bird, bacteria are the chief cause of spoilage, with the intestines a primary source of these organisms. The limited amount of work done on the bacterial spoilage of poultry has indicated that most of the bacterial growth takes place on the surfaces, that is, the skin, the lining of the body cavity, and any cut surfaces, and the decomposition products diffuse slowly into the meat. A surface odor has been noted when the bacterial count on the skin was about 2,500,000 per square centimeter. This took about 4 weeks at 32 F (0 C) and 5 weeks at 30 F (−1.1 C) in one set of experiments. Eviscerated poultry held at 10 C (50 F) or below is spoiled mostly by *Pseudomonas*, although *Achromobacter* grows to some extent, and to a lesser degree yeasts, such as *Torulopsis* and *Rhodotorula*, may be concerned. At above 10 C, micrococci usually predominate, and there also is growth of *Achromobacter* and *Flavobacterium*. Eventually the surface of the meat usually becomes slimy. Small amounts (1 to 5 ppm) of iron in the wash water used on poultry favor bacterial growth on the surface and production of the fluorescent pigment, pyoverdine, by pseudomonads; more iron will reduce pigmentation. About 100 ppm of magnesium is optimal for pigment production by *P. fluorescens*.

Iced, cut-up poultry often develops a slime that is accompanied by an odor described as "tainted," "acid," "sour," or "dish-raggy." This defect is caused chiefly by species of *Pseudomonas*, although *Alcaligenes* also may be concerned. Similar bacteria grow, whether the temperature is as low as 32 F (0 C) or as high as 50 F (10 C), and enormous numbers must be present, about 10^8 per square centimeter, before the odor becomes evident. *Pseudomonas-Achromobacter* organisms spoil poultry in polyethylene bags.

Birds treated with tetracyclines may be spoiled by resistant strains of bacteria and yeasts, e.g., pigmented pseudomonads and certain achromobacteria, and yeasts of the genera *Torulopsis, Rhodotorula, Trichosporon,* and *Candida*. Molds also may develop.

It should be kept in mind that chemical changes in poultry meat other than those caused by microorganisms occur during refrigerated storage and will, in time, reduce the quality.

REFERENCES

AYRES, J. C. 1959. Effect of sanitation, packaging and antibiotics on the microbial spoilage of commercially processed poultry. Iowa State J. Sci. 54:27–46.

AYRES, J. C., W. S. OGILVY, and G. F. STEWART. 1950. Post mortem changes in stored meats: I, Microorganisms associated with development of slime on eviscerated cut-up poultry. Food Technol. 4:199–205.

COLEBY, B., M. INGRAM, and H. J. SHEPHERD. 1960. Treatment of meats with ionising radiation: III, Radiation pasteurization of whole eviscerated chicken carcasses. J. Sci. Food Agr. 11:61–71.

EKLUND, M. W., J. V. SPENCER, E. A. SAUTER, and M. H. GEORGE. 1961. The effect of different methods of chlortetracycline application on the shelf-life of chicken fryers. Poultry Sci. 40:924–928.

ELLIOTT, R. P., R. P. STRAKA, and J. A. GARIBALDI. 1964. Polyphosphate inhibition of growth of pseudomonads from poultry meat. Appl. Microbiol. 12:517–522.

ESSARY, E. O., W. E. C. MOORE, and C. Y. KRAMER. 1958. Influence of scald temperatures, chill time, and holding temperatures on the bacterial flora and shelf-life of freshly chilled, tray-pack poultry. Food Technol. 12:684–687.

ESSELEN, W. B., and A. S. LEVINE. 1954. Bacteriological investigations on frozen stuffed poultry. J. Milk Food Technol. 17:245–250, 255.

GUNDERSON, M. F., H. W. McFADDEN, and T. S. KYLE. 1954. The bacteriology of commercial poultry processing. Burgess Publishing Company, Minneapolis.

HAMDY, M. K., N. D. BARTON, and W. E. BROWN. 1964. Source and portal of entry of bacteria found in bruised poultry tissue. Appl. Microbiol. 12:464–469.

KOTULA, A. W., and J. A. KINNER. 1964. Airborne microorganisms in broiler processing plants. Appl. Microbiol. 12:179–184.

KRAFT, A. A., and J. C. AYRES. 1965. Development of microorganisms and fluorescence of poultry chilled in water containing iron or magnesium. J. Food Sci. 30:154–159.

MAY, K. N. 1962. Bacterial contamination during cutting and packaging chicken in processing plants and retail stores. Food Technol. 16:89–91.

NAGEL, C. W., K. L. SIMPSON, H. NG, R. H. VAUGHN, and G. F. STEWART. 1960. Microorganisms associated with spoilage of refrigerated poultry. Food Technol. 14:21–23.

WALKER, H. W., and J. C. AYRES. 1958. Antibiotic residuals and microbial resistance in poultry treated with tetracyclines. Food Res. 23:525–531.

WALKER, H. W., and J. C. AYRES. 1959. Characteristics of yeasts isolated from processed poultry and the influence of tetracyclines on their growth. Appl. Microbiol. 7:251–255.

WETZLER, T. F., P. MUSICK, H. JOHNSON, and W. A. MacKENZIE. 1962. The cleaning and sanitizing of poultry processing plants. Amer. J. Public Health 52:460–471.

WILKERSON, W. B., J. C. AYRES, and A. A. KRAFT. 1961. Occurrence of enterococci and coliform organisms on fresh and stored poultry. Food Technol. 15:286–292.

CHAPTER TWENTY

CONTAMINATION, PRESERVATION, AND SPOILAGE OF MILK AND MILK PRODUCTS

Dairy products include market milk and cream, butter, frozen desserts, cheese, fermented milks, and condensed and dry milk products.

CONTAMINATION

As has been stated in Chapter 4, milk contains a few bacteria when it leaves the udder of a healthy cow, and these bacteria generally do not grow well in milk under the usual conditions of its handling. Next, the milk becomes subject to contamination from the exterior of the animal, especially the exterior of the udder and adjacent parts. Bacteria of manure, soil, and water may enter from this source, although fewer organisms enter in this way when a milking machine is employed than when milking is by hand. Such contamination is reduced by clipping the cow, especially the flanks and udder, grooming the cow, and washing the udder with water or a germicidal solution. Contamination of the cow with soil, water, and manure is reduced by paving and draining barnyards, keeping cows from stagnant pools, and cleaning manure from the barns or milking parlors. Use of a small-top milking pail reduces contamination during hand milking. Contamination from the air of barn or milking parlor may contribute organisms of feed, soil, or manure to milk, but the numbers added in this manner are negligible unless an abnormal amount of dust has been raised.

The next source of contamination is the milk pail or milking machine, as the case may be, strainers, then the milk cans, or the pipelines and milk coolers. Unless these dairy utensils are adequately cleansed and sanitized they can be a most important source of contamination in that they may add not only considerable numbers of bacteria to the milk but also some of the most undesirable kinds. If utensils are poorly cleansed, sanitized, and dried, bacteria may develop in large numbers in the dilute, milklike residue and enter the next milk to contact the utensils. These bacteria are the kinds that grow best in milk and hence endanger its keeping quality. Undesirable bacteria from utensils include lactic streptococci, coliform bacteria, psychrophilic Gram-negative rods, and "thermodurics," those that survive pasteurization (e.g., micrococci, bacilli, and brevibacteria). Application of quaternary ammonium compounds as sanitizing agents tends to increase the percentage of Gram-negative rods on the utensils (psychrophiles, coliforms), whereas hypochlorites favor Gram-positive bacteria (micrococci, bacilli).

Other possible sources of contamination are the hand milker or other dairy workers, who normally would contribute very few bacteria but might be a source of pathogens; and flies, which may add spoilage organisms or pathogens.

The numbers of bacteria per milliliter of milk added from the various sources will depend on the care taken to avoid contamination. The udder may add from several hundred to several thousand per milliliter. Quiet air adds very few organisms, but very dusty air may add several thousand per milliliter. The exterior of the cow would contribute comparatively few organisms if precautions were taken and a milking machine used, but under very bad conditions thousands per milliliter could enter the milk. Utensils can be so well cleansed and sanitized as to add only a few bacteria per milliliter of milk, but under very poor conditions may increase the bacterial count of milk by millions per milliliter. A number of flies would have to drop into a tank of milk to increase the bacterial count appreciably although the average number of bacteria per fly is about a million.

Other utensils may add contaminants after the milk leaves the farm: the tanker truck and various utensils and equipment at the market-milk plant, cheese factory, condensery, or other processing plant. In the plants, vats, tanks, pumps, pipelines, valves, separators, clarifiers, homogenizers, coolers, strainers, stirrers, bottle fillers, and bottles are possible sources of bacteria, and the amount of contamination from them will depend upon methods of cleansing and sanitizing employed. Here, too, employees are possible sources of pathogens.

Any of the dairy products made from milk may be subject to contamination additional to that already in the milk. Butter may receive organisms from the churn, from the water used in its washing, from old cream in which much growth has taken place, or from packaging materials. Dry milk, evaporated milk, and sweetened condensed milk may be contaminated by the special equipment used in their preparation. Cheese is contaminated from air, brine tanks, shelves, and packaging materials, etc. Ice cream also may have organisms added by its ingredients.

It should be kept in mind that the number of microorganisms in milk or other dairy products may be increased by the growth of the organisms, unless methods of production and handling prevent it.

PRESERVATION

Milk and milk products, which serve to illustrate most of the principles of preservation and spoilage of foods, have had more research done on them than most foods, so much that numerous text and reference books have been published on the subject. These books, some of which are listed at the end of this chapter, should be consulted for more detail than can be presented here. Milk is such a delicately flavored, easily changed food that rigorous preservative methods cannot be used on it without changing it in an undesirable manner or, at best, making a different food product. In fact, most of the products made from milk or cream at first were for the purpose of improving the keeping quality. Thus nomadic tribes found that milk that had undergone a lactic acid fermentation could be kept for long periods, although now we produce fermented milks for their characteristic body and flavor. Treatment of the curd from milk in various ways was found to lengthen the keeping time; now we produce different kinds of cheese for their individual characteristics.

ASEPSIS

The prevention, as far as is practicable, of the contamination of milk is important in its preservation, in that its keeping qualities are improved, usually, with smaller numbers of microorganisms present, especially of those that can grow readily in milk. Low numbers are indicative of sanitary precautions and careful handling during production and hence of fewer pathogens, as well as fewer spoilage organisms. Therefore the bacterial content of milk is used to measure its sanitary quality, and most grading of milk is on the basis of some method for estimating numbers.

The coat of the animal and the surfaces of utensils have been emphasized as possible sources of both high numbers and undesirable kinds of microorganisms. Especially undesirable in market milk are bacteria that grow well in milk, such as the lactics and the coliform bacteria; the psychrophiles, that can grow at the chilling temperatures at which milk usually is stored; the thermodurics, that survive the pasteurization treatment of milk; and, of course, human pathogens. When the milk is to be used for the production of a product by microbial fermentation, as in the manufacture of fermented milks or cheese, the kinds of microorganisms that are able to compete with the starter bacteria and produce defects in the product are important. Thus the coliform bacteria, anaerobes, and yeasts can cause gassiness and off-flavors unless they are kept out or are eliminated if present, and other organisms may inhibit the desirable ones or cause defects in body, texture, and flavor. Heat-resistant microorganisms may survive the heat-treatments given some cheeses during manufacture and canned products like evaporated or sweetened condensed milk and bring about spoilage, or they may be the cause of the lowering of the grade of dry milk, which has bacteriological standards.

Packaging serves to keep microorganisms from bottled milk, fermented milks, packaged butter, canned milk, dry milk, and packaged cheese, as do coatings of plastics, wax, or other protective substances on finished cheese, or on "rindless" cheeses during ripening.

REMOVAL OF MICROORGANISMS

After microorganisms have entered milk, it is difficult to remove them effectively. It is true that centrifugation, as in clarifying or separating, removes some of the organisms. High-speed centrifugation (at 10,000 g) removes about 99 percent of the spores and more than half of the vegetative cells of bacteria. Molds are removed from the surfaces of some kinds of cheese during the curing process by periodic washing, and their growth is held down.

USE OF HEAT

Pasteurization

Because milk and cream are so readily changed by heat, the milk heat-treatment called pasteurization (Chapter 6) usually is used for their preservation. These products are then cooled immediately to 45 F (7.2 C) or lower. The phosphatase test is employed commonly to test for efficiency of pasteurization and is based on the fact that this enzyme is destroyed if the minimal treatments mentioned above have been applied.

The objectives of the pasteurization of market milk are (1) to kill all pathogens that may have entered the milk, and (2) to improve the keeping quality of the milk. All this is to be accomplished without harm to the flavor, appearance, nutritive properties, and creaming of the milk. When milk is pasteurized for the manufacture of cheese, or cream is pasteurized for making butter, there is another objective: (3) to destroy microorganisms that would interfere with the activities of desirable organisms such as starter bacteria or that would cause inferiority or spoilage of the product. The cheese maker also wishes a heat-treatment that will not harm the curdling properties of the milk. The heat-treatment of cream also destroys lipases that might cause deterioration of butter during storage.

The percentage of reduction of numbers of microorganisms in milk during pasteurization will depend upon the proportion of thermoduric bacteria present, and may range from 90 to 99 percent.

Cream for buttermaking is given a greater heat-treatment during pasteurization than market cream. Heating is at 160 F (71.1 C) or above for 30 min by the holding method, or 190 to 200 F (87.8 to 93.3 C) for a few seconds by the high-temperature–short-time method. Cream is more protective to organisms than is milk, and cream for buttermaking is likely to contain a higher population of microorganisms than most lots of milk. Rapid heating of cream may be by means of injection of steam or by a combination of steam injection and evacuation in a process known as **vacreation.**

The forewarming at 160 to 212 F (71 to 100 C) for 10 to 30 min and evaporation processes at 120 to 135 F (48.9 to 57.2 C) in the manufacture of sweetened condensed milk are, in effect, pasteurization processes that kill all pathogens and should destroy all organisms that could spoil the final canned product.

What amounts to a high-temperature pasteurization is applied to bulk condensed milk. Forewarming the milk at 150 to 170 F (65.6 to 76.7 C) before evaporation kills many of the organisms present, and superheating of the condensed product, which is concentrated more than evaporated milk, at higher temperatures (180 to 200 F, or 82.2 to 93.3 C) destroys still more of the organisms. Such a product must be cooled rapidly, stored at a chilling temperature, and used fairly soon. The preheating of milk before drying, as mentioned in a later section, is, in effect, pasteurization.

The cooking at 150 F (65.6 C) or higher, in the melting and blending of cheese during the manufacture of process cheese is in effect a pasteurization process that kills most of the microorganisms present in the original cheeses. Phosphates, citrates, and tartrates are added as emulsifiers.

Pasteurization should kill all yeasts and molds and most vegetative cells of bacteria in the milk. The surviving bacteria, termed thermodurics, belong to a number of different groups of bacteria, of which only a few more important ones will be mentioned. Most important of the non-spore-forming bacteria are (1) the high-temperature lactics, e.g., the enterococci, *Streptococcus thermophilus*, high-temperature lactobacilli, such as *Lactobacillus bulgaricus*, and *L. lactis*, and species of *Microbacterium*, and (2) certain species of *Micrococcus*. Some species of

Streptococcus and *Lactobacillus* are thermophilic as well as thermoduric. The spore-forming thermodurics fall into two main groups: (1) species of *Bacillus*, that is, aerobic to facultative spore-forming bacilli, of which *B. cereus* (proteolytic) usually is the most numerous, but *B. licheniformis, B. subtilis* (proteolytic), *B. coagulans* (thermophilic), *B. polymyxa* (gas-forming), *B. calidolactis* (thermophilic), and other species sometimes are of importance; (2) species of *Clostridium,* anaerobic, spore-forming rods, some of which are saccharolytic, e.g., *C. butyricum,* and others proteolytic and saccharolytic, e.g., *C. sporogenes.* Most of those that grow in milk also form gas. Miscellaneous other bacteria may survive pasteurization but do not grow well in milk.

Boiling

Boiling of milk or heating in flowing steam destroys all microorganisms except the spores of bacteria, yet changes the appearance, palatability, digestibility, and nutritive properties of the milk. Formerly boiling was much used in the home, especially with milk for babies, but now home methods of pasteurization are more widely employed.

Steam under pressure

Evaporated milk is canned and then heat-processed by steam under pressure, often with accompanying rolling or agitation. The forewarming of the milk at about 200 to 212 F (93.3 to 100 C) or higher before evaporation kills all but the more resistant bacterial spores. The sealed cans of evaporated milk are processed usually at about 240 to 245 F (115.6 to 118.3 C) for 14 to 18 min to make the contents sterile or commercially sterile. Some of the heat-cool-fill methods, such as the Martin process, mentioned in Chapter 6, have been applied to the processing of canned whole milk, cream, and whole-milk concentrate.

USE OF LOW TEMPERATURES

With the exception of canned milk and dry milk, most dairy products require the use of low temperatures as one factor in their preservation, and often it is a most important factor.

Cooling

For the production of milk of good quality it is essential that it be cooled promptly after it is drawn from the cow. The Grade "A" Pasteurized Milk Ordinance of the United States Public Health Service stipulates that Grade A raw milk for pasteurization shall be cooled to 50 F (10 C) or less within two hours after drawn and kept that cold until processed. Newly pasteurized milk is to be cooled to 45 F (7.2 C) or less and maintained thereat. It is preferable, of course, to cool to temperatures well below 50 F, an objective that is readily attained when mechanical refrigeration or ice is available. Cooling tanks for the bulk handling of milk on the farm cool the milk rapidly to 38 to 40 F (3.3 to 4.4 C) or lower and hold the temperature there except for short periods when fresh milk is entering, and then the temperature usually does not exceed 45 F (7.2 C).

Chilling

The recommended temperatures to which freshly drawn or newly pasteurized milk should be cooled are really chilling temperatures. These temperatures are to be maintained in milk during storage on the farm, in the truck or tank during transportation to plant or receiving station, and during storage there. Chilling temperatures are recommended for the bottled milk or related product during storage in plant or in retail market and during delivery, and in the home or restaurant until consumption.

Some cheese makers maintain that milk to be used for cheese making should not be cooled as much as milk for other purposes and prefer cooling to a temperature of 55 to 60 F (12.8 to 15.6 C) and some permit morning's milk to be brought in without cooling. In general, however, milk for cheese should be cooled as adequately as milk for other purposes.

Cream for butter should be cooled to 60 F (15.6 C) or lower after separation and kept cool until picked up or delivered. The cooling is especially important on farms with low production of cream, where it may accumulate for 3 or 4 days or even longer, and unless cooled may undergo deterioration by microorganisms. Butter is stored at chilling temperatures for short periods, as in retail stores or in the home, but at freezing temperatures during warehouse storage.

Fermented milks and unripened cheeses are chilled after their manufacture and kept chilled until they reach the consumer. Most kinds of ripened cheese also are chilled and stored at chilling temperatures after their ripening is completed.

Freezing

Ice cream and other frozen dairy desserts are frozen as part of the manufacturing process and are stored at low temperatures in the frozen state, where microbial multiplication is impossible. The microbial content of the ingredients—milk, cream, sugar, eggs, stabilizers, and flavoring and coloring materials—along with contamination picked up during processing will determine the numbers and kinds of microorganisms in the mix and also will determine the microbial content after pasteurization of the mix and freezing. Pasteurization, of course, reduces numbers and kinds of microorganisms, but freezing kills few if any of the organisms, and storage in the frozen state permits survival of most of the microorganisms for long periods.

Butter in storage is held at 0 F (−17.8 C) or lower, where no microbial growth can take place. Frozen cream is stored in considerable amounts at a similar temperature. Milk, concentrated to one-third its volume, can be frozen at 0 F (−17.8 C) and stored at −10 F (−23.3 C) or lower, and can be held for several weeks without deterioration. Frozen milk can be concentrated by freeze-drying methods. Pasteurized whole milk has been frozen at about −20 F (−28.9 C) and shipped and stored in the frozen state, especially during wartime.

DRYING

Various milk products are made by the removal of different percentages of water from whole or skim milk. Only in the manufacture of dry products is enough moisture removed to prevent the growth of microorganisms in the product, although the reduction in moisture and consequent increase in the concentration of dissolved substances in liquid condensed milk products is such as to inhibit the growth of some kinds of bacteria.

Condensed products

Evaporated milk is made by removing about 60 percent of the water from whole milk, so that about 11.5 percent lactose would be in solution, plus double the amount of soluble inorganic salts in whole milk. This high concentration of sugar is inhibitory to the growth of some kinds of bacteria and might retard or prevent the growth of some survivors of the heat-treatments described above. Bulk condensed milk is more condensed than evaporated milk and is a still poorer culture medium for organisms not tolerant of high sugar concentrations. Condensed whey, called whey semisolids, is another concentrated dairy product, as is condensed buttermilk, called semisolid buttermilk, which has its concentration of acid, as well as of other solutes, increased by the condensation process.

In the preparation of the product from whole milk, enough sugar, mostly sucrose, but occasionally glucose, is added to milk before evaporation to form sweetened condensed milk with a total average sugar content (lactose plus added sugar) of about 54 percent and in the moisture part of over 64 percent. The product from skim milk would contain about 58 percent total sugar and about 66 percent in the moisture part of the condensed product. These very high concentrations of sugar plus the increased percentage of soluble inorganic salts tie up the moisture, making it unavailable to any but osmophilic microorganisms. Therefore, drying, both by removal of water and by tying it up, is a main preservative factor. Add to this the evacuation of the can and the aseptic effect of the sealed can and a product of good keeping quality results.

Dry products

Among the dairy products prepared in the dry form are milk, skim milk, cream, whey, buttermilk, ice-cream mix, and malted milk. Since, for the most part, similar principles apply to all of these products, dry skim milk, now called nonfat dry milk solids, will be discussed as a typical example. Most dry milk is prepared by either the roller process, with or without vacuum, or the spray process, mentioned in Chapter 8. Preliminary to the final drying process, the milk is concentrated three to five times for the roller process, and two or three times for the spray process. Usually the milk also is preheated before the drying process, to 150 to 185 F (65.6 to 85 C) for the roller process and to 145 to 200 F (62.8 to 93.3 C) for 30 min or less for the spray process. This preheating process pasteurizes the milk and therefore kills the less heat-resistant microorganisms.

Roller drying without vacuum involves the use of a high temperature, up to 270 F (132.2 C) or higher, for a brief period, while the vacuum-roller process utilizes lower temperatures, e.g., 100 to 170 F (37.8 to 76.7 C). In the spray process the fine spray of milk is exposed briefly to hot, dry air that may be at 250 F (121.1 C) to 400 F (204.4 C).

The microbial content of the heat-dried dairy product will depend upon the content of the liquid product to be dried, the temperature and time of preheating, the evaporation process, if used, contamination and growth in storage tanks and pipes, and the method of drying. Preheating kills organisms as pasteurization would and hence destroys all but the thermodurics. Evaporation, especially a continuously operating process, may result in increases in thermodurics and bacteria that are thermophilic. In storage or feed tanks, increases in numbers of microorganisms may take place, since the temperature of evaporation is not high and the milk soon cools. The high temperature of the roller process without vacuum destroys almost all organisms except bacterial spores. Spray drying and vacuum-roller drying destroy all organisms except some of the thermodurics. Since the American Dry Milk Institute, Inc., and the United States Army have set the bacterial standards for dry milk given in the Appendix and the Agricultural Marketing Service of the United States Department of Agriculture has set standards for standard plate counts and direct microscopic clump counts per gram of powder, the numbers of bacteria in the final product are important. Thermoduric bacteria are most numerous in the dry product: heat-resistant streptococci, micrococci, aerobic and anaerobic spore-formers, and species of *Microbacterium.*

Milk and other liquid dairy products may be dried by lyophilization, a process in which the quick-frozen product is dried under a high vacuum. To date this process has not been used to any extent commercially.

Some types of cheese are made so dry that microbial spoilage is no problem.

The moisture content of dry dairy products should be low enough to prevent the growth of microorganisms, but usually is not low enough to do more than slow down chemical changes like the oxidation of fats. The lower the moisture, the slower will be the chemical deterioration.

USE OF PRESERVATIVES

Added preservatives

The addition of so-called chemical preservatives to dairy products is not permitted except for holding samples for analysis. As has been stated, added sugar acts as a preservative of sweetened condensed milk, but its effect is mainly to make moisture unavailable to microorganisms and to draw moisture from them. Sodium chloride has been suggested as a preservative of cream for buttermaking but has not been used to any extent. Common salt is added in the manufacture of various kinds of cheese, but usually is more for flavor or for controlling the growth of microorganisms during manufacture and curing than for preservation of the finshed product. The sodium chloride in salted butter is in a concentration in the liquid phase sufficient to prevent the growth of most bacteria and, in fact, to cause a decrease in numbers of those that are not salt-tolerant. In butter with 16.34 percent moisture and 2.35 percent salt, the water phase would be about a 12.6 percent salt brine. The carbonation of milk, butter, and ice cream has been tried as an aid in preservation, but without much success. The smoking of cheese is primarily for the addition of flavor, although the drying, especially of the rind, and the chemical preservatives from the smoke may improve the keeping quality. Mold spoilage of cheese is delayed or prevented by addition of sorbic acid, sorbates, propionic acid, or propionates, or by incorporation in the wrapper.

The addition of hydrogen peroxide combined with a mild heat-treatment is now permitted for pasteurization of milk for certain kinds of cheese (e.g., Swiss and Cheddar). The excess of peroxide is destroyed by added catalase.

Developed preservatives

Fermented milks and cheese are dairy products preserved partly by acid produced by bacterial activity. Fermented milks, often called buttermilks, include cultured buttermilk, yoghurt, Bulgarian buttermilk, acidophilus milk, kefir, kumiss, skyr, taette, and numerous others very similar to or identical with the ones listed. Cultured sour cream is a similar product. In all of these fermented milks lactic acid bacteria carry on the main fermentation to produce chiefly lactic acid.

Cultured buttermilk and sour cream result from the action of *Streptococcus cremoris* and *S. lactis* at about 70 F (21.1 C) to produce most of the acidity; and aroma-forming bacteria, *Leuconostoc dextranicum*, *S. diacetilactis*, and *L. citrovorum*, add flavoring substances like diacetyl and acetic acid. The first two aroma-formers produce lactic acid. Usually the fermentation is continued until the acidity is about 0.7 to 0.9 percent as lactic acid. Sometimes some Bulgarian buttermilk is mixed in to improve the body of the cultured buttermilk.

Bulgarian buttermilk is made by growth of a pure culture of *Lactobacillus bulgaricus* at about 98.6 F (37 C), and yoghurt (Figure 20–1) by a mixed culture of *Streptococcus thermophilus* and *L. bulgaricus*. The starter for kefir is kefir grains, which are cauliflowerlike aggregates of a mixture of microorganisms, chiefly *Streptococcus lactis*, *S. cremoris*, *Leuconostoc dextranicum*, *Bacillus kefir*, and several yeasts. In addition to acid, a small amount of alcohol, 0.5 to 1.0 percent is produced, and enough carbon dioxide to charge the drink if it is kept tightly sealed during fermentation. Kumiss, ordinarily made from mare's milk, results from fermentation by a mixture of lactics and yeasts carried over from a previous lot. Acidophilus milk, prepared for its therapeutic properties for intestinal disorders, utilizes a pure culture of *Lactobacillus acidophilus* grown in milk that has been sterilized or nearly sterilized. Taette is a ropy buttermilk made by means of a ropy variety of *Streptococcus lactis*, and skyr is a semisolid fermented milk in which chiefly *S. thermophilus* and *Lactobacillus bulgaricus* have been active.

The acidity of the fermented milks is sufficient to prevent spoilage

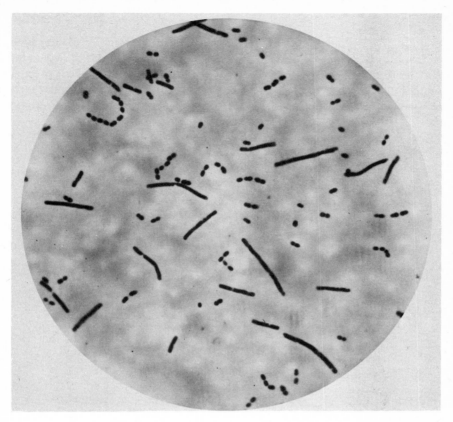

Figure 20–1. *Lactobacillus bulgaricus* and *Streptococcus thermophilus* in yoghurt. (*Courtesy of General Biological Supply House, Inc., Chicago.*)

by proteolytic or other bacteria that are not acid-tolerant. Chilling is necessary to stop acid formation by the starter bacteria at the desired stage, and bottling and sealing to avoid mold growth.

A lactic acid fermentation is involved in the making of most kinds of cheese. Unripened cheeses, such as cottage cheese and cream cheese, are made by starters similar to those used for cultured buttermilk. They must be chilled and kept cold until consumed, and they have a comparatively short keeping time. Shelf life of cottage cheese can be lengthened by incorporation of antibiotic products of the aroma-formers in the finished product. Ripened cheeses have an initial acid fermentation by lactic acid bacteria followed by action of their enzymes and those of other microorganisms; in some cheeses there is in addition action of enzymes of the rennet, and in certain Italian cheeses action of added animal lipases. The chief compounds contributing to the flavor of ripened cheeses are salt, lactic acid, fatty acids, amino acids, and carbonyl compounds (e.g., aldehydes and ketones). The soft, ripened cheeses, such as Limburger, are more perishable than the harder cheeses, e.g., Cheddar and Swiss, but all completely cured cheeses require storage at chilling temperatures to aid in their preservation. Cheeses with hard rinds, such as natural Cheddar and Swiss, are protected by these rinds to some extent from drying and spoilage. Chilling also is necessary for most kinds of cheese, as well as packaging for perishable cheeses and cut pieces of larger, more stable cheeses. Packaging is to minimize losses of moisture and penetration of oxygen, which otherwise would permit growth of molds.

In general, the curing or ripening processes do not greatly improve the keeping quality, although losses in moisture occur during the aging of long-cured hard cheeses, and a protective rind forms on most of them. The chemical products formed during ripening have little preservative effect for the most part, although fatty acids repress the anaerobes; in fact most cheeses become more alkaline as they age and hence more susceptible to spoilage by bacteria. The propionate formed in the ripening of good Swiss cheese delays the growth of most molds on the cheese; and the surface "smear" of organisms on surface-ripened cheeses such as brick or Limburger may produce products that are inhibitory to other organisms.

USE OF RADIATIONS, SOUND WAVES, AND ELECTRIC CURRENTS

Although an effect equivalent to pasteurization can be obtained by the treatment of milk with ultraviolet rays, this method is not used in the preservation of milk, because only a thin layer of milk can be successfully irradiated, and, unless great care is taken, a "burnt" flavor will result. Irradiation of milk to increase the vitamin-D content is not for its preservation. Other uses of ultraviolet light in the dairy industry include irradiation of rooms to reduce the numbers of microorganisms in the air in processing rooms where sweetened condensed milk is being prepared or cut cheese is being packaged, and in cheese curing rooms. The rays inhibit mold growth on the sides of the curing cheese exposed to them directly, but are not effective on the shaded side.

X-rays, cathode rays, and gamma rays have been tried experimentally to destroy most of the microorganisms in milk and lengthen the keeping time. Dosages large enough to kill all or most of the microorganisms present have resulted in off-flavors in the milk, but methods are being developed to minimize these effects.

Sound waves, especially ultrasonic vibrations with a frequency of about 8,900 cycles, have been combined with a temperature of 40 to 50 C (104 to 122 F) to kill most of the vegetative cells in milk and at the same time soften the curd. Ultrasonic treatment of cheese during ripening is primarily to hasten that process rather than to kill bacteria. These methods are not in practical use at present.

Alternating electric currents have been applied to milk as a method for rapid heating in the high-temperature–short-time pasteurization process. Little killing effect is claimed for the electric current itself, for heating is the main effect produced. The neutralization of milk or cream by an electric current has been reported to reduce numbers of bacteria.

It has been indicated that several preservative factors are involved in the preservation of milk and each of its products. Milk for market sale or for the manufacture of dairy products is produced as aseptically as is practicable, is cooled promptly, and is kept chilled. Often it is pasteurized, and market milk is packaged in bottles or other containers to keep out microorganisms. Cream is treated similarly. Fermented milks owe most of their keeping qualities to the acid formed during the fermentation, but require chilling and packaging for their preservation. Butter is preserved primarily by low temperatures, chilling for short-time storage, or freezing temperatures for long-time storage. The low moisture and the salt content also aid in the preservation, as does packaging or sealing to prevent contamination. Cheese is preserved by the acidity produced during its making, by chilling, and by impervious rinds or by packaging. Dry milk, if properly prepared, has too little moisture for microbial growth, but requires packaging to prevent contamination. Evaporated milk is processed by steam under pressure to kill all or most microorganisms and is sealed in cans to keep out contaminants. Sweetened condensed milk undergoes a pasteurization during its preparation, contains a high concentration of sugars, and is protected by the sealed can.

SPOILAGE

As has been indicated milk and products made from it may be preserved in a number of different ways, some of which involve killing only part of the microorganisms present and inhibiting the growth of the remainder. Some products, therefore, have a limited keeping time, and many spoil readily if the methods of preservation are inadequate.

MILK AND CREAM

Milk is an excellent culture medium for many kinds of microorganisms, for it is high in moisture, nearly neutral in pH, and rich in microbial foods. A plentiful supply of food for energy is present in the form of milk sugar (lactose), butterfat, citrate, and nitrogenous compounds; nitrogenous food is there in many forms—proteins, amino acids, ammonia, urea, and other compounds; and the accessory foods and minerals required by microorganisms are available. Some inhibitory substances (lactoperoxidase and agglutinins) are present in freshly drawn milk but soon become comparatively ineffective. Because of the fermentable sugar, an acid fermentation by bacteria is most likely under ordinary conditions, but other changes may take place if conditions are unfavorable to the acid-formers or if they are absent. The chief types of spoilage are as follows:

Souring or acid formation

When milk "sours," it often is considered spoiled, especially if it curdles, although the lactic acid fermentation of milk is utilized in the manufacture of fermented milks and cheese. The evidences of acid formation are first a sour flavor and then coagulation of the milk to give a solid jellylike curd or a weaker curd that releases clear whey. The lactic acid fermentation is most likely to take place in raw milk held at ordinary room temperatures. The "lactics" that cause this fermentation may be **homofermentative,** producing mostly lactic acid with only small amounts of acetic acid, carbon dioxide, and other volatile products, or **heterofermentative,** producing appreciable amounts of the volatile products in addition to lactic acid. In raw milk at temperatures from 10 to 37 C (50 to 98.6 F) the homofermentative *Streptococcus lactis* is most likely to cause the souring, with possibly some assistance from coliform bacteria, enterococci, lactobacilli, and micrococci. At higher temperatures, e.g., from 37 to 50 C (98.6 to 122 F), S. *thermophilus* and S. *faecalis* may produce about 1 percent acid and be followed by lactobacilli, such as *Lactobacillus bulgaricus,* to make the milk very acid. Some of the lactobacilli can grow at temperatures above 50 C (122 F), but produce less acid there. Thermophilic bacteria can grow at still higher temperatures, e.g., *Bacillus calidolactis* and *Lactobacillus thermophilus*. Little formation of acid takes place in milk held at temperatures near freezing, but proteolysis may take place. The pasteurization of milk kills the more active acid-forming bacteria but may permit the survival of heat-resistant lactics (e.g., enterococci, *Streptococcus thermophilus,* and lactobacilli), which will cause a lactic acid fermentation if the holding temperature is high enough. In the refrigerator, proteolysis by low-temperature bacteria is likely. If acid-formers do not survive, other types of bacteria may cause proteolysis at room temperature.

Many bacteria other than those termed lactics can cause an acid fermentation in milk, especially if conditions are unfavorable for the lactic acid bacteria. The coliform bacteria produce some lactic acid and considerable amounts of volatile products, such as hydrogen, carbon dioxide, acetic acid, formic acid, alcohol, etc. Species of *Micrococcus, Microbacterium,* and *Bacillus* can produce acid in milk, mostly lactic, but ordinarily cannot compete with the lactics.

Butyric acid may be produced in milk by action of *Clostridium* species under conditions that prevent or inhibit the normal lactic acid formation. Thus milk, given a heat-treatment which destroys all vegetative cells of bacteria but allows the survival of spores of *Clostridium,* may undergo the butyric acid fermentation with the production of hydrogen and carbon dioxide gas.

Gas production

Gas production by bacteria usually is accompanied by acid formation, and, with few exceptions, is undesirable in milk and milk products. The chief gas-formers are the coliform bacteria, *Clostridium* species, the aerobacilli (gas-forming *Bacillus* species) that yield both hydrogen and carbon dioxide, and the yeasts, propionics, and heterofermentative lactics that produce only carbon dioxide. The production of gas in milk is evidenced by foam at the top if the milk is liquid and is supersaturated with the gas, by gas bubbles caught in the curd or furrowing it, by floating curd containing gas bubbles, or by a ripping apart of the curd by rapid gas production, causing the so-called "stormy fermentation" of milk.

The likelihood of gas formation and the type of microorganisms to cause it will depend upon the pretreatment of the milk and the temperature of holding. In raw milk from icebox temperatures to blood heat, the coliform bacteria are most apt to be the main gas-formers because they can compete well with other acid-formers. Heterofermentative lactics also may produce gas but usually not enough to be evident in the milk. Yeasts, which must be lactose-fermenting to make gas in milk, usually are absent or in low numbers in milk and do not compete well with the bacteria. They sometimes cause gassiness of cream held on the farm for periodic collection by the creamery. The souring of the cream by bacteria favors the growth of the yeasts, which do best at an acid pH.

Gas-forming *Clostridium* and *Bacillus* (*Aerobacillus*) species seldom grow at refrigerator temperatures and do not compete well with acid-formers at higher temperatures, but may function if the acid-formers are absent or comparatively inactive. Thus in milk heated at pasteurizing temperatures or above, the chief acid-formers will be killed, the spores of *Clostridium* and *Bacillus* species will survive, and gas formation by the spore-formers may take place. These organisms sometimes are harmful to cheese from pasteurized milk. Propionic acid-forming bacteria are not active in milk.

Proteolysis

The hydrolysis of the proteins of milk by microorganisms usually is accompanied by the production of a bitter flavor caused by some of the peptides. Proteolysis is favored by storage at a low temperature, by the destruction of lactics and other acid-formers by heat, and by the destruction of formed acid in the milk by molds and film yeasts or the neutralization of acids by products of other organisms.

The types of change produced by proteolytic microorganisms include (1) acid proteolysis, in which acid production and proteolysis occur together; (2) proteolysis with little acidity or even with alkalinity; (3) sweet curdling, which is caused by renninlike enzymes of the bacteria at an early stage of proteolysis; and (4) slow proteolysis by endoenzymes of bacteria after their autolysis.

Acid proteolysis causes the production of a shrunken curd and the expression of much whey. This is followed by a slow digestion of the curd, which changes in appearance from opaqueness to translucency and may be completely dissolved by some kinds of bacteria. Sometimes, separate curd particles are formed that shrink to such a small size that they barely are visible in the large amount of whey. Acid proteolysis may be caused by several species of *Micrococcus,* some of which grow in the udder of the cow and cause acid proteolysis of aseptically drawn milk. One of the intestinal streptococci or enterococci, *Streptococcus faecalis* var. *liquefaciens* is a lactic acid organism that also is actively proteolytic. It is thermoduric, like the other enterococci, and may cause acid proteolysis in pasteurized milk. Spores of lactose-fermenting, proteolytic strains of some species of *Bacillus,* e.g., of *B. cereus,* can survive pasteurization or a more rigorous heat-treatment of milk and cause acid proteolysis.

Proteolysis by bacteria unable to ferment lactose varies with the bacterium involved, from obvious digestion of the casein to slight proteolysis that is detectable only by chemical tests. Little if any acidity is produced by most of these bacteria; in fact the milk usually becomes alkaline in time from products of protein decomposition. Many of these bacteria "sweet-curdle" the milk (curdle it with rennin) before digesting the casein, but others hydrolyze the protein so rapidly that no curdling is evident and finally a fairly clear liquid remains with no sign of casein or curd. Actively proteolytic bacteria are found among the species of *Micrococcus, Alcaligenes, Pseudomonas, Proteus, Achromobacter, Flavobacterium,* and *Serratia,* all of which are genera of non-spore-forming bacteria, and of the genera *Bacillus* and *Clostridium* of the spore-formers. It will be observed that some of these bacteria, notably some species of the genera *Micrococcus, Pseudomonas, Achromobacter,* and *Flavobacterium,* can grow well at low temperatures and hence are likely to cause some proteolysis and bitterness of milk held at chilling temperatures. None of these bacteria, except some species of *Micrococcus,* is thermoduric and therefore should not be present in pasteurized milk. In fact, however, a few of the psychrophiles are present in most pasteurized milk and grow in milk in the refrigerator. Most proteolytic species of *Bacillus* and *Clostridium,* on the other hand, are unable to grow at refrigerator temperatures and do not compete well with acid-formers at atmospheric temperatures. They are most likely to cause their proteolysis in milk that has been heated enough to destroy most or all of the acid-formers.

Slow proteolysis by endoenzymes of bacteria after their autolysis is of no significance in milk under ordinary circumstances but is significant when a long time is allowed for their action as in curing cheese.

Ropiness

Ropiness or sliminess can occur in milk, cream, or whey, but is important mostly in market milk and cream. Nonbacterial ropiness or sliminess may be due to the following factors: (1) Stringiness caused by mastitis and in particular by fibrin and leucocytes from the cow's blood. In contrast to ropiness produced by bacteria, it is present when the milk is drawn, rather than developed during holding of the milk. (2) Sliminess resulting from the thickness of cream, e.g., at the top of a bottle. (3) Stringiness due to thin films of casein or lactalbumin during cooling, as sometimes observed on surface coolers. This effect is only temporary.

Bacterial ropiness is caused by slimy capsular material from the cells, usually gums or mucins, and ordinarily develops best at low storage temperatures. The ropiness usually decreases as the acidity of the milk or cream increases. There are two main types of bacterial ropiness, one in which the milk is most ropy at the top and the other in which the milk becomes ropy throughout. Surface ropiness is caused most often by *Alcaligenes viscolactis* (*viscosus*) (Figure 20–2), an organism chiefly from water or soil that can grow fairly well in the vicinity of 10 C (50 F). Some of the thermoduric micrococci, e.g., *Micrococcus freudenreichii,* can cause surface ropiness. Ropiness throughout the milk may be caused

Figure 20–2. Ropy milk. *Alcaligenes viscolactis* has made the milk so viscous that it can be strung out with a forceps. (*Courtesy of Harper & Row, Publishers, Incorporated. From W. B. Sarles et al., Microbiology: General and Applied, 2nd ed., copyright, 1956.*)

by any of a number of kinds of bacteria: (1) Certain strains of coliform bacteria such as of *Aerobacter aerogenes*, *A. cloacae*, and rarely *Escherichia coli*. Ropiness caused by *Aerobacter* usually is worse near the top of the milk. (2) Certain strains of some of the common species of lactic acid bacteria. *Streptococcus lactis* var. *hollandicus* causes ropiness in milk, and is used in making a Scandinavian fermented milk. *Lactobacillus casei*, *L. bulgaricus*, and *L. plantarum* occasionally produce ropiness, as do strains of *Streptococcus cremoris*. Most of these lactic bacteria can grow in long chains, a characteristic that supposedly contributes to the stringy condition of the milk. (3) Miscellaneous other bacteria among the alkali-formers, micrococci, tetracocci, streptococci, and bacilli. Ordinarily these bacteria would be suppressed by the acid-formers.

Since the sources of the bacteria causing ropiness are water, manure, utensils, and feed, the reduction or elimination of contamination from these sources helps prevent ropiness. Adequate pasteurization of milk readily destroys most of the above kinds of bacteria.

Changes in butterfat

Butterfat may be decomposed by various bacteria, yeasts, and molds that do not constitute distinct groups on the basis of other characteristics. The bacteria are, for the most part, aerobic or facultative, proteolytic, and non-acid-forming. The following changes in the butterfat take place: (1) Oxidation of the unsaturated fatty acids which, coupled with other decomposition, yields aldehydes, acids, and ketones and results in tallowy odors and tastes. The reaction is favored by metals, sunlight, and oxidizing microorganisms. (2) Hydrolysis of the butterfat to fatty acids and glycerol by the enzyme, lipase. The lipase may have been in the original milk or may be microbial. (3) Combined oxidation and hydrolysis to produce "rancidity." Species of lipase-forming bacteria are found in many of the bacterial genera, e.g., *Pseudomonas, Proteus, Achromobacter, Alcaligenes, Bacillus, Micrococcus, Clostridium,* and others. Many of the molds are lipolytic and some species of yeasts.

Alkali production

The group of alkali-formers includes those bacteria which cause an alkaline reaction in milk without any evidence of proteolysis. The alkaline reaction may result from the formation of ammonia, as from urea, or of carbonates, as from organic acids like citric acid. Most of these bacteria grow at from moderate to low temperatures, and many can survive pasteurization. Examples of alkali-formers are *Pseudomonas fluorescens* and *P. trifolii, Alcaligenes faecalis* and *A. viscolactis,* and *Micrococcus ureae.*

Flavor changes

The flavor of milk as drawn is low, delicate, and easily altered. The types of changes in milk previously discussed produce changes in taste and odor, some of which will be mentioned here again. Since individuals differ in their ability to detect flavors and disagree in their description of them, a number of more or less descriptive terms have been suggested to name the off-flavors.

Milk, as drawn from the cow, may be abnormal in flavor because of the individual cow, mastitis, the stage of lactation of the cow, or feed. Flavors that develop later may be nonmicrobial in cause, e.g., absorbed flavors, tallowy flavors due to light or metals, or rancidity caused by milk lipase; or they may be caused by microorganisms.

Some of the off-flavors caused by microorganisms are described as follows:

SOUR OR ACID FLAVOR. The acidity may be described as "clean," as produced by *Streptococcus lactis* and other lactics; as "aromatic" when lactic streptococci and aroma-forming *Leuconostoc* species are growing together; and as "sharp" when appreciable amounts of volatile fatty acids—formic, acetic, or butyric—are produced by coliform bacteria, *Clostridium* species, and other organisms. Clean and aromatic flavors are desired in fermented milk products, but the sharp flavors are undesirable.

BITTER FLAVORS. Bitterness usually results from proteolysis, but may follow lipolysis or even fermentation of lactose. Milk from cows late in their lactation period sometimes is slightly bitter. Microorganisms causing proteolysis and hence bitterness have been mentioned in a previous paragraph. Other organisms causing bitterness are certain strains of coliform bacteria and of asporogenous yeasts. Some cocci cause very bitter milk, and actinomycetes sometimes give bitter-musty flavors.

BURNT OR CARAMEL FLAVOR. Certain strains of *Streptococcus lactis* (var. *maltigenes*) produce this flavor, which resembles the cooked flavor of overheated milk.

MISCELLANEOUS OTHER FLAVORS. Other flavors that are found less commonly make up a long list, only part of which will be mentioned: a barny flavor by *Aerobacter oxytocum;* soapiness by ammonia-formers like *Pseudomonas sapolactica;* a turniplike flavor by *Escherichia coli* and *P. fluorescens;* a malty flavor by yellow micrococci from the udder; a potatolike flavor by *P. graveolens* or *P. mucidolens;* fishiness by *P. ichthyosmia* or various cocci that produce trimethylamine from lecithin; earthy or musty flavors by actinomycetes; fruity, esterlike, and alcoholic flavors by yeasts; an amyl alcohol flavor by white and orange micrococci; and putrefaction by species of *Clostridium* and other putrefactive bacteria. Other flavors have been termed unclean, stale, astringent, oily, weedy, carroty, etc.

Inadequately cooled raw milk that has been held in a tightly covered can, so that volatile products of bacterial metabolism have collected above the milk, has an undesirable odor that varies in nature. Milk showing this condition is termed "smothered."

Color changes

The color of milk or cream is affected by its physical and chemical composition, for example, by the amount and yellowness of the butterfat, the thinness of the milk, the content of blood and pus, and the feed of the animal. Color changes caused by microorganisms may occur along with other changes previously discussed, but there are some special changes in color that should be mentioned. The color may be due to the surface growth of pigmented bacteria or molds in the form of a scum or ring or may be throughout the milk.

BLUE MILK. *Pseudomonas syncyanea* produces a bluish-gray to brownish color in milk in pure culture, but when growing with an acid-former like *Streptococcus lactis* causes a deep-blue color. This defect and the blue color produced by actinomycetes or species of the mold *Geotrichum* are rare in occurrence.

YELLOW MILK. *Pseudomonas synxantha* may cause a yellow color in the cream layer of milk, coincident to lipolysis and proteolysis. Species of *Flavobacterium* also can give yellowness to milk.

RED MILK. Red milk usually is caused by species of *Serratia*, e.g., *S. marcescens*, but is rare because other bacteria ordinarily outgrow the red-pigmented species. *Brevibacterium erythrogenes* produces a red layer at the top of milk, followed by proteolysis. *Micrococcus roseus* may grow and produce a red sediment, and the yeast *Torula glutinis* may produce pink or red colonies on the surface of sour milk or cream. Blood in milk will give it a red color. The red blood cells settle out or can be centrifuged out.

BROWN MILK. A brown color may result from the enzymic oxidation of tyrosine by *Pseudomonas fluorescens*.

Spoilage of milk at different temperatures

At any given storage temperature most samples of raw milk will undergo a typical series of changes caused by a succession of microorganisms. At refrigerator temperatures, proteolysis may be initiated by psychrophilic bacteria such as *Pseudomonas*, and molds may then appear. At room temperatures, an acid fermentation is most probable, first by lactic streptococci and coliform bacteria, then by the acid-tolerant lactobacilli. Then molds or film yeasts on the surface will lower the acidity thus permitting the formation of more acid. Eventually, when most of the acid has been destroyed, proteolytic or putrefactive bacteria complete the decomposition. Pasteurization kills the yeasts, molds, most psychrophilic bacteria, the coliforms, and *Streptococcus lactis*. Nevertheless, in most samples of pasteurized, refrigerated market milk there are enough psychrophiles (contaminants entering after heating) to result in a slow development of bitterness and other off-flavors. At room temperatures, acid formation by thermoduric bacteria is likely, involving acid production by bacteria like *S. thermophilus, S. faecalis, Microbacterium lacticum,* and heat-resistant micrococci, followed by further acid formation by lactobacilli. Only bacterial spores should survive the boiling of milk. Acid-forming bacteria of the genera *Bacillus* or *Clostridium* are likely to grow, but in their absence proteolytic species of these genera may be active.

CONDENSED AND DRY MILK PRODUCTS

Under this heading are included evaporated milk (unsweetened), bulk condensed milk, frozen milk, sweetened condensed milk, condensed whey or buttermilk, and condensed sour skim milk and dry milk. The quality of all of these products depends upon the quality of the material dried or condensed, for defects in the raw material will carry over to the condensed or dried product. All of the condensed products have a fairly high concentration of solutes that inhibits the growth of some kinds of bacteria. Dry milk is so low in moisture that it should offer no microbial spoilage problems when properly handled. Rise in moisture to over 8 percent might permit some mold growth. The only type of spoilage of condensed buttermilk and sour skim milk is by molds when the surface is exposed to air. The high concentration of acid and solutes prevents the growth of bacteria or yeasts.

Bulk condensed milk

The forewarming temperatures employed in the making of plain condensed milk are equivalent to no more than pasteurization, and the evaporating process is at a temperature low enough to permit the growth of thermophiles. Therefore this product has only a short storage life, although refrigerated, and, like pasteurized milk, is subject to spoilage by thermoduric bacteria that survive, except that these spoilage bacteria must be able to tolerate the increased concentration of solutes in the condensed product. In superheated condensed milk the temperature of the milk is raised to 150 to 170 F (65.6 to 76.7 C) during the introduction of steam, a process that probably destroys most of the vegetative cells of bacteria but not the spores. This product, too, has a short storage life.

Evaporated milk

Unsweetened evaporated milk is canned and heat-processed under steam pressure in an attempt to destroy all of the microorganisms present. Spoilage can take place only when the heat process is inadequate or defects in the can permit the entrance of organisms. Bacterial spores that survive the heat process may be the cause of swelling of the can, coagulation of the milk, or a bitter flavor.

Swelling of the can is caused primarily by gas-forming, anaerobic spore-formers (*Clostridium*), although overfilling of the can with cold milk may cause swelling, or action of the acid constituents of milk on the iron of the can to produce hydrogen gas may bulge the can on long storage.

Coagulation of the milk in the can may vary from a few flakes to a solid curd. Species of *Bacillus* usually are to blame, either mesophiles like *B. cereus, B. subtilis,* or *B. megaterium,* a facultative thermophile, *B. coagulans,* or an obligate thermophile, *B. calidolactis.* The extent of the curd in the milk depends to some extent on the amount of air in the can. Spoilage by the thermophiles should cause no trouble if the milk is cooled promptly and kept cool, but it can cause trouble in the tropics.

Bitterness usually results from proteolysis by species of *Bacillus* and less commonly by species of *Clostridium.* Some of the latter may cause putrefaction in rare instances.

Spoilage resulting from leakage, as indicated by the presence of non-spore-formers, may result in gas and swelling caused by coliform bacteria or yeasts, coagulation by streptococci, or bitterness caused by cocci.

Sweetened condensed milk

Sweetened condensed milk has been subjected to a fairly high temperature (160 to 212 F, or 71.1 to 100 C) during forewarming and to a milder heat-treatment (120 to 130 F, or 48.9 to 54.4 C) during condensing, so that the yeasts, molds, and most of the vegetative cells of bacteria have been destroyed. In addition a high concentration of sugar is present, about 55 to 60 percent of total sugar (lactose plus added sugar). Also, the can is evacuated and sealed. Spoilage, then, is due primarily to organisms that have entered after the heat-treatments, especially if air is present. The chief types of spoilage are (1) gas formation by sucrose-fermenting yeasts such as *Torula lactis-condensi* and *T. globula* or, more rarely, by coliform bacteria; (2) thickening caused by micrococci, which probably produce rennin; and (3) "buttons," which are mold colonies growing on the milk surface. The size of these buttons is determined by the amount of air in the can. Species of *Aspergillus*, e.g., *A. repens*, or of *Penicillium* have been incriminated.

FROZEN DESSERTS

Frozen desserts include ice cream, ice milk, frozen custards, sherbets, and ices. The ingredients may be various combinations of the following: milk, cream, evaporated milk, condensed milk, dried milk, coloring materials, flavors, fruits, nuts, sweetening agents, eggs and egg products, and stabilizers. Any of these may contribute microorganisms to the product and affect the quality of the dessert as judged by its bacterial content or its content of specific kinds of bacteria such as the coliforms. The desserts are not ordinarily subject to spoilage, however, as long as they are kept frozen. The only important types of spoilage take place in the ingredients before they are mixed or in the mix before it is frozen. Since the mix is pasteurized before it is frozen, no spoilage problems should be encountered, unless it is held at temperatures above freezing for a considerable time, when souring by acid-forming bacteria can take place.

BUTTER

Many of the defects of butter originate in the cream from which it is made, especially when the cream has been held for several days on the farm before collection by the creamery. During this time lactic acid bacteria, gas-formers, and other spoilage organisms may grow and be followed by molds, e.g., *Geotrichum candidum*. Lactose-fermenting yeasts, which are present only occasionally, may develop high gas pressures within the can of cream. In fact, most of the types of spoilage described for milk in the early part of this chapter could occur in the cream and affect the butter made from it.

The likelihood of spoilage and the kind of change will depend upon the kind of butter and the environment in which it is kept. Salted butter, because of the high salt concentration in the small amount of moisture present, is less likely to support microbial growth than unsalted butter. In general, sweet-cream butter keeps better than sour-cream butter. Today most butter in this country is made from pasteurized cream, in which most of the spoilage organisms have been destroyed. Also, butter commonly is kept refrigerated, and during commercial storage is kept at about 0 F (-17.8 C), where no microbial growth can take place. For these reasons bacteria usually do not grow in butter, and when they do their growth is not extensive. The flavor of good butter is so delicate, however, that relatively small amounts of growth may cause appreciable damage to the flavor.

Flavor defects

As has been indicated, undesirable flavors may come from the cream, which may receive such flavors from the feed of the cow, absorb them from the atmosphere, or develop them during microbial growth. Feeds like onions, garlic, French weed, peppergrass, and poor silage contribute off-flavors to the cream. Volatile products that may be absorbed from the air are odors from the barn and from the chemicals used there, e.g., kerosene, gasoline, fly sprays, disinfectants, etc. Growth of microorganisms in the cream and in the milk from which it is separated may result in any of the following bad flavors: (1) cheesiness, caused by lactobacilli; (2) rancidity, resulting from lipolysis by fat-splitting bacteria and molds, and, perhaps, by lipase in the cream; (3) barny flavor, produced by species of *Aerobacter*; (4) malty flavor, produced by *Streptococcus lactis* var. *maltigenes*; (5) yeasty flavor, produced by yeasts; (6) musty flavors, caused by molds and actinomycetes; (7) metallic flavors, caused by dissolved metals in highly acid cream; (8) a flat flavor, resulting from the destruction of diacetyl by bacteria like some of the *Pseudomonas* species; (9) highly acid flavor, when the cream has excessive acidity; and (10) "unclean" flavor, caused by coliform bacteria.

Unsatisfactory methods in the creamery may cause a cooked flavor due to overpasteurization of the cream or a "neutralizer" flavor if too much of the neutralizing compound is used, if it is unevenly distributed in the cream, or if pasteurization takes place before a balance is reached.

Like cream, butter readily absorbs volatile materials from the air. Microorganisms in the butter can cause the following defects: (1) Surface taint, also called "rabbito" and "putridity," which is blamed on *Pseudomonas putrefaciens*, introduced usually by the wash water, churns, or equipment. It is worse in unsalted or low-salt butter. The "sweaty-feet" odor is due chiefly to isovaleric acid. (2) Fishiness, caused by *P. ichthyosmia*. (3) Esterlike flavors, resulting from the action of *P. fragi*. (4) Skunklike flavors, caused by *P. mephitica*. (5) Roquefort-like flavors, produced by molds.

Chemically produced flavors include (1) rancidity produced by lipase in the cream; (2) tallowiness from oxidations of unsaturated fats catalyzed by copper and bacterial enzymes and favored by a low pH, low-temperature pasteurization, salt, air, and ozone; (3) fishiness, where trimethylamine is produced from lecithin. This defect is favored by high acidity, salt, overworking of the butter, and the presence of copper.

Color defects

Some color defects, not caused by microorganisms, are mottling because of improper working, a pink color caused by the action of sulfur dioxide refrigerant on the butter color, surface darkening resulting from the loss of water from surface layers, and bleaching that accompanies tallowiness.

Discolorations, chiefly at the surface, may be caused by molds, yeasts, or bacteria that come from churns, wrappers, liners, circles, tubs, the air, or the cream if it is unpasteurized. Colored growths of molds result in the smudged or *Alternaria* type of discoloration, with dark, smoky, or, rarely, greenish areas where *Alternaria* or *Cladosporium* species have grown, or small black spots of *Stemphylium*. *Penicillium* produces green coloration, *Phoma* or *Alternaria* molds brown areas, and *Oöspora* (*Geotrichum*) species produce orange or yellow spots. *Fusarium culmorum* can cause bright, reddish-pink areas. Yeasts sometimes grow in pink colonies. *Pseudomonas nigrifaciens* causes the reddish-brown "black smudge" in mildly salted butter.

FERMENTED MILK PRODUCTS

Fermented milks and cheese depend upon a desired fermentation or succession of fermentations for their manufacture. Therefore any abnormality in these fermentations will affect the quality of the product and may even spoil it. The finished product, too, may be subject to spoilage by microorganisms.

Fermented milks

In the manufacture of most fermented milks a starter containing the appropriate microorganisms is added to pasteurized milk, which is incubated until the desired acidity is attained. The chief product of the fermentation is lactic acid, but lesser amounts of flavoring substances may be produced or added. If the starter bacteria are inactive, other bacteria may grow and damage the curd and the flavor. Proteolytic bacteria, which ordinarily cannot compete with the lactics, may cause a poor curd and off-flavors. Coliform bacteria and lactose-fermenting yeasts should not be present but may enter from equipment and other sources to produce bad flavors and gas. The finished product is susceptible to spoilage by molds from air or equipment, provided that air is available at the surface.

Cheese

Defects of cheese may have mechanical or biological causes, but only the latter will be discussed. The types of spoilage can be divided into those developing during the manufacturing process, those appearing during the ripening process of cured cheese, and those occurring in the finished product.

SPOILAGE DURING MANUFACTURE. During the manufacture of most types of cheese, or while the cheese is draining, a lactic acid fermentation is encouraged. If the lactics are ineffective or contamination with other microorganisms is unduly heavy, abnormal changes may take place that adversely affect the quality of the cheese. In cheese made from raw milk the gas-forming organisms may be producing off-flavors, as well as gas holes in the curd (Figure 20–3). Very dangerous are the coliform bacteria, especially those resembling *Aerobacter aerogenes*. Lactose-fermenting yeasts, although not commonly present in appreciable numbers, also can cause gassiness. Spore-forming gas-formers, especially species of *Clostridium*, can cause trouble in either raw milk or pasteurized milk cheese if the lactic starter bacteria are not functioning properly; and less commonly the aerobacilli, spore-forming species of *Bacillus* like *B. polymyxa*, may produce gas and other defects. These spore-formers also can cause defects in ripening cheese.

Other bacteria may compete with the starter organisms with results that do not become evident until during the curing process, where body and flavor may be affected. Thus acid-proteolytic bacteria may produce a bitter flavor, or *Leuconostoc* species may cause holes or openness in Cheddar cheese from pasteurized milk.

Cottage cheese especially is subject to spoilage during manufacture or storage prior to consumption. If the starter bacteria produce insufficient acid or yield it too slowly, a usable curd will not result or the cheese will be inferior because of the growth of undesirable organisms. Proteolysis, gas production, or sliminess and off-flavors may ruin the product. Cheese with too low an acidity because of the addition of

Figure 20–3. Gassy Swiss cheese. Upper: early gas. Lower: late gas, with rough eyes and cracks.

cream or the failure of the starter often is made gelatinous or slimy by soil or water bacteria such as *Pseudomonas viscosa, P. fragi,* or *Alcaligenes metalcaligenes.*

SPOILAGE DURING RIPENING. During ripening or curing, cheeses normally undergo physical and chemical changes resulting from the action of enzymes released from the autolyzed cells of bacteria that grew during manufacture and from the action of microorganisms that increase during the ripening period. The growth of organisms other than the desired ones results in inferior or, in extreme cases, even worthless cheese because of alterations in texture, body, general appearance, or flavor. The most important kinds of spoilage differ enough, however, with the kind of cheese concerned to discourage generalizations. Most kinds may be subject to "late gas," usually caused by lactate-fermenting *Clostridium* species (Figure 20–3), but possibly by aerobacilli, propionic bacteria, or heterofermentative lactics. Gas holes, or eyes, are desirable in Swiss and related cheeses but not to any extent in other varieties. Especially bad in Swiss and similar cheeses is the cracking or splitting of the cheese by gas or the production of too many, too small, or misshapen eyes. Gas formation by the spore-formers is accompanied by the production of undesirable flavors, butyric acid, for example, by the anaerobes.

Bitterness may be caused by proteolytic bacteria, such as the acid-proteolytic types; coliforms; micrococci; various other bacteria; and rarely by yeasts, which usually give a sweet, fruity, or "yeasty" flavor.

Putrefaction may occur locally or generally in cheese where insufficient acidity has been produced by the lactics or the acid has been destroyed by a lactate-fermenter, for example *Clostridium tyrobutyricum*. Putrefactive anaerobes, such as *C. sporogenes* or *C. lentoputrescens*, may be concerned. Foul-smelling gray rot may occur locally, especially at the surface of Swiss cheese. The dark color, due to sulfides, and bad flavors are caused by *Bacterium proteolyticum*.

Discolorations of the ripening cheese may result from the action of microorganisms on compounds produced during curing or on added coloring material, such as the annatto used in Cheddar cheese, or may be due to the development of pigmented colonies of organisms on or in the cheese. Blue, green, or black discoloration may be produced by the reaction of hydrogen sulfide produced by organisms with metals or metallic salts. Bacterially formed sulfhydryl groups give a pink to muddy appearance to annatto. Reddish-brown to grayish-brown colors sometimes result from the oxidation of tyrosine by bacteria growing in soft cheeses. Rusty spot of Cheddar and similar cheeses is due to colonies of *Lactobacillus plantarum* var. *rudensis* or *L. brevis* var. *rudensis*, while yellow, pink, or brown spots in Swiss cheese, mostly on the surface of the eyes, are colonies of pigmented species of *Propionibacterium*, e.g., *P. zeae*, *P. rubrum*, and *P. thoenii*.

SPOILAGE OF THE FINISHED CHEESE. In general, the perishability of cured cheeses increases with their moisture content. Therefore, soft cheeses, like Limburger and Brie, are most perishable, and hard cheeses like Cheddar and Swiss most stable. Most feared among the spoilage organisms are the molds that tend to grow on the cheese surfaces and into cracks or trier holes. Even cheeses that depend partly on a specific mold for ripening may be damaged by other molds. Most natural cheeses have rinds that serve as some protection to the anaerobic interior, but usually are not dry enough to prevent mold growth. The acidity of the cheese is no deterrent to growth, and the storage temperature is not too low for such growth. Most molds grow in colored colonies on the surface or in crevices, without much penetration into the cheese, but some kinds produce actual rots. Not only are discolorations evident, but also off-flavors are produced locally. Among the molds that grow on cheese surfaces are the following: (1) Species of *Oöspora*

(*Geotrichum*). *Oöspora* (*Geotrichum*) *lactis,* called the "dairy mold," grows on soft cheeses, and during ripening sometimes suppresses other molds, as well as surface-ripening bacteria. The curd gradually becomes liquefied under the felt. *O. rubrum* and *O. crustacea* produce a red coloration, and *O. aurianticum* forms orange to red spots. *O. caseovorans* causes "cheese cancer" of Swiss and similar cheeses. Bumps of growth become filled with a white, chalky mass. (2) Species of *Cladosporium.* The mycelium and spores of these molds are dark or smoky and give dark colors to the cheese. Most common is *C. herbarum,* characterized by dark-green to black colors. Other species cause green, brown, or black discolorations. (3) Species of *Penicillium. P. puberulum* and other green-spored species grow in cracks, crevices, and trier holes of Cheddar and related cheeses to give a green coloration because of their spores. They may act on annatto to cause mottling and discoloration. *P. casei* causes yellowish-brown spots on the rind, and *P. aurantio-virens* discolors Camembert cheese. (4) Species of *Monilia. M. nigra* produces penetrating black spots on the rind of hard cheeses. Species of many other genera may discolor cheeses and give off-flavors, e.g., the genera *Scopulariopsis, Aspergillus, Mucor, Alternaria,* etc.

If the surface is sufficiently moist, yeasts may form colored colonies or areas, and film yeasts may pave the way for the yellow to red growth of *Brevibacterium linens.* The latter is desired in some surface-ripened cheeses but not in others.

REFERENCES

American Public Health Association. 1960. Standard methods for the examination of dairy products. 11th ed. American Public Health Assn., New York.

COOK, D. J. 1963. Radiations for the dairy industry: II, Effects of radiation in milk and milk products. Dairy Ind. 28:465–470, 536–540.

CUTHBERT, W. A. 1964. The significance of thermoduric organisms in milk. Int. Dairy Fed. Annu. Bull. 1964(4):10–22.

ELLIKER, P. R., 1949. Practical dairy bacteriology. McGraw-Hill Book Company, New York.

FELL, L. R., 1964. Machine milking and mastitis—a review. Dairy Sci. Abstr. 26:551–569.

FORSS, D. A. 1964. Fishy flavor in dairy products. J. Dairy Sci. 47:245–250.

FOSTER, E. M., F. E. NELSON, M. L. SPECK, R. D. DOETSCH, and J. C. OLSON, JR. 1957. Dairy microbiology. Prentice-Hall, Inc., Englewood Cliffs, New Jersey.

HARPER, W. J. 1959. Chemistry of cheese flavors. J. Dairy Sci. 42:207–213.

HOURAN, G. A. 1964. Utilization of centrifugal force for removal of microorganisms from milk. J. Dairy Sci. 47:100–101.

MABBITT, L. A. 1961. Reviews of the progress of dairy science: Section B (Bacteriology), The flavour of cheese. J. Dairy Res. 28:303–318.

MARTH, E. H. 1963. Microbiological and chemical aspects of Cheddar cheese ripening; a review. J. Dairy Sci. 46:869–890.

PLASTRIDGE, W. N. 1958. Bovine mastitis: a review. J. Dairy Sci. 41:1141–1181.

SPECK, M. L. (Convenor). 1962. Symposium on lactic starter cultures. J. Dairy Sci. 45:1262–1294.

THOMAS, S. B. 1964. Investigations on the bacterial content and microflora of farm dairy equipment. J. Soc. Dairy Technol. 17:210–215.

THOMAS, S. B., R. G. DRUCE, and K. ELSON. 1960. Ropy milk. Dairy Ind. 25:202–207.

THOMAS, S. B., PHYLLIS M. HOBSON, and K. ELSON. 1964. The microflora of milking equipment cleansed by chemical methods. J. Appl. Bacteriol. 27:15–26.

THOMAS, S. B., MENA JONES, PHYLLIS M. HOBSON, G. WILLIAMS, and R. G. DRUCE. 1963. Microflora of raw milk and farm dairy equipment. Dairy Ind. 28:212–219.

CHAPTER TWENTY-ONE

MISCELLANEOUS FOODS

Foods not in previously discussed groups are included in this chapter: fatty foods, salad dressings, essential oils, bottled soft drinks, spices and other condiments, salt, and nutmeats.

Food products compounded from combinations of the different groups of foods also would combine their microbial contents, and the new product may furnish a good culture medium for microorganisms that previously had little chance to grow. Thus yeasts from sugar added to bottled soft beverages may spoil the product. The bottlers of carbonated beverages have bacteriological standards for sugar: not over 200 mesophilic bacteria per 10 g and not more than ten yeasts or molds (see Appendix for standards for liquid sugar). The water and flavoring materials also are potential sources of contamination. Spices and other condiments added to foods may be important sources of microorganisms, although spices may be treated with propylene oxide gas to give them a low microbial content. Microorganisms are added to salad dressings by ingredients such as the spices, condiments, eggs, and pickles. Salt (sodium chloride), especially solar salt, may add halophilic and salt-tolerant bacteria to salted fish and other salted or brined products.

Most of the foods to be discussed in this chapter are of plant origin, although part of the fats and oils comes from animals. The sugar, acids, and flavoring materials used in soft drinks are chiefly of plant origin, but some synthetic flavoring or coloring compounds may be employed.

FATTY FOODS

Fats and oils

The fatty parts of foods, the foods made up chiefly of fats and oils, and the fats and oils themselves are subject more often to chemical spoilage than to microbial. Besides the fatty glycerides, natural fats and oils usually contain small amounts of fatty acids, glycerol or other liquid alcohols, sterols, hydrocarbons, proteins and other nitrogenous compounds, phosphatides, and carotinoid pigments. The chief types of spoilage result from hydrolysis, oxidation, or combinations of the two processes. The terms applied to the different types of spoilage often are used rather loosely, although when applied to the deterioration of a specific kind of fat or oil they may have a definite meaning. The term **rancidity** sometimes is used for the result of any change in fats or oils that is accompanied by undesirable flavors, regardless of the cause. The spoilage due to oxidation, chemical or microbial, is termed **oxidative rancidity,** as distinguished from changes resulting from hydrolysis, by lipases originally present or by those from microorganisms, causing **hydrolytic rancidity.** Extensive oxidation, usually following hydrolysis and the release of fatty acids, can result in **ketonic rancidity. Flavor reversion** is defined as the appearance of objectionable flavors from less oxidation than is needed to produce rancidity. Oils that contain linolenic acid, fish and vegetable oils, for example, are subject to flavor reversion.

The oxidation of fats and oils may be catalyzed by various metals and rays and by moisture, as well as by microorganisms; such oxidation is prevented or delayed by natural or added antioxidants. Hydrolysis by lipases results in fatty acids and glycerol or other alcohols. Fats subjected to either or both of these types of changes may contain fatty oxy- and hydroxy- acids, glycerol and other alcohols, aldehydes, ketones, and lactones, and, in the presence of lecithin, may include trimethylamine, with its fishy odor.

Butterfat and meat fats become "tallowy" as the result of oxidation, but butterfat is called rancid when only hydrolysis to fatty acids and glycerol has taken place.

Some of the pigments produced by microorganisms are fat-soluble and therefore can diffuse into fat, producing discolorations ranging through yellow, red, purple, and brown. Best known is the "stamping-ink" discoloration of meat fat that Jensen and coworkers have shown to be caused by yellow-pigmented micrococci and bacilli. The fat-soluble pigment is an oxidation-reduction indicator that changes from yellow to green to blue and finally to purple as the fat becomes more oxidized by the peroxides formed by the bacteria. Yellow, pink, and red fat-soluble pigments may be produced by various bacteria, yeasts, and molds.

Bacteriostatic and bactericidal properties have been claimed for many of the fixed vegetable and animal oils, but most of these, as well as the fats, can be hydrolyzed and oxidized by microorganisms. These fatty materials ordinarily are very low in moisture, a condition that favors molds more than other microorganisms. Molds have been found to cause both oxidative and hydrolytic decomposition that results in rancidity. Bacteria causing rancidity of butter have been shown to cause a similar defect in olive oil. Among the bacteria that can decompose fats are species of *Pseudomonas, Micrococcus, Bacillus, Serratia, Achro-*

mobacter, and *Proteus;* and among the molds, species of *Geotrichum, Penicillium, Aspergillus, Cladosporium,* and *Monilia.* Some yeasts, especially film yeasts, are lipolytic. Copra and cocoa butter may be spoiled by molds.

Salad dressings

Salad dressings contain oil, which may become oxidized or hydrolyzed, and enough moisture to permit microbial growth. For the most part, however, their acidity is too great for most bacteria, ranging from about pH 3 to 4, but favorable for yeasts or molds. Egg or egg products, pickles, relish, pimientos, sugar, starch, gums, gelatin, spices, and other ingredients may add microorganisms, sometimes in appreciable numbers, and may make the dressings better media for microbial growth. The three types of spoilage of mayonnaise and similar dressings are (1) separation of the oil or water from the emulsion, (2) oxidation and hydrolysis of the oils by chemical or biological action, and (3) growth of microorganisms to produce gas, off-flavors, or other defects. Darkening often takes place.

The decomposition of salad dressings and related products can be caused by bacteria, yeasts, or molds. The acidity, coupled with the sugar content, about 4.5 percent on the average in the moisture of mayonnaise, is most favorable to yeasts, which have been reported to cause gassiness. Species of *Zygosaccharomyces* and *Saccharomyces* have spoiled mayonnaise, salad dressing, and French dressing. Bacteria would have to be acid-tolerant to spoil most types of dressing. Therefore it is not unexpected to learn of a heterofermentative lactobacillus resembling *Lactobacillus brevis* causing gas in a salad dressing. More surprising is the report of species of *Bacillus,* e.g., *B. subtilis* and *B. megaterium,* as organisms causing gas, rancidity, and separation, since they are not tolerant of much acidity. Yeasts growing with *B. megaterium* could account for the gas. Darkening and separation of Thousand Island dressing with a pH of 4.2 to 4.4 by *B. vulgatus* from the pepper and paprika have been reported. Molds can grow on salad dressings if air is available and are favored by the addition of starch or pectin to the dressing.

ESSENTIAL OILS

Essential oils or volatile oils are products obtained from the plant kingdom in which the odoriferous and flavoring characteristics are concentrated. These present no spoilage problems, but on the contrary may have some preservative effect as ingredients of foods, e.g., mustard, cinnamon, garlic, and onion oils. Most of them do not affect the heat resistance of microorganisms.

BOTTLED BEVERAGES

Bottled beverages may be alcoholic or nonalcoholic, carbonated or uncarbonated, and acid or nonacid. The spoilage of alcoholic beverages will be discussed in Chapter 24. The ease with which the nonalcoholic beverages will spoil and the type of spoilage to be expected will depend upon the composition of the soft drink. Carbonation is inhibitory or even germicidal to some microorganisms but not to others, and the acidity resulting from carbonation and the addition of acids, e.g., citric, lactic, phosphoric, tartaric, and malic, inhibits the growth of organisms not tolerant to acidity. Also benzoic acid may be added as a preservative. Nonacid drinks like root beer and sarsaparilla are better culture media for spoilage organisms than acid drinks, such as the cola drinks, ginger ale, and fruit-flavored drinks. The ingredients of soft drinks not only affect the suitability for microbial growth but also can affect the kinds and numbers of microorganisms present and hence the likelihood of spoilage organisms being added. In addition, the bottles and closures are possible sources of contamination. The water for soft drinks is purified in regard to carbonate and mineral content and filtered. The filtration process may remove microorganisms, or, in the instance of a badly contaminated filter, add them. Ultraviolet irradiation sometimes is used to destroy microorganisms in the water. Discoloration of water and a flocculent precipitate may be caused by growth of algae. Treatment with chlorine or chlorine dioxide has been recommended to kill the algae, and filtration, e.g., through diatomaceous earth, to remove the flocculent dead cells.

Yeasts, chiefly *Torulopsis* and *Candida,* are the most likely causes of spoilage of soft drinks, for most such beverages are acid and contain sugar. One worker found 85 percent of 1,500 spoiled samples of carbonated beverages had been spoiled by yeasts. Since the sugars are a possible source of yeasts, the American Bottlers of Carbonated Beverages have set a standard of not more than ten yeasts per 10 g of dry sugar. Fruit concentrates are another possible source of yeasts.

Cloudiness and ropiness are types of spoilage of soft drinks. Cloudiness will result from marked growth of various yeasts or bacteria, and ropiness from the development of capsulated bacteria, most of which seem to be of the genus *Bacillus.* Bacteria may enter from ingredients, bottles, or closures. An *Achromobacter* sp. was found responsible for a musty odor and taste in root beer.

Since molds must have air, they cannot grow on carbonated beverages, but may develop at the surface of uncarbonated ones containing air above the liquid. They may come from sugar, coloring materials or flavoring materials, from the air, or from bottles or closures.

SPICES AND OTHER CONDIMENTS

The dry spices are not subject to spoilage normally, although mold growth during their drying may give them a heavy load of mold spores. Chip dips flavored with vegetables or spices usually have much higher total, coliform, and mold counts than those flavored with cheese. As has been mentioned, treatment of the spices with propylene oxide greatly reduces their content of microorganisms. Spices can be purchased with guaranteed low numbers of organisms.

Prepared mustard has been reported spoiled by yeasts and by species of *Proteus* and *Bacillus*, usually with a gassy fermentation. Horseradish seldom spoils, but on the contrary is bacteriostatic to bactericidal. The spoilage of vinegar is discussed in Chapter 24.

SALT

The three kinds of salt used in foods are: (1) *solar salt* from the evaporation of surface salt water; (2) *mined* or *rock salt;* and (3) *welled salt,* from salt dissolved from subterranean salt deposits. Solar salt contains halophiles, such as *Halobacterium* species and *Serratia salinaria.* About three-fourths of the bacteria are *Bacillus* organisms, and the rest mainly *Micrococcus* and *Sarcina.* Mined salt has been found to contain about 70 percent *Micrococcus,* 20 percent coryneforms, and 4 percent *Bacillus;* and putrefactive anaerobes have been found. Wet salt used on fish averages about 10 to 1,000 organisms per gram. Most purified salt, however, adds few organisms to foods.

NUTMEATS

Nutmeats in the shell are usually sterile or nearly so. Shelled nuts to be used as ingredients of foods, e.g., in frozen desserts, may be contaminated with bacteria, yeasts, and molds. The test for coliform bacteria is used most often to indicate possible contamination with fecal matter during handling. *Escherichia coli* does not survive long, however, on untreated English walnuts, unless they are kept refrigerated. Roasting, and heating in oil or sugar solution reduce the load of microorganisms.

REFERENCES

ANDERSON, E. E., W. B. ESSELEN, JR., and A. R. HANDLEMAN. 1953. The effect of essential oils on the inhibition and thermal resistance of microorganisms in acid food products. Food Res. 18:40–47.

APPLEMAN, M. D., E. P. HESS, and S. C. RITTENBERG. 1949. An investigation of a mayonnaise spoilage. Food Technol. 3:201–203.

BAIN, NORA, W. HODGKISS, and J. M. SHEWAN. 1958. The bacteriology of salt used in fish curing: the microbiology of fish and meat curing brines. II International Symposium on Food Microbiology, Proc. (1957), p. 1–11.

CLAYTON, W. 1931, 1932. The bacteriology of common salt. Food Manufacture 6:133, 257; 7:76–77, 109–110, 172–173.

CONNOR, J. W., 1950. Mayonnaise spoilage. Canad. Food Ind. 21:27.

CORRAN, J. W., and S. H. EDGAR. 1933. Preservative action of spices and related compounds against yeast fermentation. J. Soc. Chem. Ind. 52:149–152T.

EAGON, R. G., and C. R. GREEN. 1957. Effect of carbonated beverages on bacteria. Food Res. 22:687–688.

FABIAN, F. W., and MARY C. WETHINGTON. 1950. Spoilage in salad and French dressing due to yeasts. Food Res. 15:135–137.

FOTER, M. J., and A. M. GORLICK. 1938. Inhibitory properties of horse-radish vapors. Food Res. 3:609–613.

HARMON, L. C., C. M. STINE, and G. C. WALKER. 1962. Composition, physical properties and microbiological quality of chip-dips. J. Milk Food Technol. 25:7–11.

INSALATA, N. F. 1952. Balking algae in beverage water. Food Eng. 24(12):72–74, 197.

INSALATA, N. F. 1956. These bacteria checks prevent beverage spoilage. Food Eng. 28(4):84–86.

JACOBS, M. B. 1959. Manufacture and analysis of carbonated beverages. Chemical Publishing Company, Inc., New York.

JENSEN, L. B., and D. P. GRETTIE. 1937. Action of microorganisms on fats. Food Res. 2:97–120.

LEA, C. H. 1961. Some biological aspects of fat deterioration. Food Technol. 15(7):33–40.

LEHMANN, D. L., and B. E. BYRD. 1953. A bacterium responsible for a musty odor and taste in root beer. Food Res. 18:76–78.

MEYER, LILLIAN H. 1960. Food chemistry. Reinhold Publishing Corporation, New York.

NICOL, H. 1937. Watch your salt. Food Manufacture 12:111–113.

PERIGO, J. A., B. L. GIMBERT, and T. E. BASHFORD. 1964. The effect of carbonation, benzoic acid and pH on the growth rate of a soft drink spoilage yeast as determined by a turbidostatic continuous culture apparatus. J. Appl. Bacteriol. 27:315–332.

PEDERSON, C. S. 1930. Bacterial spoilage of a Thousand Island dressing. J. Bacteriol. 20:99–106.

RAINBOW, C., and A. H. ROSE. 1963. Biochemistry of industrial microorganisms. Academic Press Inc., New York.

SCHULTZ, H. W. 1960. Food enzymes. Avi Publishing Co., Inc., Westport, Conn.

SULTZER, B. M. 1961. Oxidative activity of psychrophilic and mesophilic bacteria on saturated fatty acids. J. Bacteriol. 82:492–497.

VAKIL, J. R., and J. V. BHAT. 1958. The microbiology of coconut oil. J. Univ. Bombay, Sect. B 27(3):83–89.

VOLLRATH, R. E., L. WALTON, and C. C. LINDEGREN. 1937. Bactericidal properties of acrolein. Soc. Exp. Biol. Med., Proc. 36:55–58.

WITTER, L. D., J. M. BERRY, and J. F. FOLINAZZO. 1958. The viability of *Escherichia coli* and a spoilage yeast in carbonated beverages. Food Res. 23:133–142.

WRIGHT, W. J., C. W. BICE, and J. M. FOGELBERG. 1954. The effect of spices on yeast fermentation. Cereal Chem. 31:100–112.

CHAPTER TWENTY-TWO

SPOILAGE OF HEATED CANNED FOODS

It has been pointed out in Chapter 6 that the heat-treatments given foods for their preservation may vary from a mild pasteurization treatment, as for milk or fruit juice, to an attempt to sterilize by steam under pressure, as for canned vegetables or soups. Obviously, the milder the heat-treatment administered, the less heat-resistant the microorganisms will need to be to survive and the greater will be the numbers and kinds of survivors. There is always the possibility that a surviving organism may grow and cause spoilage if environmental conditions permit.

CAUSES OF SPOILAGE

Spoilage of heated foods may have a chemical cause or a biological cause, or both. The most important kind of chemical spoilage of canned foods is the **hydrogen swell,** resulting from the pressure of hydrogen gas, released by the action of the acid of a food on the iron of the can. Hydrogen swells are favored by (1) increasing acidities of foods, (2) increasing temperatures of storage, (3) imperfections in the tinning and lacquering of the interior of the can, (4) a poor exhaust, and (5) the presence of soluble sulfur and phosphorus compounds. Other defects, caused by interaction between the steel base of the can and the contained food include (1) discoloration of the inside of the can, (2) discoloration of the food, (3) production of off-flavors in the food, (4) cloudiness of liquors or sirups, (5) corrosion or perforation of the metal, and (6) loss in nutritive value.

Biological spoilage of canned foods by microorganisms may result from either or both of two causes: (1) survival of organisms after the administration of the heat-treatment, and (2) leakage of the container after the heat process, permitting the entrance of organisms. Mild heat-treatments may be only enough to permit the successful storage of the foods for limited periods with aid of another preservative method such as refrigeration. Surviving microorganisms are likely to be of several kinds and may even include vegetative cells. Processing of meat loaves and pasteurization of milk are examples of such mild heat processes.

Acid foods, such as fruits, usually are filled into the can hot or are processed at temperatures approaching 100 C (212 F), treatments which result in the killing of all vegetative cells of bacteria, yeasts, and molds and their spores, and some bacterial spores. The only survivors ordinarily are spores of bacteria, which cannot grow in a very acid food. Any survivors of heat-treatments by steam under pressure are very heat-resistant bacterial spores, usually only one or two kinds.

Microorganisms entering through leaks in containers may be of various kinds, and are not necessarily heat-resistant. Leakage and subsequent spoilage of cans of food may be the result of mechanical damage of the empty cans so that side and end seams are defective; rough handling of filled cans may also result in damage. Microorganisms may enter from outer surfaces of filled cans that have become contaminated from equipment, especially if the cans are wet, or they may enter from contaminated cooling water after the heat process. Leakage also may cause a loss in can vacuum, thus encouraging chemical and microbial deterioration of the food. Spoilage resulting from leakage of the container, then, will be caused by the organism or combination of organisms that happens to enter. The presence of organisms known to be of low heat resistance, and especially of more than one kind of such organisms, is evidence of leakage.

APPEARANCE OF THE UNOPENED CONTAINER

Normally the ends of a can of food are termed "flat," which means that they are actually slightly concave; and a partial vacuum exists

Figure 22–1. Normal can of food and swells: (A) normal can with concave ends; (B) swollen can with ends bulged out: (C) swollen can with leakage where seams are breaking; (D) a burst can. (*Courtesy of Harper & Row, Publishers, Incorporated. From W. B. Sarles et al., Microbiology: General and Applied, 2nd ed., copyright, 1956.*)

within the container. If pressure develops inside, the can goes through a series of distortions as the result of increasing pressures (Figure 22–1) and is called successively a flipper, springer, soft swell, or hard swell. A **flipper** has flat ends, one of which will become convex when the side of the can is struck sharply or the temperature of the contents is increased. A **springer** has both ends of the can bulged, but one or both ends will stay concave if pushed in; or if a swollen end is pushed in, an opposite flat end will pop out. The terms flipper and springer are used by some to designate slight pressures in the can not caused by gas production but by such things as a poor exhaust, overfilling, denting of the can, changes in temperature, etc., but the can may have the same outward characteristics at the start of gas production from either a microbial or a chemical cause or both. A **soft swell** has both ends of the can bulged, but the gas pressure is low enough that the ends can be dented by pressure of the fingers. A **hard swell** has such high gas pressure from within that the ends are too hard to dent by hand. Often the high gas pressures distort or buckle the ends or side seam of the cans. The final step is the bursting of the can, usually through the side seam, but sometimes through the seals at the ends. A **breather** is a can with a minute leak that permits air to move in or out, but does not necessarily allow microorganisms to enter.

Other defects in the general appearance of the can are noted before and after it is opened: dents, which may be responsible for a flipper; rust; perforations; defective side seam or end seals; and corrosion.

The glass container of food under gas pressure may have its cover bulged or popped off, or may show leakage of food through the broken seal. Of course, it is possible to see any evidences of microbial growth through the glass sides, such as gas bubbles, cloudiness, and films of growth.

GROUPING OF CANNED FOODS ON THE BASIS OF ACIDITY

The acidity of canned foods is important in determining the heat process necessary for their sterilization and the type of spoilage to be expected if the process is inadequate or leakage takes place. Various groupings of canned foods have been made by workers of the National Canners Association, always with a division into the less-acid foods, with the pH above 4.5, and a more-acid group, with the pH below 4.5 (see Chapter 6). Cameron classified canned vegetables into four groups on the basis of their pH. For convenience other foods have been included in the groups, which follow:

Group I.   Low-acid foods, with a pH above 5.3. Peas, corn, and lima beans are vegetables in this group. Meats, fish, and poultry also are in this pH range.

Group II. Medium-acid foods, with a pH between 5.3 and 4.5. Spinach, asparagus, beets, and pumpkin usually are within this range.

Group III. Acid foods, with a pH between 4.5 and 3.7. Tomatoes, pears, and red cabbage have this acidity.

Group IV. High-acid foods, with a pH below 3.7. Sauerkraut and berries are examples.

Some of the vegetables and fruits have pH values near the border lines of divisions between groups. Pumpkin, carrots, spinach, asparagus, green beans, and beets might be in Group I or II, and fruit cocktail, peaches, apricots, and sliced pineapple might be in Group III or IV. Some foods may be artificially acidified before canning, e.g., onions and artichokes.

TYPES OF BIOLOGICAL SPOILAGE OF CANNED FOODS

Types of spoilage of canned foods by microorganisms usually are divided into those caused by thermophilic bacteria and those caused by mesophilic microorganisms. Other methods of classification of kinds of spoilage are based on the kinds of changes produced in the food, e.g., putrefaction, acid production, gas formation, blackening, etc. Types of spoilage also may be grouped on the basis of the kinds of foods involved.

The three most important kinds of biological spoilage of commercially canned foods (described later) are flat sour spoilage, "T.A." spoilage, and putrefaction. A fourth important kind of spoilage, caused by action of food acid on the iron of the can, results in a "hydrogen swell."

Types of spoilage by thermophilic spore-forming bacteria

Most spoilage of commercially heat-processed canned foods resulting from underprocessing is caused by thermophilic bacteria, because their spores are more heat-resistant than those of most mesophilic bacteria. The three chief types of spoilage by thermophiles are flat sour spoilage, "T.A." spoilage, and sulfide spoilage.

FLAT SOUR SPOILAGE. This kind of spoilage derives its name from the fact that the ends of the can of food remain "flat," that is, have a normal concavity, during "souring," or the development of lactic acid in the food by the flat sour bacteria. Because the can retains a normal outward appearance, this type of spoilage cannot be detected by examination of the unopened can but must be detected by cultural methods. The spoilage occurs chiefly in low-acid foods, such as peas or corn (Group I), and is caused by species of *Bacillus*. Flat sour spoilage of acid foods, e.g., tomatoes or tomato juice, is caused by a special, facultatively thermophilic species, *B. coagulans*. The various species of *Bacillus* that are able to form acid without gas in food may be mesophiles, facultative thermophiles, or obligate thermophiles. The spores of the mesophiles are the least heat-resistant and are usually killed by the heat processing and hence are rarely concerned with flat sour spoilage of low-acid foods. The spores of the thermophiles, on the other hand, are considerably more heat-resistant and may survive the heat process to cause flat sour spoilage. Surviving obligate thermophiles, such as *B. stearothermophilus* and *B. pepo*, would not cause spoilage unless the food were held hot for a while, as in slow cooling or storage in the tropics, but facultative thermophiles could grow at ordinary temperatures. The immediate source of the flat sour bacteria is usually the plant equipment, e.g., the blanchers, but they may come originally from sugar, starch, or soil.

The ability of *B. coagulans* to grow in tomato juice depends upon the number of spores present, the availability of oxygen, and the pH of the juice. The more spores there are present, the lower will be the minimal pH for germination (ca. 5.0) and for growth (ca. 4.2 to 4.3). The organism, which is homofermentative under almost anaerobic conditions and heterofermentative under aerobic conditions, can grow in low concentrations of oxygen (e.g., 0.1 mm of mercury).

"T.A." SPOILAGE. The bacterium causing this type of spoilage has been nicknamed "T.A.," which is short for "thermophilic anaerobe not producing hydrogen sulfide," or for the species *Clostridium thermosaccharolyticum*. This is a sugar-splitting, obligately thermophilic, spore-forming anaerobe that forms acid and gas in low- and medium-acid foods (Groups I and II). The gas, a mixture of carbon dioxide and hydrogen, swells the can if it is held long enough at a high temperature, and may eventually cause bursting. The spoiled food usually has a sour odor. Since the organism does not form colonies readily in agar, it is detected usually by the inoculation of liquid media, such as liver corn mash, liver broth with lumps of liver, or thioglycolate broth, incubation at 55 C, and examination for gas production and a sour odor. Sources are the same as for flat sour bacteria.

SULFIDE, OR "SULFUR STINKER," SPOILAGE. This spoilage, caused by *Clostridium nigrificans*, is found, and then uncommonly, in low-acid foods like peas and corn. The spores of this bacterium have considerably less heat resistance than those of flat sour and T.A. bacteria and hence their appearance in canned foods is indicative of gross underprocessing. The organism is an obligate thermophile and therefore also requires poor cooling of the heat-processed foods or hot storage for its development. It is detected by means of black (FeS) colonies it forms in an iron-sulfite agar at 55 C. Hydrogen sulfide, formed in the canned peas or corn, is evident by odor when the can is opened. In corn, a bluish-gray liquid is evident in which blackened germs and gray kernels of corn float. Peas usually give the H_2S odor, but without any marked discoloration. Sources of the spores are similar to those for flat sour and T.A. bacteria, but manure also can be an original source.

Types of spoilage by mesophilic spore-forming bacteria

Most spoilage by mesophilic microorganisms that results from underprocessing is caused by spore-forming bacteria of the genera *Bacillus* and *Clostridium*, but lightly heated foods, e.g., some acid ones, may permit the survival of and spoilage by non-spore-forming bacteria or even yeasts or molds.

SPOILAGE BY MESOPHILIC *Clostridium* SPECIES. Species of *Clostridium* may be sugar fermenting, e.g., *C. butyricum* and *C. pasteurianum*, and cause the butyric acid type of fermentation in acid or medium-acid foods, with swelling of the container by the carbon dioxide and hydrogen gas produced. Other species, like *C. sporogenes*, *C. putrefaciens*, and *C. botulinum*, are proteolytic or putrefactive, decomposing proteins with the production of malodorous compounds such as hydrogen sulfide, mercaptans, ammonia, indole, and skatole. The putrefactive anaerobes also produce carbon dioxide and hydrogen, causing the can to swell. The spores of some of the putrefactive anaerobes are very heat-resistant; therefore putrefaction joins flat sour and T.A. spoilage to comprise the chief types of biological spoilage of canned foods resulting from underprocessing.

Because the spores of the saccharolytic clostridia, sometimes called "butyrics," have a comparatively low heat resistance, spoilage by these anaerobes takes place most commonly in canned foods which have been processed at 212 F (100 C) or less, as are many of the commercially canned acid foods, and the home-canned foods processed by the hot water, flowing stream, or oven method. Thus the canned acid foods, such as pineapple, tomatoes, and pears, have been found spoiled by *C. pasteurianum*. Such spoilage is more likely when the pH of the food is above 4.5. Home-canned foods heated to about 212 F may be spoiled by the saccharolytic bacteria with the production of butyric acid, carbon dioxide, and hydrogen.

The putrefactive anaerobes grow best in the low-acid canned foods (Group I), such as peas, corn, meats, fish, and poultry, but on rare occasions may spoil medium-acid foods (Group II). One of the putrefiers, *C. botulinum,* is a cause of food poisoning and will be discussed in Chapter 26.

SPOILAGE BY MESOPHILIC *Bacillus* SPECIES. Spores of various species of *Bacillus* differ considerably in their heat resistance, but in general the spores of the mesophiles are not as resistant as those of the thermophiles. Spores of many of the mesophiles are killed in a short time at 212 F or less, but a few kinds can survive the heat-treatments employed in steam-pressure processing. The spores that survive do not necessarily cause spoilage, for conditions in the can of food may be unfavorable for germination and growth. Many species of *Bacillus* are aerobic and therefore cannot grow in a well-evacuated container. The food may be too acid for the bacteria, or the medium may be unfavorable otherwise. There have been reports, however, of *B. subtilis, B. mesentericus,* and other species growing in low-acid home-canned foods that had been given a heat processing at 212 F. Commercially canned foods have been spoiled by *Bacillus* species, especially in poorly evacuated cans. Foods so spoiled have been mostly canned seafood, meats, and evaporated milk. Recently, the aerobacilli, or gas-forming *Bacillus* species (*B. polymyxa* and *B. macerans*), have been reported to cause spoilage of canned peas, asparagus, spinach, peaches, and tomatoes, but there is some doubt whether they survived the heat process. Entrance may have been through a leak in the container. Spores of these bacteria have about the same heat resistance as those of *Clostridium pasteurianum.*

Spoilage by non-spore-forming bacteria

If viable non-spore-forming bacteria are found in canned foods, either a very mild heat-treatment was used or the bacteria entered the container through a leak. Vegetative cells of some kinds of bacteria are fairly heat-resistant in that they can withstand pasteurization. Among these "thermoduric" bacteria are the enterococci, *Streptococcus thermophilus,* some species of *Micrococcus* and *Lactobacillus,* and *Microbacterium.* Acid-forming *Lactobacillus* and *Leuconostoc* species have been found growing in underprocessed tomato products, pears, and other fruits. The heterofermentative species may release enough CO_2 gas to swell the can. Micrococci have been reported in meat pastes and

in similar products with very poor heat penetration, and S. *faecalis* or
S. *faecium* is often present in canned hams that are only partially sterilized, and may be responsible for spoilage on storage.

Usually, however, the presence of viable non-spore-forming bacteria
in heat-processed canned foods is indicative of leakage of the container.
Since the cooling water is most frequently the source of contamination,
types of bacteria commonly found in water usually cause spoilage of
the leaky cans. Some of these, e.g., the coliform bacteria, produce gas
which swells the cans. The tiny orifice through which the bacteria enter
apparently becomes tightly plugged by a food particle, permitting the
accumulation of gas pressure in the can. It should be noted that spore-
forming bacteria also can enter the can through a leak, so that the aerobacilli (*B. macerans* and *B. polymyxa*) or the clostridia could be responsible for the gas formation. Often non-gas-forming bacteria are
found growing in the food in leaky cans, along with the gas-former
or by themselves. It is possible, of course, that only one kind of bacterium may enter, so that an apparently pure culture would be growing.
Non-spore-forming, non-gas-forming bacteria that may enter include
those in the genera *Pseudomonas, Achromobacter, Micrococcus, Flavobacterium, Proteus,* and others. Less common non-spore-formers other
than those of water may enter through leaks in the cans and cause
spoilage.

Spoilage by yeasts

Since yeasts and their spores are killed readily by most pasteurization treatments, their presence in canned foods is the result of either
gross underprocessing or leakage. Canned fruits, jams, jellies, fruit juices,
sirups, and sweetened condensed milk have been spoiled by fermentative
yeasts with swelling of the cans by the CO_2 produced. Film yeasts
may grow on the surface of jelled, pickled pork, repacked pickles or
olives, and similar products, but their presence indicates recontamination
or lack of heat processing, plus poor evacuation.

Spoilage by molds

Molds probably are the most common cause of the spoilage of
home-canned foods, which they enter through a leak in the seal of
the container. Jams, jellies, marmalades, and fruit butters will permit
mold growth when sugar concentrations are as high as 70 percent and
in the acidity usually present in these products. It has been claimed
that adjustment of the soluble extract of jam to 70 to 72 percent sugar
in the presence of a normal 0.8 to 1.0 percent acid will practically remove

the risk of mold spoilage. Strains of *Aspergillus, Penicillium,* and *Citromyces,* found growing in jellies and candied fruits, have been found able to grow in sugar concentrations up to 67.5 percent. Acidification to pH 3 prevented growth of the first two molds, and heating the foods at 90 C for 1 min killed all strains. Some molds are fairly resistant to heat, especially those forming the tightly packed masses of mycelium called sclerotia. *Byssochlamys fulva* is a pectin-fermenting mold that has ascospores that have resisted the heat processing of bottled and canned fruits and have caused spoilage. The high sugar content of sirups about fruits helps protect microorganisms against heat, and this protective effect is increased if the sugar and fruit are added to the can separately, localizing the sugar during processing.

Spoilage of canned foods of different acidities

The previous discussion has indicated that the grouping of commercially canned foods on the basis of pH divides them also on the basis of the chief types of spoilage to be expected. Thus the low-acid foods of Group I, with a pH above 5.3, are especially subject to flat sour spoilage and putrefaction. The medium-acid foods of Group II, with a pH between 5.3 and 4.5, are likely to undergo T.A. spoilage. Acid foods of Group III, with a pH between 4.5 and 3.7, usually are spoiled by a special flat sour bacterium (*Bacillus coagulans*) or by a saccharolytic anaerobe. High-acid foods of Group IV, with a pH below 3.7, ordinarily do not undergo spoilage by microorganisms, but in cans may become hydrogen swells (as may canned foods in Group III).

It has been noted that some canned foods may fall into either of two adjacent pH groups. Therefore some lots of pumpkin, carrots, spinach, asparagus, green beans, and beets may be low-acid foods (Group I) and therefore most likely to undergo flat sour spoilage, while other lots may be medium-acid foods (Group II) and therefore subject usually to T.A. spoilage. When the pH of canned pineapple, pears, pimientos, and chili is above 4.5, spoilage by saccharolytic anaerobes like *Clostridium pasteurianum* is much more likely than when the pH is below 4.5.

Spoilage of canned meats and fish

In general, canned meats and fish exhibit two chief types of spoilage: (1) by *Bacillus* species, resulting in softening and souring; and (2) by *Clostridium* species (e.g., *C. sporogenes*), producing putrid swells. Less commonly, bacilli of the *Aerobacillus* type produce acid and gas and swell the cans. Even non-spore-formers have been found in some pastes. Besides spoilage by these organisms, mostly spore-formers that have survived the heat process, there may be spoilage by organisms entering through leaks.

Canned cured meats which are given a heat process insufficient for sterilization, such as ham or luncheon meats, may be subject to production of carbon dioxide, nitrogen oxides, or nitrogen gas by species of *Bacillus* (e.g., *B. licheniformis, B. coagulans, B. cereus,* or *B. subtilis*) from the nitrate, sugar, and meat; or they may be subject to putrefaction with gas produced by *Clostridium* species (e.g., *C. sporogenes*). Such spoilage ordinarily is prevented by adequate refrigeration. Gas also may be produced by heterofermentative lactic acid bacteria (e.g., leuconostocs), but only after inadequate processing. Spoilage without gas production, but with souring and changes in color and texture, may be caused by species of *Bacillus* and by homofermentative lactic acid bacteria (e.g., *Streptococcus faecium* or *faecalis*).

Unusual types of spoilage of canned foods

Certain types of spoilage seem to be limited to one or two kinds of food. Sulfide spoilage has been reported only for peas and sweet corn. Black beets are caused by the mesophilic *Bacillus betanigrificans* in the presence of a high content of soluble iron. This is distinct from the darkening resulting from a deficiency of boron in the soil. *Bacillus* species can cause bitterness, acidity, and curdling in canned milk, cream, and evaporated milk. The only important alkaline canned vegetable, hominy (pH 6.8 to 7.8), undergoes flat sour spoilage which is characterized by a sweetish taste. Canned poultry is more often spoiled by putrefactive than by saccharolytic clostridia, chiefly because of the lower heat resistance of spores of the latter.

Canned sweetened condensed milk may become gassy because of growth of yeasts or coliform bacteria, may become thickened by a *Micrococcus* species, or may exhibit "buttons," which consist of small masses of mold mycelium and coagulated milk usually on the surface of the milk. The size of the buttons is limited by the quantity of free oxygen in the head space of the can.

Sometimes, as the result of **autosterilization,** no viable organisms can be found in cans of food that have undergone obviously biological spoilage. All vegetative cells have died, including those of spore-formers that did not sporulate.

A scheme for the diagnosis of the cause of spoilage of a canned food is outlined in Figure 22–2. This scheme applies only to the commonly encountered types of spoilage.

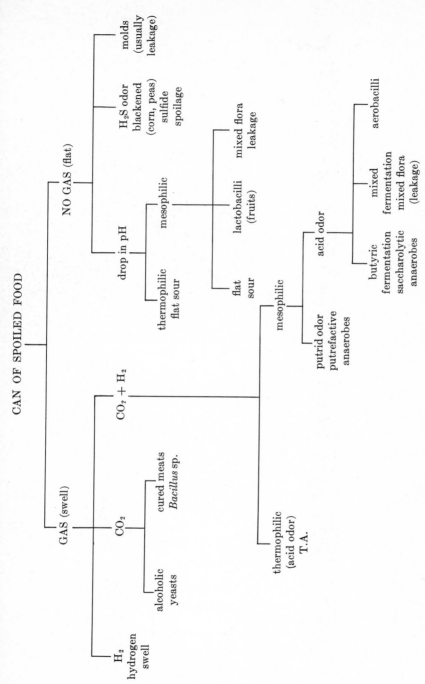

Figure 22–2. Scheme for diagnosis of cause of spoilage of a canned food.

REFERENCES

ARTIFEX. 1935. Further investigations in jam manufacture. Food 4:412–414.

ASCHEHOUG, V., and E. JANSEN. 1950. Studies on putrefactive anaerobes as spoilage agents in canned foods. Food Res. 15:62–67.

AYRES, J. C., and A. T. ADAMS. 1953. Occurrence and nature of bacteria in canned beef. Food Technol. 7:318–323.

BOWEN, J. F., C. C. STRACHAN, and A. W. MOYLS. 1954. Further studies of butyric fermentations in canned tomatoes. Food Technol. 8:471–473.

BUTTIAUX, R., and H. BEERENS. 1955. Gas-producing mesophilic clostridia in canned meats, with improved techniques for their identification. J. Appl. Bacteriol. 18:581–590.

CAMERON, E. J., and J. R. ESTY. 1940. Comments on the microbiology of spoilage in canned foods. Food Res. 5:549–557.

DAKIN, J. C., and PAMELA M. DAY. 1958. Yeasts causing spoilage in acetic acid preserves. J. Appl. Bacteriol. 21:94–96.

GOLDBLITH, S. A., M. A. JOSLYN, and J. T. R. NICKERSON (Eds.) 1961. An introduction to the thermal processing of foods. Avi Publishing Co., Inc., Westport, Conn. (Contains reproductions of early papers on bacteriological spoilage of canned foods.)

HERSON, A. C., and E. D. HULLAND. 1964. Canned foods: an introduction to their microbiology. Chemical Publishing Company, Inc., New York.

HUNWICKE, R. F., 1928. Bacteria and the canning industry. Food Manufacture 1:19–20; 2:179–180, 184.

JARVIS, N. D., 1940. Spoilage in canned fishery products. Canning Age 21:434–436, 444, 476–477.

JENSEN, L. B., 1954. Microbiology of meats, chaps. 11, 12, 13. 3rd ed. The Garrard Press, Champaign, Ill.

POTTER, R. S., 1935. Jam troubles. Food Manufacture 10:232–233.

RANGASWAMI, G., and R. VENKATESAN. 1959. Studies on the microbial spoilage of canned food: I, Isolation and identification of some spoilage bacteria. Indian Acad. Sci., Proc., Sect. B. 50:349–359.

RIEMANN, H., 1957. Bacteriology of canned fish. Food Manufacture 32:265–267, 333–335.

TANIKAWA, E. 1958. Studies on the technical problems in the processing of canned salmon. Mem. Fac. Fish. Hokkaido Univ. 6:67–138.

VAUGHN, R. H., I. H. KREULEVITCH, and W. A. MERCER. 1952. Spoilage of canned foods caused by the Bacillus macerans-polymyxa group of bacteria. Food Res. 17:560–570.

PART FOUR

FOODS AND ENZYMES PRODUCED BY MICROORGANISMS

Microorganisms themselves may serve as food or feed, may be employed in the preparation of special nutrients, such as organic acids or vitamins, to be added to foods, may be used in the production of special foods by fermentation, or may serve as sources of enzyme mixtures or single enzymes for the treatment of foods during processing. The appropriate cultures for these various purposes must be maintained, usually in pure culture, in a stable yet active condition, and must be built up to considerable volume for use as mass or bulk cultures for a particular process.

CHAPTER TWENTY-THREE

PRODUCTION OF CULTURES
FOR FOOD FERMENTATIONS

Microorganisms necessary in food fermentation may be added as pure cultures or mixed cultures, or, in some instances, no cultures may be added if the desired microorganisms are known to be present in sufficient numbers in the original raw material. In the food fermentations discussed in Chapter 15 for the manufacture of sauerkraut, fermented pickles, and green olives, and those in Chapter 24 for the processing of cocoa, coffee, poi, and citron, the original raw product carries a sufficiency of the desired organisms, and these will act in proper succession if favorable environmental conditions are provided and maintained. Therefore, the addition of pure or mixed cultures of the organisms responsible for the fermentations has not been found necessary or even advantageous. On the other hand, controlled "starter" cultures, pure or mixed, usually are employed in the manufacture of certain dairy products (Chapter 20), such as fermented milks, some kinds of butter, and most types of cheese, and in most of the food fermentations discussed in Chapter 24, e.g., bread, malt beverages, wines, distilled liquors, and vinegar.

GENERAL PRINCIPLES OF CULTURE MAINTENANCE
AND PREPARATION

Selection of cultures

Cultures for food fermentations are selected primarily on the basis of their stability and their ability to produce desired products or changes efficiently. These cultures may be established ones obtained from other laboratories or may be selected after the testing of numerous strains. Stability is an important characteristic; yields and rates of changes must not be erratic. Some cultures may be improved by breeding, e.g., the sporogenous yeasts, but selection is the most commonly used method for the improvement of strains.

Maintenance of activity of cultures

Once a satisfactory culture has been obtained, it must be kept pure and active. Usually this objective is attained by periodic transfer of the culture into the proper culture medium, incubation until the culture reaches the maximal stationary phase of growth, and then storage at temperatures low enough to prevent further growth. Too frequent transfer of an unstable culture may lead to undesirable changes in its characteristics.

Also, stock cultures should be prepared for storage of cultures over long periods without transfer. Such cultures tend to remain stable and serve as a source of culture if the active culture deteriorates or is lost. Lyophilization (freeze drying) is now frequently used to prepare stock cultures, although some use still is made of a paraffin-oil seal over ordinary tube cultures. Bacterial cultures have been preserved for months to years at room temperature on slants of agar in which one percent NaCl had been incorporated. A dry spore stock on sterilized soil can be used to preserve spores of bacteria or molds for long periods.

Maintenance of purity of cultures

To ensure the purity of cultures they should be obtained periodically from a culture laboratory or be checked regularly for purity. Methods for testing a culture for purity will vary with the type of culture being tested. Microscopic examination will indicate contamination only if the contaminant differs from the desired organism in appearance and is high in numbers. Another method is to plate the culture with an agar medium that will grow contaminants but not the desired organism. Tests may be made for the presence of substances not produced by the desired organism, e.g., for catalase in a culture of catalase-negative lactic acid bacteria as indicative of the presence of catalase-positive contaminants.

Preparation of cultures

Mother culture is prepared daily, usually, from a previous mother culture and originally from the stock culture. These mother cultures can be used to inoculate a larger quantity of culture medium to produce the mass or bulk culture to be used in the fermentation process. Often, however, the fermentation is on such a large scale that several intermediate cultures of increasing size must be built up between the mother culture and the final bulk or mass culture. The culture maker attempts to produce and maintain a culture that (1) contains only the desired microorganism(s); (2) is uniform in microbial numbers, proportions (if a mixed culture), and activity from day to day; (3) is active in producing the products desired; and (4) has adequate resistance to unfavorable conditions, if necessary, e.g., heat resistance, if it has to take heating in a cheese curd. He tries to maintain uniformity by standardizing methods of preparation and sterilization of the culture medium, of inoculation, and of incubation temperature and time. The stage of growth to which he will grow his culture will depend upon the purpose for which it is to be used. If he wishes prompt and rapid growth, he uses a culture that is late in its logarithmic phase of growth. If he wishes more resistance to heat or other unfavorable conditions, he uses a culture that has just entered the maximal stationary phase. The temperature of incubation usually is somewhere near the optimal temperature for the organism, although there are exceptions. Temperature and time of incubation often are adjusted so that the culture will be ready at the time it is needed. Otherwise it may have to be cooled to stop further development.

Activity of culture

The activity of a culture is judged by its rate of growth and production of products. It should be good if the mother or the intermediate culture is satisfactory and culture medium, incubation time, and temperature are optimal. Deterioration of cultures may result from wrong nandling and cultivation, frequent transfer over long periods in an inadequate culture medium, selection, variation or mutation, or attack of bacteria by bacteriophage.

Mixed cultures

Known mixtures of pure cultures sometimes are prepared, being either grown together continuously or grown separately and mixed at the time of use. The so-called butter, or lactic, culture used in the dairy industry is an example of a mixture of several species of bacteria growing together and sometimes several strains of individual species. When different strains of the same species or different species are grown together, these organisms must be compatible, that is, grow well together without one or more causing the elimination of others. The maintenance of the desired balance of kinds of organisms within these mixed cultures is difficult.

Unknown mixtures of organisms are present in "starters" used in some food products. An example is the dough carried over from one lot of special French bread to a succeeding lot, or the mixture of yeasts and bacteria carried from the surface smear of one Limburger cheese to the surface of another.

BACTERIAL CULTURES

Most of the bacterial cultures that are employed as "starters" for dairy products, sausage, and bread are pure or mixed cultures of lactic acid bacteria, exceptions being the propionic acid bacteria added to Swiss cheese and the surface-smear organisms on some types of cheese. Acetic acid bacteria are used in vinegar making, and special bacteria are used for the manufacture of certain enzymes.

Lactic acid cultures

The most commonly used of the dairy starters is the common butter, or lactic, starter, which ordinarily consists of a mixture of strains of *Streptococcus lactis* and *S. cremoris* for the production of lactic acid and *Leuconostoc dextranicum* and *L. citrovorum* or *Streptococcus diacetilactis* for the production of flavor and aroma. Several strains of lactic streptococci of different sensitivities to bacteriophage usually are included in a culture as some insurance against phage trouble. *Leuconostoc* strains often are included in cultures for making ripened cheeses because the combination of lactic streptococci and aroma-formers seems to yield a better acid-producing culture than a pure culture of a lactic streptococcus, although the aroma-formers may tend to cause an open texture in the cheese. The lactic culture usually is incubated at about 70 to 72 F (21.1 to 22.2 C), although 75 to 80 F (23.9 to 26.7 C) has been employed in making starters for cheese; and the culture is ripened to a titratable acidity that will vary with the purpose for which the culture is to be used.

The mixed lactic culture is used in the manufacture of cultured buttermilk, butter, and most types of cheese in which the curd is heated at a comparatively low temperature, e.g., cottage, cream, Limburger, Cheddar, blue, and brick cheese. The aroma bacteria are especially important in flavor production in cultured buttermilk, butter, and the un-cured cheeses, and the culture must be handled so as to keep these bacteria numerous and active, as well as the lactic streptococci. This is accomplished by ripening the starter to a fairly high acidity in both mother and bulk starters, 0.85 percent lactic acid or higher. On the other hand, a lower acidity that would favor the aroma-formers less would be better for a starter for Cheddar cheese.

Although some dairy plants have their own cultures that they have carried successfully for years, most operators obtain new cultures periodically from a culture laboratory. Most laboratories distribute dry cultures prepared from milk cultures by lyophilization or freeze drying. A few cultures still are distributed as liquid milk cultures, but such cultures are shorter lived than the dry kinds and, unless fresh, are less active. The user of the culture activates it by successive transfer, if necessary, and then prepares daily mother cultures to be used in inoculating the larger bulk cultures for making a dairy product. He tries to maintain a uniform culture that is active in acid production and in aroma production, if that is desired.

High-temperature lactic acid bacteria, growing best at 37 C (98.6 F) or above are used as starters for cheeses such as Swiss, whose curd is cooked at a fairly high temperature, and for manufacture of lactic acid. Commonly used in cheese are *Streptococcus thermophilus* and a thermoduric lactobacillus, *Lactobacillus bulgaricus*, *L. lactis*, or *L. helveticus*. These may be used as separate pure cultures, or the streptococcus may be grown with one of the lactobacilli in milk or in whey. Or the cheese maker may prepare an impure, mixed culture by pasteurizing and incubating whey from the cheese vat or kettle. Dry cultures of some of these species now are available. *Lactobacillus delbrueckii* or *Bacillus coagulans* is employed for manufacture of lactic acid from sucrose, invert sugar, or glucose, and *L. bulgaricus* for production from whey.

As mentioned in Chapter 24, the "sour" used by the maker of rye bread is a varying mixture of lactic acid bacteria ordinarily grown in mixed and impure culture in flour paste, dough, or other medium. It is claimed that *L. bulgaricus* must grow if enough acid is to be produced and that heterofermentative lactics are desirable from the standpoint of flavor production. Some makers grow pure cultures of various lactic acid bacteria to inoculate the dough: *Streptococcus lactis*, *Leuconostoc* species, *Lactobacillus plantarum*, *L. casei*, *L. brevis*, *L. bulgaricus*, *Streptococcus thermophilus*, and others. Some makers also add yeasts.

Pediococcus cerevisiae cultures are used as starters for summer, Thuringer, and similar fermented sausages. Broth cultures or, preferably, dried cultures have been prepared.

Propionic culture

Cultures of *Propionibacterium shermanii* have been added as lactose, peptone broth cultures, or dried cultures to Swiss cheese to improve the flavor and assist eye formation. Cultures have been dried by the spray-drying process or by lyophilization.

Cheese smear organisms

Most cheese makers trust to contamination of the surfaces of smear-ripened cheese from previous cheeses, shelves, cloths, brine tank, hands, and other sources in the plant. The micrococci, *Brevibacterium linens,* and film yeasts important in the smear have been isolated, however, and used in pure or mixed cultures to wash the cheese surface and thus to inoculate it.

Acetic acid bacteria

Pure cultures of vinegar bacteria have not proved efficient in the production of acetic acid. Therefore, in vinegar making, impure, mixed cultures are allowed to develop naturally as in the home method, are added by means of raw vinegar from a previous run, as in the Orleans process or the quick process, or are transferred from a vinegar cask or generator.

Other bacterial cultures

Pure cultures of selected strains of bacteria are employed for specific purposes in other parts of the food industries: *Acetobacter suboxydans* for the oxidation of D-sorbitol to L-sorbose in the manufacture of ascorbic acid (vitamin C); *Leuconostoc mesenteroides* in the production of dextran; *Bacillus subtilis* in the production of α-amylase and protease; *Brevibacterium glutamicum* and an array of other organisms for the production of glutamic acid and other amino acids.

YEAST CULTURES

Most yeasts of industrial importance are of the genus *Saccharomyces,* and mostly of the species *S. cerevisiae.* These ascospore-forming yeasts are readily bred for desired characteristics. A yeast for a given purpose may be improved for that use but must also be guarded against possible undesirable changes.

Bakers' yeast

Strains of *S. cerevisiae* used in the manufacture of bakers' yeast are usually single-cell isolates that have been selected especially for the purpose. They should give a good yield of cells in the mash or medium chosen for their cultivation, should be stable in their characteristics, should remain viable in the cake or dried form for a reasonably long period during storage before use, and should produce carbon dioxide rapidly in the bread dough when used for leavening.

As with other cultures used on a large scale, this culture is built up from the original mother culture through several intermediate cultures of increasing size to the final "seed" culture. The cells from the seed-culture tank are concentrated into a "cream" by centrifugation, and this heavy suspension is added to the large volume of mash in which the yeast is to be grown, so that about 3 to 5 lb of yeast are added per 100 gal of medium.

The most commonly used medium for the build-up of cultures and the production of bakers' yeast is a molasses–mineral-salts mash that contains molasses, nitrogen foods in the form of ammonium salts, urea, malt sprouts, etc., inorganic salts as phosphates and other mineral salts, and accessory growth substances in the form of extracts of vegetables, grain, or yeast, or small quantities of vitamin precursors or vitamins. The pH is adjusted to about 4.3 to 4.5, and the incubation temperature is around 30 C (86 F). During growth of the yeast the medium is aerated at a rapid rate, and molasses is added gradually to maintain the sugar level at about 0.5 to 1.5 percent. After four or five budding cycles, the yeast is centrifuged out in the form of a "cream," which is put through a filter press to remove excess liquid. The mass of yeasts is made into cakes of different sizes after incorporation of small amounts of vegetable oils.

Active dry yeast now is made by drying the yeast cells to less than 8 percent moisture. Cells so dried are grown especially for the purpose and are dried carefully at low temperatures so that most of the cells will survive and during storage at room temperature will retain for some months their ability to actively leaven dough.

Bakers' yeast also can be prepared from grain mashes, waste sulfite liquor from paper mills, wood hydrolyzate, and other materials.

Yeasts for malt beverages

Yeasts for malt beverages may be carried in pure culture in brewery laboratories or obtained when needed from specialized laboratories. The strain selected is one intended for the product to be made, a special bottom yeast for beer, usually a top yeast for ale but sometimes a bottom yeast, and a top yeast for stout and porter. The top yeasts used are strains of *Saccharomyces cerevisiae,* and the bottom yeast used is S. *carlsbergensis.* When the yeast is started from a pure culture it must be built up in wort from a laboratory culture to a final large seed or "pitching" culture. In practice, however, the pitching yeast is nearly always yeast recovered from a previous fermentation. The recovered yeast is concentrated and may or may not be washed. It is obvious that such a yeast culture will always be contaminated with other organisms, which ordinarily include bacteria and wild yeasts that build up during successive fermentations and recoveries. Fortunately, most of the contaminants, although able to grow in the yeast culture, are inhibited by the hops and low pH in the wort and do not damage the malt beverage appreciably. It is possible, however, for organisms causing "beer diseases" (see Chapter 24) to build up in the pitching yeast.

Wine yeasts

For wine making, a special strain of S. *cerevisiae* var. *ellipsoideus*, adapted to the making of the specific type of wine, is selected. Famous types are the Burgundy, Tokay, and champagne cultures so widely used. The cultures are grown and built up in volume in must (juice of the grape or other fruit) like that to be used in the main fermentation.

Distillers' yeast

Distillers' yeast ordinarily is a high-alcohol-yielding strain of *Saccharomyces cerevisiae* var. *ellipsoideus*, usually one adapted to growing in the medium or mash to be employed. The medium or mash would be malted grain, usually corn or rye for whiskey, molasses for rum, or juices or mashes of fruits for brandy. The liquors are distillates of the fermented mashes.

MOLD CULTURES

Stock cultures of molds usually are carried on slants of a suitable agar medium, e.g., malt-extract agar, and may be preserved in the spore state for long periods by lyophilization (freeze drying) or as soil stocks (see page 371). There are a number of different ways of preparing spore or mycelial cultures for use on a plant scale. These include (1) surface growth on a liquid or agar medium in a flask or similar container; (2) surface growth on media in shallow layers in trays; (3) growth on loose, moistened wheat bran which may be acidified or may have liquid nutrient added, e.g., corn-steep liquor; (4) growth on previously sterilized and moistened bread or crackers; or (5) growth by the submerged method in an aerated liquid medium, usually resulting in pellets composed of mycelium, with or without spores. The mold spores are re-covered in different ways, depending upon the method of production. They may be washed or drawn from dry surfaces, may be left in dry material that is ground up or powdered, or for convenience in use may be incorporated in some dry powder, e.g., flour. The pellets, of course, are used as such.

Spores of *Penicillium roqueforti* for blue cheeses, Roquefort, Stilton, Gorgonzola, etc., usually are grown on cubes of sterilized, moistened, and usually acidified bread; whole wheat or bread of a special formula may be employed. After the sporulation of the mold is complete, bread and culture are dried, powdered, and packaged, commonly in cans.

P. camemberti spores are prepared by growing the mold on moist-ened sterile crackers. A spore suspension is prepared for the inoculation of the surface of the Camembert, Brie, or similar cheese.

Mold starters to be used as inoculum in industrial fermentations where products are to be produced by the submerged method usually are prepared in the form of pellets or masses of mycelium that are produced during submerged growth while the culture is being actively shaken. When surface growth is desired on liquid or agar medium or on bran, mold spores, produced by the methods listed above, ordinarily serve as the inoculum.

The koji, or starter, for soy sauce, described in the next chapter, usually is a mixed culture carried over from a previous lot, although pure cultures of *Aspergillus oryzae*, together with a yeast and *Lactobacillus delbrueckii*, have been used. The mold culture is grown on cooked, sterilized rice.

REFERENCES

AMERINE, M. A., and W. V. CRUESS. 1960. The technology of wine making. Avi Publishing Co., Inc., Westport, Conn.

CHANCE, H. L. 1963. Salt—a preservative for bacterial cultures. J. Bacteriol. 85:719–720.

FOSTER, E. M., F. E. NELSON, M. L. Speck, R. N. DOETSCH, and J. C. OLSON, JR. 1957. Dairy microbiology. Prentice-Hall, Inc., Englewood Cliffs, N.J.

FRY, R. M. 1954. The preservation of bacteria. *In* R. J. C. Harris (*Ed.*) Biological applications of freezing and drying. Academic Press Inc., New York.

HALES, M. W. 1963. The care of cultures. J. Dairy Sci. 46:1439–1440.

HARTSELL, S. E. 1956. Maintenance of cultures under paraffin oil. Appl. Microbiol. 4:350–355.

McCOY, ELIZABETH. 1954. Selection and maintenance of cultures. *In* L. A. Underkofler and R. J. Hickey (*Eds.*) Industrial fermentations. Chemical Publishing Company, Inc., New York.

PERLMAN, D., et al. 1955. Symposium on the maintenance of cultures of microorganisms. Bacteriol. Rev. 19:280–283.

PRICE, W. V., M. DEAN, and N. F. OLSON. 1964. Handbook on lactic starter. Univ. Wis., Coll. Agr., Circ. 383 (May 1964 revision).

PRESCOTT, S. C., and C. G. DUNN. 1959. Industrial microbiology. 3rd ed. McGraw-Hill Book Company, New York.

SIMMONS, J. C., and D. M. GRAHAM. 1959. Maintenance of active lactic cultures by freezing as an alternative to daily transfer. J. Dairy Sci. 42:363–364.

SPECK, M. L. (Convenor). 1962. Symposium on lactic starter cultures. J. Dairy Sci. 45:1262–1294.

STEEL, K. J., and HELEN E. ROSS. 1963. Survival of freeze dried bacterial cultures. J. Appl. Bacteriol. 26:370–375.

WAGMAN, J., and E. J. WENECK. 1963. Preservation of bacteria by circulating-gas freeze drying. Appl. Microbiol. 11:244–248.

CHAPTER TWENTY-FOUR

FOOD FERMENTATIONS

The preservation of cabbage, cucumbers, and green olives by means of a lactic acid fermentation has been discussed in Chapter 15. These fermentations not only aid in the preservation of the foods but also result in distinctive new food products that are in demand because of their new characteristics of appearance, body, and flavor. Most of the fermentations to be discussed here are to make new and desired products, and preservative effects are incidental. The fermentations may be by yeasts, bacteria, molds, or combinations of these organisms. In the making of the first group of food products, bread, beer, wine, and distilled liquors, fermentation by yeasts is of primary importance. Yeasts and bacteria are involved in the manufacture of vinegar from sugar-bearing materials, and bacteria chiefly in the production of fermented milks. Molds are important in the preparation of some of the Oriental foods.

BREAD

Microorganisms are useful in two chief ways in breadmaking: (1) they may produce gas to "leaven" or raise the dough, giving the bread the desired loose, porous texture; and (2) they may produce desirable flavoring substances. They also may function in the "conditioning" of the dough, to be discussed later.

Leavening

Leavening of dough usually is accomplished by means of bread yeasts (Chapter 23), which ferment the sugars in the dough and produce mainly carbon dioxide and alcohol. However, other actively gas-forming microorganisms, such as wild yeasts, coliform bacteria, saccharolytic *Clostridium* species, heterofermentative lactic acid bacteria, and various naturally occurring mixtures of these organisms have been used instead of bread yeasts for leavening. Leavening also may be accomplished by the direct incorporation of gas (CO_2) in the dough, or by the addition of chemicals that yield gas.

LEAVENING BY BREAD YEASTS. The primary function of yeasts added to the bread dough is the fermentation of sugar to produce carbon dioxide to leaven or raise the dough. There is little or no growth during the first 2 hr after the yeast is added to the dough, but some growth in 2 to 4 hr, if that much time is allowed before baking, and then a decline in growth in 4 to 6 hr. Fermentation by the yeast begins as soon as the dough (or sponge) is mixed and continues until the temperature of the oven inactivates the yeast enzymes. The professional baker adds a considerable amount of yeast and has a comparatively short making time. Modern trends in home baking are toward the addition of an excess of yeasts so that the fermentation may be even shorter than in commercial practice. These short-time processes encourage little or no growth of yeast during the fermentation process. Older home methods involved the use of less yeast or less effective yeast and therefore resulted in a longer making time and some opportunity for yeast and bacterial growth. During the fermentation, "conditioning" of the dough takes place when the flour proteins (gluten) mature, that is, become elastic and springy and therefore capable of retaining a maximal amount of the carbon dioxide gas produced by the yeasts. The conditioning results from action on the gluten by (1) proteolytic enzymes in the flour, from the yeast, from the malt, or added otherwise; and (2) the reduction in pH by acids added and formed. Fungal enzymes have been employed to aid this action. Dough conditioners, sometimes called yeast foods, that are added include ammonium salts to stimulate the yeasts and various salts, e.g., $KBrO_3$, KIO_3, CaO_2, and $(NH_4)_2S_2O_8$, to improve dough characteristics.

Although the sugar in the flour added to the bread mix plus that produced by the action of the flour amylase may provide enough sugar for the yeast fermentation, most formulas call for the addition of more sugar or of amylase-bearing malt. Rate of gas production by the yeasts is increased by the addition of (1) more yeast; (2) sugar, or amylase-bearing malt; and (3) yeast food, within limits. It is decreased by (1) the addition of salt; (2) the addition of too much yeast food; and (3) the use of too high or too low temperatures. The main objectives of the baker during leavening are to have enough gas produced and to have the dough in such a condition that it will hold the gas at the right time.

In the sponge method of breadmaking, part of the ingredients are mixed at 74 to 76 F (23.3 to 24.4 C) and allowed to ferment to the desired maturity. Then the rest of the ingredients are added and the fermentation is continued until the dough is in the desired condition. In the straight-dough method of making, all the ingredients are mixed at 78 to 82 F (25.6 to 27.8 C). The fermentation room, where the dough is held for most of the leavening process, is usually held at about 80 F (26.7 C). These temperatures are favorable to the yeast fermentation. Recently, a "liquid-ferment" process has been described to replace the conventional sponge operation (see McGhee et al.). In this process most of the yeast is added to part of the ingredients of the dough, is allowed to ferment there, and is kept active. Later the rest of the ingredients are added, and the leavening proceeds later in the newly formed loaf.

LEAVENING BY OTHER MICROORGANISMS. As has been stated, leavening can be accomplished by gas-forming organisms other than the bread yeasts. Breads leavened by dough carried over from a previous making, as for certain special breads and sour-dough bread, carry a mixture of coliform bacteria and wild yeasts, the former being the more effective. Heterofermentative lactic acid bacteria and saccharolytic anaerobes also can take part in the leavening. Salt-rising bread is leavened by "salt-rising yeast," as well as by microorganisms from the ingredients; it also may utilize bakers' yeast. In some cases a succession of organisms leaven, flavor, and modify dough, such as in the production of soda crackers, where a 3- to 4-hr fermentation is followed by action by lactic acid bacteria. Soda-cracker sponge also may be fermented by added yeast and by bacteria that are present.

LEAVENING BY CHEMICALS. Leavening of dough may be accomplished by chemical agents instead of by microorganisms, but the product cannot be called bread, according to standards of identity which specify yeast leavening. Carbon dioxide gas may be incorporated directly into the dough; or baking powders, which are combinations of chemical compounds that release gas when mixed into the dough, may be employed for leavening. Self-rising flour contains both the acidic and basic components of baking powder, which react upon moistening.

Continuous breadmaking

Continuous breadmaking processes usually involve growth and fermentation by the yeast in part of the ingredients to get a large yield of active yeast—or at least the addition of more yeast than usual—before the final dough is formed. Leavening may take place in the pans just prior to baking.

Flavor production

Yeasts are reported to contribute to the flavor of bread through products released during the fermentation of sugars, although most workers believe that yeasts add little or no flavor, especially in bread made by the rapid methods now employed. Alcohol, acids, esters, and aldehydes are products that might add desirable flavors. Most experts maintain, however, that bacteria growing in the dough can contribute the most to flavor. Too little time is allowed in the usual industrial leavening and working process for the bacteria to grow enough to appreciably affect the flavor, but the longer time available during the older methods of making in the home permits enough bacterial growth for a considerable production of desirable flavors. Dough leavened by means of a previous lot of dough may receive a good inoculum of desirable flavor-producing bacteria in this way. Some special brands of bread, made in this manner, are famous for their characteristic flavors.

Most of the flavor in bakers' bread, then, comes from the ingredients and the changes in them during baking. If enough time is given previous to baking for the growth of bacteria, they may add to the flavor, as may the yeasts to a lesser extent. Flavoring substances so developed may include alcohol, diacetyl, aldehydes, acetoin, and isoalcohols, and lactic, acetic, and succinic acids and their esters.

The baking process

Although the interior of the loaf does not quite reach 212 F (100 C) during baking, the heat serves to kill the yeasts, inactivate their enzymes and those of the flour and malt, expand the gas present, and set the structure of the loaf. Baking, besides producing the appearance of the loaf, also contributes desirable flavors. The heat also drives off most of the alcohol and other volatile substances formed by the yeasts, but contributes substances such as furfural, pyruvic and other aldehydes, and other compounds that add to the flavor. The most important change in bread during baking is gelatinization of starch. "Set" of bread results from this process, in which gluten gives structural support in the dough, but starch supports the structure of baked bread.

Rye bread

Rye bread may be made with or without a starter, or "sour," but only the type made with starter is of interest here. The old method of preparing sour depended upon the bacteria naturally present in a mixture of rye flour and water. The mixture was allowed to ferment for 5 to 10 hr; then more flour and water were added and the fermentation was continued for an additional 5 or 6 hr; then this was repeated several times. Half of the sour thus produced was incorporated in the sponge or dough for the bread and the rest carried over to start a new sour. This sour was modified by some bakers by the addition of yeasts and of lactic acid bacteria from cultured buttermilk or Bulgarian buttermilk to a sour that was made anew daily. It is obvious that the sours described would lack uniformity.

Modern methods involve adding considerable amounts of cultures of acid-forming bacteria to the dough mass to be used as a sour and controlling the time of fermentation (18 to 24 hr) and the temperature of incubation (about 25 C, or 77 F). An excessively high incubation temperature, e.g., 32 to 35 C (89.6 to 95 F) favors the growth of undesirable gas-formers, such as coliform and butyric bacteria. Some bakers prefer to use low-temperature lactic acid bacteria, e.g., *Lactobacillus plantarum*, *L. brevis*, and *Leuconostoc mesenteroides*, whereas others use high-temperature lactics, such as *Lactobacillus bulgaricus*, *L. fermenti*, and *Streptococcus thermophilus*, and adjust the incubation temperature accordingly. The growth of heterofermentative lactic acid bacteria is considered desirable. The starter imparts a desired tangy or sour flavor to the rye bread that is not given by the addition of lactic and acetic acids.

MALT BEVERAGES

Beer and ale are the principal malt beverages produced and consumed in this country and will be the ones discussed here. They are made of malt, hops, yeasts, water, and malt adjuncts. The malt is prepared from barley grains which have been germinated and dried, and then the sprouts or germs removed. Hops are the dried flowers of the hop plant. The malt adjuncts are starch- or sugar-containing materials added in addition to the carbohydrates in the malt. Starch adjuncts include corn and corn products, rice, wheat, barley, sorghum grain, soybeans, cassava, potatoes, etc., with corn and rice used most frequently. Saccharine adjuncts are materials like sugars and sirups.

Brewing of beer

The manufacture of beer will be outlined briefly as an example of the brewing process.

MALTING. In the preparation of malt, barley grains are soaked, or "steeped," at 50 to 60 F (10 to 15.6 C), germinated at 60 to 70 F (15.6 to 21.1 C) for 5 to 7 days, and dried to about 5 percent moisture. Most of the sprouts or germs are removed, and the malt remains. The malt, a source of amylases and proteinases, is crushed before use.

MASHING. The purpose of the mashing process is to make soluble as much as possible of the valuable portions of the malt and malt adjuncts, and especially to cause hydrolysis of starches and other polysaccharides and of proteins and products of their hydrolysis. First, the main malt mash is prepared by mixing the ground malt with water at 38 to 50 C (100.4 to 122 F). To this are added the cooked, starchy malt adjuncts in water, which are at about 100 C (212 F) after a boiling or cooking under steam pressure. This brings the temperature of the resulting cereal-malt mash to about 65 to 70 C (149 to 158 F), at which temperature saccharification (production of sugars from the starch) takes place within a short time. Then, the temperature is increased to about 75 C (167 F), which inactivates the enzymes. Insoluble materials that settle to the bottom of the container serve as a filter, so that the liquid that emerges, called **wort,** is clear. A special "lauter" tub may be used for this filtration. Rinsings from the filtering material are added to the wort. Next, hops are added to the wort to constitute the liquid from which the final wort is to be prepared for fermentation.

BOILING THE WORT WITH HOPS. The liquid containing wort and hops is boiled for about 2.5 hr, after which it is filtered through the hop residues. In this way the hop solids and precipitated proteins are removed. The precipitate is washed with hot water to remove most of the soluble material, and the washings are added to the original filtrate. The resulting mash or wort is ready for fermentation.

The boiling of the wort with hops has a number of purposes: (1) to concentrate it, (2) to practically sterilize it, (3) to inactivate enzymes, (4) to extract soluble substances from the hops, (5) to coagulate and precipitate proteins and other substances, (6) to caramelize the sugar slightly, and (7) to contribute antiseptic substances (chiefly the alpha resins, humulone, cohumulone, and adhumulone) to the wort and the beer. These resins are effective against Gram-positive bacteria. Extracted from the hops are the bitter acids and resins, which aid in flavor, stability, and head retention of the beer; essential oil, which adds a little flavor; and tannins, which are removed as much as possible, because they may be responsible for poor flavors and haziness in the beer.

FERMENTATION. A special beer yeast of the bottom type, a strain of *Saccharomyces carlsbergensis,* is used for the inoculation or "pitching" of the cooled wort. The pitching yeast ordinarily has been recovered from a previous fermentation. A fairly heavy inoculum, about 1 lb per barrel of beer, is employed. The temperature of the wort during the fermentation differs in different breweries but is low, usually being in the range from 38 to 57 F (3.3 to 14 C). Some brewers maintain the temperature at about 38 to 40 F (3.3 to 4.4 C), while others start with a low temperature and raise it later. The fermentation is complete within 8 to 14 days, usually in 8 to 10 days.

During the fermentation, the yeast converts the sugar in the wort chiefly to alcohol and carbon dioxide, plus small amounts of glycerol and acetic acid. Proteins and fat derivatives yield small amounts of higher alcohols and acids, and organic acids and alcohols combine to form aromatic esters. As the carbon dioxide is evolved in increasing amounts the foaming increases; then later it decreases to none when the fermentation has concluded. At a later stage the bottom yeasts "break," that is, flocculate and settle. Bacterial growth is not desired during the fermentation and the subsequent aging of the beer.

AGING, OR MATURING. The young, or "green," beer is stored, or "lagered," in vats at about 32 F (0 C) for from several weeks to several months, during which period precipitation of proteins, yeast, resin, and other undesirable substances takes place and the beer becomes clear and mellowed or matured. Esters and other compounds are produced to add to the taste and aroma, and the body changes from harsh to smooth.

FINISHING. After aging, the "lager" beer is carbonated to a CO_2 content of about 0.45 to 0.52 percent, mostly by means of gas collected during the fermentation. Then the beer is cooled, clarified, or filtered, and packaged in bottles, cans, or barrels. The beer for cans or smaller bottles is pasteurized briefly at about 60 to 61 C (140 to 141.8 F), or filtered through membranes or other materials to remove all yeasts.

CONTINUOUS PROCESSES. Continuous malting to shorten germination time of the barley involves application of aerated water to the grain as a finely divided spray, after the barley has been wetted and washed. Continuous brewing employs nonstop mashing and boiling of wort, as well as flow-through fermentation. The beer may be bottled without aging.

MICROBIOLOGY. The procedures during the brewing process have a great influence on the ability of microorganisms to survive or grow. Little is known about the growth of organisms during malting, in the main malt mash, or in the cereal-malt mash, although growth must take place in the first two. The boiling of the wort and hops for 2.5 hr, however, provides sufficient heat to destroy all but the most resistant bacterial spores, such as those of some species of *Bacillus* or *Clostridium*, and the combined action of heat and hop antiseptics might destroy most of those and inhibit any survivors. The yeast (Figure 24–1,A) used in pitching should be a pure culture (but usually is not) and hence should contribute no contaminating organisms. The wort is unfavorable to some organisms because of its acid pH and its content of antiseptics extracted from the hops. Then, too, temperatures are low during both fermentation and aging, and conditions are anaerobic. The alcohol produced also may be inhibitory to organisms.

Beer should hinder the growth of microorganisms because of its low pH, its content of antiseptics in the form of CO_2, alcohol, and hop extracts, and its low temperature of storage. Also, conditions are anaerobic throughout its processing and storage, and much of the beer sold has been pasteurized or filtered.

Yet, despite all these reasons why beer should be free of spoilage organisms, it is subject not only to defects from physical and chemical causes but also to "diseases" caused by microorganisms. Since the microorganisms that are most important in causing diseases of beer are readily killed by temperatures below boiling, they must enter after the boiling of the wort with hops.

Beer defects and "diseases"

As has been indicated, the term **defects** will be applied here to undesirable characteristics with causes that are not microbial, such as (1) turbidity due to unstable protein, protein-tannin complexes, starch, and resin; (2) off-flavors caused by poor ingredients or contact with metals; and (3) poor physical characteristics. This discussion will be limited to the troubles caused by microorganisms and therefore termed **beer infections** or **beer diseases.** The mash in the brewhouse may undergo butyric acid fermentation by *Clostridium* species or lactic acid fermentation by lactics if the mash is held too long at temperatures favoring these bacteria. Off-flavors so produced may carry over into the beer. It was pointed out in Chapter 23 that the pitching yeast, recovered from a previous fermentation, ordinarily is contaminated with bacteria and wild yeasts and may be a source of spoilage organisms (Figure 24–1,D). Yeasts and bacteria produce turbidity when they grow in beer; and beer yeasts carried over from the fermentation may be responsible for cloudiness. Likewise, wild yeasts, e.g., *Saccharomyces pastorianus* (Figure 24–1,C), can cause cloudiness in beer. Yeasts can be inhibited or excluded by keeping out air, fermenting most of the sugar in the wort to produce a "dry" beer, using good cultures of beer yeasts, and sanitizing the plant adequately. Yeasts also may be responsible for off-tastes and -odors. Thus, for example, bitterness may be caused by *S. pastorianus,* and an esterlike taste by *Hansenula anomala.* Most yeasts produce fruity odors, and some produce hydrogen sulfide from the hop extract in the beer. Yeasts able to utilize the dextrins in beer (e.g., *Saccharomyces diastaticus*) are potential spoilage organisms.

The bacteria causing beer diseases are mostly from the genera *Pediococcus, Lactobacillus, Achromobacter, Flavobacterium,* and *Acetobacter.*

"Sarcina sickness," characterized by sourness, turbidity, and ropiness of beer, is caused by *Pediococcus cerevisiae* (Figure 24–1,*E*). Because the cocci often aggregate in fours or tetrads, they were first thought to be sarcinae.

Some lactobacilli, being tolerant to acid and hop antiseptics, can grow in beer. *Lactobacillus pastorianus* (Figure 24–1,*F*) causes sourness and a silky turbidity. This bacterium produces lactic, acetic, and formic acids, alcohol and carbon dioxide from sugars, and is especially bad in top fermentations such as are used for ales.

Achromobacter anaerobium, when growing in beer, causes a silky turbidity and produces an odor reminiscent of hydrogen sulfide and of apples. It forms carbon dioxide and alcohol.

Flavobacterium proteus is responsible for a parsniplike odor and taste in wort and in beer. It produces alcohol and acid and is not tolerant of a pH as low as 4.2. It has been found as a common contaminant of pitching yeast.

Species of *Acetobacter,* the vinegar bacteria, which are tolerant of acid and hop antiseptics, can cause sourness of wort or beer under aerobic conditions. Exposure to oxygen can occur in cooks that are stored too long, in empty beer barrels, and in pitching yeast. A number of species can cause sourness; *A. capsulatum* and *A. viscosum* may produce ropiness; and *A. turbidans* has been blamed for turbidity and sourness.

Other incompletely described, unidentified bacteria have been blamed for defects. *Micrococcus, Streptococcus, Bacterium,* and *Bacillus* species have been accused of causing trouble, but in some instances probably merely were present. *Bacterium termo* organisms (Figure 24–1,*B*), so-called "apparatus bacteria," have been blamed for putrefaction, turbidity, and off-tastes and -odors, and sometimes for a musty odor. *Streptococcus mucilaginosus,* which probably is a pediococcus, has been reported to cause ropiness.

It should be reemphasized that all of the yeasts and bacteria that cause infections or diseases in wort and beer are killed by the boiling of the wort and hops and must enter thereafter from equipment, the air, the water, or the pitching yeast, and that aseptic and sanitary precautions will help prevent these troubles.

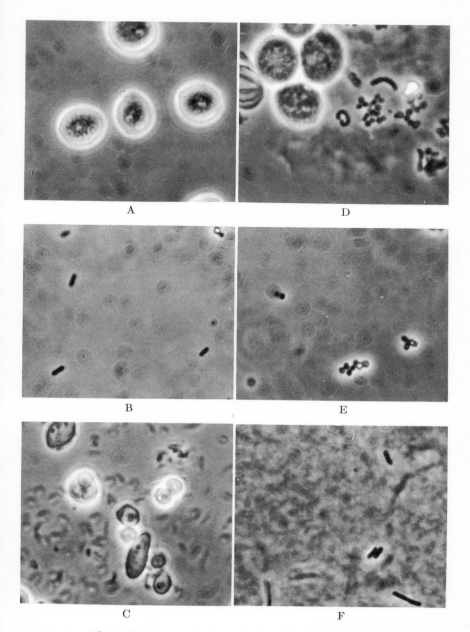

Figure 24–1. Phase photomicrographs of brewery microorganisms: (A) brewers' yeast; (B) termobacteria that give wort an off-color and -flavor; (C) wild yeast (*Saccharomyces pastorianus*), showing ascospores; harmful to beer; (D) brewers' yeast with contaminating organisms; (E) *Pediococcus cerevisiae*, causing "sarcina sickness"; (F) *Lactobacillus pastorianus*, a common contaminant of brewers' yeast, causing sourness. (×1,200.) (*Courtesy of Chas. Pfizer & Co., Inc.*)

Other malt beverages

Ale usually is made with a top yeast, a strain of *Saccharomyces cerevisiae,* instead of the bottom yeast employed for beer, although a bottom yeast sometimes is used. The primary fermentation takes place at 54 to 76 F (12.2 to 24.4 C), a higher temperature than for beer, and therefore the fermentation is more rapid, taking 5 to 7 days. The top yeast forms clumps which collect carbon dioxide gas and are carried to the top of the wort. At intervals the yeast scum is skimmed off. More hops are used in ale than in beer, and a higher alcohol content usually is attained. Ale usually is pale in color and tart in taste.

Weiss beer, porter, and **stout** are ales in that top yeasts are employed in their manufacture. Weiss beer is a light, tart ale made chiefly from wheat. Porter and stout are dark, heavy, sweet ales.

Related beverages

Sake is a yellow rice beer or wine of Japanese origin with an alcohol content of about 14 to 17 percent. A starter, or koji, for sake is made by growing *Aspergillus oryzae* on a soaked and steamed rice mash until a maximal yield of enzymes is obtained. This koji contains amylases which cause the hydrolysis of rice starch to sugars available to yeasts plus other hydrolytic enzymes such as proteinases. The koji is mixed with more rice mash, starch is converted to sugar, and several species of yeast of the genus *Saccharomyces* carry out the alcoholic fermentation. The liquor filtered from the fermented mass after 10 to 14 days is the sake. **Sonti** is a rice beer or wine of India. The mold *Rhizopus sonti* and yeasts are active in the fermentation.

Pulque is a Latin-American beerlike beverage containing about 6 percent of alcohol that results from a natural yeast fermentation of the juice of the agave, or century plant.

Ginger beer is a mildly alcoholic, acid beverage made by the fermentation of a sugar solution flavored with ginger. The starter is the "ginger-beer plant," in which a yeast, *Saccharomyces pyriformis,* and a capsulated bacterium, *Lactobacillus vermiformis,* are enclosed in the gelatinous capsular material of the lactobacillus.

WINES

Unless otherwise specified the term **wine** is applied here to the product made by the alcoholic fermentation of grapes or grape juice by yeasts and a subsequent aging process. Wines, however, can be produced by the fermentation of the juices of fruits, berries, rhubarb, dandelions, honey, cereals, etc.

Grape wine

Grape wines are, for the most part, either red or white. The red wines, fermented on the skins, contain the red pigment from the skins of purple or red varieties of grapes, whereas white wines are made from white grapes or the expressed juice of other grapes, fermented free of the skins. The manufacture of red wine will be outlined briefly as an example of the wine-making process.

PREPARATION OF JUICE. Grapes of a variety especially adapted to wine making are harvested at a stage when they have the desired sugar content. They are stemmed and crushed by machine and then treated with sulfur dioxide (75 to 200 ppm) or potassium metabisulfite in equivalent amounts to inhibit the growth of undesirable competitors of the wine yeast.

FERMENTATION. About 2 to 5 percent of a special wine yeast, a strain of *Saccharomyces cerevisiae* var. *ellipsoideus*, is added to the crushed grapes, or **must**, rather than to trust to yeasts already present. At first the contents of the tank are mixed twice a day by punching the "cap" of floating skins, stems, etc., pumping juice over the skins, or mixing in some other way to aerate and hence encourage growth of the yeast and aid in the extraction of color from the skins (for red wines). Or the red pigments may be extracted from the skins by heat and added back to the juice. Later the mixing is discontinued, for anaerobic conditions are most favorable to the alcoholic fermentation. It is very important that the temperature be maintained within an optimal range, i.e., beween 75 and 80 F (23.9 and 26.7 C) for red wines, during the active fermentation, which takes about 3 to 5 days for red wines and 7 to 14 days (at 50 to 70 F, or 10 to 21.1 C) for white wines. An excessively high temperature inhibits the wine yeasts and permits competing organisms, the lactobacilli, for example, to grow and cause defects; and too low a temperature slows down action of the wine yeasts and permits wild yeasts, lactic acid bacteria, and other organisms to grow. Heat is liberated during the fermentation and this, coupled with high atmospheric temperatures, may necessitate artificial cooling of the must.

After the primary or active fermentation has advanced sufficiently, the fermented juice is drawn off from the residues (pomace) and placed in a storage tank under a light pressure of carbon dioxide for the secondary fermentation for 7 to 11 days at about 70 to 85 F (21.1 to 29.4 C). Here the remaining sugar is fermented if a dry wine is desired. Clear wine is drawn off, or "racked," from the sediment at the bottom of the tank.

STORING AND AGING. The wine may be flash-pasteurized before aging (but usually is not) to precipitate proteins. It is cooled, held for a few days, filtered, and transferred into wooden tanks (usually of white oak or redwood) or plastic-coated concrete tanks for aging. The tanks are filled completely and sealed to keep out air. Periodically the wine is racked from bottom sediment. Final aging may be in the bottle. Aging for months or years results in desirable changes in body and flavor of the wine, giving it the aroma or bouquet that should be one of its characteristics. Esters and alcohols are considered important contributors to bouquet and taste. During aging some fermentation of the malic acid of grape juice by lactobacilli or micrococci (malolactic fermentation) may take place with the production of lactic acid and carbon dioxide and reduction in acidity.

After aging, the wine is filtered or otherwise clarified, barreled or bottled, and stored. Some wines are pasteurized after aging, usually in the bottle.

VOLATILE ACIDITY OF WINES. A high content of volatile acid in wines is indicative of a faulty fermentation. In the United States the legal limit for volatile acid content is 0.14 g per 100 ml, expressed as acetic acid, for red wine and 0.12 g for white wine.

MICROBIOLOGY. The grapes when crushed have a variety of microorganisms on their surfaces, including yeasts and bacteria. Not only is the surface flora of the grape present but also an array of contaminants from the soil. To suppress these organisms the wine maker adds sulfur dioxide or sulfite, or less commonly pasteurizes the must (expressed juice). During the primary fermentation the added wine yeast predominates. During early stages, growth of the yeast is favored by aeration of the must; then later anaerobic conditions favor the alcoholic fermentation by the yeasts, liberating carbon dioxide and ethyl alcohol, both of which help inhibit organisms other than the wine yeasts. The atmosphere of carbon dioxide above the wine during the secondary fermentation prevents the growth of aerobic contaminants, such as the acetic acid bacteria. The pasteurization that follows, although not for that purpose, reduces the numbers of microorganisms that later might cause spoilage ("diseases") of the wine. There should be little or no growth of organisms during the aging and storage of the wine, but organisms that can grow then may be introduced by contamination from the tanks, barrels, or bottles, and changes such as the malolactic fermentation may take place.

Kinds of wine

No attempt will be made to list the scores of names applied to different types of wines. Instead a few general descriptive terms will be defined. Most wines are **still** wines, that is, they retain none of the carbon dioxide produced during the fermentation, in contrast to **sparkling** wines, which do contain considerable amounts. Other wines may be artificially **carbonated.**

Dry wines contain little or no unfermented sugar, as contrasted to **sweet** wines, which have sugar left or added. Wines usually contain from 11 to 16 percent of alcohol by volume but may go as low as 7 percent. **Fortified** wines, however, to which distillate of wine called "wine spirits" or "brandy" has been added, contain about 19 to 21 percent of alcohol by volume. **Table** wines have a comparatively low content of alcohol and little or no sugar, while **dessert** wines are fortified, sweet wines.

French dry sherry is of interest because it is made from grapes which have a high sugar content as a result of being dried out by an infecting gray mold, *Botrytis cinerea;* they therefore yield a wine with a high content of alcohol. Spanish (Jerez) sherry supports the growth of a yeast film, presumably of one or more species of *Saccharomyces*, while the wine is being racked in partially filled barrels following the main fermentations. This yeast growth, or "flor," imparts a special bouquet and flavor to the wine.

Wine defects and microbial spoilage

Like beer, wine has its defects from nonmicrobial causes and its spoilage caused by microorganisms. Defects include those due to metals or their salts, enzymes, and agents employed in clearing the wine. Iron, for example, may produce a sediment known variously as gray, black, blue, or ferric casse, or in white wine may be responsible for a white precipitate of iron phosphate termed white casse. Tin and copper and their salts have been blamed for cloudiness. White wines may be turned brown, and red wines may have their color precipitated by an oxidizing enzyme, peroxidase, of certain molds. Gelatin, used in clarifying wines, may cause cloudiness.

The microorganisms causing wine spoilage are chiefly wild yeasts, molds, and bacteria of the genera *Acetobacter, Lactobacillus, Leuconostoc* and, perhaps, *Micrococcus* and *Pediococcus.*

FACTORS AFFECTING GROWTH OF MICROORGANISMS IN WINE. Factors that are known to influence the susceptibility of wines to microbial spoilage are:

1. Acidity or pH. The lower the pH, the less likely there is to be spoilage. The minimal pH permitting the growth of microorganisms varies with the organism, the type of wine, and the alcoholic content. Molds, yeasts, and acetic acid bacteria would not be stopped by any pH normal to wines, but most lactic acid bacteria will tolerate acidity down to about pH 3.3 to 3.5, a pH lower than that of most wines (most California table wines have a pH of about 3.5 to 4.0).
2. Sugar content. Dry wines (about 0.1 percent or less of sugar), with their low sugar content, are rarely spoiled by bacteria, but 0.5 to 1.0 percent or more of sugar will favor spoilage.
3. Concentration of alcohol. Tolerance of alcohol varies with the spoilage organism. Thus acetic bacteria spoiling musts and wines are inhibited by over 14 to 15 percent of alcohol by volume, but deacidifying cocci are stopped by about 12 percent, *Leuconostoc* species by over 14 percent, heterofermentative lactobacilli by about 18 percent (except *Lactobacillus trichodes*, which grows in fortified wines in over 20 percent of alcohol), and homofermentative lactobacilli by about 10 percent.
4. Concentration of accessory growth substances. *Acetobacter* species can make their own vitamins, but the lactic acid bacteria must have most of them provided. The chief source of these substances in wines is the wine yeast, which releases the accessory growth factors on autolysis. The more of these substances there are present, the greater will be the likelihood of spoilage by lactic acid bacteria.
5. Concentration of tannins. Tannins added with gelatin for clarification retard bacteria, but usually not enough are added to be of much practical importance as inhibitors.
6. Amount of sulfur dioxide present. The more sulfur dioxide added, the greater will be the retardation of the spoilage microorganisms. The 75 to 200 ppm customarily added to the must usually is adequate. Effectiveness depends upon the kind of organism to be suppressed and increases with a lowering of pH and sugar content.
7. Temperature of storage. Spoilage usually is most rapid at 20 to 35 C (68 to 95 F) and slows down as the temperature is dropped toward freezing.
8. Availability of air. Absence of air prevents the growth of aerobic organisms, such as molds, film yeasts, and *Acetobacter,* but the lactic acid bacteria grow well anaerobically.

SPOILAGE BY AEROBIC MICROORGANISMS. Film yeasts, which can oxidize alcohol and organic acids, may grow on the surface of must and wines exposed to air, producing a heavy pellicle called "wine flowers."

They should cause no trouble if the must is mixed periodically and if air is kept away from the wine.

In the presence of air the aerobic acetic acid bacteria, usually *Acetobacter aceti* or *A. oxydans*, oxidize alcohol in must or wine to acetic acid, an undesirable process called **acetification.** They also may oxidize glucose in the must to gluconic acid and may give a "mousy" or "sweet-sour" taste to the must.

Molds, such as *Mucor, Penicillium, Aspergillus,* and others, may grow on plant walls, barrels, tanks, hose lines, and corks, and may also grow on the grapes or on cold must. Molds are kept down by adequate cleansing of walls and equipment.

SPOILAGE BY FACULTATIVE MICROORGANISMS. Wild yeasts, which include all yeasts but the wine yeast added as starter, may bring about abnormal fermentations that result in low alcohol content, high volatile acidity, undesirable flavors, and cloudiness in the wine. These yeasts, which come chiefly from the grapes from which the must is prepared and usually are predominantly of the apiculate type, are suppressed or eliminated by use of an active starter of the wine yeast, sulfiting or pasteurization of the must before the fermentation, and control of the temperature of the must during the fermentation. Low temperatures, below 70 F (21.1 C), will favor some of the wild yeasts and slime-producing bacteria.

Lactic acid bacteria are the principal causes of the bacterial spoilage of musts and wines. There has been some confusion in the application of names to the various types of bacterial spoilage of wines, probably because several different kinds of bacteria may be able to cause the same defect or because the same organism may cause different defects under different conditions. Probably most commonly occurring is **tourne** (turned or soured) spoilage, in which acid is formed from sugars, glucose, and fructose in the wine, chiefly by heterofermentative *Lactobacillus* species, such as *L. brevis, L. hilgardii, L. trichodes,* and perhaps *L. fermenti* and *L. buchneri.* The growth of the lactobacilli produces silky cloudiness, increases lactic and acetic acid, yields carbon dioxide, sometimes gives "mousy" or other disagreeable flavors, and damages the color of the wine. When the fermentation of fructose results in the bitter product mannitol, the fermentation sometimes is termed "mannitic"; bitterness (**amertume**) also may result from the fermentation of the glycerol in the wine. Gassiness, resulting from any cause, such as from the liberation of carbon dioxide by heterofermentative lactics, is called **pousse.** The homofermentative *L. plantarum* forms mostly lactic acid from sugars in table wines, increasing the fixed acidity and giving a "mousy" flavor.

The acidity of the wine may be reduced by the spoilage bacteria through the oxidation of the malic, citric, and tartaric acids by *Acetobacter* (aerobic) or through fermentation of malic and tartaric acids by species of *Lactobacillus, Leuconostic,* or *Pediococcus* or by other cocci.

Sliminess or ropiness of young white wines, accompanied by cloudiness and increased volatile acidity, has been blamed on *Leuconostoc* species, *L. mesenteroides,* and *L. dextranicum,* on micrococci and on lactobacilli. Addition of sucrose, when permitted, favors the production of dextran and therefore of sliminess by *Leuconostoc.*

Any bacteria or yeasts growing in wine are likely to cause an undesirable cloudiness, and any acetic bacterium or heterofermentative lactic growing in the wine may increase the volatile acidity to an extent that will make the product unsalable. Fermentation of sugars usually results in an increase in acidity, in fixed acid if the lactic is homofermentative, and in fixed and volatile acid if the organism is heterofermentative. Oxidation or fermentation of the organic acids of the grape results in a decrease in amounts of fixed acid. It should be reemphasized that the composition of the wine is important in determining its susceptibility to bacterial spoilage. Thus white wines of low alcohol content are more readily subject to sliminess and spoilage by cocci than are other wines; musts and table wines support the growth of *Lactobacillus hilgardii, L. brevis,* and *Leuconostoc mesenteroides;* while *Lactobacillus trichodes* (Figure 24–2) is the only species known to spoil California dessert wines and does not grow in musts.

It is interesting to note that the formation of lactic acid from malic acid in very sour wines might improve their quality by reducing the acidity. Lactic acid is a weaker acid than malic, and only two molecules of lactic acid are formed from three of malic. Some bacteria can also form lactic acid from tartaric acid and glycerol in wine.

Other wines

WINES FROM OTHER FRUITS. Wines can be made from most kinds of fruits, including apples (**hard cider**), peaches, apricots, plums, pears (**perry**), cherries, berries, and many others. Berries and most other fruits (except wine grapes) contain insufficient sugar to make a good wine and must be "ameliorated" by the addition of sugar before the fermentation. Otherwise, the manufacture of the wine is similar to that of grape wine. The products may be dry, sweet, fortified, sparkling, or carbonated. Wines for consumption as such or for distillation to produce brandies may also be made from dried fruits such as raisins, dates, figs, and prunes.

The manufacture and diseases of apple wine or hard cider, prepared from juice expressed from apples, have received considerable attention.

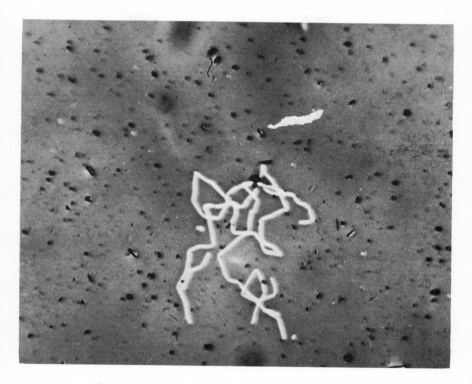

Figure 24–2. Photomicrograph of *Lactobacillus trichodes,* an organism which can cause spoilage of wine (×1,200). (*Courtesy of R. H. Vaughn.*)

Much American hard cider is made locally from apples not especially suited to its manufacture and is fermented by the yeasts that happen to be present. Therefore, often only 4 to 6 percent of alcohol is produced, there is a residue of sugar, and flavor and bouquet may be defective or lacking. British and French ciders and some industrially produced American hard ciders are made from apples that are high in sugars and in tannin. Manufacturing methods may include sulfiting, the addition of sugars and yeast food, and inoculation with a proven yeast culture. Heterofermentative lactobacilli have been reported to be active during cider making, fermenting malic and citric acids, sugars, and glycerol to produce carbon dioxide, lactic, acetic, propionic, and succinic acids, and mannitol.

Most of the sicknesses or diseases of hard cider are similar to those of grape wine: cloudiness, low alcohol content, and off-flavors caused by wild yeasts; cloudiness and bad flavors produced by lactobacilli; sliminess or ropiness due to bacteria; acetic acid production, etc. The low acidity and low nitrogen content of ciders encourage the growth of the spoilage organisms.

WINES FROM OTHER AGRICULTURAL PRODUCTS.   Theoretically any edible product that contains sufficient moisture, sugar, and other foods for yeasts can be used to make wine. Honey wine, or **mead**, is made from diluted honey to which minerals and nitrogenous food for the yeasts have been added. Dandelion wine is a homemade product made by alcoholic fermentation of a water extract of flowers of the dandelion to which sugar, flavoring substances, and yeast have been added.

DISTILLED LIQUORS

Distilled liquors or spirits of interest here are those produced by distillation of an alcoholically fermented product. **Rum** is the distillate from alcoholically fermented sugar-cane juice, sirup, or molasses. **Whiskeys** are distilled from saccharified and fermented grain mashes, e.g., rye whiskey from rye mash, bourbon or corn whiskey from corn mash, wheat whiskey from wheat mash, etc. Rums and whiskeys are made from mashes fermented by special distillers' yeasts, strains of *Saccharomyces cerevisiae* var. *ellipsoideus,* which give high yields of alcohol. The grain mashes usually are acidified to favor the yeasts. The aging of the distilled liquors in charred oaken barrels or tuns is a chemical rather than a biological process. **Brandy** means the distillate from grape wine, unless a qualifying word is added, as is done in naming apple brandy (applejack), peach brandy, apricot brandy, etc.

There should be no problems of spoilage of distilled liquors by microorganisms.

VINEGAR

It has been pointed out in Chapter 15 that the normal course of changes in fruit juices at ordinary temperatures is an alcoholic fermentation by yeasts, followed by oxidation of the alcohol to acetic acid by acetic acid bacteria. If enough acetic acid were produced the product would be vinegar. **Vinegar** is defined as a condiment made from sugary or starchy materials by an alcoholic fermentation followed by an acetous one. It must contain at least 4 g of acetic acid per 100 ml to be legal vinegar.

Kinds of vinegar

Vinegars may be classified on the basis of the materials from which they have been made: (1) those from the juices of fruits, e.g., apples, grapes, oranges, pears, berries, etc.; (2) those from starchy vegetables, potatoes or sweet potatoes, for example, whose starch must first be hydrolyzed to sugars; (3) those from malted cereals, such as barley, rye, wheat, and corn; (4) those from sugars, such as sirups, molasses, honey, maple skimmings, etc.; and (5) those from spirits or alcohol, e.g., from waste alcoholic liquor (beer) from yeast manufacture or from dilute, denatured ethyl alcohol. Anything, in fact, that contains enough sugar or alcohol and is in no way objectionable as food may be used to make vinegar. The vinegar usually derives its descriptive name from the material from which it was made: cider vinegar from apple juice, alegar from ale, malt vinegar from malted grains, spirit vinegar from alcohol, etc. In the United States most table vinegar is cider vinegar, and therefore the term vinegar by itself usually means cider vinegar. Vinegar from grapes (wine) is most popular in France, and vinegar from malt liquors (alegar) in the British Isles.

The fermentation

As has been indicated, the manufacture of vinegar from saccharine materials involves two steps: (1) the fermentation of sugar to ethyl alcohol and (2) the oxidation of alcohol to acetic acid. The first step is an anaerobic process carried out by yeasts, either those naturally present in the raw material or, preferably, added cultures of high-alcohol-producing strains of *Saccharomyces cerevisiae* var. *ellipsoideus*. A simplified equation for the process is:

$$C_6H_{12}O_6 \rightarrow 2CO_2 + 2C_2H_5OH$$
$$\text{glucose} \qquad\qquad \text{alcohol}$$

Actually a series of intermediate reactions takes place, and small amounts of other final products are produced, such as glycerol and acetic acid. Also, there are small amounts of other substances, produced from compounds other than sugar, including succinic acid and amyl alcohol.

The second step, oxidation of the alcohol to acetic acid, is an aerobic reaction carried out by the acetic acid bacteria:

$$C_2H_5OH + O_2 \rightarrow CH_3 \cdot COOH + H_2O$$
$$\text{alcohol} \quad \text{oxygen} \quad\quad \text{acetic acid} \quad \text{water}$$

Acetaldehyde is an intermediate compound in this reaction. Among the final products are small amounts of aldehydes, esters, acetoin, etc.

"Bergey's Manual of Determinative Bacteriology," sixth edition, lists some nineteen suggested species of *Acetobacter* and six species of *Bacterium* that are acetic acid bacteria, and the seventh edition lists seven species so far agreed upon, all in the genus *Acetobacter*. The seven species of *Acetobacter* are divided into those that oxidize acetic acid to carbon dioxide and water and those that do not. Some species, e.g., *Bacterium scheutzenbachii* and *B. curvum,* have been found active in the "quick" vinegar process to be described. *Acetobacter aceti, A. rancens, A xylinum,* and others have been isolated from active vinegar starters for the "slow" methods. *B. orleanense* can function in either type of process. It should be emphasized that pure cultures of acetic acid bacteria are not used in practice because they are less efficient than the mixed cultures.

Methods of manufacture

The ways of making vinegar may be divided into the "slow" methods, such as the home, or "let-alone," method and the French, or Orleans, method, and "quick" methods, like the generator process or the fogging procedure. In slow methods the alcoholic liquid is not moved during acetification, while in quick methods the alcoholic liquid is in motion. For the most part the slow methods utilize fermented fruit juices or malt liquors for acetic acid production, whereas the quick methods are applied mostly to the production of vinegar from spirits (alcohol). Fruit or malt liquors are well supplied with food for the vinegar bacteria, but to maintain active vinegar bacteria in generator methods using alcohol, denatured with ethyl acetate or vinegar, it must be supplemented with a "vinegar food," which is a combination of organic and inorganic compounds that varies with the compounder. Combinations of substances such as dibasic ammonium phosphate, urea, asparagine, peptones, yeast autolysate, glucose, malt, starch, dextrins, salts, and other substances have been reported in use.

SLOW METHODS.   In the home, or let-alone, method, a fruit juice, like apple juice, is allowed to undergo a spontaneous alcoholic fermentation, preferably to about 11 to 13 percent of alcohol, by yeasts originally present, after which a barrel is partially filled with the fermented juice and placed on its side with the bunghole upward and open. Then the alcoholic solution is allowed to undergo an acetic acid fermentation, called acetification, carried out by vinegar bacteria naturally present until vinegar is produced. A film of vinegar bacteria, called "mother of vinegar," should grow on the surface of the liquid and oxidize the alcohol to acetic acid. Unfortunately, the yield may be low because of a poor yield of alcohol during the yeast fermentation; because of the absence of productive strains of vinegar bacteria; because of the oxidation of acetic acid by the vinegar bacteria if there is a shortage of alcohol; or because of the competitive growth of film yeasts and molds on the surface, which destroy alcohol and acids, and of undesirable bacteria in the liquid, which produce undesirable flavors. The process is very slow, and the product often is of inferior quality.

In contrast to the batch process just described, the Orleans, or French, method, employed considerably in Europe, is a continuous process, although both processes usually are carried out in barrels. In the Orleans process raw vinegar from a previous run is introduced to fill about one-fourth to one-third of the barrel and serves to introduce an inoculum of active vinegar bacteria and to acidify the added wine, hard cider, or malt liquor so as to inhibit competing microorganisms. Enough of the alcoholic liquor is added to the vinegar to fill about half the barrel, leaving an air space above that is open to the outside air through the bunghole at the top and a hole in each end of the barrel above the level of the liquid. These holes are protected by screening. The acetic acid bacteria growing in a film on top of the liquid carry out the oxidation of alcohol to acetic acid for weeks to months at about 70 to 85 F (21.1 to 29.4 C), after which part of the vinegar so formed is drawn off for bottling and is replaced in the barrel by an equal quantity of alcoholic liquor. This operation is repeated a number of times, so that the process in this way becomes more or less continuous. Vinegar of high quality can be produced by this rather slow process.

One difficulty encountered in this method is the dropping of the gelatinous film of vinegar bacteria and the resulting retardation of acetification. To avoid this difficulty a raft or floating framework sometimes is provided to support the film. It is claimed that too heavy a bacterial film will result in reduced acetification.

QUICK METHODS. As has been indicated, quick methods of vinegar manufacture involve the movement of the alcoholic liquid during the process of acetification. Most commonly this liquid is trickled down over surfaces on which films of the vinegar bacteria have grown and to which a plentiful supply of air is provided.

The generator method is the one in common use at present. The simple generator is a cylindrical tank that comes in different sizes and usually is made of wood. The interior is divided into three parts: an upper section where the alcoholic liquid is introduced; the large middle section where the liquid is allowed to trickle down over beechwood shavings, corncobs, rattan shavings, charcoal, coke, pomace, or some other material that will give a large total surface yet not settle into a compact mass; and the bottom section where the vinegar collects. The alcoholic liquid is fed in at the top through an automatic feed trough or a sprinkling device (sparger) and trickled down over the shavings or other material on which has developed a slimy growth of acetic acid bacteria, which oxidize the alcohol to acetic acid. Air enters through the false bottom of the middle section and, on becoming warm, rises, to be vented above. Since the oxidation process here releases considerable heat, it usually is necessary to control the temperature so that it does not rise much above 85 F (29.4 C). This can be done by means of cooling coils, by adjustment of the rate of feeding air and alcoholic liquid, and by cooling the alcoholic liquid before it enters the generator or by cooling the partially acetified liquid that is returned to the top from the bottom section of the tank for further action.

In starting a new generator the slime of vinegar bacteria must be established before vinegar can be made. First, the middle section of the tank is filled with raw vinegar that contains active vinegar bacteria to inoculate the shavings with the desired bacteria, or this material is circulated through the generator. Then an alcoholic liquid, acidified with vinegar, is slowly trickled through the generator to build up bacterial growth on the shavings and then is recirculated. Some makers acidify all of the alcoholic liquid with vinegar before introducing it into the generator or leave some vinegar to acidify the new lot of liquid.

The vinegar may be made by one run of the alcoholic liquid through the generator, or the vinegar collected at the bottom may be recirculated through the generator if insufficient acid has been produced at first or too much alcohol is left. Sometimes generators are operated in tandem, the liquid from the first tank going through a second or even a third generator.

The Frings generator (Figure 24–3) is a large, cylindrical, airtight tank, equipped with a sparger (sprinkler) at the top, cooling coils about the lower part of the middle section containing the shavings, and facilities for the recirculation of the vinegar from the bottom collection chamber through the system. Modern types of these generators are equipped with automatic controls for feeding the alcoholic liquid, for introducing filtered air, for controlling temperature, and for recirculating the liquid collected at the bottom. These generators give high yields of acetic acid and leave little residue of alcohol.

In the Mackin process, a fog or fine mist of a mixture of vinegar bacteria and nutrient alcohol solution is sprayed through jet nozzles into a chamber. The mist is kept in circulation by filtered air for a

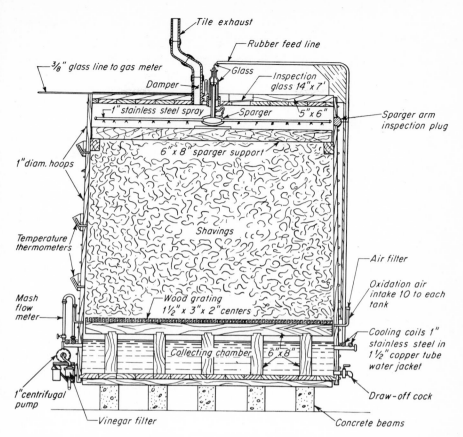

Figure 24–3. Diagram of a quick-method vinegar generator. (*Courtesy of Food Engineering.*)

while and then is allowed to fall to the bottom for collection, to be cooled, reatomized, and returned to the chamber. This process is continued until oxidation of the alcohol is almost complete.

The dipping generator consists of a tank containing a basket filled with beechwood shavings that can be raised out of or lowered into dilute alcohol solution in the lower part of the tank. While the basket is out of the liquid, aeration permits rapid acetification by vinegar bacteria on the shavings, and lowering the basket into the liquid adds more culture medium and removes some of the acetic acid made.

SUBMERGED METHOD. In the submerged (Hromatka and Ebner) fermentation a stirred medium containing 8 to 12 percent alcohol (hard cider, wine, fermented malt mash, or spirits) is inoculated with *Aceto-*

bacter acetigenum and is held at 76 to 85 F (24.4 to 29.4 C) with controlled aeration by means of finely divided air. The Frings "Acetator," shown in Figure 24–4, is an example of equipment for this method.

Bacteria grow in a suspension of fine air bubbles and fermenting liquid. The suspension is achieved by a specially designed aerator. It consists of a rotor (*a*) driven by an electric motor (*c*) mounted below the tank (*d*). The rotor sucks in the air, accelerates it after thoroughly mixing it with water, and distributes the suspension uniformly over the cross section of the tank.

To ensure uniform distribution of the air bubbles, the self-priming rotor is surrounded by a stator (*b*). An air pipe (*e*) connected to the rotor leaves the inside of the tank near the top and continues in two branches outside of the tank. One branch leads to a rotameter (*f*), which measures the amount of air entering the tank at any minute.

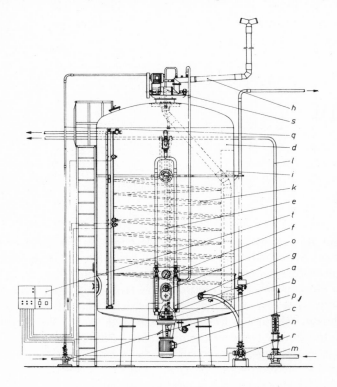

Figure 24–4.   Vinegar by submerged oxidation fermentation (Frings Acetator).   (*By permission from Dr. H. Ebner and Food Engineering, Chilton Company, Philadelphia, Pa., copyright, 1965.*)

The other branch goes to a condensate cylinder (g) connected to the exhaust air pipe (h).

Four stabilizer boards (i) are fastened to the circumference of the inside of the tank. To these the stainless steel cooling coil (k) is fastened. Cooling water enters the coil through a pipe (l) coming from a pump (m) and a flowmeter (n). The pump is operated by a regulating thermometer (o) to maintain constant fermentation temperature. The charging pump (p), close to the charging vat, pumps the mash through the charging pipe (q). Entering the tank close to the center of the top, the mash flows directly into the rotor.

The discharging pump (r), near the Acetator, half empties the tank after the completion of each fermentation cycle. Attached to the top center of the tank is the defoamer (s), which destroys the foam mechanically [Morton generators do not have defoamer]. The liquid part of the foam is pumped back into the tank by the defoamer. Air leaves through the exhaust pipe. Controller (t) charges and discharges the generator.

Finishing

The composition of vinegar depends, of course, on the material from which it was made. Vinegars from fruits and malt liquors carry flavors characteristic of these materials. The method of manufacture also influences the character of the product. Vinegars made by slow methods are less harsh than those made by quick methods because of the aging undergone during the long time of preparation. Quickly made vinegars, when aged in tanks or barrels, improve in body, taste, and bouquet. Filtration and "fining," which is clarification by the settling out of added suspended materials, are employed to clarify the vinegar, which should be very clear. Most market vinegar now is pasteurized in bulk or in the bottles. Times and temperatures vary, but heating at 140 to 150 F (60 to 65.6 C) for a few seconds is an example.

The strength of the vinegar is expressed in grains, that is, ten times the number of grams of acetic acid per 100 ml of vinegar. Thus 40-grain vinegar contains 4 g of acetic acid per 100 ml of vinegar at 20 C.

Vinegar defects and diseases

As in wines, metals and their salts may cause cloudiness and discoloration of vinegar. Ferrous iron may be oxidized to ferric iron and combine with tannins, phosphates, or proteins to produce a haze. Cloudiness also may be caused by salts of tin or copper. Iron acting upon tannin or oxidase activity may be responsible for the darkening of vinegar.

ANIMAL PESTS.   Mites and vinegar flies (*Drosophila*) breed readily around the vinegar factory and are objectionable from an esthetic point of view. The vinegar eel, *Anguillula aceti*, is a small, barely visible nematode worm that causes trouble. These worms have been said to attack the film of acetic acid bacteria in the slow methods and cause it to sink, hindering acetification and deteriorating the vinegar. They are harmless to human beings, but the presence of their bodies, alive or dead, is objectionable from an esthetic standpoint. They usually come from fruits or their juices, but once established they will continue to breed in barrels or generators. Recently it has been reported that they aid in acetification in a synthetic medium and in the manufacture of spirit vinegar by the quick method. They are killed by pasteurization and readily removed by filtration.

MICROBIAL DEFECTS.   Defects caused by microorganisms may result in inferior materials from which the vinegar is made or in inferiority of the condiment itself. Wine and hard cider, for example, are subject to the troubles listed in the discussion of diseases of wine. *Lactobacillus* and *Leuconostoc* species in fruit juice not only may be responsible for off-flavors, e.g., the "mousy" taste, but also may produce enough acetic acid to interfere with the alcoholic fermentation by yeasts. Under anaerobic conditions butyric acid bacteria may produce their undesirable acid. These difficulties may be reduced by the addition of sulfur dioxide to the juices, but this chemical is inhibitory to the vinegar bacteria.

The defects of vinegar itself are confined for the most part to the production of excessive sliminess in the mass of vinegar bacteria and the destruction of acetic acid in the product. It has been mentioned that an especially heavy, thick, slimy film of bacteria in the slow process of vinegar manufacture reduces the rate of acetification. Excessive sliminess is much more harmful, however, in the generator process, for it interferes with aeration. Sliminess is favored by an alcoholic liquid that is a good culture medium, e.g., cider, wine, or a medium to which too much rich vinegar food has been added, but is not ordinarily trouble-some in the acetification of a poor medium like that used in making vinegar from spirits (alcohol). Several species of vinegar bacteria can cause sliminess, but *Acetobacter xylinum* is probably the most important one.

Oxidation of acetic acid in vinegar to carbon dioxide and water can be brought about by the acetic acid bacteria themselves during the vinegar-making process if there is a shortage of alcohol or an excessive amount of aeration. Other organisms that can oxidize acetic acid under aerobic conditions are the film yeasts ("wine flowers"), molds, and algae.

FERMENTED DAIRY PRODUCTS

The preparation and spoilage of fermented milks and cheese have been discussed in Chapter 20.

TEA, COFFEE, CACAO, VANILLA, CITRON

Tea

Tea can be classified as (1) fermented, or black, (2) unfermented, or green, and (3) semifermented, or oolong. The experts agree that the "fermentation" of tea leaves is the result of the activity of enzymes of the leaves rather than of microorganisms present, although action of microorganisms may harm the flavor and reduce the quality of black tea. Molds of the genera *Aspergillus, Penicillium,* and *Rhizopus* have been reported to cause spoilage.

Coffee

The two chief methods of curing coffee are: (1) the dry method in which the berries or "cherries" are spread out and air-dried in the sun or artificially; and (2) the wet, or "washed coffee," method in which the berries, after removal of outer skin, are soaked in water. The removal of pulpy material in either method is accomplished primarily by pectinolytic bacteria, mostly coliforms, although pectinolytic bacilli and fungi may be present. This is followed by an acid fermentation by lactic acid bacteria such as *Leuconostoc mesenteroides, Lactobacillus brevis* and *plantarum,* and *Streptococcus faecalis.* The acids produced may be degraded by oxidizing organisms. After the pulp and its residues have been washed away, the beans are dried and hulled.

There are patents on the fermentation of coffee beans by certain kinds of bacteria before the roasting process to improve taste and aroma of the roasted product. Also a pectinolytic enzyme, derived from molds, has been utilized in a method for removal of the pulpy layer from the beans, thus avoiding the production of off-flavored coffee that sometimes results from the uncontrolled natural fermentation.

Deterioration of the released coffee beans can occur during their drying if this procedure is too prolonged.

Cacao

Cacao, or cocoa, beans (seeds) when taken from the pod are covered with a slimy or fruity pulp, which is removed by means of a fermentation process. The pulp-covered seeds are placed in piles, in pits, or in a "sweating box," where the beans are covered with banana or plantain leaves. During the 3 to 13 days of fermentation the beans are stirred and turned to aerate them and to keep down the temperature. The fermentation has the following purposes: (1) to remove the adhering pulp from the bean; (2) to kill the embryo in the seed; and (3) to give aroma, flavor, and color to the bean. Here again it is difficult to say how much is accomplished by the plant enzymes and how much by the growing microorganisms. According to some reports the fermentation is in several stages: (1) the fermentation of sugars in the pulp to ethanol and carbon dioxide by a variety of yeasts, especially *Candida krusei* and apiculate yeasts, and to acid by lactic acid bacteria; (2) the oxidation of the ethanol to acetic acid by acetic bacteria, with a rise in pH from an original 3.9 to about 7.1 in 7 days; (3) chemical changes brought about by the heat of the fermentation, where the temperatures may rise to 44 to 50 C (111.2 to 122 F), and by plant enzymes; (4) further chemical changes during curing or drying of the beans. Molds and actinomycetes can cause spoilage of the beans during curing.

Vanilla

Vanilla beans also are subjected to fermentation by being placed in piles where they undergo alternate sweating and drying processes. No reports are available upon the microbiology of the fermentation.

Citron

Halved citron fruits are held for 6 to 7 weeks in sea water or a 5 to 10 percent brine. It has been reported that yeasts improve the aroma by esterification of the essence of citron and that alcoholic fermentation by yeasts followed by an acetic acid fermentation combine to improve the flavor, color, and texture of citron. A yeast, *Saccharomyces citri medicae,* and a bacterium, *Bacillus citri medicae,* have been found to predominate during the fermentation.

ORIENTAL FERMENTED FOODS

Most of the Oriental fermented foods mentioned below have molds concerned in their preparation. In the starter, termed **koji** by the Japanese and **chou** by the Chinese, molds serve as sources of hydrolytic enzymes, such as amylases to hydrolyze the starch in the grains, proteinases, lipases, and many others. For the most part the starters are mixtures of molds, yeasts, and bacteria, but for a few products pure cultures have been employed.

Soy sauce

The chief Oriental fermented food imported into the United States and also made here is soy sauce, a brown, salty, tangy sauce used on dishes such as chop suey or as a constituent of other sauces. The methods of preparation of the starter and of manufacture of soy sauce have many variations and may result in different types of product.

THE STARTER. The starters (koji or chou) may be mixed cultures carried over from previous lots or pure cultures grown separately. The substrate on which the starter is grown varies, although most often it is an autoclaved mixture of soybeans, cracked wheat and wheat bran, a mixture of wheat bran and soybean flour, or rice. This moistened material is inoculated with spores of *Aspergillus oryzae* (*A. soyae*), spread in small boxes or trays, and held at 25 to 30 C (77 to 86 F) until the mold growth on the surfaces of the mash is adjudged to have attained a maximal content of enzymes (usually after about 3 days). A flora of lactic acid bacteria, streptococci and lactobacilli, also develops in the koji and produces lactic acid, and some growth of *Bacillus* species takes place. The starter may be used at once as it is, may be dried and used later, or may be dried and extracted and the extract used.

MANUFACTURE OF SOY SAUCE. The mash may consist of autoclaved soybeans or defatted, chemically hydrolyzed soybeans, roasted and crushed wheat, and steamed wheat bran. The mash is inoculated with the koji and incubated in trays for three days at about 30 C (86 F). Then it is soaked with sterile, 24 percent sodium chloride brine (sometimes the koji is mixed directly with an equal amount of saline water). The brined mash is held for from 2.5 months to a year or longer, depending upon the temperature.

THE FERMENTATION. The proteinases, amylases, and other enzymes of the koji continue to act throughout the holding period. There are three stages in the curing: (1) lactic acid fermentation by lactic acid bacteria from the koji, followed by more acid production by *Pediococcus soyae;* (2) alcoholic fermentation by yeasts, such as *Saccharomyces rouxii* and *Zygosaccharomyces soyae;* and (3) completion of the fermentation and aging.

The various microorganisms important in the making of soy sauce may be added in pure culture, or may come from previous lots of koji and from the ingredients. The chief organisms are: *Aspergillus soyae* (*oryzae*), the most important organism, which grows in the koji to yield proteinases, amylases, and other enzymes for soy-sauce brewing and contributes aromas and flavors; lactic acid bacteria, e.g., *Lactobacillus delbrueckii,* which makes the koji acid enough to prevent spoilage and acidifies the mash; *Bacillus subtilis* and other bacilli, which grow in the koji to improve flavor and make the soy sauce less turbid; *Pediococcus soyae,* which increases the acid in the mash, thereby stimulating the yeasts, contributing to essential aromas and flavors, decreasing color intensity, and reducing the activity of the mold proteinases; and *Hansenula* sp., *Saccharomyces rouxii,* and other yeasts, which produce alcohol and help the flavor.

Tamari sauce

This is a Japanese sauce, similar to soy sauce, made by a short fermentation process from a soybean mash to which rice may have been added. A different mold, *Aspergillus tamarii,* is the principal mold concerned in the manufacture of tamari sauce.

Miso

The koji for **miso** is a culture of *Aspergillus oryzae* grown at about 35 C (95 F) on a steamed polished-rice mash in shallow trays until the grains are completely covered (Figure 24–5) but the mold has not sporulated. The koji is mixed with a mash of crushed, steamed soybeans, salt is added, and the fermentation is allowed to proceed for a week at 28 C (82.4 F), then for two months at 35 C (95 F), after which the mixture is ripened for several weeks at room temperature. Involved in the main fermentation are the enzymes of the koji, yeasts (*Saccharomyces rouxii* and *Zygosaccharomyces* species), lactic acid bacteria, and bacilli. The final product is ground into a paste to be used in combination with other foods.

Figure 24–5. Rice covered with white mycelium during manufacture of miso. (*Courtesy of C. W. Hesseltine, Northern Utilization Research and Development Division, A.R.S.; U.S.D.A. Photograph.*)

Tempeh

In the manufacture of **tempeh**, an Indonesian food, soybeans are soaked overnight at 25 C (77 F), the seed coats are removed, and the beans, split into halves, are boiled in water for 20 min, dried on mats, cooled, and inoculated with a previous lot of tempeh or with mold spores of species of *Rhizopus* (*R. stolonifer, R. oryzae, R. oligosporus,* or *R. arrhizus*). The mash is packed into a plastic container or a hollow tube, or is rolled in banana leaves. It then is incubated at about 32 C (89.6 F) for 20 hr until there is good growth of mycelium but little sporulation (Figure 24–6). The product is sliced thin, dipped into salt water, and fried in vegetable fat to a golden brown.

Ang-khak

Ang-khak, or Chinese red rice, is produced by growth of *Monascus purpureus* on autoclaved rice and is used for coloring and flavoring fish and other food products.

Natto

In the manufacture of **natto,** boiled soybeans are wrapped in rice straw and fermented for 1 or 2 days. The package becomes slimy on

Figure 24–6. Tempeh in form of cakes (cracked soybeans) and in plastic containers (dehulled whole beans). Observe white surface growth of *Rhizopus*. (*Courtesy of C. W. Hesseltine, Northern Utilization Research and Development Division, A.R.S.; U.S.D.A. Photograph.*)

the outside. *Bacillus natto,* probably identical with *Bacillus subtilis,* grows in natto, releasing trypsinlike enzymes that are supposed to be important in the ripening process.

Soybean cheese

Soybean cheese, or **tou-fu-ru,** is a Chinese fermented food made by soaking soybeans, grinding them to a paste, and then filtering through linen. The protein in the filtrate is curdled by means of a magnesium or calcium salt, after which the curd is pressed into blocks. The blocks, arranged on trays, are held in a fermentation chamber for a month at about 14 C (57.2 F), during which period white molds, probably *Mucor* species, develop. Final ripening takes place in brine or in a special wine.

Minchin

Fermented **minchin** is made from wheat gluten from which the starch has been removed. The moist, raw gluten is placed in a closed jar and allowed to ferment for 2 to 3 weeks, after which it is salted. A typical specimen was found to contain seven species of molds, nine of bacteria, and three of yeasts. The final product is cut into strips to be boiled, baked, or fried.

Idli

Idli, a fermented food of India, is made from rice and black gram mungo in equal parts. The ingredients are washed and soaked separately, ground, mixed, and finally allowed to ferment overnight. When the batter has arisen enough, it is cooked by steaming and served hot. *Leuconostoc mesenteroides* grows first in the batter, leavening it, and is followed by *Streptococcus faecalis* and finally *Pediococcus cerevisiae,* all of which contribute to the acidity.

Fermented fish

The Japanese prepare a fermented fish by cutting it into strips, cooking it, and then encouraging fermentation by molds, chiefly species of *Aspergillus.* The strips then are dried. The Chinese have a fermented fish product preserved in **lao-chao,** a fermented rice product. Molds and yeasts are concerned in the fermentation of the steamed rice and some alcohol is produced.

Preserved eggs

Pidan, or the Chinese preserved egg, is made from duck eggs coated with a slurry of soda, burned straw, salt, and slaked lime and covered with rice husks. The eggs then are kept in sealed clay jars for a month. An assortment of bacteria grows in the egg, but coliform bacteria and species of *Bacillus* apparently are predominant.

Poi

Poi, although Hawaiian rather than Oriental, will be mentioned here. In the preparation of poi, the corms (bulblike, fleshy stems) of the taro plant are steamed at 70 to 100 C (158 to 212 F) for 2 to 3 hr, cooled, washed, peeled, trimmed and scraped, and then finely ground. This ground mass, after being mixed with water to attain a desired consistency, is fresh poi that is ready for consumption. The sour or fermented poi is prepared by the incubation of barrels or other containers of fresh poi at room temperatures for at least 1 day and at most not more than 6 days. During the first 6 hr of the fermentation the poi swells or puffs slightly and the color changes. During this period a mixture of soil and water microorganisms are prominent, organisms such as coliforms, *Pseudomonas* species, chromogenic bacteria, and yeasts. Between 6 hr and 4 days the flora consists predominantly of acid-forming bacteria, among which are *Lactobacillus pastorianus, L. delbrueckii, L. brevis, Streptococcus lactis,* and S. *kefir* (*Leuconostoc*). These lactic acid bacteria carry on the main fermentation in the poi, although during the latter part of the process, yeasts, film yeasts, and the mold *Geotrichum candidum* increase in numbers and probably contribute pleasing fruity odors and tastes (bouquet). The fermentation

products are chiefly lactic, acetic, and formic acids, alcohol, and carbon
dioxide. Abnormal fermentations are likely to be of the butyric acid
type.

REFERENCES

ALLEN, O. N., and ETHEL K. ALLEN. 1933. The manufacture of poi from
taro in Hawaii. Hawaii Agr. Exp. Sta. Bull. 70.
ALLGEIER, R. J., and F. M. HILDEBRANDT. 1960. Newer developments in
vinegar manufacture. Advances Appl. Microbiol. 2:163–182.
AMERINE, M. A., and W. V. CRUESS. 1960. The technology of wine making.
Avi Publishing Co., Inc., Westport, Conn.
AMOS, A. J. 1942. Microbiology and baking. Chem. Ind. 61:117–119.
CAMARGO, R. DE, J. LEME, JR., and A. MARTINELLI, FILHO. 1963. General
observations on the microflora of fermenting cocoa beans (Theobroma
cacao) in Bahia (Brazil). Food Technol. 17: 1328–1330.
CARR, J. G. 1962. The microbiology of wines and ciders. Rep. Progr. Appl.
Chem. 47:645–657.
CARR, J. G., and J. L. SHIMWELL. 1961. Acetic acid bacteria, 1941–1961:
a critical review. Antonie van Leeuwenhoek 27:386–400.
CRUESS, W. V. 1958. Commercial fruit and vegetable products. 4th ed. Mc-
Graw-Hill Book Company, New York.
DJIEN, K. S., and C. W. HESSELTINE. 1961. Indonesian fermented foods.
Soybean Dig., Nov., p. 14–15.
DOMERCQ, SIMONE. 1957. Etude et classification des levures de vin de la
Gironde. Annu. Inst. Nat. Rech. Agron., Ser. "E" Annu. Technol. Agr.
6:5–58.
FELL, G. 1961. Etude sur la fermentation malolactique du vin et les possibili-
tes de la provoquer par ensemencement. Landw. Jahrb. Schweiz. 10:249–
264.
Food Engineering Staff. 1962. Bread-dough process. Food Eng. 34(2):51.
GINI, B., and R. H. VAUGHN. 1962. Characteristics of some bacteria associated
with the spoilage of California dessert wines. Amer. J. Enol. Viticult.
13:20–31.
HAAS, G. J. 1960. Microbial control methods in the brewery. Academic Press
Inc., New York.
HALCROW, R. M. 1963. Biological considerations of brewery water supplies.
Brewers Dig. 28(8):39–45.
HESSELTINE, C. W. 1961. Research at Northern Regional Research Laboratory
on fermented foods. Proceedings of the Conference on Soybean Products
for Protein in Human Foods. Agr. Res. Serv., U.S. Dep. Agr.
HOGGAN, J. 1963a. Recent developments in brewing technology. Food Manu-
facture 38:308–314, 331.
HOGGAN, J. 1963b. Brewing, malting, and allied processes. Rep. Prog. Appl.
Chem. 48:524–532.
HOYNAK, S., T. S. POLANSKY, and R. W. STONE. 1941. Microbiological studies
of cacao fermentation. Food Res. 6:471–479.
HROMATKA, O., and H. EBNER. 1959. Vinegar by submerged oxidative fermen-
tation. Ind. Eng. Chem. 51:1279–1280.

Joslyn, M. A., and M. W. Turbovsky. 1954. Commercial production of table and dessert wines, Vol. I, chap. 7. *In* L. A. Underkofler and R. J. Hickey (*Eds.*) Industrial fermentations. Chemical Publishing Company, Inc., New York.

Martinelli, Filho, A., and C. W. Hesseltine. 1964. Tempeh fermentation: package and tray fermentations. Food Technol. 18:761–765.

Matz, S. A. (*Ed.*) 1960. Bakery technology and engineering. Avi Publishing Co., Inc., Westport, Conn.

McGhee, H., W. C. Drews, and C. Kandlbinder. 1956. Liquid ferment system perks up bakery performance. Food Eng. 28(7):56–58, 162–165.

Micka, J. 1955. Bacterial aspects of soda cracker fermentation. Cereal Chem. 32:125–131.

Mukherjee, S. K., M. N. Albury, C. S. Pederson, A. G. Van Veen, and K. H. Steinkraus. 1965. Role of *Leuconostoc mesenteroides* in leavening the batter of idli, a fermented food of India. Appl. Microbiol. 13:227–231.

Palo, M. A., Luz Vidal-Adeva, and Leticia M. Maceda. 1962. A study of ang-kak and its production. Philippine J. Sci. 89:1–22.

Pederson, C. S., and R. S. Breed. 1946. Fermentation of coffee. Food Res. 11:99–106.

Prescott, S. C., and C. G. Dunn. 1959. Industrial microbiology, chaps. 6, 7, 15. 3rd ed. McGraw-Hill Book Company, New York.

Rose, A. H. 1961. Industrial microbiology. Butterworth & Co. (Publishers), Ltd., London.

Sakaguchi, K. 1959. Studies on the activities of bacteria in soy sauce brewing: V, The effects of *Aspergillus sojae, Pediococcus soyae, Bacillus subtilis* and *Saccharomyces rouxii* in purely cultured soy sauce brewing. Rep. Noda Inst. Sci. Res. 3:23–29.

Samuel, O. C. 1963. Continuous brewhouse operation. Food Processing 24(9):60–63.

Schulz, A. 1952. The microbiology of sour dough. Baker's Dig. 26(5):27–29, 37.

Shimwell, J. L. 1960. The beer acetic acid bacteria. Brewers Dig. 25(7): 38–40.

Steinkraus, K. H., Yap B. Hwa, J. P. Van Buren, M. I. Provvidenti, and D. B. Hand. 1960. Studies on tempeh—an Indonesian fermented soybean food. Food Res. 25:777–788.

Vaughn, R. H. 1954. Acetic acid—vinegar, Vol. I, chap. 6. *In* L. A. Underkofler and R. J. Hickey (*Eds.*) Industrial fermentations. Chemical Publishing Company, Inc., New York.

Vaughn, R. H. 1955. Bacterial spoilage of wines with special reference to California conditions. Advances Food Res. 6:67–108.

Wiles, A. E. 1961. The wild yeasts; a review. Brewers Dig. 36(1):40–46.

Windisch, S. 1962. Microbiological problems in brewing. Brewers Dig. 37(2):47–51.

Yokotsuka, T. 1960. Aroma and flavor of Japanese soy sauce. Advances Food Res. 10:75–134.

CHAPTER TWENTY-FIVE

FOODS AND ENZYMES FROM MICROORGANISMS

Microorganisms may serve as food for human beings or feed for animals, as a source of enzymes to be used in the processing of foods, or as manufacturers of products to be added to foods.

MICROORGANISMS AS FOOD

Although a number of kinds of microorganisms have been recommended for human consumption, including yeasts, molds, and algae, to date only yeasts have been used as food to any extent and then under unusual conditions. During World War II, when there were shortages in proteins and vitamins in the diet, the Germans produced yeasts and a mold (*Geotrichum candidum*) in some quantity for food. After the war the British established a plant in Jamaica for the production of food yeast. Food- and feed-yeast plants are reported in operation now in Germany, Switzerland, Finland, the Union of South Africa, Jamaica, Formosa, and the United States. Production is of special interest to areas having plentiful supplies of cheap carbohydrate and shortages of proteins and vitamins. In the past most of the yeast used in the United States for feed and pharmaceuticals has been recovered by brewers and distillers after the alcoholic fermentation, but now increasing quantities of yeast are being grown directly for such purposes. Organisms may be termed **primary** when grown directly for the purpose in mind and **secondary** when they are recovered as a by-product of a fermentation.

Organisms used

Secondary yeasts recovered as a by-product of an alcoholic fermentation are strains of brewers' yeasts, *Saccharomyces cerevisiae* or *S. carlsbergensis*, or of distillers' yeasts, *S. cerevisiae* var. *ellipsoideus*. Brewers' yeast is debittered by washing with caustic soda solution and water, adjusted to pH 5.5 to 5.7 with phosphoric acid, salted, and dried, usually on a drum dryer. Before drying it may be fortified with thiamine, riboflavin, and niacin. A type of primary yeast may be produced by growing harvested yeast cells from breweries or distilleries for a few generations in a molasses or similar medium.

The yeast most often used for direct production is *Candida utilis*, called **torula yeast**. This yeast grows rapidly, utilizes pentose as well as hexose sugars, and synthesizes its accessory foods for growth from simple compounds, enabling its production from raw materials that are comparatively poor culture media. This is in contrast to *S. cerevisiae*, which utilizes only hexose sugar and otherwise is more fastidious in its growth requirements. Two special varieties of *Candida utilis* have been used, *C. utilis* var. *major*, which has larger cells than the original strain, and *C. utilis* var. *thermophila*, a high-temperature strain which can be grown at 36 to 39 C (96.8 to 102.2 F) rather than at 32 to 34 C (89.6 to 93.2 F). *C. arborea* (*Monilia candida*), *C. pulcherrima*, and other yeasts also have been used. A lactose-fermenting yeast is required, of course, if yeast is to be grown in whey. *Geotrichum candidum* (*Oöspora lactis*) is the only mold that has been grown for food on an industrial scale. Wide use of this or a similar mold is not indicated at present. The growth and use of algae for food is still in the experimental stage.

Raw materials

Materials that have been employed to produce primary yeasts include (1) molasses from sugar manufacture or hydrolysis of starch; (2) spent sulfite liquor, which is a waste product of the sulfite-pulping process in the paper industry; (3) the acid hydrolyzate of wood; and (4) agricultural wastes (e.g., whey from the dairy industry), hydrolyzed starchy foods (e.g., grains and cull potatoes), fruit wastes (e.g., fruit juice or citrus-peel hydrolyzate), etc. Molasses, spent sulfite liquor, wood hydrolyzate, and whey are the chief materials to be used thus far on a plant scale.

Production of primary yeast

Since most of the yeast produced directly is *Candida utilis*, this discussion will be limited to its growth in commonly utilized raw materials. For the most part, such torula yeast is produced by a continuous process which requires (1) establishment of active yeast growth in the fermentor; (2) feeding of carbohydrates and sources of nitrogen, phosphorus, and potassium at increasing rates until a maximal level of yeast growth is maintained continuously; (3) application of optimal aeration and agitation; and (4) withdrawal of liquor (beer) containing the yeast cells at rates and volumes equaling the additions of fresh medium.

Optimal conditions for yeast production vary with the yeast employed and the raw material used. Aeration should be considerable but at an optimal level; too little encourages alcohol production rather than growth, and too much favors increased respiration and heat production and hence lowered yields of the yeast cells. The optimal temperature depends upon the yeast strain, as has been indicated earlier. The pH should be kept on the acid side, usually between 4.5 and 6.0. The concentration of fermentable sugar is maintained at a level no higher than that necessary for a good yield of cells. The amounts and kinds of inorganic nutrients to be added depend upon the substrate. Cane- or beet-sugar molasses, for example, usually is high in potassium and fairly well supplied with available phosphorus and nitrogen, but spent sulfite liquor is deficient in all three of these elements, which are needed in relatively large amounts. Nitrogen usually is added in the form of ammonia or ammonium salts.

Yields are 45 percent or more of dry yeast on the basis of the sugar fed. The yeast cells are centrifuged from the medium, washed, concentrated, and dried. All food yeasts are killed before use.

Special pretreatments are required before some of the raw materials can be used to grow yeast. Most of the sulfur dioxide must be removed from the spent sulfite liquor by means of steam stripping or by aeration and treatment with lime. The treated liquor remains selective enough to favor the yeasts over competing organisms so that troubles with contamination are negligible throughout a long period of continuous yeast production. Wood hydrolyzate similarly requires pretreatment.

Nutritive value of yeast

Food yeast is high in protein and in most of the B complex of vitamins but may be deficient in methionine and perhaps cystine. Also, the thiamine content may be lower than in secondary yeast, and there is a deficiency of vitamin B_{12}. Furnished by food yeast in varying amounts are thiamine, riboflavin, biotin, niacin, pantothenic acid, pyridoxine, choline, streptogenin, glutathione, and perhaps folic acid and p-aminobenzoic acid.

Uses of food yeast

Thus far, food yeasts have been used to fortify human diet with protein, vitamins, and other nutritional factors chiefly in time of war emergency and in areas where these factors are in short supply and cheap carbohydrates are abundant. Yeasts have not displaced plentiful, inexpensive plant and animal sources of these factors. Food yeasts also are used in pharmaceutical products as sources of vitamins, amino acids, nucleic acids, and other protein fractions. It also is possible that yeasts may serve as raw material for fractionation processes to yield products useful in pharmaceutical and food industries.

FATS FROM MICROORGANISMS

Fats, or more properly lipids, are synthesized in appreciable amounts by certain yeasts, yeastlike organisms, and molds, but production in this manner has been employed only during periods of emergency, e.g., a state of war, when cheaper and more easily available animal and plant lipids were not available in sufficient quantity.

Organisms used

Among the yeasts, *Candida pulcherrima, Torulopsis lipofera, Saccharomyces cerevisiae,* and *Rhodotorula glutinis* have been studied for their fat production, and the first-named was used for this purpose in Germany and in Sweden during World War II. The yeastlike *Trichosporon pullulans* was used by the Germans during World War I, as were special strains of the mold, *Geotrichum candidum.*

Raw materials

Media for fat production should, in general, have a high carbon-to-nitrogen ratio, a good supply of phosphates, and for most organisms a pH on the acid side. In fat production by *Trichosporon pullulans,* good growth first is obtained in a mash with a high ratio of nitrogen to carbon, and fat production is carried out in a medium with a high ratio of carbon to nitrogen. Molasses, cellulose waste, hydrolyzed wood, and spent sulfite liquor are among the sources of carbohydrate employed. Nitrogen can be in the form of ammonium salts, urea, urine, yeast water, molasses slop, or extracts of grains. Salts to be added include potassium chloride, monopotassium phosphate, and magnesium sulfate. Addition of small amounts of alcohol or sodium acetate increases yields of lipids. The choice of a yeast will depend upon its ability to utilize both pentoses and hexoses or only the latter. *Geotrichum candidum* is able to utilize the lactose in whey, to which some additional nitrogenous food and salts may be added.

Production of fat

All of the above-mentioned organisms grow best under aerobic conditions. Their growth for fat production, therefore, has been in thin layers on trays or other flat areas when *Trichosporon pullulans* or a mold was employed, and in well-aerated tanks when yeasts were grown submerged. The optimal temperature varies with the organism employed, ranging from 15 to 20 C (59 to 68 F) for *T. pullulans* to 25 to 30 C (77 to 86 F) or slightly higher for molds and yeasts. A fairly long period is required for maximal fat production by most of the organisms mentioned. *T. pullulans*, for example, requires 2 to 3 days to attain growth and 6 or 8 days more for maximal yields of lipids. Yields of lipids vary widely with the organisms and with methods of production. Recovery of lipids is by extraction with solvents, with or without autolysis.

VITAMINS FROM MICROORGANISMS

Microorganisms may be able to absorb vitamins from the substrate, synthesize vitamins from precursors added or present, synthesize them from simple compounds, or manufacture provitamins which are readily convertible into vitamins.

Absorption of vitamins

Yeasts can absorb thiamine, niacin, biotin, and, to a lesser extent, pyridoxine, inositol, and other vitamins. Therefore, the level of these vitamins in the substrate in which the yeasts are grown is a factor in determining the vitamin content of the yeast cells. Brewers' and distillers' yeasts ordinarily are grown in mashes comparatively rich in vitamins and therefore become enriched themselves; or vitamin-rich rice polishings, malt sprouts, etc., or even purified thiamine, niacin, or biotin may be added to the mash to encourage absorption. On the other hand, yeasts grown in comparatively simple substrates, such as molasses-salts medium or spent sulfite liquor, have to synthesize most of their vitamins.

Synthesis of vitamins and provitamins

Microorganisms can synthesize many of the vitamins and some provitamins. However, only certain ones are prepared industrially in this manner, for the production of some of the vitamins is better accomplished by other means.

VITAMIN-B COMPLEX.   Yeasts are most often used to synthesize (and absorb) vitamins of the B complex. They have been reported to synthesize thiamine, riboflavin, p-aminobenzoic acid, and others. A pyrimidine and a thiazole added to the medium are said to favor synthesis of thiamine, and added β-alanine to serve as a precursor to pantothenic acid. Also present, at least in some yeasts, are niacin, pyridoxine, folic acid, biotin, inositol, and choline.

Riboflavin may be made by chemical synthesis or produced by microorganisms. It may be recovered from wastes of the butanol-acetone fermentation of molasses or grain or from whey in which *Clostridium acetobutylicum* has been grown. An optimal iron content in the foregoing media is important for its effect on the yield of riboflavin. Direct industrial production of riboflavin is chiefly by means of two very similar plant parasites, *Eremothecium ashbyii* and *Ashbya gossypii. E. ashbyii* is grown mostly by the submerged method in various proteinaceous foods, e.g., wastes such as stillage or slops from the butanol-acetone fermentation, fortified variously with milk, malt extract, corn-steep liquor, peptones, yeast, sugars, lipids, minerals, etc. Many other media have been suggested. An incubation temperature between 20 and 34 C (68 and 93.2 F) is recommended. *A. gossypii* is grown in media containing glucose, sucrose, or maltose, peptones, corn-steep liquor or yeast extract, and lipids. These organisms need i-inositol, thiamine, and biotin for good growth and produce best at about 26 to 28 C (78.8 to 82.4 F). Riboflavin is recovered in a crude form by dehydration or in purer form by precipitation by means of a reducing substance and subsequent crystallization.

Vitamin B_{12} (cyanocobalamin), the anti-pernicious-anemia factor, has been recovered from waste products from the production of some of the antibiotics by *Streptomyces* cultures, and is found in appreciable amounts in activated sludge. It also is synthesized by various bacteria, especially the propionibacteria.

FAT-SOLUBLE VITAMINS.   Fat-soluble vitamins (A, D, E, and K) or provitamins for them are synthesized by microorganisms but are not produced industrially in this manner. An exception is provitamin D, or ergosterol, which is made by special strains of yeast, *Saccharomyces carlsbergensis* for example, or by molds, such as *Aspergillus fischeri.* Industrially, most microbial ergosterol is synthesized by yeasts. This synthesis is favored by aeration, presence of nontoxic oxidizing agents, a low content of available nitrogenous food and lipids, addition of ethanol, presence of available di- or trisaccharides, and a fairly high temperature during the latter part of the fermentation. Conversion of ergosterol to vitamin D is by means of irradiation with ultraviolet light, either of yeast cells containing ergosterol or of the purified compound.

VITAMIN C. Ascorbic acid, or vitamin C, is obtained by chemical synthesis or from plant sources. In the synthesis of the vitamin, D-sorbitol, obtained by the chemical reduction of glucose, is oxidized by an acetic acid bacterium, *Acetobacter suboxydans,* to L-sorbose, which is then used in the chemical synthesis of ascorbic acid.

Production of amino acids

Lysine, aspartic acid, threonine, and glutamic acid are now being produced commercially by means of microorganisms. *Brevibacterium glutamicum,* various enteric bacteria, and a yeast have been employed for the production of lysine, *Bacillus* species and *Pseudomonas fluorescens* for aspartic acid from fumarate, *Escherichia coli* for threonine from sorbitol or mannitol, and *Brevibacterium glutamicum* and other organisms for glutamic acid. Organisms and methods are available for production of other amino acids, to be used when a demand occurs (e.g., for alanine, aminobutyric acid, diaminopimelic acid, homoserine, isoleucine, tryptophan, and valine).

PRODUCTION OF OTHER SUBSTANCES ADDED TO FOODS

Microorganisms may be employed to produce dextran, lactic and citric acids, and perhaps other substances to be added to foods.

Dextran

Dextran, a gummy polysaccharide, made by *Leuconostoc mesenteroides* from a molasses or refined sucrose medium, may serve as a stabilizer for sugar sirups, ice cream, or confections.

Lactic acid

For the most part, lactic acid is produced industrially by means of homofermentative lactic acid bacteria or bacteria resembling them. A simplified equation for the production of lactic acid from glucose by such organisms is as follows:

$$C_6H_{12}O_6 \xrightarrow[\text{bacteria}]{\text{lactic acid}} 2CH_3{\cdot}CHOH{\cdot}COOH$$
$$\text{glucose} \qquad\qquad\qquad \text{lactic acid}$$

Actually a series of steps is involved, and small amounts of other products are produced. It has been estimated that edible lactic acid makes up about half of the commercial acid produced by fermentation.

MICROORGANISMS USED. The microorganism used will depend upon the raw material to be fermented. *Lactobacillus delbrueckii* has been most often employed in the past for lactic acid production from glucose, maltose, or sucrose, although increasing use in being made of a flat sour bacterium such as *Bacillus coagulans*. *Lactobacillus bulgaricus* has been selected for making lactic acid from whey. The pentose-fermenting *L. pentosus* (*L. plantarum*) is recommended for use in spent sulfite liquors, and *L. brevis* (*L. pentoaceticus*) for hydrolyzed corncobs, cottonseed hulls, etc. Molds, e.g., *Rhizopus oryzae*, have been used experimentally to produce lactic acid from a glucose-salts medium.

RAW MATERIALS. It is advantageous in the manufacture of lactic acid by fermentation to start with a relatively simple medium or mash in order to facilitate the recovery of the product. When glucose, sucrose, or maltose is the carbohydrate fermented, as with a molasses or a starch hydrolyzate mash, *Lactobacillus delbrueckii* or a flat sour bacillus usually is employed. Nitrogenous food, minerals, and various growth factors required by the bacteria may be added in the form of malt sprouts, corn-steep liquor, milk, etc. Similar addition must be made to spent sulfite liquor for growth of *L. plantarum*. The production of lactic acid from whey by means of *L. bulgaricus* also has been carried out on an industrial scale. Other raw materials that have been suggested are cull fruits, Jerusalem artichokes, molasses, beet juice, hydrolyzed starchy materials, e.g., potatoes or grains, and hydrolyzed corncobs, corn stalks, cottonseed hulls, and straw.

PRODUCTION OF LACTIC ACID. The heat-processed mash is held at a temperature favorable to the organism with which it is inoculated: about 45 C (113 F) for *L. delbrueckii,* 45 to 50 C (113 to 122 F) for *L. bulgaricus* or *Bacillus coagulans,* or 30 C (86 F) for *L. plantarum.* The optimal sugar content of the mash will vary from 5 to 20 percent, depending upon the raw material and the fermenting organism. Anaerobic conditions are maintained in the mash. The pH is kept slightly on the acid side, the lactic acid being neutralized as formed by the periodic addition of calcium hydroxide or calcium carbonate during the several days of the fermentation. The calcium lactate may be crystallized as such or converted to lactic acid by the addition of sulfuric acid. The edible grade of lactic acid requires considerably more refining than the technical grade.

USES OF EDIBLE LACTIC ACID. In the food industries, lactic acid is used to acidify jams, jellies, confectionery, sherbets, soft drinks, extracts, and other products. It is added to brines for pickles and olives and to horse-radish and fish to aid in preservation. Its addition makes milk more digestible for infants. Calcium lactate is an ingredient of some baking powders.

Citric acid

In the United States most of the citric acid is produced by fermentation.

ORGANISMS USED. *Aspergillus niger* is the principal mold used in citric acid production, although various other molds are known to be able to make the acid and many have been tried experimentally, molds such as *A. clavatus, A. wentii, Penicillium luteum, P. citrinum, Mucor pyriformis,* and others. Apparently different strains of *A. niger* are preferred for surface methods of citric acid production than for methods involving submerged growth. A simplified, theoretical equation for the production of citric acid from glucose is as follows:

$$C_6H_{12}O_6 + 1\tfrac{1}{2}O_2 \rightarrow \quad C_6H_8O_7 \quad + 2H_2O$$
$$\text{glucose} \qquad\qquad \text{citric acid}$$

A series of intermediate steps are involved, and other products are produced.

RAW MATERIALS. Most citric acid is made from molasses, with beet molasses preferred over cane. Cane blackstrap and cane invert molasses also have been tried, as well as solutions of glucose or sucrose plus sources of nitrogen and minerals. The source of nitrogen usually is some simple compound such as ammonium salts or urea. The concentration of mineral ions is very important in obtaining good yields, especially levels of iron, zinc, and manganese. A slight deficiency in nitrogenous food and phosphate also favors production. Factors that tend to reduce the formation of mycelium below the maximum, therefore, appear to increase the yield of citric acid. Conditions favoring citric acid production tend to suppress the formation of oxalic acid.

PRODUCTION OF CITRIC ACID. The older method of citric acid manufacture was by means of surface growth; the mold mycelium was cultured on shallow layers of medium in trays or similar containers. Modern methods involve the submerged growth of the mycelium. It should be noted again that the strain of mold selected and the medium for production differ with the method of cultivation of the mold. Most methods utilize a sugar concentration between 14 and 20 percent; a fairly low pH, that is more acid for the surface method; a temperature of about 25 to 30 C (77 to 86 F); and a fermentation period of 7 to 10 days for the surface method and a shorter time for the submerged method.

USES OF CITRIC ACID. In the food industries, citric acid is added to flavoring extracts, soft drinks, and candies. It has been added to fish to adjust the pH to about 5.0 to aid in its preservation, to artichokes to enable the processor to utilize a milder heat-treatment in their canning than would otherwise be employed, to crab meat to prevent discoloration, as a synergist with antioxidants for oils, and as a dip for sliced peaches to delay browning. A large portion of the citric acid produced is used for medicinal purposes.

Monosodium glutamate

Monosodium glutamate (MSG), which is added to meats and fish and their products, sauces, soups, and other foods to accentuate their flavor, is made from L-glutamic acid, which may be produced by action of *Brevibacterium glutamicum*, other species of *Brevibacterium*, or other organisms on fumaric acid, glucose or various sugars (see Production of Amino Acids).

PRODUCTION OF ENZYMES

The production of enzymes from microorganisms is an important and rapidly growing industry, but one about which there still is considerable secrecy. However, some information is available from patents and from a few general articles. Because the separation and purification of the enzymes is difficult and costly and because too much purification often reduces their effectiveness, most of the enzymes are utilized in a crude form, usually in a preparation containing chiefly the desired enzyme but also other enzymes in lesser quantity, and, perhaps, considerable amounts of inert material. Most of the enzymes prepared on an industrial scale are hydrolytic.

Amylases

The amylases, which hydrolyze the starches, are classified in various ways, depending on how they act on the starch molecules. Starch is composed of two glucans: the linear **amylose** containing D-glucose units joined by alpha-1,4-linkages and the branched **amylopectin** containing in addition 1,6-linkages at the branching points. The amylases include: **phosphorylase** which can completely degrade amylose but only partially break down amylopectin, leaving "limit dextrins"; **isoamylase,** the debranching enzyme, which splits the terminal 1,6-glucosidic linkages of amylopectin; **alpha-amylase** which randomly hydrolyzes the alpha-1,4-glucoside linkages of amylose or amylopectin but not the 1,6-glucosidic linkages of amylopectin; **beta-amylase** which splits only the second alpha-1,4-glucoside linkage from the nonreducing chain ends, detaching one molecule of maltose at a time from the chain and finally leaving "limit dextrins"; and **amyloglucosidase** which hydrolyzes both 1,6- and 1,4-glucoside linkages to produce glucose without intermediate dextrins and maltose. **Maltase** or **α-glucosidase** (not an amylase) hydrolyzes maltose to glucose. Much of the amylase used in industry is provided by grain malt, which contains amylases in varying proportions. Increasing amounts of amylase are being produced, however, from molds and bacteria, and they may yield primarily one kind of amylase or a mixture of amylases.

FROM MOLDS. Molds are used as sources of amylases as well as of other hydrolytic enzymes, the species and strain of mold being selected especially for the purpose. *Rhizopus delemar, Mucor rouxii,* and related species have been employed in the "Amylo" process, in which a starchy grain mash is saccharified by amylases produced by the mold growing on it. This process has been applied principally to the preparation of mashes for alcoholic fermentation by yeasts. For the production of preparations rich in amylases, *Aspergillus oryzae* has been used most, although *A. niger* has been recommended for submerged methods of production. The preparation of koji, the *A. oryzae* starter for soy sauce that is rich in amylases, as well as other hydrolytic enzymes, was described in the preceding chapter.

Moistened, steamed wheat or rice bran is used for the production of amylases from *A. oryzae* by the tray method, in which the mold is grown on thin layers of the medium in trays, or the drum method, in which the bran is tossed loosely in a rotating drum. A maximal yield is obtained by the tray method in 40 to 48 hr at about 30 C (86 F) in an atmopshere with a high humidity and adequate ventilation. The amylases are extracted from the mycelium and may be purified by precipitation and washing or may be concentrated as desired. Other hydrolytic enzymes also are present in such preparations.

FROM BACTERIA. *Bacillus subtilis* has been the principal bacterium used for the production of amylases, although other species of bacteria are known to yield these enzymes. Most bacteria produce more α-amylase than β-amylase. Production may be by means of the surface-growth method on shallow layers of mash in trays or by the submerged method.

Various culture media or mashes have been recommended for the production of amylases by means of B. *subtilis,* ranging from a complex medium such as thin stillage, a by-product of alcohol production from grains, to a simpler medium consisting of soluble or hydrolyzed starch, ammonium salts, and buffers. Another mash utilizes hydrolyzed soybean cake, peanut cake, or casein as a source of nitrogen, hydrolyzed starch for energy, and various mineral salts. Also, wheat bran has been used in a bran–salts medium. Incubation temperatures from 25 to 37 C (77 to 98.6 F) and times of from 2 to 6 days have been employed in the tray method and 24 to 48 hr at 30 to 40 C (86 to 104 F) in the submerged method.

Bacterial amylase may be purified and concentrated by dialysis, condensation, and fractional precipitation.

USES OF AMYLASES. In the food industries, fungal amylases have been employed to remove starch from fruit extracts, e.g., in the production of pectin from apple pomace, to clarify starch turbidities in wines, beer, and fruit juices, to convert acid-modified starches to sweet sirups, to substitute for malt in breadmaking in aiding leavening and improving dough consistency and hence gas retention (proteinases also are involved), and to saccharify starch in mashes for alcoholic fermentation. Bacterial amylase, chiefly the alpha type, has been used in brewing to produce dextrins of low fermentability and also to clarify the beer, and in the manufacture of corn sirup and chocolate sirup to prevent thickening by dextrinizing the starch present.

Invertase

Invertase catalyzes the hydrolysis of sucrose to glucose and fructose. The invertase of yeasts is a fructosidase in that it attacks the fructose end of the sucrose molecule, in contrast to the glucosidase of molds that attacks the glucose end. Industrially, invertase is produced mainly by growth of special strains of Saccharomyces cerevisiae (bottom type) in a medium that contains sucrose, an ammonium salt, and phosphate buffer and other minerals, and is adjusted to about pH 4.5. Incubation is for about 8 hr at 28 to 30 C (82.4 to 86 F). For recovery of the invertase the yeast cells are filtered off, compressed, plasmolyzed, and autolyzed. The invertase extracted from the cells may be dried with sugar or held in a sucrose sirup; or the enzyme may be purified by dialysis, ultrafiltration, adsorption, and elution. Most commercial preparations of invertase are not highly purified.

USES. Invertase is used in the confectionery industry to make invert sugar for the preparation of liqueurs and ice creams in which the crystallization of sugars from high concentrations is to be avoided. In soft-center, chocolate-coated candies, e.g., maraschino cherries, invertase incorporated in the centers softens the fondant after it has been coated with chocolate. Invertase is added to sucrose sirups to hydrolyze that sugar and in this way to prevent crystallization on standing. It also has been used in the manufacture of artificial honey.

Pectolytic enzymes

Pectin, which is methylated polygalacturonic acid, is important in food industries because of its ability to form gels with sugar and acid. In jellies this characteristic is desirable, but not in fruit juices. Most authorities agree that chiefly two enzymes are involved in the hydrolysis of pectin: **pectinesterase**, to hydrolyze the pectin to methanol and polygalacturonic acid (pectic acid), and **polygalacturonase**, to hydrolyze the polygalacturonic acid to monogalacturonic acid. Further hydrolysis would yield sugars and other products.

The mixture of pectolytic enzymes sometimes is called **pectinase** (pectase, pectinols, or filtragols), a term which will be used here for a mixture of pectolytic enzymes such as is produced by microorganisms. Pectinase is yielded by a number of molds and by various bacteria, including the clostridia involved in retting. Only the fungal type of pectinase is produced industrially to any extent, and this from molds such as species of *Aspergillus, Penicillium,* and other genera. The mycelium is developed on a medium containing pectin or a pectinlike compound, a nitrogen source, such as plant, yeast, or malt extract, ammonia, peptone, etc., and mineral salts. The mycelium is harvested, macerated, and extracted, and the crude enzyme mixture thus obtained may be precipitated and concentrated.

USES. Pectinases from extracts of plant materials or from fungi are used in the food industries for the clarification of fruit juices, wines, vinegars, sirups, and jellies that may contain suspended pectic material. Treatment of fruit juices with pectinase helps prevent jelling of the juices upon concentration. The addition of pectinase to crushed fruit, e.g., grapes, aids in the expression of the juice and results in wines that clarify readily. Partial deesterification by means of pectinesterase to yield modified pectins which set slowly is employed in the manufacture of candy jellies of high sugar content.

Proteolytic enzymes

The proteolytic enzymes, or **proteases,** include the **proteinases,** which catalyze the hydrolysis of the protein molecule into large fragments, and the **peptidases,** which hydrolyze these polypeptide fragments as far down as amino acids. The proteolytic enzyme preparations from microorganisms are proteases, that is, mixtures of proteinases and peptidases. Proteases also are prepared from plant or animal sources. Papain, for example, from the papaya fruit, is injected into meat animals before slaughter, so that the meat will be tenderized by the enzyme during cooking.

FROM BACTERIA. For the most part, bacterial protease is prepared from cultures of *Bacillus subtilis,* although many other bacteria yield proteases. A high-yielding strain is selected, special culture media are employed, and temperature and degree of aeration are adjusted to favor the production of protease over that of amylase. The medium or mash has a fairly high content of carbohydrate (2 to 6 percent), as well as of protein, and also contains mineral salts. Incubation is for 3 to 5 days at about 37 C (98.6 F) with adequate ventilation. The filtrate from the culture is concentrated and the enzymes are used in this form, or purified further, or adsorbed onto some inert material, such as sawdust. The enzyme mixture also contains varying amounts of amylases.

FROM MOLDS. Protease preparations from molds also contain other enzymes. Thus the koji for soy sauce or the Taka-Diastase for pharmaceutical purposes contains a variety of enzymes. It is possible, however, to select strains of molds that give high yields of proteases and comparatively low yields of other enzymes. The mold also can be chosen for its ability to produce proteases which are active under acid conditions or active under alkaline conditions. The methods of preparation of mold proteases are similar to those for the production of amylases. *Aspergillus oryzae* is a good source of proteases, although other molds have been recommended. Many different media have been suggested, including those containing wheat bran, soybean cake, alfalfa meal, middlings, yeast, and other materials. Recovery of the enzyme is by extraction, concentration, and precipitation, as for other hydrolytic enzymes.

USES. The proteases from microorganisms are used primarily for their proteinase activity. Bacterial proteases have been applied to the digestion of fish livers to liberate fish oil, to the tenderization of meat, and to the clarification and maturing of malt beverages. Fungal proteases are active in the manufacture of soy sauce and other Oriental mold-fermented foods and may be added to bread dough, where, along with amylase, they help improve the consistency of the dough. They also may be used for chill-proofing beer and ale by removal of protein haze (the fungal tannase present also may be helpful), for the tenderization of meats, for thinning egg white so that it can be filtered before drying, and for the hydrolysis of the gelatinous protein material in fish waste and press water to facilitate concentration and drying.

Catalase

Catalase usually is obtained from animal organs but can be prepared from a mold grown in an aerated, agitated culture or from *Micrococcus lysodeikticus*. It is used to neutralize excess hydrogen peroxide in processes involving the combined action of mild heat and the peroxide as applied to milk and other foods.

Glucose-oxidase

Glucose-oxidase is produced by the submerged growth of *Aspergillus niger* or another mold. It is used to remove glucose from egg white or whole eggs to facilitate drying, prevent deterioration, and improve the whipping properties (of the reconstituted dried whites). It has been employed also to extend the shelf life of canned soft drinks by retarding the pickup of iron and the fading of color. Oxidation of the glucose by glucose-oxidase forms gluconic acid and hydrogen peroxide, the latter then being decomposed by the catalase in the same preparation. A combination of glucose-oxidase and catalase is used to remove small residues of oxygen in packaged foods.

Other enzymes

Cellulase, catalyzing the hydrolysis of cellulose to cellulodextrins and glucose, has been recommended for producing more fermentable sugar in brewers' mashes, clarifying orange and lemon juices and concentrates, and tenderizing green beans. Microbial **lipase** removes fat from yolk residues in dried egg albumin, assists mold spores in production of blue-cheese flavor in spreads, and adds to the flavor of milk chocolate. **Dextransucrase** increases viscosity by production of dextran in sucrose-containing foods. Flavor-producing enzymes, both from raw foods and from microorganisms, are receiving special attention. **Lactase**

from *Saccharomyces fragilis* may find use in hydrolysis of the lactose in whey to glucose and galactose, which are less laxative sugars. Undoubtedly more kinds of enzymes will be added to those mentioned.

REFERENCES

ANONYMOUS. 1964. What you can do with enzymes. Food Eng. 36(5):80–81.

BARTON, R. R., and C. E. LAND, JR. 1961. How latest enzymes sharpen your process control. Food Eng. 33(9):85–88.

DEINDOERFER, F. H., R. I. MATELES, and A. E. HUMPHREY. 1963. 1961 fermentation process review. Appl. Microbiol. 11:273–303.

FAWNS, H. T. 1943, 1944. Food production by micro-organisms. Food Manufacture 18:194–198, 200, 333–337; 19:394–400.

HUMPHREY, A. E., and F. H. DEINDOERFER. 1962. 1960 fermentation process review. Appl. Microbiol. 10:359–385.

JOSLYN, M. A. 1962. The chemistry of protopectin: a critical review of historical data and recent developments. Advances Food Res. 11:1–107.

MEYER, LILLIAN H. 1960. Food chemistry. Reinhold Publishing Corporation, New York.

PORTER, J. R. 1946. Bacterial chemistry and physiology. John Wiley & Sons, Inc., New York.

PRESCOTT, S. C., and C. G. DUNN. 1959. Industrial microbiology. 3rd ed. McGraw-Hill Book Company, New York.

ROSE, A. H. 1961. Industrial microbiology. Butterworth & Co. (Publishers), Ltd., London.

SCHOCH, T. J. 1962. Recent developments in starch chemistry. Brewers Dig. 37(2):41–46, 66–67.

SCHULTZ, H. W. 1960. Food enzymes. Avi Publishing Co., Inc., Westport, Conn.

SIZER, I. W. 1964. Enzymes and their applications. Advances Appl. Microbiol. 6:207–226.

UNDERKOFLER, L. A., R. R. BARTON, and S. S. RENNERT. 1958. Production of microbial enzymes and their applications. Appl. Microbiol. 6:212–221.

UNDERKOFLER, L. A., and R. J. HICKEY (*Eds.*) 1954. Industrial fermentations, Vol. I: chaps. 10, 12, 13; Vol. II: chaps. 3, 4, 5, 6, 10. Chemical Publishing Company, Inc., New York. 2 vols.

WALLERSTEIN, L. 1938. Enzyme preparations from microorganisms. Ind. Eng. Chem. 31:1218–1224.

PART FIVE

FOODS IN RELATION TO DISEASE

Every food microbiologist should know how disease may be spread by foods and how such transmission can be prevented. Of special interest is so-called "food poisoning," which may be true food poisoning or be food infection, and the investigation of outbreaks of food poisoning.

CHAPTER TWENTY-SIX

FOOD POISONINGS AND INFECTIONS

Gastrointestinal disturbances resulting from the ingestion of food can have a variety of causes, such as overeating, allergies and nutritional deficiencies, actual poisoning by chemicals, toxic plants or animals, toxins produced by bacteria, infestation by animal parasites, or infection by microorganisms. These illnesses often are grouped together because they have rather similar symptoms at times and sometimes are mistaken one for the other.

POISONING BY CHEMICALS

Poisoning by consumption of chemicals is rather uncommon and usually is characterized by appearance of the symptoms within a short time after the poisonous food is eaten. Antimony, arsenic, cadmium, lead, and zinc in foods have been blamed for food poisoning. Poisonous chemicals may enter foods from utensils, e.g., from cadmium-plated ware or cheap enameled ware containing antimony. The insecticide sodium fluoride has been accidentally added to food in place of baking powder, flour, dry milk, or starch. Lead and arsenic residues from fruit sprays may be on the surfaces of fruits, but usually in harmless amounts, especially after washing. Often the source of poison is wrongly attributed to food; methyl chloride poisoning from a leaking mechanical refrigerator, for example, or poisoning from materials that are hazards in some industries has been mistaken for food poisoning.

POISONOUS PLANTS AND ANIMALS

Gastrointestinal disturbances or even death may result from the consumption of plants or plant products. Favism is poisoning from eating young fava beans or even from smelling the blossoms of the plant. Snakeroot poisoning results from drinking milk from cows that have fed on this or a similar weed. Rhubarb greens have been reported responsible for oxalic acid poisoning. The eating of poisonous varieties of mushrooms, mistaken for the edible kinds, causes mushroom poisoning.

Ocean mussels and clams during certain seasons of the year contain a poisonous alkaloid, apparently from plankton consumed by the shellfish. Some fish found in tropical waters are poisonous.

"POISONING" BY MICROORGANISMS AND THEIR PRODUCTS

Ordinarily, the term "food poisoning," as applied to diseases caused by microorganisms, is used very loosely to include both illnesses caused by the ingestion of toxins elaborated by the organisms and those resulting from infection of the host through the intestinal tract. Here, **food illness** will be employed as a general term for any disease caused by eating food. A **food poisoning** will refer to an illness caused by a poison present in the food when it is consumed; and a **food infection** will mean an illness caused by an infection produced by invasion, growth, and damage to the tissue of the host by pathogenic organisms carried by the food.

There are two chief kinds of food poisoning caused by bacteria: (1) botulism, caused by the presence in food of toxin produced by *Clostridium botulinum,* and (2) staphylococcus poisoning, caused by a toxin in the food from *Staphylococcus aureus.*

Food infections may be divided into two types: (1) those in which the food does not ordinarily support growth of the pathogens but merely carries them, pathogens such as those causing tuberculosis, diphtheria, the dysenteries, typhoid fever, brucellosis, cholera, infectious hepatitis, Q fever, etc.; and (2) those in which the food can serve as a culture medium for growth of the pathogens to numbers that will increase the likelihood of infection of the consumer of the food; these are chiefly pathogens of the genus *Salmonella.* Outbreaks of food infections caused by species of this genus are likely to be more explosive than outbreaks caused by other intestinal pathogens.

Table 26–1 shows numbers of cases and deaths in one year and the average numbers of cases and deaths during a ten-year period in the United States from botulism, salmonellosis, hepatitis, and trichinosis, much of which may have come from foods. No reliable figures are available on staphylococcus and *Clostridium perfringens* poisoning. Experts have estimated, however, that there are from several hundred thousand to a million cases of food illnesses annually in the United States.

There was an upsurge in numbers of cases of botulism reported for 1963, when twelve outbreaks, forty-six cases, and fourteen deaths were recorded. Four of these outbreaks, with twenty-four cases and

TABLE 26–1. Cases (1964) and deaths (1963), notifiable illnesses in the United States, and averages for ten-year periods.†

Illness	No. of cases		No. of deaths	
	1964	*Avg. 1955–64*	*1963*	*Avg. 1954–63*
Botulism	23	19.3	14	12.3
Salmonellosis (except typhoid)	17,144	8,950	72	65
Hepatitis (infectious and serum)	37,740	35,403	852*	883.6*
Trichinosis	198	217.3	5	3.7

  * Infectious only.
  † From Public Health Service, U.S. Department of Health, Education, and Welfare. 1965. Morbidity and Mortality; Weekly Report 13(54): Sept. 30.

nine deaths, were from commercially processed foods. Three outbreaks were of type E botulism (see following section) from processed fish, one from canned tuna (three cases, two deaths), and two from smoked fish (nineteen cases, seven deaths). Number of cases of salmonellosis reported for 1963 also rose sharply.

TRUE FOOD POISONING

Botulism

Botulism is a true food poisoning, caused by the ingestion of food containing the exotoxin produced by *Clostridium botulinum* during its growth in the food.

THE ORGANISM. This rod-shaped soil bacterium (Figure 26–1,A,B) is saprophytic, spore-forming, gas-forming, and anaerobic. In the seventh edition of "Bergey's Manual of Determinative Bacteriology" it is suggested that the name *C. botulinum* be applied to the nonovolytic (albumin not liquefied) type of organism and *C. parabotulinum* to the ovolytic type. In the following discussion only the name *C. botulinum* will be applied to all types of clostridia that cause botulism. Six types are distinguished on the basis of the serological specificity of their toxins:

Type A (ovolytic) is the one that most commonly causes human botulism in the western part of the United States. It usually is more toxic than type B.

Type B (some ovolytic and others not) is found more often than type A in most soils of the world and is less toxic to human beings.

Type C (nonovolytic) causes botulism of fowls, cattle, mink, and other animals, but not of human beings so far as is known.

Type D (nonovolytic) is associated with forage poisoning of cattle in the Union of South Africa and is rare in human beings.

Type E (nonovolytic), which is toxic for man, has been obtained chiefly from fish and fish products.

Type F, which except for its toxin is similar to types A and B, has been isolated in Denmark.

Type A strains and most cultures of type B are proteolytic and are putrefactive enough to give an obnoxious odor to proteinaceous foods, but some strains of type B and those of type E are not. Even the first two types fail to give marked indications of putrefaction in

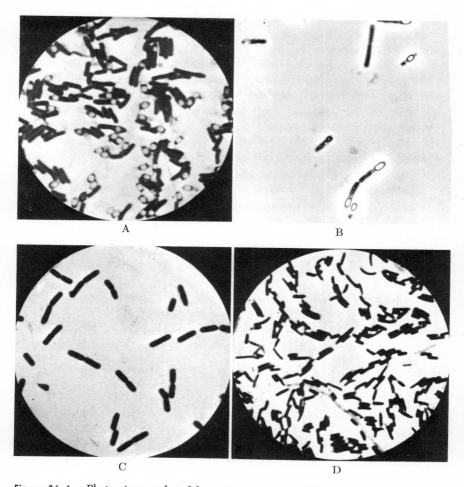

Figure 26–1. Photomicrographs of bacteria causing food illnesses: (A) *Clostridium botulinum* with spores (*From J. Nowak*); (B) *C. botulinum* type E with spores (*Courtesy of Janet S. Deffner*); (C) *C. perfringens* (*From J. Nowak*); (D) *Salmonella enteritidis* (*From J. Nowak*).

low-protein foods such as string beans and corn, although they produce toxin. The organism ferments carbohydrates with gas production, but sometimes this, too, is not evident.

GROWTH AND TOXIN PRODUCTION.   Toxin production by *C. botulinum* depends upon the ability of its cells to grow in a food and to autolyze there, for the types A and B toxins apparently are synthesized as large, comparatively inactive molecules which become fully toxic after some hydrolysis. Therefore, the factors that influence spore germination, growth, and hence toxin production are of special interest. These factors include the composition of the food or medium, especially its nutritive properties (e.g., glucose or maltose is known to be essential for toxin production), moisture content, pH, oxidation-reduction potential, and salt content, and the temperature and time of storage of the food. It is the combination of these factors that determines whether growth can take place and the rate and extent of that growth. Thus the nutritive properties of the food are likely to determine the minimal pH or temperature and the maximal concentration of sodium chloride for growth and toxin production. Results will differ with the serological type of organism and the particular strain. The characteristics given in Table 26–2 are, for the most part, extremes.

Although foods are known to differ as culture media for *C. botulinum,* much of the evidence is empirical. Most of the studies have been on toxin production in various foods. Meats, fish, and low- or medium-acid canned foods have been shown to support toxin production and to differ in the potency of the toxin formed. Even good culture media may differ in the relative potency of the toxin formed in them.

TABLE 26–2.   Some characteristics* of bacteria causing food illnesses
(from numerous sources)

Organism	Growth temperatures, C			Lowest pH for growth reported	Killing by heat†	Killing by gamma rays† (12D), Mrad	Salt toler- ance, %
	Lowest reported	Optimal	Highest reported				
Clostridium botulinum:							
Types A, B	10	35	48	4.7	15 min, 121 C	4.0	8.9
Type E	3.3	30	45	5.0	15 min, 80 C	2.5	ca. 5.0
Staphylococcus aureus	6.6	35	47	4.8	30 min, 62.8 C	0.5	17.0
Salmonella	6.6	37	45	5.5	30 min, 60 C	0.5	10.0
C. perfringens	10 (?)	43–47	50	5.0	1–4 hr, 100 C	2.75	5.0

* All characteristics listed vary with substrate and strain of organism.
† For killing most resistant strains.

It has been reported, for instance, that media containing milk or casein, glucose or maltose, and corn-steep liquor yield more potent type A toxin than other media, and that potencies of toxin from the following canned foods are, in descending order: corn > peas > string beans > spinach. Dissolved tin from cans has been shown to inhibit growth and toxin production in canned vegetables. Experiments on dehydrated meat have shown that toxin was produced more slowly when the moisture content was 40 percent than when it was 60 percent, and that reduction to 30 percent prevented toxin production.

The concentrations of sodium chloride necessary to prevent growth and toxin production in foods depend upon the composition of the food and the temperature. The presence of sodium nitrate in sausage or of disodium phosphate in cheese spread reduces the level of sodium chloride necessary to prevent toxin production. More salt is needed at a higher temperature, such as 37 C, than at a lower one, e.g., 15 C. Under favorable conditions for growth 8 percent or more of salt is needed to inhibit *C. botulinum*.

A pH near neutrality favors *C. botulinum*. The minimal pH at which growth and toxin production will take place will depend upon the kind of food and the temperature. Workers in the laboratories of the National Canners Association have tested spore germination, growth, and toxin production by strains of both types A and B botulinum cultures in a variety of foods adjusted to different pH levels. These foods included green peppers, pimientos, eggplant, fruit puddings, spaghetti with sauce or with meat, vegetable-juice cocktails, etc., that often are acidified before the heat process to permit a less rigorous heating than would be required otherwise. They obtained different results with the various foods and the different strains, although type A strains would, with one exception, produce toxin at a lower pH than type B strains. They agree with other workers that a pH of 4.5 or lower will prevent toxin production in most foods, but that the lowest pH for spore germination is considerably higher. Minimal pH values reported are: 4.87 for vegetative cells and 5.01 for spore germination in a veal infusion broth, 4.8 to 5.0 in bread, and 4.8 in pineapple-rice pudding. A maximal pH of 8.89 was found for vegetative growth. *C. botulinum* has been found growing and producing toxin, however, in foods that normally are too acid for it, when other microorganisms also were growing in the food and presumably raising the pH locally or generally.

Temperature is an important factor in determining whether toxin production will take place and what the rate of production will be. Vegetative growth will take place at a lower temperature than the minimum for spore germination. Different strains of *C. botulinum* types A and B vary in their temperature requirements. A few strains have been reported able to grow at 10 or 11 C (50 or 51.8 F), but about 15 C (59 F) has been claimed to be the lowest temperature for spore germination. The maximal temperature for growth is about 48 C (118.4 F) for these types, and about 45 C (113 F) for type E (Table 26–2). Type E organisms have been found to produce gas and toxin within 31 to 45 days at as low as 38 F (3.3 C). Although 37 C (98.6 F) often is given as the optimal temperature for the organism, more toxin usually is produced at lower temperatures, possibly because the toxin is less stable at the higher temperatures. Obviously, the slower the rate of toxin production, the longer it will take to obtain appreciable amounts. The optimal temperature for type E organisms is about 30 C (86 F).

As previously stated, the growth of *C. botulinum* in some foods results in such a foul, rancid odor that they would be rejected. Meats and proteinaceous, low-acid vegetables develop an especially obnoxious odor. More acid foods, however, and those low in proteins may become just as toxic without much evidence of putrefaction. Moreover, the non-ovolytic strains of *C. botulinum* give less evidence of spoilage than the ovolytic ones. Also, gas production is not always evident and therefore is not a reliable indication of spoilage by this organism. Certainly it is advisable to reject all foods, raw or canned, that give evidences of spoilage and to do as the canners have done for years—reject canned foods that exhibit any pressure in the container.

THE TOXIN. The toxin of *C. botulinum*, a protein that has been purified and crystallized, is so powerful that only a tiny amount is sufficient to cause death. It is absorbed mostly in the small intestine, and paralyzes the involuntary muscles of the body. An important characteristic is its comparative thermolability. The heat-treatment necessary to destroy it depends upon the type of organism producing the toxin and the medium in which it is heated. In the laboratory, heat-treatments of from a few minutes to 30 min at 80 C (176 F) have been reported to inactivate the toxin, but in practice it is recommended that suspected foods be kept at a full boil for at least 15 min. It is destroyed in cheese by 7.3 Mrad of gamma rays, and in broth by 4.9 Mrad. The toxin has been known to persist in foods for long periods, especially when storage has been at low temperatures. It is unstable at pH values above 6.8. As has been indicated, the toxins of the six types of bacteria responsible for botulism are antigenic, causing the production of antitoxin specific for the type of toxin injected. Toxoids have been prepared for the active immunization of workers who might be exposed to accidental poisoning by the toxin of botulism.

HEAT RESISTANCE OF SPORES. Compared with the spores of most other *Clostridium* species, those of some of the putrefactive anaerobes, including *C. botulinum*, have a comparatively high resistance to heat. The heat-treatment necessary to destroy all of the spores in a food will depend upon the kind of food, the type and strain of *C. botulinum*, the medium in which the spores were formed, the temperature at which they were produced, the age of the spores, and the numbers of spores present. The reader should refer to Chapter 6 for a discussion of the factors that influence the heat resistance of spores, Esty has recommended the following heat-treatments to destroy all spores of *C. botulinum* in a food:

100 C (212 F)	360 min
105 C (221 F)	120 min
110 C (230 F)	36 min
115 C (239 F)	12 min
120 C (248 F)	4 min

In general, spores of organisms of types C, D, and E are less heat-resistant than those of types A and B, type E spores being inactivated in 15 min at 80 C (176 F).

Minimal heat processes for canned foods recommended by the National Canners Association and other agencies are sufficient to destroy all spores of *C. botulinum* and allow a good margin of safety. As has been indicated previously, one of the questions raised concerning new methods for processing foods, such as heat plus antibiotics or irradiation, is whether these processes will guarantee the destruction of spores of *C. botulinum* or at least prevention of their germination.

RAY RESISTANCE OF SPORES. Spores of *C. botulinum*, types A and B, differ considerably in their resistance to gamma rays; the D values for most of them in meat ranges from 0.224 to 0.336 Mrad, but some are more sensitive. The less resistant spores of type E in beef stew have D values from 0.125 to 0.138 Mrad.

DISTRIBUTION OF SPORES. The habitat of *C. botulinum* is believed to be the soil, for spores have been found in both cultivated and virgin soils all over the world. Tests have shown that type A spores are found more in western soils in this country and type B spores elsewhere. Plant crops may become contaminated from the soil, and intestinal contents and hence manure of animals after consumption of such plants. Type E spores are found in soil, in sea and lake mud, and in fish, primarily in their intestinal tracts.

INCIDENCE OF BOTULISM. Fortunately, botulism occurs only rarely; but it always receives much attention because of the high mortality.

In the United States from 1938 to 1953, inclusive, there was an average of less than seven outbreaks reported per year. The incidence has decreased with increased use of pressure cookers in home canning.

Foods INVOLVED. In this country, inadequately processed home-canned foods are most often the cause of botulism; and in Europe the main causes are preserved meats and fish. Of the canned foods, those most often responsible for botulism have been string beans, sweet corn, beets, asparagus, spinach, and chard, but each of many other kinds of food has been responsible for one or several outbreaks (fifty kinds of canned fruits and vegetables have been involved between 1899 and 1947). Table 26–3 lists home-canned foods responsible for six or more outbreaks of botulism in the United States in the period 1899 to 1947. In general, the low- and medium-acid canned foods are most often incriminated, but there have been exceptional instances of poisoning from acid foods, such as tomatoes, apricots, pears, and peaches. These acid foods had been grossly underprocessed, and the underprocessing had permitted the growth of other microorganisms to aid growth and toxin production by *C. botulinum.*

Meats, fish and seafood, and milk and milk products also have been responsible for outbreaks of botulism. Recent outbreaks of type E botulism in the United States from smoked fish have been mentioned. Such poisoning was favored by inadequate refrigeration. The single outbreak from canned tuna resulted from contamination of the fish through defective seams in the can. Type E botulism from fish, other seafoods, and meat of sea mammals is reported periodically in many of the northern countries of the world.

Sausage and ham often are involved; in fact the name botulism is derived from the Latin word for sausage, *botulus,* because the first recognized European outbreaks were caused by spoiled sausages. Outbreaks from some nineteen kinds of meat products and nine kinds of

TABLE 26–3.   Home-canned foods causing six or more outbreaks of botulism during the period 1899–1947 (U.S.)

Food	Number of Outbreaks	Food	Number of Outbreaks
String beans	94	Beet greens	9
Corn	46	Chili peppers	9
Beets	22	Beans	7
Asparagus	21	Tomatoes	7
Spinach and chard	12	Mushrooms	6
Peas	10	Sausages	9
Figs	10	Fish	10

seafood were reported in the United States in the period 1899 to 1947, and a few cases have been caused by milk and milk products.

Investigators have shown that spores of *C. botulinum* will survive long storage periods in raw and precooked frozen foods, and can grow and produce toxin if these foods are held for a long enough time at a high enough temperature after thawing. To date, no such outbreak of botulism has been reported.

THE DISEASE. Man is so susceptible to botulism that if appreciable amounts of toxin are present everyone who eats the food becomes ill; and consumption of very small pieces of food, a pod of a string bean or a few peas, can cause illness and death. The typical symptoms of botulism usually appear within 12 to 36 hr, although a longer or shorter time may be required. The earliest symptoms usually are an acute digestive disturbance followed by nausea and vomiting and possibly diarrhea, together with fatigue, dizziness, and a headache. Later there is constipation. Double vision may be evident early, and difficulty in swallowing and speaking may be noted. Patients may complain of dryness of the mouth and constriction of the throat, and the tongue may become swollen and coated. The temperature of the victim is normal or subnormal. Involuntary muscles become paralyzed, paralysis spreads to the respiratory system and heart, and death usually results from respiratory failure. Symptoms are similar for types A, B, and E poisoning, although nausea, vomiting and urinary retention usually are more severe with type E toxin. In fatal cases, death usually comes within 3 to 6 days after the poisonous food has been ingested, but the period may be shorter or longer.

The mortality averages over 65 percent in the United States, but is considerably lower in Europe, about 20 percent in Germany, for example.

The only known method for the successful treatment of botulism is the administration of antitoxin. Unfortunately, this injection usually is not successful if made after the symptoms of botulism have appeared, but it should always be used at the earliest possible moment, for it might prove helpful. Other treatments include artificial respiration, keeping the patient quiet, maintaining the fluid balance in the body, and, perhaps, elimination treatments.

CONDITIONS NECESSARY FOR AN OUTBREAK. The following conditions are necessary for an outbreak of botulism: (1) presence of spores of *C. botulinum* of type A, B, or E in the food being canned or being processed in some other way; (2) a food in which the spores can germinate and the clostridia can grow and produce toxin; (3) survival of the spores of the organism, e.g., because of inadequate heating in canning or inadequate processing otherwise; (4) environmental conditions after processing that will permit germination of the spores and growth and toxin production by the organism; (5) insufficient cooking of the food to inactivate the toxin; and (6) ingestion of the toxin-bearing food.

PREVENTION OF OUTBREAKS. The methods and precautions for the prevention of botulism that have been mentioned in the preceding discussion include (1) use of approved heat processes for canned foods; (2) rejection of all gassy (swollen) of otherwise spoiled canned foods; (3) refusal even to taste a doubtful food; (4) avoidance of foods that have been cooked, held, and not well reheated; and (5) boiling of a suspected food for at least 15 min. To this list might be added avoidance of raw or precooked foods that have been frozen, thawed, and held at room temperatures. To prevent botulism from smoked fish it has been recommended: (1) that good sanitation be maintained throughout production and handling; (2) that during smoking or thereafter the fish be heated to at least 180 F (82 C) for 30 min in the coldest part; (3) that the fish be frozen immediately after packaging and kept frozen; and (4) that all packages be marked "Perishable—Keep Frozen."

Staphylococcus food poisoning

This most commonly occurring true food poisoning is caused by the ingestion of the enterotoxin formed in food during growth of certain strains of *Staphylococcus aureus*. The toxin is termed an **enterotoxin** because it causes gastroenteritis or inflammation of the lining of the stomach and intestines.

THE ORGANISM. The organism is a typical staphylococcus, occurring in masses like clusters of grapes or in pairs and short chains (Figure 26–2). Growth on solid media usually is golden or yellow but may be unpigmented in some strains. All enterotoxin-producing *S. aureus* cultures are coagulase-positive (coagulating oxalated blood plasma) and are facultative in their oxygen requirements in a complex glucose medium, but grow better aerobically than anaerobically. However, not all coagulase-positive staphylococci are necessarily enterotoxigenic. Some of the toxigenic cocci are very salt-tolerant, growing in sodium chloride solutions that approach saturation, and also tolerate nitrites fairly well and therefore can grow in curing solutions and on curing and cured meats if other environmental conditions are favorable. They also are fairly tolerant of dissolved sugars. They are fermentative and proteolytic, but usually do not produce obnoxious odors in most foods or make

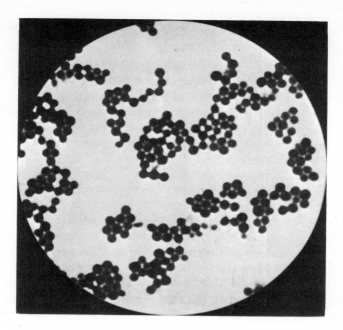

Figure 26–2. Photomicrograph of *Staphylococcus aureus*. (*From J. Nowak*).

them appear unattractive. Staphylococci may produce four serological types of enterotoxin, A, B, C, and D, that differ in toxicity; most food poisoning is from type A. The enterotoxin-producing strains of S. *aureus* yield several other toxins as well. The range of conditions permitting growth of the staphylococcus, and hence toxin production, varies with the food involved. In general, the better medium the food is for the coccus, the wider will be the range of temperature, pH, or a_w over which growth can take place. The minimal temperature in custard, condensed milk, and chicken à la king is about 44 to 46 F (6.67 to 7.78 C), but is about 50 F (10 C) in beef-heart infusion, and higher in ham salad. Of course growth at these minimal temperatures is very slow, and the time to attain numbers sufficient for production of detectable levels of toxin is longer than the storage period for most refrigerated foods. The maximal temperature for growth is about 112 to 116 F (44.4 to 46.6 C), depending upon the food. The minimal pH for growth is lower under aerobic conditions than under anaerobic; for example, in meat the minimal pH is 4.8 in aerobic conditions and about 5.5 in anaerobic. The minimal a_w is about 0.86 under aerobic conditions and about 0.90 under anaerobic conditions. A sublethal heat-treatment decreases tolerance to salt.

Other kinds of food bacteria, competing with the staphylococcus, may repress its growth enough to delay or prevent production of toxin, or the spoilage bacteria may make the food inedible before it is dangerous. The effectiveness of the repression varies with the kinds and numbers of competing organisms, the type of food, temperature, and time. Ordinarily staphylococci enter foods in low numbers and usually are outnumbered by competing bacteria in raw foods. This competition may not occur, however, in heated foods, and unrestrained growth of *Staphylococcus* contaminants may ensue.

A million staphylococci per milliliter, or gram of perishable foods will be inactivated by 150 F (66 C) maintained for at least 12 min, or by 140 F (60 C) for 78 to 83 min. Heat resistance varies some with the food and the strain of staphylococcus. *D* values at 140 F (60 C) in custard are about 7.7 to 7.8 min and in chicken à la king about 5.2 to 5.4 min. About 0.37 to 0.48 Mrad of gamma rays on moist foods will kill most of the staphylococci.

The sources from which the food-poisoning staphylococci enter foods are, for the most part, human or animal. The nasal passages of many persons are laden with these organisms, which are a common cause of sinus infections. Also, boils and infected wounds may be sources. The human skin apparently is a source of these bacteria only when they have come from nasal passages or local infections. Staphylococci are becoming increasingly important in causing mastitis of cows, and some of these cocci can form enterotoxin in milk or milk products. Ordinarily, air is a relatively unimportant source of the cocci, except when they are being introduced there from human sources.

THE ENTEROTOXIN. Appreciable levels of enterotoxin are produced only after considerable growth of the staphylococcus; usually a population of at least several millions per milliliter or gram must be attained. Therefore, the conditions that favor toxin production are those best for growth of the staphylococcus. It has been reported that toxin is produced at an appreciable rate at temperatures between 60 and 115 F (15.6 and 46.1 C); and production is best at 70 to 97 F (21.1 to 36.1 C). Under the best conditions enterotoxin may become evident within 4 to 6 hr. The lower the temperature during growth, the longer it will take to produce enough enterotoxin to cause poisoning. Enterotoxin has been demonstrated in a good culture medium in 3 days at 18 C (64.4 F), and in 12 hr at 37 C (98.6 F), but not in 3 days at 15 C (59 F), 7 days at 9 C (48.2 F), or 4 weeks at 4 to 6.7 C (39.2 to 44 F). It has been observed that production of enterotoxin by the staphylococci is more likely when competing microorganisms are absent, few, or inhibited for some reason. Therefore, a food that had been contaminated with the staphylococci after a heat process would be favorable for toxin production. There is evidence that toxin is produced by staphylococci growing in the intestinal tracts of patients when treatment with antibiotics has destroyed or inactivated other competitive bacteria there.

The type of food evidently has an influence on the amount of enterotoxin produced: little is produced in canned salmon, for example, and much in meat products and custard-filled bakery goods. The presence of starch and protein in considerable amounts is supposed to encourage toxin production by the staphylococci.

An important characteristic of the enterotoxin is its stability toward heat. It has been shown to withstand boiling for 20 to 60 min or even autoclaving, although it gradually loses its potency during such heating. The cooking usually given most foods will not destroy the toxin formed therein prior to the heat process. Such foods might cause poisoning, although no live staphylococci could be demonstrated.

INCIDENCE OF THE DISEASE. There are no reliable figures on the numbers of cases of staphylococcus poisoning in the United States or in any of the states for any given period. The poisoning usually is not reported or publicized unless the outbreak is fairly large, as at a picnic, large dinner, or convention. It is known, however, that a large percentage of all cases reported as "food poisoning" or food infection actually are staphylococcus poisoning, and that most of us encounter this illness a number of times during our lifetime.

FOODS INVOLVED. Of the many kinds of food that have been involved in causing staphylococcus food poisoning, custard- and cream-filled bakery goods, ham, tongue, and poultry have caused the most outbreaks. Other foods incriminated include other meats and meat products, fish and fish products, milk and milk products, cream sauces, salads, puddings, custards, pies, and salad dressings. The fillings in bakery goods usually are good culture media in which the staphylococci can grow during the time that these foods are held at room temperatures. Toxin production has even been reported in imitation-cream filling. Tongue and mildly cured, rapidly cured, tenderized, or precooked hams, although perishable, are often held without adequate refrigeration, as had been done without difficulty with the old-style country-cured hams. If contaminated leftover turkey, or other fowl, along with the gravy and dressing, is kept out of the refrigerator, it may cause poisoning. Foods that ordinarily are too acid for good growth of the staphylococci may have this acidity reduced by added ingredients, such as eggs or cream, and then become dangerous. Growth and toxin production by staphylococci may take place in the steam tables in cafeterias and restaurants and in food-vending machines that keep foods heated for extended periods.

THE DISEASE. Individuals differ in their susceptibility to staphylococcus poisoning, so that of a group of people eating food containing toxin some may become very ill and a lucky few may be affected little or not at all. Because most animals are not susceptible, it is difficult to test for the toxin without human volunteers, although cats and monkeys have been used as test animals with some success. Presently gel-diffusion tests are being perfected for specific types of enterotoxin.

The incubation period (time between consumption of the food and appearance of the first symptoms) for this kind of poisoning usually is a brief 2 or 3 hr (ranging from 1 to 6 hr), differing in this respect from the other common food poisonings and infections, which usually have longer incubation periods.

The most common human symptoms are salivation, then nausea, vomiting, retching, abdominal cramping of varying severity, and diarrhea. Blood and mucus may be found in stools and vomitus in severe cases. Headache, muscular cramping, sweating, chills, prostration, weak pulse, shock, and shallow respiration may occur. Usually a subnormal body temperature is found rather than fever. The duration is brief, usually only a day or two, and recovery ordinarily is uneventful and complete. The mortality is extremely low. For the most part no treatment is given, except in extreme cases, when saline solutions may be given parenterally to restore the salt balance and counteract dehydration.

Outbreaks of food poisoning often are attributed to staphylococci on the basis of the type of food involved, the short incubation period and, perhaps, the demonstration of the presence of staphylococci in the food. An actual diagnosis of the poisoning would depend, however, upon isolation of staphylococci and demonstration that these produce enterotoxin. The latter procedure presumably would involve the use of cats as test animals, or, for more certainty, human volunteers.

CONDITIONS NECESSARY FOR AN OUTBREAK. The following conditions are necessary for an outbreak of staphylococcus food poisoning: (1) the food must contain enterotoxin-producing staphylococci; (2) the food must be a good culture medium for growth and toxin production by the staphylococci; (3) the temperature must be favorable to growth of the cocci and enough time must be allowed for production of enterotoxin; and (4) the enterotoxin-bearing food must be ingested.

PREVENTION OF OUTBREAKS. The means of prevention of outbreaks of staphylococcus food poisoning include (1) prevention of contamination of the food with the staphylococci; (2) prevention of the growth of the staphylococci; and (3) killing staphylococci in foods. Contamination of foods may be reduced by general methods of sanitation, by use of ingredients free from the cocci, e.g., pasteurized rather than raw milk, and by keeping employees away from foods when these workers have staphylococcal infections in the form of colds, boils, carbuncles, etc. Growth of the cocci can be prevented by adequate refrigeration of foods, and, in some instances, by adjustment to a more acid pH. Also the addition of a bacteriostatic substance, such as serine or an antibiotic, has been suggested. Some foods may be pasteurized to kill the staphylococci before these foods are exposed to ordinary temperatures; the pasteurization of custard-filled puffs and éclairs for 30 min at 190.6 to 218.3 C (375 to 425 F) oven temperature has been so used. The United States Public Health Service in Bull. 280 recommends an oven temperature of at least 425 F (218.3 C) for at least 20 min, followed by cooling to 50 F (10 C) or less within an hour after rebaking; or heating all parts of the filling at 190 F (88 C) or over for at least 10 min before filling of the shells and similar cooling.

Clostridium perfringens food poisoning

Food poisoning by Clostridium perfringens is being detected and reported more often than formerly, although it probably is no more prevalent. Incidence apparently is high in some European countries.

THE ORGANISM. The bacteria causing this illness for the most part resemble C. perfringens (welchii), type A (Figure 26-1,C), except for low levels of lecithinase and hemolysin. Maximal temperature for growth is about 50 C (122 F), and optimal temperature about 43 to 47 C (109.4 to 116.6 F). Growth is restricted at 15 to 20 C (59 to 68 F). The organism will not grow below pH 5.0 or above pH 9.0. It is inhibited by 5 percent NaCl (a_w 0.97) (see Table 26-2), and some strains are held back by 2.5 percent sodium nitrate.

The spores of food-poisoning strains differ considerably in their heat resistance, many of them requiring 1 to 4 hr at 100 C (212 F) for their destruction, whereas others are killed within a few minutes. Meat cultures are toxic, but sterile filtrates usually are not. Large numbers of actively multiplying organisms are necessary for food poisoning.

FOODS INVOLVED. The spores have been found in part of the samples of most raw foods examined, as well as in soil, sewage, and animal feces. Most commonly involved in the poisoning are meats that have been cooked, allowed to cool slowly, and then held for some time before consumption. Fish paste and cold chicken also have been incriminated.

THE DISEASE. The symptoms, which appear usually in 10 to 12 hr after eating, consist of nausea and more or less severe abdominal pains and diarrhea, usually without fever or vomiting. Recovery is rapid.

PREVENTION OF OUTBREAKS. Obviously, outbreaks can be prevented by cooling flesh foods promptly and refrigerating them adequately until use.

Poisoning by other organisms

Many microorganisms other than *Clostridium botulinum*, *Staphylococcus aureus*, and *C. perfringens* have been suspected of causing food poisoning, but not many have actually been proved to be involved, and some of the organisms so accused may cause infections rather than true poisoning. *Escherichia coli* has been reported to cause poisoning, as have other bacteria of the coliform group. Species of *Proteus*, e.g., *P. vulgaris* and *P. mirabilis*, have been found in large numbers in foods that caused intestinal upsets, but it has not been definitely demonstrated that these actually produce an enterotoxin or cause food poisoning. *Bacillus cereus* and *B. subtilis*, growing in starchy foods, also have been reported to cause food poisoning by means of an endotoxin with symptoms similar to those from *Clostridium perfringens*. Japanese workers report a halophile, *Vibrio parahemolyticus*, responsible for food illness. Also, viruses may cause enteric disturbances, primarily as infective agents.

Aflatoxin (B_1, B_2, G_1, and G_2) produced by molds (*Aspergillus flavus*, *A. parasiticus*, and *Penicillium puberulum*) growing on peanuts, wheat, and various cereals, and in culture media, has been shown to cause illness or even death in animals and cancer in rats and trout. To date such illness has not been reported in man. During World War II there were many cases of toxic alimentary aleukia and many deaths in Russia resulting from the consumption of grain that had overwintered in the field. Chiefly species of *Fusarium*, *Cladosporium*, *Alternaria*, *Penicillium*, and *Mucor* were incriminated. Similar toxic molds have been found on rice.

FOOD INFECTIONS

Infections by certain species of *Salmonella* sometimes are called "food poisoning" because the symptoms in general resemble those of staphylococcus poisoning and the outbreaks commonly are explosive.

Usually the infecting bacterium has grown in the food to attain high numbers, increasing the likelihood of infection, and often resulting in outbreaks in families or larger groups. By contrast, other intestinal pathogens, such as organisms causing the dysenteries and typhoid and paratyphoid fevers, usually have a longer incubation period before symptoms and, except under epidemic conditions, occur in only scattered cases.

Salmonella infections

The *Salmonella* infections that are called "food poisoning" may be caused by any of a large number of species or serological types of species of that genus, most of which bear the name of the location where the first identified culture was obtained, e.g., S. *newport*, S. *panama*, S. *sandiego*, S. *montevideo*, etc., although a few bear definite species names such as S. *typhimurium* and S. *enteritidis* (Figure 26–1,D) Included also would be the closely related Arizona group of bacteria. The differentiation of species and types is chiefly by agglutination tests for somatic (O) and flagellar (H) antigens. Surveys of human outbreaks have indicated S. *typhimurium* to be the species most often encountered.

THE ORGANISM. The salmonellae are Gram-negative, non-spore-forming rods that ferment glucose, usually with gas, but do not ferment lactose or sucrose. As has been stated, they are typed on the basis of their antigen content. As with other bacteria, they will grow over a wider range of temperature, pH, and a_w in a good culture medium than in a poor one. For example, minimal temperatures for growth range from 44 or 46 F in chicken à la king to over 50 F in custard or ham salad. Their maximal temperature is about 114 F (45.6 C). They grow well at room temperatures, but their optimum is about 37 C (98.6 F). They grow best in low-acid foods; salad at pH 5.5 to 5.7 has been found unfavorable for growth. The lowest a_w for growth varies with the food, but is about 0.93 to 0.95. The species and strains of *Salmonella* differ, too, in heat resistance and in the effect of environmental factors on growth. Recommendations for thermal destruction of salmonellae in perishable foods are similar to those for staphylococci, viz., heating to 150 F (66 C) and holding all parts at that temperature for at least 12 min (or 78 to 83 min at 140 F or 60 C). F_{140} values (minutes at 140 F necessary to reduce an inoculum to an undetectable level) found for two species were 78 and 19 min, respectively, in custard, and 81.5 and 3.1 min in chicken à la king. These results illustrate how the required heat-treatment differs with the species of *Salmonella* and the food heated.

The likelihood of infection by consumption of a food containing

salmonellae depends upon the resistance of the consumer, the infective-
ness of the particular strain of *Salmonella,* and the number of organisms
ingested. Less infective species like S. *pullorum* must be ingested in
hundreds of millions or in billions to bring about infection, but considerably fewer (about a million) organisms of more infective species,
e.g., S. *enteritidis,* usually would be sufficient. Salmonellae apparently
can attain considerable numbers in foods without causing detectable
alterations in appearance, odor, or even taste. Of course, the more of
any of these pathogens the food contains, the greater will be the likelihood of infection developing in the person who eats the food, and the
shorter will be the incubation time.

Sources of *Salmonella.* Human beings and animals are directly
or indirectly the source of the contamination of foods with salmonellae.
The organisms may come from actual cases of the disease or from carriers. Most frequently S. *typhimurium,* S. *montevideo,* S. *oranienburg,* and
S. *newport* cause human gastroenteritis, but any of many other types may
be responsible. The organisms also may come from cats, dogs, swine, and
cattle, but more important sources for foods are poultry and their eggs
and rodents (see Figure 26–3). Chickens, turkeys, ducks, and
geese may be infected with any of a large number of types of *Salmonella,*
which are then found in the fecal matter, in eggs from the hens, and
in the flesh of the dressed fowl. Considerable attention is now being
given to shell eggs and to liquid, frozen, and dried eggs as sources
of *Salmonella.* Infected rodents, rats and mice, may contaminate unprotected foods with their feces and thus spread *Salmonella* bacteria. Flies
may play an important role in the spread of *Salmonella,* especially from
contaminated fecal matter to foods. Roaches apparently also can spread
the disease.

Changes in processing, packaging, and compounding of foods and
feeds in recent years have resulted in an apparent increase in salmonellosis from these products. Salmonellae have been introduced by the incorporation of cracked and dried eggs in baked goods, candy, ice cream,
and convenience foods such as cake and cookie mixes. The compounding
of new food products may make possible the growth of salmonellae
or other food-poisoning organisms, introduced by means of an ingredient
in which they had been unable to grow; or these organisms may be
in a product when sold and become able to grow in this food as it
is modified for use. Large-scale handling of foods, as by commissaries
or institutions, tends to increase the spread of trouble, and food vending
machines add to the risk, as do precooked foods.

Feeds, especially those from meat or fish by-products, may carry
salmonellae to poultry or meat animals. Even pet feeds have been known

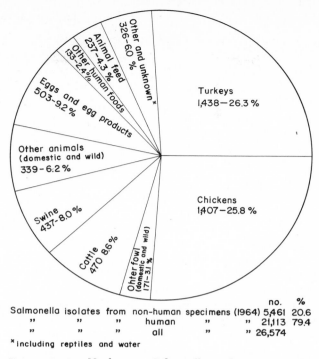

Turkeys
1,438—26.3 %

Chickens
1,407—25.8 %

Other and unknown*
326-6.0 %

Animal feed
237-4.3 %

Other human foods
133-2.4%

Eggs and egg products
503-9.2 %

Other animals
(domestic and wild)
339-6.2 %

Swine
437-8.0 %

Cattle
470 8.6 %

Ohter fowl
(domestic and wild)
171-3.1 %

				no.	%
Salmonella isolates from non-human specimens (1964)				5,461	20.6
,, ,, ,, human ,,				,, 21,113	79.4
,, ,, ,, all ,,				,, 26,574	

*including reptiles and water

Figure 26–3. Nonhuman *Salmonella* isolations in the United States in 1964. (*From Salmonella Surveillance, Annual Summary, 1964 (Oct., 1965), Communicable Disease Center, Public Health Service, U.S. Department of Health, Education, and Welfare.*)

to transmit salmonellae to domestic animals, from which children have been infected.

INCIDENCE OF THE DISEASE. Numbers of reported human cases of salmonellosis (other than typhoid fever) in the United States have been increasing in recent years, with 21,113 cases reported for 1964, as compared with 18,000 in 1963. Canada also reports increases in numbers of cases. This increase has been attributed partially to changing food habits and food-processing methods, although improved methods of detection and reporting may be involved.

FOODS INVOLVED. A large variety of foods has been found to be involved in causing outbreaks of *Salmonella* infections. Most commonly incriminated are various kinds of meats, poultry, and products from them, especially if these foods are held unrefrigerated for long periods. Fresh meats may carry *Salmonella* bacteria that caused disease in the slaughtered animals or may be contaminated by handlers. Meat products, such as meat pies, hash, sausages, cured meats (ham, bacon, and tongue), sandwiches, chili, etc., often are allowed to stand at room temperatures, permitting the growth of salmonellae. Poultry and its dressing and gravy should not give trouble if properly handled and cooked, but often are mishandled, as are fish and other seafood and products from them. Milk and milk products, including fresh milk, fermented milks, ice cream, and cheese, have caused infections. Since eggs may carry the salmonellae, foods made with eggs and not sufficiently cooked or pasteurized may carry live organisms, e.g., pastries filled with cream or custard, cream cakes, baked Alaska, and eggnog.

THE DISEASE. As with other infectious diseases, individuals differ in their susceptibility to *Salmonella* infections, but, in general, morbidity is high in any outbreak. As has been stated, the susceptibility of man varies with the species and strain of the organism and the total numbers of bacteria ingested.

A longer incubation period usually distinguishes salmonellosis from staphylococcus poisoning: usually 12 to 24 hr for the former and about 3 hr for the latter. Shorter (as little as 3 hr), or longer (up to 72 hr) incubation periods may occur in some cases of *Salmonella* infections.

The principal symptoms of a *Salmonella* gastrointestinal infection are nausea, vomiting, abdominal pain, and diarrhea that usually appear suddenly. This may be preceded by a headache and chills. Other evidences of the disease are watery, greenish foul-smelling stools, prostration, muscular weakness, faintness, usually a moderate fever, restlessness, twitching, and drowsiness. The mortality is low, being less than 1 percent. The severity and duration vary not only with the amount of food eaten and hence the numbers of *Salmonella* bacteria ingested but also with the individual. Intensity may vary from slight discomfort and diarrhea to death in 2 to 6 days. Usually the symptoms persist for 2 to 3 days, followed by uncomplicated recovery, but they may linger for weeks or months. About 0.2 to 5 percent of the patients may become carriers of the *Salmonella* organism.

The laboratory diagnosis of the disease is difficult unless *Salmonella* can be isolated from the suspected food and from the stools of individuals. Often, however, the incriminated foods are no longer available, and the organisms disappear rapidly from the intestinal tract. The development of specific antibodies in the blood usually is variable, and therefore tests for them are not dependable.

CONDITIONS NECESSARY FOR AN OUTBREAK. The following conditions are necessary for an outbreak of a food-borne *Salmonella* gastrointestinal infection: (1) the food must contain or become contaminated with the

Salmonella bacteria; (2) these bacteria must be there in considerable numbers, either because of contamination or more often because of growth; these high numbers mean that the food must be a good culture medium, the temperature must be favorable, and enough time must be allowed for appreciable growth; and (3) the organisms must be ingested.

PREVENTION OF OUTBREAKS. Three main principles are involved in the prevention of outbreaks of food-borne *Salmonella* infections: (1) avoidance of contamination of the food with salmonellae from sources such as diseased human beings and animals and carriers, and ingredients carrying the organisms, e.g., contaminated eggs; (2) destruction of the organisms in foods by heat (or other means) when possible, as by cooking or pasteurization, paying special attention to held-over foods; and (3) prevention of the growth of *Salmonella* in foods by adequate refrigeration or by other means. In the prevention of contamination, care and cleanliness in food handling and preparation are important. The food handlers should be healthy (and not be carriers) and clean. Rats and other vermin and insects should be kept away from the food. Ingredients used in foods should be free of salmonellae, if possible. Of course foods should not be allowed to stand at room temperature for any length of time, but if this happens, thorough cooking will destroy the *Salmonella* organisms (but not staphylococcus enterotoxin). Warmed-over leftovers, held without refrigeration, often support the growth of *Salmonella*, as may canned foods that have been contaminated and held after the cans were opened. Inspection of animals and meats at packing houses may remove some *Salmonella* infected meats, but is not in itself a successful method for the prevention of human salmonellosis.

Streptococcus infections

On primarily circumstantial evidence, *Streptococcus faecalis* and closely related species have been blamed for food illnesses, and since no enterotoxin could be demonstrated the organisms were presumed but not proved to cause an infection. Foods reported to be involved in a limited number of outbreaks included barbecued beef, beef croquettes, Vienna sausage, ham bologna, turkey dressing, turkey à la king, Albanian cheese and other cheeses, charlotte russe, and evaporated milk. Enterococci were isolated from most of these foods. The incubation period has been given as from 2 to 18 hr, and symptoms as nausea, vomiting, and diarrhea. Formerly it was thought that only certain strains caused poisoning, but presently most authorities maintain that all of the accused organisms are innocuous. Volunteers have been fed massive doses of the organisms, including some supposedly incriminated in food-poisoning outbreaks, and also their products, without any resulting enteritis.

OTHER FOOD-BORNE INFECTIONS

The food infections just discussed have been those involving bacteria able to grow in foods and hence to increase the dosage of the pathogen delivered to the person eating the food. Pathogenic organisms unable to grow in the food, as it ordinarily is handled, may be carried by it, the food in this instance serving merely as an inactive carrier of the disease organisms, much as a doorknob, handkerchief, bus strap, drinking cup, or other fomes would serve that purpose. Most of the diseases so transmitted are intestinal or respiratory. Some of the intestinal diseases so transported are typhoid and paratyphoid fevers, bacillary and amebic dysenteries, and cholera. Disease organisms from the throat and respiratory tract are those causing tuberculosis, scarlet fever, diphtheria, and other diseases. Also reportedly spread by foods are brucellosis, tularemia, Q fever, scarlet fever, septic sore throat, and infectious hepatitis.

Foods may be contaminated with disease organisms from food handlers, food utensils (eating, drinking, and kitchen), air, soil, water (e.g., oysters), the animal from which meat or milk came, and vermin, such as flies, roaches, and rodents. Especially likely to be important are the people who handle foods after pasteurization, cooking, or other processing, e.g., cooks and helpers in the kitchen, waiters in the eating place, preparers and salesmen of unwrapped foods such as bulk candies, ice cream sandwiches and cones, baked goods, hot dogs and hamburgers, and vendors of frozen desserts from counter freezers or other bulk sources.

Foods consumed raw, of course, are possible sources of pathogens. Thus, fresh fruits or vegetables may carry pathogens from a diseased handler to a healthy consumer, although few instances of such a transfer have been proven. Where night soil is used as fertilizer, there is great risk of the presence of intestinal pathogens on the surface of fresh salad greens. Depending upon the heat-treatment administered, the cooking of foods may or may not destroy all pathogens present. Usually all on the surface will be killed, but not always those in the interior. Instances have been reported where even comparatively non-heat-resistant pathogens have survived cooking and caused an outbreak of disease.

The principle to be applied in the prevention of food-borne infections is to prevent the transfer of the pathogen from its source to the food, preferably by elimination of that source, but this objective is not always readily attained. Contamination from vermin can be prevented by their eradication; use of night soil for fertilizing soil for growing plant foods to be consumed raw can be prohibited; fruits or vegetables to be eaten raw can be washed thoroughly with water or with chlorine solution; shellfish from polluted waters can be rejected; and milk from diseased cows can be refused. However, meat animals or food handlers that are diseased or are carriers are not always easy to detect. Methods for the detection of disease or disease organisms are, for the most part, laborious, difficult, and not always reliable; and usually the tests are impractical to apply to meat animals at the packing plant and cannot be applied to food handlers as often as would be desirable. Of course food handlers should not be permitted to work while they are ill or recovering, but they may be only mildly ill or be carriers and therefore be allowed to work while giving off pathogens.

Many municipalities have ordinances which state that no person who is affected with any disease in a communicable form or is a carrier of such disease shall work in any eating or drinking establishment or be hired there, but the difficulty lies in enforcement of the ordinances. In most plants where foods are handled and processed, however, any check on the health of the workers is at present primarily the responsibility of the employer.

TRICHINOSIS

Trichinosis, although caused by a nematode worm, *Trichinella spiralis,* usually is discussed along with bacterial food poisoning and infection because all have similar symptoms and all are food-borne. Most human trichinosis results from the consumption of raw or incompletely cooked pork containing the encysted larvae. The larvae are released into the intestinal tract during digestion and invade the mucous membranes of the first parts of the small intestine, where they develop into adults. The fertilized females give birth to numerous larvae, which travel through the blood vessels and lymphatics to skeletal muscle tissue, where they encyst (Figure 26-4). The worm goes through a similar cycle in hogs and rats, and other hosts, such as mice, rabbits, cats, dogs, and bears.

SYMPTOMS. The incubation period, between ingestion of the pork and the first symptoms, varies widely, being reported as short as a day or two and as long as several weeks. The first symptoms that may be confused with food poisoning appear when the larvae, freshly released from their cysts in the ingested and digested pork, invade the mucosa.

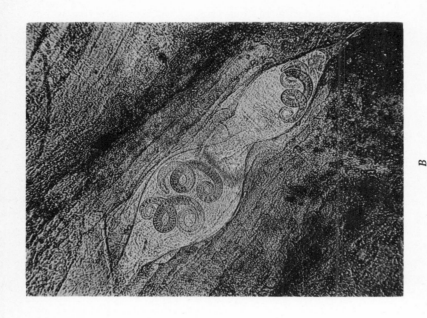

B

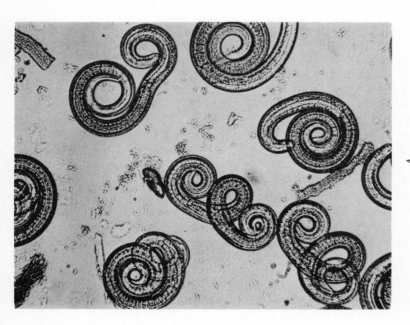

A

Figure 26–4. *Trichinella spiralis*: (*A*) pepsin-HCl digest, showing free trichinae; (*B*) encysted worms in muscle. (*Courtesy of L. B. Jensen, Swift and Co., Chicago.*)

Symptoms may include nausea, vomiting, diarrhea, profuse sweating, colic, and loss of appetite and may continue for days. Later symptoms, resulting from the migration of the newborn larvae to the muscles and their encystment there, are mostly related to muscular soreness and swelling and would not be confused with food poisoning. Death may follow in severe cases.

DIAGNOSIS. Diagnosis often is based primarily on symptoms, for other methods of diagnosis are difficult. The suspected food may be examined for encysted larvae, but they usually are hard to find, as are adult worms in stools of patients. Muscle strips from the patient may be examined for cysts, or intradermal and precipitin tests may be positive several weeks after the first symptoms have appeared.

PREVENTION. The chief method for the prevention of trichinosis is the treatment of pork (or other meat) to ensure the destruction of any trichinae that may be present. This can be accomplished by (1) the thorough cooking of all pork so that every part reaches at least 137 F (58.3 C), (2) quick freezing or storage at 5 F (−15 C) or lower for not less than 20 days, (3) treatment with 20,000 rep of ionizing rays (see Chapter 10), or (4) processing sausage or similar pork products according to a recommended schedule of salting, drying, smoking, and refrigeration as formulated originally by the former Bureau of Animal Industry of the United States Department of Agriculture. Directions for drying specify 1 part of salt per 30 parts of meat and holding in the drying room for over 20 days at 45 F (7.2 C) or above. Smoking should be for 40 hr or more at 80 F (26.7 C) or above, followed by 10 days or more in the drying room at 45 F or above.

Inspection of animals and meats for trichinae at the packing plant is laborious and not always successful, although formerly practiced. Some attempt has been made to reduce the incidence of trichinosis in swine by control of rats, which may carry the worm, and by the cooking of garbage fed to the hogs.

SUMMARY ON FOOD POISONING AND INFECTIONS

A comparison of botulism, staphylococcus poisoning, *Clostridium perfringens* poisoning, *Salmonella* infections, and infestation with *Trichinella spiralis* is made in Table 26–4. Figures for incubation times, duration of illness, and mortality in individual outbreaks may vary from those given. Not all of the symptoms listed may be evident always, and, of course, the list of foods involved is not complete.

TABLE 26–4. Comparison of food poisonings and infections

Disease	Cause	Incubation time, hr	Symptoms	Duration	Mortality	Foods commonly involved
Botulism	Toxin— *C. botulinum*	12–36+	Early nausea, vomiting, (diarrhea). Fatigue, dizziness, headache. Dry skin, mouth, throat. Constipation, no fever, paralysis of muscles, double vision, respiratory failure.	1–10 days	65%+	Low- or medium-acid canned foods; meats, sausage, fish and other seafood. Aquatic mammals.
Staphylococcus poisoning	Toxin— *Staphylococcus aureus*	3 (1–6)	Nausea, vomiting, abdominal cramping, diarrhea. Headache, sweating, prostration. Low temperature.	1–2 days	Very low	Custard- or cream-filled baked goods, ham, tongue, fowls, their dressing and gravy, head cheese, meat sandwiches, salads, cream sauces, cakes, dairy products, etc.
Clostridium perfringens poisoning	Toxin— *C. perfringens*	10–12	Nausea, abdominal pains, diarrhea.	1–2 days		Poorly cooled cooked meats, poultry, fish.
Salmonellosis	Infection— *Salmonella* sp.	12–24 (7–30)	Nausea, vomiting, abdominal pain, diarrhea—sudden onset. Usually *fever*. Headache, chills, prostration.	2–3 days	<1%	Meat products, warmed-up leftovers. Salads, meat pies, hash, sausage.
Trichinosis	Infestation— *Trichinella spiralis*	48	Nausea, vomiting, diarrhea, sweating, colic, loss of appetite. Later, muscular pains.	Weeks to months	1–30%	Raw pork or pork products.

The symptoms of botulism differ from those for the other four illnesses. Botulism and staphylococcus and *C. perfringens* poisoning are caused by toxins produced in the food. *Salmonella* bacteria cause gastrointestinal infections. Trichinosis results from infestation with a nematode worm. Most useful in distinguishing these food illnesses are the times of incubation: usually only 2 to 3 hr for staphylococcus poisoning, about 10 to 12 hr for *C. perfringens* poisoning, and over 12 hr for botulism and *Salmonella* infections. Duration usually is brief and recovery uneventful for all but botulism and trichinosis. The food involved also may give a hint of the type of illness: botulism would be suspected if moderately acid foods had been canned; trichinosis usually would come from inadequately cooked or cured pork; but many of the same foods could cause the other three illnesses. It should be noted that growth of the causative organism in the food usually is involved in all the illnesses but trichinosis. Therefore, bacterial food poisoning or infection can be prevented by keeping the causative organisms out of the food or preventing their growth if they enter.

REFERENCES

ANGELOTTI, R., ELIZABETH WILSON, M. J. FOTER, and K. H. LEWIS. 1959. Time-temperature effects on salmonellae and staphylococci in foods: I, Behavior in broth cultures and refrigerated foods. U.S. Department of Health, Education, and Welfare. Robert A. Taft Sanit. Eng. Cent. Tech. Rep. F59-2.

ANGELOTTI, R., M. J. FOTER, and K. H. LEWIS. 1960. Time-temperature effects on salmonellae and staphylococci in foods: II, Behavior at warm holding temperatures; thermal-death-time studies. U.S. Department of Health, Education, and Welfare. Robert A. Taft Sanit. Eng. Cent. Tech. Rep. F60-5.

ANONYMOUS. 1965. Government agencies step up surveillance of *Salmonella* in foods. Food Processing 26(3):21–23.

AYERST, G., and P. K. C. AUSTWICK. 1964. Toxin production by moulds. Chem. Ind. 1964:1222 (July 4).

AYRES, J. C., A. A. KRAFT, H. E. SNYDER, and H. W. WALKER (*Eds.*) 1962. Chemical and biological hazards in food. The Iowa State University Press, Ames, Iowa.

BOWMER, E. J. 1965. Salmonellae in food—a review. J. Milk Food Technol. 28:74–86.

COCKBURN, W. C., JOAN TAYLOR, E. S. ANDERSON, and BETTY C. HOBBS. 1962. Food poisoning. Royal Society of Health, London.

DACK, G. M. 1956. Food poisoning. 3rd ed. The University of Chicago Press, Chicago.

DEIBEL, R. H., and J. H. SILLIKER. 1963. Food-poisoning potential of the enterococci. J. Bacteriol. 85:827–832.

DEWBERRY, E. B. 1959. Food poisoning. 4th ed. Leonard Hill, Ltd., London.

EDWARDS, P. R., MARY A. FIFE, and CAROLYN H. RAMSEY. 1959. Studies on the Arizona group of Enterobacteriaceae. Bacteriol. Revs. 23:155–174.

FOSTER, E. M. 1965. Food-borne illnesses—minor problem or hidden epidemic? Food Processing 26(2):56–58, 108.

GEORGALA, D. L., and A. HURST. 1963. The survival of food poisoning bacteria in frozen foods. J. Appl. Bacteriol. 26:346–358.

GOULD, S. E. 1945. Trichinosis. Charles C Thomas, Publisher, Springfield, Illinois.

HALL, H. E., and R. ANGELOTTI. 1965. *Clostridium perfringens* in meat and meat products. Appl. Microbiol. 13:352–357.

HAUGE, S. 1955. Food poisoning caused by aerobic spore-forming bacilli. J. Appl. Bacteriol. 18:591–595.

HOBBS, BETTY C. 1965. *Clostridium welchii* as a food poisoning organism. J. Appl. Bacteriol. 28:74–82.

HODGES, F. A., J. R. ZUST, H. R. SMITH, A. A. NELSON, B. H. ARMBRECHT, and A. D. CAMPBELL. 1964. Mycotoxins: aflatoxin isolate from *Penicillium puberulum*. Science 146:1439.

HORWOOD, M. P., and V. A. MINCH. 1951. The numbers and types of bacteria found on the hands of food handlers. Food Res. 16:133–136.

LEWIS, K. H., and K. CASSEL, JR. (*Eds.*) 1964. Botulism. U.S. Department of Health, Education, and Welfare. Public Health Serv. Pub. 999-FP-1.

National Research Council. 1964. An evaluation of public health hazards from microbiological contamination of foods. Rep. Food Protection Comm., Food and Nutrition Bd. Pub. 1195.

NIVEN, C. F., JR., and J. B. EVANS. 1955. Popular misconceptions concerning staphylococcus food poisoning. 7th Res. Conf. Amer. Meat Inst., Proc., p. 73–77.

OSHEROFF, B. J., G. G. SLOCUM, and W. M. DECKER. 1964. Status of botulism in the United States. Public Health Rep. 79:871–878.

ROBERTS, T. A., and M. INGRAM. 1965. The resistance of spores of *Clostridium botulinum* type E to heat and radiation. J. Appl. Bacteriol. 28:125–138.

SADLER, W. W., and R. E. CORSTVET. 1965. Second survey of market poultry for *Salmonella* infection. Appl. Microbiol. 13:348–351.

SHEWAN, J. M., and J. LISTON. 1955. A review of food poisoning caused by fish and fishery products. J. Appl. Bacteriol. 18:522–534.

SLANETZ, L. W., C. O. CHICHESTER, A. R. GAUFIN, and Z. J. ORDAL (*Eds.*) 1963. Microbiological quality of foods. Academic Press Inc., New York.

United States Department of Health, Education, and Welfare, Division of Environmental Engineering. 1963. Conference on viruses and rickettsia in foods.

United States Department of Health, Education, and Welfare, Public Health Service. 1965. Proceedings, national conference on salmonellosis. Public Health Serv. Pub. 1262.

United States Department of Health, Education, and Welfare, Communicable Disease Center. 1965. Salmonella surveillance; annual summary, 1964.

WILSON, ELIZABETH, R. S. PAFFENBARGER, JR., M. J. FOTER, and K. H. LEWIS. 1961. Prevalence of salmonellae in meat and poultry products. J. Infect. Dis. 109:166–171.

WOGAN, G. N. (*Ed.*) 1965. Mycotoxins in foodstuffs. The M.I.T. Press, Cambridge, Mass.

CHAPTER TWENTY-SEVEN

INVESTIGATION OF FOOD-BORNE DISEASE OUTBREAKS

A manual entitled "Procedure for the Investigation of Foodborne Disease Outbreaks" has been written by the Committee on Communicable Diseases Affecting Man of the International Association of Milk and Food Sanitarians and has been published and recommended by that Association (1957). Much of the discussion to follow is based on the contents of this manual, to which the reader is referred for a more complete discussion of the subject, for typical report forms to be employed, and for details that must be omitted from a textbook discussion.

FOOD-BORNE DISEASES

Food-borne diseases include those resulting from consumption of any solid food or of milk, water, or other beverage. The more important diseases and their causes have been mentioned in the preceding chapter: staphylococcus poisoning, caused by *Staphylococcus aureus;* botulism, by *Clostridium botulinum;* poisoning, by *C. perfringens;* salmonellosis, including typhoid and paratyphoid fevers, by *Salmonella* species; bacillary dysenteries, by *Shigella* species; amebic dysentery, by *Endamoeba histolytica;* and infection with alpha-type streptococci (*Streptococcus faecalis*). These microorganisms or evidences of their presence and growth are sought by the investigators of outbreaks of food-borne diseases. Less commonly, poisoning by *Bacillus cereus, Proteus* spp., and other organisms may be encountered. Other food-borne infections are brucellosis, diphtheria, scarlet fever, septic sore throat, tuberculosis, infectious heptatitis, and tularemia. Various nonbacterial parasitic infections, and poisoning by chemicals, plants, and animals, and by ionizing radiations may be encountered.

OBJECTIVES OF INVESTIGATION

From the public health viewpoint the main purposes of the investigation of an outbreak of food-borne disease are to determine how the foodstuff became contaminated, and, if growth of a toxigenic or infectious organism was involved, to find how such growth could take place, so that measures may be taken to prevent a repetition of the same set of conditions. This requires the location and identification of the causative agent, the establishment of the means of transmission, demonstration of the opportunity for growth of the pathogen, and, in instances of infections, proof that the pathogen has infected the victims. Prompt investigation also may aid in limiting the spread of the outbreak and sometimes may be of assistance to physicians treating the victims. Publicity given an outbreak and the explanation of its cause may be helpful in educating and warning the public and therefore in avoiding future outbreaks.

PERSONNEL INVOLVED IN INVESTIGATION

The organization of the "team" to investigate an outbreak of foodborne disease will vary with the public health department concerned, but this team will consist ordinarily of a person in charge, a field group, and a laboratory group. The field group interviews persons, both ill and healthy, who consumed the suspected foods, physicians and nurses who are treating the victims, and personnel at the place of exposure to the disease; collects samples of suspected foods and transmits them to the laboratory; collects specimens from patients or food handlers when such sampling is indicated; inspects the premises where the foods were stored, prepared, and served; ascertains where suspected foods were purchased and the conditions there, and fills out appropriate reports on these activities, to be made available to the person in charge and to the laboratory staff. The laboratory group makes such microbiological and chemical tests as are indicated by the reports of the field group and the nature of the suspected food and records its findings on appropriate report blanks. The person in charge or a qualified epidemiologist then can interpret the data from all sources to determine the cause and the source of the disease outbreak. The manual cited at the beginning of this chapter states that, at times, cooperation may be needed between the epidemiologist, the health officer, physicians, nurses, veterinarians, sanitarians, sanitary engineers, laboratory technicians, and statisticians in an investigation.

MATERIALS AND EQUIPMENT REQUIRED

The equipment and supplies needed to equip a field kit to take food samples and transmit them to the laboratory, as listed by Geidt (see reference at end of chapter), include sterile containers and sampling devices for sampling, a thermometer, an alcohol lamp, sterile swabs in diluent, sterile wrapping paper, tape for sealing samples, sterile paper towels, an insulated chest for conveying samples, forms for recording data, etc. The laboratory should be prepared, of course, to conduct the necessary microbiological and chemical tests and record results. Prepared materials would include sterile glassware, water blanks, appropriate culture media, test solutions, serological materials and equipment, hypodermic syringes and needles, etc., as well as experimental animals. If possible, the laboratory staff should be warned far enough ahead to be ready for the samples and specimens on their arrival.

THE FIELD INVESTIGATION

Prompt reporting of an outbreak to the health department is most important if the investigation is to be successful, yet the information often is delayed. Reports usually come from physicians, hospitals, news agencies, or even rumors. Promptness in initiating the investigation also is important because samples or specimens may be available for only a short time, information obtained promptly is usually more reliable than delayed reports, and additional cases in the outbreak may be prevented.

Gathering of information

A complete inspection is made immediately of the place or places where the suspected meal, meals, or beverages were prepared and consumed, and the results are recorded on appropriate forms. Information sought includes: the menu for the meal or meals, the source and method of preparation of each item of food on the menu, methods of storage of perishable foods, sources of purchase of foods served, and health of employees serving or preparing foods and their health history. Observations are made on infections on exposed surfaces of the bodies of employees and on the sanitary condition of the establishment.

Theoretically, all persons present at the time the questionable meal was eaten, including those who prepared, served, or ate the meal, and physicians treating victims should be interviewed. Actually, it usually is not practicable to interview all victims of a large outbreak. Geidt has suggested that in outbreaks where no more than twenty persons are involved, an attempt should be made to question all of them; if about fifty persons were involved, about half of them should be questioned, including proportionate numbers of those who became ill and those who did not; and about 25 percent of a total of one hundred or more persons should be interviewed. The information obtained may be recorded on a form such as that entitled "Case History Questionnaire" in the previously mentioned manual, where there is a place to record age and sex, whether the individual partook of the suspected meal, if so the exact time, whether he became ill and if so the incubation time and symptoms, and which of the list of the foods and beverages served were consumed. The case histories are then summarized on a second form, and by means of these data the suspected food is located.

Collection of food samples

Samples of all leftover foods and beverages served at the suspected meal or meals should be taken aseptically, and samples of perishable foods should be refrigerated immediately and kept cold in transit to the laboratory. Samples of food should be taken aseptically by means of sterilized sampling devices into sterile containers. Entire packages of foods in small, unopened containers may be taken when available. It is essential to label each sample to give information as to the type of food, the place and time of sampling, the reason for its collection, the organism or chemical suspected, and any other pertinent information. Each sample must be sealed, both inner and outer container, with the date and time of sealing and the name of the person who collected and sealed the sample written on the tape. All the information available at the time of sampling (see preceding section) should accompany the samples, and the reasons for suspecting one or more of the foods should be given.

Unfortunately, the food samples sought often are no longer available, such as food scraps from plates upon which the food was served or samples from serving containers. Then the collector of samples must settle for what he can get—rinsings, garbage, or food handled in the same manner as the suspected food. If a canned food is involved, part of the used can of food is first choice, but only a sample from the same lot may be available. Brand and lot number should be obtained from the can of the commercially canned food, and the method of canning and heating of the home-canned food should be ascertained. Food samples should be taken, labeled, and sealed in the presence of witnesses if legal action is likely to be involved.

Collection of specimens from human sources

Specimens may be obtained from patients with food illnesses or from food handlers, sometimes for the purpose of finding the causative organism of the outbreak, but more often to ascertain the ultimate source of the pathogen that entered the food. The type of specimen to be taken will depend, of course, on the illness concerned. Cultures from the nose or throat or from skin lesions of food handlers are made to test for staphylococci able to produce enterotoxin. Fecal samples are used to test for organisms, e.g., *Salmonella* or *Shigella* species, capable of causing enteric infections; the tests might be to find carriers among food handlers or to identify the cause of illness in patients. Blood specimens from patients may be used for serological tests for the identification and typing of certain pathogens, e.g., *Salmonella* spp. Vomitus may be tested when chemical poisoning is suspected.

LABORATORY TESTING

The procedure to be followed in testing the samples of food or specimens from human sources upon receipt in the laboratory will depend upon the type of food and the information available about the outbreak of food illness. The more complete that information is, the better the laboratory men can select the type of examination to be used. Especially helpful will be reliable information on the symptoms of those made ill and the incubation period (see Chapter 26).

The first act in most laboratories is to make a microscopic examination of a preparation of the food stained by the Gram method. The smear is made from liquid or from the sediment from homogenized, centrifuged food. The microscopic examination may give a clue to the causative organism and may indicate the numbers in the original food, if the sample has been properly refrigerated. The methods of testing for the various causes of important food-borne illnesses will be discussed only briefly. Culture media commonly employed are shown in Table 27–1.

Tests for botulism

An example of one method of testing is as follows: If enough of the suspected food is available, white mice or guinea pigs are inoculated with it, two being left unprotected, two protected with type A botulinum antitoxin, and two with type B antitoxin. If the animals receiving both types of antitoxin die, protection is attempted with type E antitoxin. Type A toxin will kill animals not protected with type A antitoxin; type B toxin will kill animals not protected with type B antitoxin, etc. If the food is heavily contaminated and not filtered before injection, the numerous contaminating bacteria may cause death. If insufficient food is available for injection of animals or if results of injections are unreliable, it may be necessary to attempt to isolate *Clostridium botulinum* from the food and type it on the basis of the type of toxin produced.

Tests for staphylococcus poisoning

Staphylococcal toxin is tested for in cultures obtained from foods or directly in the foods by means of single- or double-gel diffusion procedures, in which antiserum specific for a particular type of toxin (A, B, C, D) reacts with that toxin. Most laboratories first try to isolate staphylococci from the food by means of blood agar plates, mannitol

TABLE 27–1. Examination of foods for bacteria causing food
infections or poisonings

Organism	Enrichment media	Media for direct inoculation or plating
Salmonella (and *Arizona*)	Lactose broth, selenite-F broth, selenite-cystine broth	Brilliant green sulfadiazine agar, salmonella-shigella agar
Shigella	Lactose broth, selenite-cystine broth	Salmonella-shigella agar, brilliant green sulfadiazine agar
Staphylococcus aureus	Cooked-meat–10% NaCl medium, mannitol–sorbic acid medium	Tellurite–polymyxin–egg yolk agar, tellurite-glycine agar, mannitol-salt agar, staph. medium no. 110 agar
Fecal streptococci	Azide-dextrose broth, buffered azide glucose-glycerol broth, KF-streptococcus broth	KF-streptococcus agar, sodium azide–blood agar, Parker's crystal violet agar
*Clostridium perfringens**	Fluid thioglycolate medium, cooked-meat medium, anaerobic litmus milk	Sulfite–polymyxin–sulfadiazine agar, glucose-free sulfite-iron agar
C. botulinum, types A, B†	Beef-heart medium, cooked-meat–glucose-starch medium	Blood agar, meat-infusion agar, Anderson's pork infusion agar, reinforced clostridium medium
C. botulinum, type E‡	Similar	Similar

* Preheat sample 1 min at 80 C.

† Preheat inoculated enrichment medium or serial dilutions 20 min at 80 C.

‡ After inoculation, leave enrichment medium unheated, heat 15 min at 60 C, and heat 30 min at 80 C.

salt agar, tellurite glycine agar, or another selective medium (see Table 27–1), and then demonstrate that the organism is coagulase-positive on blood plasma (i.e., clots rabbit plasma). Cultures may be tested by injection of culture filtrates into cats or by feeding of human volunteers. Some progress has been made on the use of bacteriophage typing to select toxigenic strains and the use of smooth muscle to test for enterotoxin.

The interpretation of the results must be done with caution, for not all staphylococci, nor even all coagulase-positive staphylococci, produce enterotoxin, and individual animals and human volunteers differ in their susceptibility to the enterotoxin. Usually, considerable emphasis is placed on the type of food concerned, how it was handled, the chance for contamination with staphylococci from food handlers, the symptoms, and the usually comparatively short incubation period of 2 or 3 hr.

Cultures from nose, throat, or skin lesions of food handlers are tested in a similar manner to those from foods.

Tests for *Streptococcus faecalis*

The number of enterococci in the properly collected and refrigerated sample of food is considered significant; if there are not 750,000 or more colonies per gram of food, as estimated by plating on selective agar (e.g., sodium azide agar), the presence of the enterococci is not considered significant. Isolates are examined microscopically and tested for their ability to survive 60 C for 30 min and to grow in broth containing 6.5 percent salt or at pH 9.6.

Tests for *Salmonella* and *Shigella*

Food samples or fecal samples from patients or from suspected carriers are tested. Samples may be plated directly in a selective agar medium, SS agar or MacConkey agar, for example, or first may be put through an enrichment process in a selective broth, such as selenite or tetrathionate broth, and then plated. Characteristic colonies are picked and the cultures are differentiated into *Salmonella, Shigella,* or neither. Gram-negative cultures are differentiated on the basis of physiological (biochemical) tests, such as reactions on triple sugar iron agar slants, and of agglutination tests. Other characteristics may be studied following results with agglutination tests. Final typing of species usually is not attempted by the local or state public health laboratory, but may be done at a *Salmonella–Shigella* center.

When food handlers are tested to find carriers, at least three successive stools are examined. Agglutination tests on the blood serum of patients are useful in discovering whether the pathogen causing the food illness is identical with the one found at the ultimate source.

Test for *Clostridium perfringens*

Black colonies from anaerobic plates at 37 C of sulfite-polymyxin-sulfadiazine agar are confirmed in motility nitrate medium and sporulation broth (see Angelotti et al.).

Tests for other food-borne diseases

The methods of testing for the presence of other pathogens which are less commonly encountered in foods or for their effects, or for poisoning by chemicals, plants, or animals can be found in other text and reference books.

INTERPRETATION AND APPLICATION OF RESULTS

The results of the field and laboratory investigations, if complete, can lead to incrimination of the guilty food and location and elimination of the ultimate source of the cause of the food-borne disease outbreak. For reasons previously indicated, data often are incomplete and circumstantial evidence may have to be substituted for actual proof.

When an outbreak is small, as within a family, the location of the food responsible for the illness may be fairly simple. Table 27–2 summarizes a simplified report on such an outbreak. Chocolate éclairs obviously are the food to be suspected, and because of the nature of the food and the short incubation time (2½ to 4 hr), staphylococcus poisoning is probable. The source of the staphylococci would be sought among the food handlers at the bakery from which the éclairs came.

When the number of persons involved in an outbreak is fairly large, it is difficult to obtain complete and accurate data. Not all the persons may be available for interviewing, not all persons are equally susceptible to food illnesses, and people are forgetful about just what they ate. An aid to finding the offending food is a comparison of the percentage of persons who ate each food without becoming ill with the percentage of those who became ill. A complete tabulation would include data on the group of persons who ate each of the foods served, and on the group of persons who did not eat each of the foods, and for each food the percentage of persons consuming this food who became ill. The highest percentage of ill persons should be in the group which ate the offending food, a markedly higher percentage of ill persons than

TABLE 27-2. Summary of data on a food-borne disease outbreak within a family

| Name | Age, years | Time of first symptoms, hr* | | Foods consumed | | | | | | | | |
			Ham- burger	Gravy	Pota- toes	Fruit salad	Choco- late éclairs	Fresh peas	Milk	Coffee	Canned cherries
Jones, John	35	3	✓	✓	✓	✓	✓	✓		✓	✓
Jones, Mary	33	2½	✓	✓	✓	✓	✓	✓		✓	✓
Jones, Wm.	4			✓	✓			✓	✓		
Jones, Ruth	8		✓	✓	✓	✓		✓	✓		✓
Jones, Mabel	63	4	✓	✓	✓	✓	✓	✓		✓	✓

* Nausea, vomiting, abdominal cramping, diarrhea.

for any of the other foods consumed. The lowest percentage of ill persons should be in the group that did not eat the offending food. Statistical analyses are useful in determining the food responsible for a large outbreak. Final conclusions in regard to the food involved are drawn only after the laboratory results also are considered.

As has been indicated, one of the main purposes of the investigation is to locate the source of the agent causing the food illness so that this source can be eliminated and future outbreaks prevented. The location of the source also usually requires an analysis of both field and laboratory data.

The completely successful investigation of an outbreak of food-borne disease, then, finds the food responsible for the transmission of the disease, identifies the causative agent, finds the source of contamination of the food with the agent, and eliminates that source.

PREVENTIVE MEASURES

Means of prevention of food-borne outbreaks of some of the more important intoxications and infections have been discussed in Chapter 26. The general principles are:

1. To keep foods as free as possible from contamination with pathogenic agents by selection of uncontaminated foods, by adequate pasteurization or other heat processing, by keeping away pathogen-bearing vermin, by avoiding contamination from infected food handlers or carriers, and by generally good sanitary practice throughout the handling, preparation, and serving of foods.
2. To eliminate opportunities for the growth of pathogens, toxigenic or infectious, in foods by adjustment of the composition, by prompt consumption after preparation, and by adequate refrigeration of perishable foods if they must be held for any considerable time. Keeping foods warm for long periods is especially to be avoided.
3. To reject suspected foods.
4. To educate the public better concerning the causes and prevention of food-borne illnesses and the dangers involved.

REFERENCES

American Public Health Association. 1958. Recommended methods for the microbiological examination of foods, chap. 7. American Public Health Association, New York.
ANGELOTTI, R., H. E. HALL, M. J. FOTER, and K. H. LEWIS. 1962. Quantitation of *Clostridium perfringens* in foods. Appl. Microbiol. 10:193–199.

BERQUIST, K. R. 1957. The laboratory aspects of investigating food-borne disease outbreaks. J. Milk Food Technol. 20:101–165.

CASMAN, E. P., and R. W. BENNETT. 1965. Detection of staphylococcal enterotoxin in food. Appl. Microbiol. 13:181–189.

DACK, G. M. 1956. Food poisoning. 3rd ed. The University of Chicago Press, Chicago.

DAVIS, J. G. 1963. Examination of foods for food-poisoning organisms. Dairy Ind. 28:256–263.

GEIDT, W. R. 1957. The field application of the "Suggested procedures for the investigation of foodborne disease outbreaks." J. Milk Food Technol. 10:39–43.

HALL, H. E., R. ANGELOTTI, and K. H. LEWIS. 1963. Quantitative detection of staphylococcal enterotoxin B in food by gel-diffusion methods. Public Health Rep. 78:1089–1098.

HALL, H. E., D. F. BROWN, and R. ANGELOTTI. 1964. The quantification of salmonellae in foods by using the lactose pre-enrichment method of North. J. Milk Food Technol. 27:235–240.

International Association of Milk and Food Sanitarians, Inc. 1957. Procedure for the investigation of foodborne disease outbreaks. Shelbyville, Ind.

LEWIS, K. H., and R. ANGELOTTI (Eds.) 1963. Examination of foods for enteropathogenic and indicator bacteria; review of methodology and manual of selected procedures. Milk, Food Research Section. Milk, Food Branch, Div. Environmental Eng., Food Protection, Public Health Serv. U.S. Dep. Health, Educ., Welfare.

READ, R. B., JR., J. BRADSHAW, W. L. PRITCHARD, and L. A. BLACK. 1965a. Assay of staphylococcal enterotoxin from cheese. J. Dairy Sci. 48:420–424.

READ, R. B., JR., W. L. PRITCHARD, J. BRADSHAW, and L. A. BLACK. 1965b. In vitro assay of staphylococcal enterotoxins A and B from milk. J. Dairy Sci. 48:411–419.

U.S. Public Health Service, Department of Health, Education, and Welfare. 1964. Training course manual; food microbiology. Robert A. Taft Sanitary Engineering Center, Cincinnati, Ohio.

WENTWORTH, BERTTINA B. 1963. Bacteriophage typing of the staphylococci. Bacteriol. Revs. 27:253–272.

PART SIX

FOOD SANITATION, CONTROL, AND INSPECTION

Food sanitation, control, and inspection have been considered primarily from the microbiological standpoint. The laboratory manual style has been avoided in the discussion of microbiological laboratory methods; instead, basic principles have been treated. Since the Federal Food, Drug, and Cosmetic Act is the model for much of the legislation, its bacteriological aspects have been discussed in some detail.

CHAPTER TWENTY-EIGHT

MICROBIOLOGY IN FOOD PLANT SANITATION

The food-industry sanitarian is concerned with aseptic practices in the preparation, processing, and packaging of the food products of a plant (or plants), the general cleanliness and sanitation of plant and premises, and the health of employees. His duties in connection with the food products may involve quality control and storage of raw products, the provision of a good water supply, prevention of the contamination of the foods at all stages during processing from equipment, personnel, and vermin, and supervision of packaging and warehousing of finished products. The supervision of cleanliness and sanitation of plant and premises includes not only the maintenance of clean and well-sanitized surfaces of all equipment touching the foods but also generally good housekeeping in and about the plant and adequate treatment and disposal of wastes. Duties affecting the health of the employees include provision of a potable water supply, supervision of matters of personal hygiene, regulation of sanitary facilities in the plant and in plant-operated housing units, and contact with sanitary aspects of plant lighting, heating, and ventilation. He also usually participates in the training of employees in sanitary practices. Only bacteriological aspects of plant sanitation will be discussed here. For other facets of the problem the reader is referred to the references at the end of the chapter.

For the most part, the sanitarian concerns himself chiefly with general aspects of sanitation, making inspections, consulting with personnel responsible for details of sanitation and executives directing such work, and training personnel in sanitation. He may or may not be connected with a plant laboratory.

BACTERIOLOGY OF WATER SUPPLIES

The water for drinking purposes and for plant use may be from the same source or from different sources.

Drinking water

The water that the employees drink must meet public health standards when tested by methods recommended in the latest edition of "Standard Methods for the Examination of Water and Wastewater" (see Chapter 4, and suggested standards in Appendix). Coliform bacteria must not be present at levels indicating contamination of the water by sewage. Total plate counts of the water sometimes are made to indicate when trouble may be incipient so that such trouble can be forestalled.

Plant water

All water that comes into contact with foods should meet the bacteriological standards for drinking water, and preferably all fresh water at the plant should be that good. But this water also should be satisfactory from a bacteriological standpoint for use with the particular food being processed. A water supply may be adjudged potable, yet be unsatisfactory for use with a food. Thus, for example, water containing appreciable numbers of psychrophiles of the genera *Pseudomonas*, *Achromobacter*, or *Alcaligenes* might be unsatisfactory without treatment in a dairy plant making butter or cottage cheese. The slimy growth of iron bacteria in water supplies often leads to trouble in the food plant.

More likely to be important is the chemical composition of the water, which must be suited to the use to be made of it. Thus hard water is undesirable in pea canning and in brewing; iron and manganese are bad in beet canning and in brewing; excessive organic matter may lead to off-flavors, etc.

Of special interest in canning factories is the bacteriology of the water in which the cans of processed foods are cooled after their heat-treatment. If this water contains microorganisms able to spoil the food, it will, after entering defective cans through minute leaks, increase the percentage of cans of food spoiling during storage. Many canneries routinely chlorinate the cooling water to reduce or eliminate this trouble.

The shortage of water in many food plants has necessitated reuse of part of the water, and microorganisms may build up in such reused water. Water employed for the final rinse of a food must be fresh and potable, but after use may be returned for soaking, first wash, or fluming, preferably after treatment with chlorine, chlorine dioxide, or a similar germicide.

In-plant or continuous chlorination beyond the break point (the point where the chlorine demand has been satisfied) to a residual of 5 to 7 ppm of chlorine is employed for continuous application to areas and equipment where slime bacteria may be a problem, e.g., conveyors or belts, can coolers, product washers, and flumes. The chlorinated water may be applied as a spray, or parts of equipment may be immersed. When operations cease, chlorinated water may be applied to fillers, peelers, dicers, and similar equipment. Contaminated or polluted water lines are held filled with chlorinated water containing 50 to 100 ppm of chlorine for 12 to 48 hr, the strength of chlorine and length of time depending on the extent of pollution.

Ice used in contact with foods should meet the bacteriological requirements for potable water. Much work has been done on the incorporation of bacteriostatic or bactericidal chemicals in water and in ice to aid in food preservation. It has been noted previously that a chlortetracycline or oxytetracycline dip for dressed poultry has been approved, and that these antibiotics may be incorporated in ice to be applied to fish and other seafood.

SEWAGE AND WASTE TREATMENT AND DISPOSAL

The food sanitarian is concerned directly or indirectly with the adequate treatment and disposal of wastes from his industry. Solid and concentrated wastes ordinarily are kept separate from the watery wastes, and may be used directly for food, feed, fertilizer, or other purpose, may first be concentrated, dried, or fermented (e.g., pea-vine silage), or may be carted away to available land as unusable waste. Care is taken to keep out of the waste waters as much wasted liquid or solid food material as possible, by taking precautions to avoid introduction into the watery wastes of drip, leakage, overflow, spillage, large residues in containers, foam, frozen-on food, and food dust during the handling and processing of the food. It is recommended that sewage of human origin be kept separate from other plant waters because of the possible presence of human intestinal pathogens and the necessity for a guarantee of their removal or destruction. Such sewage may be turned into a municipal system, if one is available, for adequate treatment and disposal or may be treated separately at the food plant. Other food-plant wastes should not contain human pathogens.

Wastes from food plants ordinarily contain a variety of organic compounds, which range from simple and readily oxidizable kinds to those that are complex and difficult to decompose. The strength of the sewage or food waste containing organic matter is expressed in terms of **biochemical oxygen demand** (BOD), which is the quantity of oxygen used by aerobic microorganisms and reducing compounds in the stabilization of decomposable matter during a selected time at a certain temperature. A period of 5 days at 20 C is generally used, and results are expressed as 5-day BOD. The BOD is determined by dilution of a measured quantity of waste with water that has been saturated with oxygen and incubation of the mixture at 20 C, along with a control of dilution water alone. After 5 days, the residual oxygen in both control and test sample is measured by titration. The difference represents the oxygen-consuming capacity of the waste, and is calculated to be expressed as parts per million of oxygen taken up by the waste. To calculate the strength of the waste in terms of pounds of BOD:

$$\frac{\text{ppm 5-day BOD} \times \text{gallons of waste} \times 8.34}{1,000,000} = \text{pounds BOD}$$

This value can be converted to population equivalent (P.E.) by assuming that the domestic sewage of one person is equivalent to one-sixth of a pound of BOD per day.

Whenever appreciable amounts of wastes high in oxidizable organic matter (high BOD) are emptied into natural waters, such as streams, ponds, or lakes, the 7 to 8 ppm of free oxygen normally present in the waters is used up soon by oxidation processes carried out by aerobic or facultative microorganisms. When the oxygen drops below 3 ppm, the fish either leave or die, and when anaerobic conditions have been attained, hydrolysis, putrefaction, and fermentation by microorganisms will follow, with the result that the body of water will become malodorous and cloudy and hence unsuited for recreational use and unfit for drinking and for use in the food plant. Wastes from a food plant, then, to be emptied into a body of water, must either be so greatly diluted by that water as to be innocuous or must be treated first to reduce the oxidizable compounds to a harmless level. Even the effluent from an efficiently operated sewage treatment system will encourage the growth of algae and higher aquatic plants in the water and make it less attractive for recreational purposes.

Preliminary treatments of food-plant wastes by chemical means may be employed, but most systems of treatment and disposal depend upon (1) screening out of large particles; (2) floating off of fatty and other floating materials; (3) sedimentation of as much of the remaining solids as is practicable; (4) hydrolysis, fermentation, and putrefaction of complex organic compounds; and finally (5) oxidation of the remaining solids in the water to a point where they can enter a municipal sewage treatment and disposal system, a plant disposal system, a lake or stream, or soil. The completeness of oxidation required will depend upon the disposal to be made. Thus less oxidation might be required for feeding to a municipal system or for irrigating soil than for entering a stream or lake.

Chemical treatment

In chemical pretreatments, a chemical or mixture of chemicals is added to the sewage or waste so as to cause formation of a flocculent precipitate, which, in settling, carries with it much of the suspended and colloidal material, including bacteria. The effluent then is run into a body of water, onto soil, or into a biological treatment system. The chemicals commonly used are soluble aluminum or iron salts, such as alum or ferrous sulfate, plus lime, giving a flocculent precipitate of aluminum or ferric hydroxide. Disposal of the sludge (settlings) so obtained may be difficult.

Biological treatment and disposal

The general biological methods for waste disposal and/or treatment include (1) **dilution,** by running waste waters into a large body of water; (2) **irrigation,** in which waste waters are sprayed onto fields of open-textured soil; (3) **lagooning,** by running the waste waters into shallow artificial ponds (with or without other treatments); (4) use of **trickling filters,** made of crushed rock, coke, filter tile, etc.; (5) use of the **activated sludge** method, in which waste water is inoculated heavily with sludge from a previous run and is actively aerated in tanks; and (6) use of **anaerobic tanks** of various kinds, where settling, hydrolysis, putrefaction, and fermentation take place, usually to be followed by some aerobic treatment.

The dilution method seldom is practicable because a sufficiently large or rapidly moving volume of water rarely is available or because the location is such that sewage decomposition cannot proceed without objections from nearby populations. Irrigation is increasing in popularity and is especially adaptable to use by plants located in rural areas and near open-textured soil. Lagooning has been used especially for seasonal wastes, as from canning factories. The wastes are decomposed slowly in these shallow ponds or lagoons until the liquid part can be run into a stream or other body of water during the rainy season or time of melting snow, when there is a good volume of water. Usually, sodium nitrate is added to reduce obnoxious odors. Sometimes the liquid is pumped from and returned to a lagoon, or it may be pumped from one lagoon to another in a series of lagoons. Trickling filters and activated sludge systems are probably the most effective of the systems listed, but they are expensive to run and require supervision by an expert.

TABLE 28–1. Range of 5-day BOD values for wastes from various food-processing plants

Source of waste	5-day BOD, ppm	Source of waste	5-day BOD, ppm
Dairy plants	500–2,000	String-bean cannery	160–600
Meat-packing plants	Up to 2,500	Lima-bean cannery	190–450
Poultry plants	300–7,500	Sweet-corn cannery	625–6,000
Sugar processing	500–1,500	Pea cannery	380–4,700
Fruit cannery	200–2,100	Pumpkin cannery	2,800–6,900
Tomato cannery	180–4,000	Spinach cannery	280–730
Brewery	420–1,200	Sauerkraut cannery	Up to 6,300

Anaerobic tanks yield an effluent that needs further treatment and should be either turned into a municipal system or given an aerobic treatment.

Types of food wastes

An extended discussion of the nature and composition of wastes from the different food industries cannot be given here. It should be noted, however, that each type of waste has a characteristic BOD that may be high, low, or intermediate and each presents its own problems of treatment and disposal. Dairy wastes, for example, are usually high in protein and lactose and contain many microorganisms. Such wastes, if not already acid, will turn acid if kept under anaerobic conditions and then will be more difficult to treat. Some wastes may be acid originally—wastes from fruit canneries, for instance. Malthouse, brewery, distillery, sweet-corn cannery, and corn-products plant wastes are high in carbohydrates and likely to become acid under anaerobic conditions. Wastes high in proteins, e.g., pea- or fish-cannery or packing-plant wastes, are likely to putrefy under anaerobic conditions. Other wastes may contain antiseptic chemicals, such as the sulfite in waste sulfite liquors from paper mills, and therefore may be difficult to decompose by means of microorganisms. Ranges of 5-day BOD values reported for wastes from various types of food-processing plants are shown in Table 28–1.

MICROBIOLOGY OF THE FOOD PRODUCT

To reduce contamination with microorganisms to a minimum and obtain good keeping quality of the product, the raw materials are examined, the equipment contacting the food is adequately cleansed, sanitized and tested, the preserving process is checked, and packaging and storage are supervised.

The ingredients

The raw product is inspected and tested for quality, but this does not necessarily involve bacteriological laboratory testing in all instances. Some of the ingredients of some products may contain numbers and kinds of microorganisms that can affect the keeping quality of the product or even its acceptability. Some ingredients, such as sweetening agents, starch, and spices, can be purchased on specification as to maximal allowable content of microorganisms or of numbers of certain kinds. The numbers of bacteria in ingredients are important in foods for which there are bacterial standards, e.g., ice cream, which must not contain more than 50,000 bacteria per gram in some areas. Large numbers of spores of aerobes are undesirable in dry milk to be used in breadmaking because of the increased risk of ropiness developing; heat-resistant spores in sugar and starch may add to the difficulty in adequately heat-processing canned vegetables to which sugar or starch is added; and large numbers of bacteria in spices may favor the spoilage of summer sausage.

The microbiology of the main raw product often is important. Excessive mold mycelium in the raw fruit, which is indicative of the presence of rotten parts, may lead to condemnation of the canned or frozen product. Large numbers of thermoduric bacteria in raw milk may yield a pasteurized milk that will not meet the bacterial standards for numbers as estimated by the standard plate-count method. Large numbers of bacteria on vegetables or in fruits may indicate inferiority that will carry over into the frozen product. Laboratory examination may be employed to detect these undesirable organisms and estimate their numbers.

Often there is opportunity for microorganisms to grow in a food product during handling and processing in the plant. Examples are the build-up of thermophiles where foods are kept hot, as in forewarmers and blanchers, and increases in total numbers of bacteria in vegetables between blanching and freezing. Line samples may be tested in the laboratory to ascertain where appreciable growth of microorganisms is taking place.

Packaging materials

Packaging materials are a possible source of contamination of foods with microorganisms, but ordinarily the penetrability of nonmetallic materials to moisture and to gases is of more significance in the preservation of foods than the microbiology of these materials, for they harbor mostly low numbers of innocuous microorganisms or no organisms. Also, as has been indicated previously, wrappers may be treated or impregnated with bacteriostatic or fungistatic compounds, e.g., cheese wraps with sorbic or caprylic acid.

Paper and paperboard used for milk cartons contain mostly bacilli and micrococci, and occasionally other rods, actinomycetes, and mold spores, but no organisms of public health significance. Treatment with hot paraffin kills most of the organisms present, provides an almost sterile surface, and prevents bacteria within the cardboard, mostly bacterial spores, from reaching the food. Wax paper is practically sterile as produced, as are most plastic packaging materials. All packaging materials should be protected from contamination with dust or other sources of microorganisms in handling.

According to Federal regulations a food is deemed to be adulterated "if its container is composed, in whole or in part, of any poisonous or deleterious substance which may render its contents injurious to health."

The equipment

Unless the equipment that comes in contact with foods is adequately cleansed and sanitized, it may be an important source of contamination of foods with microorganisms. Not only may organisms persist on equipment, but they may increase in numbers when treatment has been inadequate. Microbiological standards for plant sanitation have been suggested but to date have not been applied officially. Thatcher has proposed that standards be based on counts of total numbers of microorganisms per unit surface area of equipment coming in contact with foods as an indication of the effectiveness of cleaning and sanitizing, and on special tests for bacteria of human origin, e.g., coliform bacteria and streptococci of fecal origin, staphylococci from nasopharyngeal or suppurative sources, and *Streptococcus salivarius* from the mouth, to indicate contamination from personnel.

CLEANSING.   From a bacteriological viewpoint, cleansing of equipment is primarily to remove as much food for microorganisms as is practicable. Equipment may be disassembled for cleaning and sanitizing, although this is difficult with some pieces. To aid in the cleansing action of water, cleansing agents called **detergents** are employed. These agents may serve to soften or condition the water, improve the wetting ability of the cleansing solution, emulsify or saponify fats, solubilize minerals, deflocculate or disperse suspended materials, and dissolve as much soluble material as possible. At the same time the detergents should be noncorrosive and readily rinsed from the surfaces. Among the detergents used alone or in mixtures are the **alkaline** varieties, such as lye, soda ash, sodium metasilicate, trisodium phosphate, and the polyphosphates; **acid** detergents, usually organic acids, such as hydroxyacetic, gluconic, citric, tartaric, and levulinic acids; and **wetting agents,** which may be anionic (NaR), such as the hydrocarbon sulfonates; nonionic, e.g., polyether alcohol; or cationic (RCl), for example, the quaternary ammonium compounds. Cleaning may be aided by the employment of brushes (Figure 28–1) and of water under pressure.

SANITIZING. The sanitizing process is an attempt to kill most or all of the microorganisms on equipment surfaces. The kind of sanitizer, the concentration employed, the temperature of the sanitizer, and the method of application will vary with the kind of sanitizing agent, the conditions during use, the type of equipment to be treated, and the microorganisms to be destroyed. Among the sanitizing agents in common use are hot water, flowing steam or steam under pressure, halogens (chlorine or iodine) and halogen derivatives, and the quaternary ammonium compounds.

Steam under pressure is the most effective way of applying heat as a sanitizing agent, but its use is limited to closed systems that can withstand pressure. Steam jets, flowing steam, or hot water may be

Figure 28–1. General cleanup in a food-processing plant. (*Courtesy of Klenzade Products, Beloit, Wis.*)

used, but jets are ineffective except at very short distances, flowing steam may condense and drop in temperature as it passes through equipment, and hot water may undergo a similar drop in temperature. All microorganisms and their spores can be killed by adequate treatment with high-pressure steam. Effectively applied flowing steam and boiling water will kill all but some of the more resistant bacterial spores. The lower the temperature of "hot" water, the less effective it will be in killing organisms.

Chlorine, iodine, and their compounds (hypochlorites, chloramines, iodophors, etc.) are effective germicides if in proper concentrations and if given enough time to act. Usually, more sanitizer is necessary in the presence of organic matter. Bacterial spores are especially resistant to these sanitizers. Chlorine is used to destroy undesirable bacteria in water for drinking, for use in foods, for washing foods or equipment, and for cooling. Hypochlorites are more labile but more effective at acid pH values than at alkaline ones. As stated earlier, in-plant or continuous chlorination beyond the break point (where chlorine demand has been satisfied) to a residual of 5 to 7 ppm is employed for continuous application to areas where slime bacteria may be a problem, e.g., on conveyors, belts, or product washers. Chlorine (50 to 100 ppm) also is used to treat contaminated or polluted water lines.

Quaternary ammonium compounds are, in general, more effective against Gram-positive than Gram-negative bacteria. These compounds have a residual effect, that is, they adhere to equipment surfaces and deter bacterial growth; but they rinse off onto foods coming into contact with these surfaces, and, if they are present in detectable concentrations, might be considered undesirable. Many of these compounds are active under alkaline conditions. Most are affected by hardness of water.

Detergent sanitizers, which usually are a combination of an alkaline detergent and a quaternary ammonium compound, sometimes are used to cleanse and sanitize utensils or equipment in one operation.

CLEANED-IN-PLACE SYSTEMS. Some industries, especially the dairy industry, leave pipelines permanently connected and clean and sanitize them in place. Apparatus is available for accomplishing this automatically. Different sequences of treatments are recommended for different cleaned-in-place (CIP) systems. Milk pipelines, for example are rinsed first with tepid water, which is pushed or pulled through the system. Then hot (160 F, or 71 C) detergent solution may be passed through, followed by rinsing water and finally a sanitizing agent, such as hot water (170 F, or 77 C, or over), chlorine solution (200 ppm), or a quaternary ammonium compound (200 ppm). Often a sanitizing treatment is given immediately before use.

References listed at the end of the chapter should be consulted for more detail on detergents and sanitizers and their selection and use.

FUNGISTATIC PAINTS. There also is the possibility of contamination of foods and equipment from walls, ceilings, and other parts of the food plant. Most troublesome are the molds, whose spores are readily air-borne and may travel considerable distances in the plant. Cleanliness, of course, will retard mold growth, but at the present time, fungistatic paints are used on walls and ceilings likely to be subject to mold growth. Commonly used in paints are copper-containing compounds such as copper naphthanate or copper-8-quinolinolate, or phenol derivatives, e.g., pentachlorophenol or other chlorophenols, oxyquinoline sulfate, phenyl mercurial, etc. Occasionally, drip from overhead ceilings, pipes, etc., introduces organisms other than molds and must be prevented.

The preservation process

The sanitarian usually has little to do with the processing of the foods except to check through the laboratory, if one is available, for the effectiveness of the processing. The laboratory, for example, might run keeping-quality tests on canned foods and bacterial counts on frozen foods, pasteurized milk, dry milks, etc. Some of the laboratory tests on foods are discussed in Chapter 29.

Vending machines for foods and beverages

With the rapid expansion of the use of vending machines to dispense perishable foods has come increased interest in the sanitation of these machines and dispensed foods. The United States Public Health Service Ordinance and Code (The Vending of Foods and Beverages) covers sanitation of foods and machines, operation of machines, and inspection. "Readily perishable foods" are defined as those consisting in whole or in part of milk, milk products, eggs, meat, fish, poultry, etc. These are foods which can support rapid growth of microorganisms and can cause food infections or intoxications. Adequately dried or canned foods are excepted. Perishable foods include sandwiches, pastries, hot coffee, tea or chocolate, malted milk, fluid milk, ice cream, frozen desserts, and hot-food plates (meat, stews, soup, baked beans, poultry, fish, etc.). During transportation from the commissaries and in the machine, perishable foods should be kept either cold (below 50 F, or 10 C, usually at 38 to 40 F, or 3.3 to 4.4 C) or hot (150 F, or 66 C, or above). Slow growth of psychrophiles can take place at the lower temperatures and of thermophiles in the hot foods if these recommendations are barely met, and excessive heat will deteriorate many foods. All parts of vending machines in contact with readily perishable foods should be cleaned and sanitized periodically, daily if the above temperature limitations are not met. Water used in connection with the foods should be potable, and waste disposal should be adequate. Most machines dispensing perishable foods are equipped with safety devices to stop the dispensing of food when refrigeration or heating fails.

Food handling on a large scale

Food handling on a large scale by caterers, commissaries, restaurants, institutions, airlines, camps, etc., are subject to similar considerations. General recommendations and a Food Service Sanitation Ordinance and Code are included in the Food Service Sanitation Manual published in 1962 by the Public Health Service of the U.S. Department of Health, Education, and Welfare. The ordinance defines "safe" temperatures for storage of foods as 45 F (7.2 C) or below or 140 F (60 C) or above, except during necessary periods of preparation and service. It requires the washing of raw fruits and vegetables, and the thorough cooking of stuffing, poultry, stuffed meats and poultry (heating to at least 165 F, or 74 C), and pork and pork products (all parts heated to at least 150 F, or 66 C) before being served. The ordinance has regulations concerning the health and cleanliness of personnel, the cleanliness, sanitization, and protection of food utensils, and the potability of water. It specifies foods that are clean, wholesome, unspoiled, free from adulteration and misbranding, safe for consumption, and meeting any standards of quality or inspection. Also described is the handling of pastry fillings and of puddings.

Sandwiches

Sandwiches and other foods may be retailed without vending machines. Such foods may be a potential food-poisoning hazard, for they often are held at ambient temperatures for 18 to 24 hr before being sold. One survey (Adame et al.) indicated that the wrapped sandwiches examined showed signs of contamination during preparation and of growth of bacteria prior to vending. All sandwiches showed high total numbers of bacteria per gram, no salmonellae or *Clostridium perfringens,* and considerable numbers of staphylococci (most of which, however, were coagulase-negative), with higher numbers in the moist sandwiches than in the dry, and appreciable numbers in those heated to 55.5 C (132 F) and served hot. Similar results were obtained by McCroan et al., who concluded that spiced-ham sandwiches and cheese sandwiches were more hazardous than sandwiches containing mayonnaise, e.g., egg-salad and chicken-salad sandwiches, for contact with the acid dressing helped repress the staphylococci.

Microbiological standards for foods

It is evident from previous discussion that microbiological standards for foods may be of three types: (1) those on the raw product and ingredients; (2) those concerned with plant sanitation and methods of packaging, storage, and handling; and (3) those on the finished product as marketed. Examples of types (1) and (3) have been given in previous chapters and some are presented in the Appendix. Standards must be adapted to the types of food for which they are intended. They probably would be different for a food to be consumed raw than for the same food to be cooked or subjected to heating or other processing before being marketed. The type of spoilage organism to be feared and, therefore, watched for will vary with the food and the method of processing. Standards for ingredients of soft drinks, for example, include those for numbers of yeasts; for low-acid foods to be canned, numbers and kinds of heat-resistant bacterial spores are significant; and numbers of aerobic spore-formers in flour may indicate the likelihood of the development of ropiness in bread. The type of pathogen most likely to be present will be different in different foods. Tests for coliform bacteria to indicate the possible presence of intestinal pathogens are useful in setting standards for oysters but have little meaning when made on frozen orange juice. Salmonellae might be looked for in eggs or egg products and trichinae in raw pork.

The chief purposes of microbiological standards for foods are to give assurance (1) that the foods will be acceptable from the public health standpoint, i.e., will not be responsible for the spread of infectious disease or for food poisoning; (2) that the foods will be of satisfactory quality, that is, will consist of good original materials that have not deteriorated or become unduly contaminated during processing, packaging, storage, handling, or marketing; (3) that the foods will be acceptable from an esthetic viewpoint, in that the introduction of filth in the form of fecal material, parts of vermin, pus cells, mold mycelium, etc., has been prevented; and (4) that the foods will have keeping qualities that should be expected of the product.

Many difficulties are encountered in establishing and applying microbiological standards for foods. Sampling for tests is a problem, for the lack of homogeneity in most foods makes location, size, and number of samples significant. Standards usually are based on total numbers of organisms, numbers of an indicator organism, or numbers (or total absence) of pathogens; but there has been some disagreement as to what counts should be considered significant, what the indicator organism should be, and whether pathogens can be demonstrated. Counts are statistically uncertain, and dye reduction tests do not necessarily test for the important organisms. Finally the numbers and kinds of organisms in most foods decrease during storage in the dry or frozen condition. If an "average standard" is adopted, half of the food samples are eliminated. If legal action is involved, the level of a standard must be justified; and counts or results on a sample by prosecutor, defendant, and a neutral agency may not necessarily agree.

It has been recommended that: (1) testing procedure and standards be adapted to the particular kind of food; (2) a numerical relationship be demonstrated between the standard and the hazard; (3) tolerances be allowed for admitted inaccuracies of sampling and analysis, i.e., all samples would not be required to meet the standard (see Standards for Drinking Water in the Appendix), and results from successive samples, taken at stated intervals, would be considered in setting and interpreting standards; and (4) any suggested standard be tried out first on a voluntary basis.

It has been found, however, that setting a standard usually results in better plant sanitation within the industry and in lower average bacterial counts and better keeping quality for the product.

HEALTH OF EMPLOYEES

As has been pointed out, duties of the sanitarian that affect the health of employees include provision of potable drinking water, supervision of matters of personal hygiene, regulation of sanitary facilities within the plant and of sewage treatment and disposal, and supervision of sanitation in plant eating establishments and in plant-operated housing units. Most of these duties involve the sanitary aspects of plant and housing construction, selection of qualified personnel to direct operations of the facilities, and training of employees in sanitary practices.

The bacteriology of drinking water and of sewage treatment and disposal has been discussed briefly. However, sanitation in eating places in the plant deserves special mention. Special places should be designated for eating carried lunches, and such places should be kept neat and sanitary. If the plant serves meals to employees in a cafeteria or restaurant, the sanitarian should be responsible for supervision of sanitation in the preparation, handling, serving, and storage of the food so as to avoid the spread of infectious microorganisms and to prevent outbreaks of food poisoning. The prevention of food poisoning and food infections has been discussed in Chapter 26. To prevent the spread of disease, food equipment and utensils should be handled, washed, and sanitized as directed in the Food Service Sanitation Manual of the U.S. Public Health Service.

Before washing, all food equipment and utensils should be presoaked, if necessary, and preflushed or prescraped. Washing should be by means of a solution of a suitable detergent, which must be at 140 F (60 C) or higher (160 F, or 71 C, or over for single-tank conveyor machines). When hot water is employed in sanitizing it should be at 170 F (77 C) or over, with exposure for at least ½ min. The final rinse water should be at that temperature, or at 180 F (82 C) or over at the entrance of the manifold. With chemical sanitizers, immersion should be for at least 1 min at not less than 75 F (24 C), and the solutions should contain at least 50 ppm of available chlorine, or 12.5 ppm of available iodine, or have an equivalent bactericidal effect. Very large pieces of equipment may be treated with live steam, rinsed with boiling water, or sprayed or swabbed with a chemical sanitizing solution at least twice as strong as that used for immersion treatment.

Transient bacteria are removed from hands of workers by soap and water, with scrubbing aiding in the removal; but the permanent organisms remain, such as staphylococci, streptococci, coliforms, and pseudomonads. Chemicals in soaps or used as detergents, e.g., bisphenols, iodophors, and quaternary ammonium compounds, are effective in reducing numbers of bacteria, if used daily in successive hand washings.

REFERENCES

ADAME, J. L., F. J. POST, and A. H. BLISS. 1960. Preliminary report on a bacteriological study of selected commercially prepared wrapped sandwiches. J. Milk Food Technol. 23:363–366.

American Public Health Association. 1960. Standard methods for the examination of water and wastewater. 11th ed. New York.

ANONYMOUS. 1958, 1959. Some public health aspects of food and beverage vending. J. Milk Food Technol. 21:322–324, 350–352; 22:18–21.

ANONYMOUS. 1963. Sanitation; FE special report. Food Eng. 35(4):69–86.

Association of Food Industry Sanitarians, Inc. 1952. Sanitation for the food-preservation industries. McGraw-Hill Book Company, New York.

CLARY, R. K. 1963. Evaluation of chemical sanitizers. Health Quart. Bull., Wis. State Bd. Health, Oct.– Dec., p. 23–25.

CRISLEY, F. D., and M. J. FOTER. 1965. The use of antimicrobial soaps and detergents for hand washing in food service establishments. J. Milk Food Technol. 28:278–284.

DACK, G. M. 1956. Evaluation of microbiological standards for foods. Food Technol. 10:507–509.

ECKENFELDER, W. W., and E. L. BARNHART. 1965. Treatment of food processing wastes. Avi Publishing Co., Inc., Westport, Conn.

ELLIOTT, R. P., and H. D. MICHENER. 1961. Microbiological standards and handling codes for chilled and frozen foods; a review. Appl. Microbiol. 9:452–468.

HARTLEY, D. E. 1963. Inspecting automatic vending operations. J. Milk Food Technol. 26:130–133.

INGRAM, M. 1961. Microbiological standards for foods. Food Technol. 15(2):4–12, 16.

International Union of Pure and Applied Chemistry. 1963. Re-use of water in industry. Butterworth, Inc., Washington, D.C.

LACHMANN, A. 1963. International food regulations. Food Eng. 35(8):59–63.

McCROAN, J. E., T. W. McKINLEY, ALICE BRIM, and W. C. HENNING. 1964. Staphylococci and salmonellae in commercial wrapped sandwiches. Public Health Rep. 79:997–1004.

MANGOLD, W. S. (Ed.) 1946. A sanitation manual for food industries. University of California Press, Berkeley, Calif.

MERCER, W. A. 1964. Physical characteristics of recirculated waters as related to their sanitary condition. Food Technol. 18:335–344.

Microbiological and Biochemical Center, Syracuse University Research Corporation. 1964. Manual of sanitation standards for certain products of paper, paperboard, or moulded pulp. J. Milk Food Technol. 27:366–369.

PARKER, M. E., and J. H. LITCHFIELD. 1962. Food plant sanitation. Reinhold Publishing Corporation, New York.

PETERSON, G. T., J. F. FOX, and L. E. MARTIN. 1959. Problems in the preparation and handling of hot vended canned foods. Food Technol. 13(4):22 [of insert]. (Abstr.)

RACK, B.-G., and R. BINSTEAD. 1964. Hygiene in food manufacturing and handling. Food Trade Press, Ltd., London.

ROGERS, J. L. 1958. Automatic vending, merchandising—catering. Food Trade Review, London.

SENN, C. L., and P. P. LOGAN. 1955. Sanitation aspects of "take-out" type foods. Amer. J. Public Health 45:33-38.

SHIFFMAN, M. A., and D. KRONICK. 1963. The development of microbiological standards for foods. J. Milk Food Technol. 26:110–114.

STUART, L. S. 1962. Federal regulation of bactericidal chemicals used in building, industrial and institutional sanitation programs. J. Milk Food Technol. 25:308–312.

THATCHER, F. S. 1955. Microbiological standards for foods: their function and limitations. J. Appl. Bacteriol. 18:449–461.

THOMAS, S. B., R. G. DRUCE, PHYLLIS M. HOBSON, and G. WILLIAMS. 1963. Effect of steam, boiling water, caustic soda, hypochlorite and quaternary ammonium compounds on the bacterial content of farm dairy equipment. Dairy Ind. 28:390–399.

TIEDEMAN, W. D., and M. SHIFFMAN (Chairmen). 1960. Microbiological standards for foods; conference report. Public Health Rep. 75:815-822.

U.S. Department of Health, Education, and Welfare. 1962. Food service manual. Public Health Serv. Pub. 934.

U.S. Department of Health, Education, and Welfare, Public Health Service. 1958. Frozen desserts ordinance and code. [Reprinted March, 1958]

U.S. Department of Health, Education, and Welfare. 1953. Sanitary food service. Public Health Serv. NAVMED P-1333.

U.S. Department of Health, Education, and Welfare. 1957. The vending of foods and beverages. Public Health Serv. Publ. 546.

U.S. Public Health Service. 1943. Ordinance and code regulating eating and drinking establishments. Bull. 280.

WARRICK, L. F., F. J. McKEE, H. E. WIRTH, and N. H. SANBORN. 1939. Methods of treating cannery wastes. Bull. Wis. State. Bd. Health and Nat. Canners Ass.

CHAPTER TWENTY-NINE

MICROBIOLOGICAL LABORATORY METHODS

Microbiological tests on the quality of foods and food ingredients may be conducted in laboratories of food plants or of various control agencies—Federal, state, local, private, or commercial.

PURPOSES OF TESTS

The objectives of the microbiological tests employed will depend upon the type of laboratory making them. Laboratories of governmental control agencies will be concerned primarily with testing foods to ascertain whether they meet the standards—Federal, state, or local—as promulgated by law. Thus, for example, total numbers of bacteria might be estimated in ice cream or market milk, coliform bacteria might be estimated in oysters or drinking water, and mold mycelia might be counted in frozen berries. The tests would be concerned for the most part with sanitary aspects of the foods, that is, fitness and especially healthfulness for consumption.

The food-plant laboratories, which are concerned chiefly with quality control, test the raw product and ingredients and line samples during handling and processing as a check on the success of their methods and as a warning of possible troubles to come. They also ascertain whether bacteriological standards are being met (if such standards exist), the keeping quality of the product is acceptable, and no microorganisms or microbial products injurious to human health are present.

Private control agencies, e.g., the Committee on Foods of the American Medical Association or the Good Housekeeping Institute, might have laboratory tests to enable the recommendation or acceptance of certain brands of foods.

Laboratories of commercial control agencies, e.g., of the National Canners Association, the American Dry Milk Institute, the American Meat Institute, etc., may set standards of their own for the special types of food with which they are concerned and describe recommended or official methods of examination.

SAMPLING

Accepted methods for the microbiological examination of foods as described by the appropriate agencies usually include specific directions regarding the sampling of the food.

Aseptic sampling

Any packaged food product that is supposedly sterile or nearly so is sampled with aseptic precautions from the container opened in the laboratory. This type of sampling is essential when legal aspects are involved. Foods containing large numbers of microorganisms and especially those exposed normally to open air and other contamination may not require strictly aseptic precautions. All apparatus connected with sampling should, however, be previously sterilized and kept sterile until used.

Sampling devices

Liquid foods ordinarily are sampled by means of sterile pipettes or sampling tubes, preferably after stirring or mixing to homogeneity. Solid foods may be sampled by means of sterile sampling tubes on the order of cork borers or triers, or by augers, scoops, spoons, or knives. Separate small units of liquid or solid foods may be taken as samples during successive stages in handling and processing. A filterable liquid, e.g., water, may be sampled by means of a permeable, bacteriatight membrane filter, through which a measured quantity of the liquid is passed. The membrane is either stained for direct microscopic examination, or is treated with an appropriate culture medium and incubated, after which colonies are counted. Inner surfaces of equipment may be sampled by the rinse method; flat surfaces may be sampled by pressing the agar surface of a contact plate against the surface; and various surfaces of equipment may be sampled by the swab technique, in which a sterilized swab, soaked in a dilution liquid, is squeezed out partially, then rubbed over an indicated surface area, and rinsed in the liquid, which then is plated or cultured on agar slants.

Number and size of samples

Total numbers and evenness of distribution of microorganisms in the food will influence the number of samples to be taken and the size of those samples. The more bacteriologically homogeneous the food is, the fewer and smaller need be the samples. Only 1 ml of a well-mixed lot of milk is sufficient for plating, but 11 g or more of flour or of frozen peas would be taken because of the probable uneven distribution of organisms in these foods. Some foods are fairly homogeneous within a given unit or lot but vary considerably in different lots. In this event, sampling of the different lots is indicated, after which either individual or pooled samples are examined.

Part of food sampled

Because some foods carry nearly all their load of microorganisms on outer surfaces, e.g., poultry, meats, fish, and fruits, unit areas of these surfaces usually are sampled, and numbers of microorganisms are expressed as per square centimeter or per square inch. When the organisms are distributed throughout the food, a specified mass or volume is taken, and the counts are per gram or per milliliter.

Release of microorganisms from sample

Microorganisms are freely suspended in most liquid foods and are readily washed from the surfaces of some solid foods (e.g., dried fruits). In many foods the organisms adhere strongly to surfaces or are held in the interior and are loosened only by thorough comminution of the food by some means, such as grinding with sterile sand in a sterilized mortar or breaking up the food in a mechanical blender. Thus, for example, the bacteria on the surface of cabbage or lettuce leaves or in hamburger or a wiener would be released only by comminution.

Handling of sample

When perishable food samples cannot be tested for microorganisms promptly, they must be cooled immediately if not cold and must be iced or refrigerated during transportation to the laboratory and until they are tested. It is recommended that milk samples, for example, be held at 32 to 40 F (0 to 4.4 C) and tested within a few hours after collection, although little change in bacterial population is likely to occur within 24 hr at this chilling temperature. Usually, freezing of food samples is avoided.

TYPES OF SAMPLES TESTED

The samples collected for testing will depend upon the kind of food and the type of laboratory making the tests. Most often only the final product is given a microbiological test. At food plants the amount of bacteriological testing varies from none to regular testing of raw product, ingredients, samples at various stages during handling and processing of foods and from equipment surfaces, processing, and packaging, to testing the final product in some plants.

KINDS OF MICROBIOLOGICAL TESTS MADE

Microbiological tests on foods may be quantitative, to estimate total numbers of organisms or numbers of those kinds of organisms that have special significance, or may be qualitative, to detect certain kinds of organisms or products produced by them. Tests for sterility may be conducted on containers or on finished packaged products; and tests may be made on the keeping quality of foods. The methods employed vary from those that may be considered official, as those for milk and dairy products and water, to recommended or suggested methods. A summary of commonly used microbiological tests made on various foods follows.

Summary of commonly used microbiological tests made on foods

Plating for Total Numbers of Bacteria: bottled beverages, concentrated milks, condensed milk (sweetened), dry milk, eggs (pulp, dried, frozen), flour, frozen desserts, fruits (dried, frozen, juices), milk and cream, nuts, oysters (and other shellfish), salad dressings, spices, stabilizers (ice cream), vegetables (dried, frozen), water

Plating for Yeasts and Molds: bottled beverages, butter, eggs (pulp, dried, frozen), fermented foods, salad dressings, stabilizers (ice cream), sugars and other sweetening agents

Reductase Tests: egg pulp, milk, oysters, prawns, poultry

Direct Microscopic Count of Bacteria: eggs (pulp, frozen), frozen vegetables, frozen desserts, milk and cream

Microscopic Count of Mold Fragments: butter, frozen fruits, canned tomatoes, and other fruits

Microscopic Count of Living and Dead Bacterial Cells: dry milk, pasteurized milk, sweetened condensed milk

Counts of Spores of Mesophilic Bacteria: flour, sugars and other sweetening agents

Tests for Spores of Thermophiles and Putrefactive Anaerobes: dry milk, spices, starch, sugars and other sweetening agents

Tests for Coliform Bacteria (and Counts): bottled beverages, certified milk, colors and flavors, nutmeats, dry whole milk, eggs (pulp, dried, frozen), frozen desserts, fruit juices, oysters and other shellfish, pasteurized milk, salad dressings, stabilizers (ice cream), water

Tests for Enterococci: water, fruit juices, pasteurized dairy products, liquid eggs, breaded and batter-dipped foods

Tests for Pathogens (see Table 27–1)

Brucella spp.: milk and other dairy products
Clostridium botulinum: canned foods (low-, medium-acid), cured meats, fish
Hemolytic bacteria: cheese, eggs (pulp, dried, frozen), frozen desserts, milk and cream
Salmonella spp.: cheese, eggs (pulp, dried, frozen), meats, foods suspected of causing *Salmonella* infections
Staphylococcus aureus: foods suspected of causing staphylococcus poisoning
Trichinae: pork, bear meat, rabbit meat, flesh of infested patient
Tubercle bacilli: milk and cream

Quantitative methods

Agar plate counts or modifications of them are used most commonly for the estimation of total numbers of microorganisms in foods, although most probable numbers are calculated from liquid media inoculated with decimal dilutions in some instances, and the direct microscopic count has been recommended for some foods. The agar plating medium employed usually is a fairly simple and readily reproducible one, such as plain nutrient agar or plate count agar, a medium that will yield comparable results in different laboratories, although not maximal counts. Agar plate counts have been made on a number of foods, some of which have been listed above. Also, plate counts often are used in line tests in food plants during handling and processing of such foods and on equipment coming into contact with foods.

Especially selective culture media, treatments, and environments are recommended for the counting or the detection of special kinds of microorganisms. Thermoduric bacteria of raw milk are counted by the plate method after pasteurization of the sample. When bacterial spores are to be counted, the sample or a dilution of it is pasteurized, the heat-treatment depending upon the heat resistance of the spores sought. In counting spores of mesophilic bacteria, as in flour or dry milk to be used for bread, a heat-treatment of 10 to 15 min at 80 C (176 F) is recommended. When more heat-resistant spores are of interest, such as spores of bacteria spoiling canned foods, a more rigorous heat-treatment is applied, such as 5 min at 100 C (212 F) for sugar or starch to be used in canning. Oxygen is supplied for aerobes and anaerobic conditions for anaerobes. High incubation temperatures (e.g., 55 C, or 131 F) are utilized for thermophiles and low temperatures (e.g., 5 to 10 C, or 41 to 50 F) for psychrophiles. Media for growing yeasts and molds usually are acidified to about pH 3.5 to 4.0 with lactic, tartaric, or other organic acid. Halophiles are cultivated in high-salt media and osmophiles in media high in sugar. Some selective media have chemicals added that inhibit all but the desired organisms. Selective media for coliform bacteria are good examples: violet-red bile agar, desoxycholate agar, brilliant-green lactose peptone bile, and formate ricinoleate lactose peptone broth. Such media are used in estimating numbers of coliform bacteria in milk, milk products, oysters, and water or in detecting their presence. The liquid media sometimes are employed to estimate most probable numbers. Other selective media (see Table 27–1) are bismuth sulfite agar for *Salmonella*, tellurite-glycine agar for staphylococci, and antibiotic-bearing media for yeasts.

The selection of an optimal incubation temperature for quantitative counts or special tests is most important. Many food microorganisms grow poorly or not at all at the commonly employed temperatures of 35 to 37 C (95 to 98.6 F), requiring lower temperatures. For this reason the incubation temperature for the standard plate count on milk has been lowered to 32 C (89.6 F). And the temperature to which agar is cooled before use, 45 C (113 F) or higher, may even harm some of the low-temperature organisms.

Some culture media contain indicator materials to aid in the identification of colonies of definite types of organisms, e.g., suspended casein to clear up as an indication of proteolysis, suspended fat to indicate lipolysis, a pH indicator to detect acid- or alkali-forming colonies, sulfite plus citrate of iron to bring out the black colonies of sulfide spoilage bacteria in sugar or starch, etc. Other selective culture media are blood agar for hemolytic bacteria, and, of course, experimental animals for various pathogens.

The direct microscopic method of counting can be used when numbers of microorganisms in a food are high. This method also is useful, in some instances, for indicating previous growth of organisms or the addition of ingredients that once were high in microbial content. Thus the presence of appreciable numbers of fragments of mold mycelia in butter, canned tomatoes, or frozen fruits indicates an inferior or perhaps rotted original raw product; many bacterial cells in pasteurized milk show that the original raw milk was poor in quality; and large numbers of stained bacterial cells in dry milk indicate excessive growth during the handling and processing of the liquid milk.

Methylene blue and resazurin reduction tests have been used on milk, egg pulp, etc., as indications of the bacteriological quality of those products.

Qualitative methods

Selective methods like those mentioned in the preceding section also may be used for the detection of specific kinds of microorganisms without an estimate of numbers. Some food organisms are so difficult to count that tests for them are more qualitative than quantitative. For example, spores of *Clostridium thermosaccharolyticum* (T.A. spoilage of canned foods) in sugar are difficult to count and are tested for by distribution of 20 ml of pasteurized dilution containing 4 g of sugar through six tubes of liver–liver broth, and counting the number of tubes showing gas after incubation at 55 C (131 F). In addition, finished products may be tested for the presence of or amounts of one or more microbial products to indicate either a good or an inferior product. Thus, a high volatile acid content of wine indicates a defective fermentation, and too little or too much titratable acid in sauerkraut shows that the fermentation has been abnormal or the time has been insufficient or excessive. Pathogenic microorganisms commonly are detected in foods rather than counted. Tests for "food-poisoning" organisms have been mentioned in Chapter 27.

Tests for sterility

Tests for sterility of packaged foods are made by inoculation of aliquots from the aseptically opened and sampled container of food into appropriate culture media, or, if the food is a good culture medium, by incubation of the unopened containers of food at recommended temperatures, after which the contents are examined macroscopically for changes, or, if the food is apparently unchanged it may be plated or examined microscopically. Tests for sterility of containers and caps or covers may be made by pouring culture medium into container or cap, followed by incubation, or by plating rinsings made with sterile water or buffer solution.

Keeping-quality tests

These tests usually involve the very simple procedure of incubation of the food under conditions simulating those under which the food is likely to be kept after leaving the food plant. Usually several different temperatures of incubation are employed, and samples are examined periodically for signs of spoilage. Canned vegetables and evaporated milk are held at 55 C (131 F) for growth of thermophiles and at 35 to 37 C (95 to 98.6 F) for mesophiles; waxed vegetables would be stored at room and refrigerator temperatures. The British formerly employed keeping-quality tests on market milk, storing it overnight at "atmospheric shade temperature" before making a methylene blue reduction test.

AGENCIES RECOMMENDING METHODS

The American Public Health Association (APHA) and its committees are responsible for many of the recommendations of microbiological methods for the examination of foods. Most of these methods have practically an official standing, e.g., those in "Standard Methods for the Examination of Dairy Products" (in collaboration with the Association of Official Agricultural Chemists) and "Standard Methods for the Examination of Water and Wastewater." The manual on dairy products deals with methods for the examination of milk, products made from milk, and ingredients that go into these products, such as flavoring and coloring materials, sweetening agents, eggs and egg products, and stabilizers. Included also are methods for the examination of equipment and packaging materials. The Association also has published "Recommended Methods for the Microbiological Examination of Foods" with directions for the examination of the chief kinds of raw and processed foods except dairy products and seafoods. Another publication recommends methods for the bacteriological examination of sea water and shellfish.

Governmental agencies also may be concerned in the recommendation of methods. The United States Public Health Service recommends microbiological methods in its "Ordinance and Code Regulating Eating and Drinking Establishments" (Bull. 280) and its "Milk Ordinance and Code" (Bull. 220, revised). The Armed Forces have their own recommended methods for the microbiological examination of foods, and governmental control agencies may recommend techniques for the bacteriological examination of foods for the purpose of grading such foods or testing for acceptability. Thus the Agricultural Marketing Service recommends a method for the direct microscopic examination of dry milk, and the Food and Drug Administration a method for counting mold filaments in foods.

The Association of Official Agricultural Chemists describes some microbiological methods in various editions of its "Official (and Tentative) Methods of Analysis," including methods for the examination of sugars and of canned foods for spoilage bacteria, the Howard method for examining fruits (fresh, frozen, and canned) for mold fragments, examination of eggs and egg products, etc.

Organizations representing the various food industries sometimes present methods and may recommend standards. Thus the American Dry Milk Institute, Inc., publishes recommended methods for the microbiological examination of dry milks, as well as for grading these milks. Likewise, the American Bottlers of Carbonated Beverages recommends test methods and standards for granulated and liquid sugar to be used in its products. The laboratories of the National Canners Association have worked out many of the methods employed in the testing of canned foods and ingredients used in them. The American Meat Institute laboratories worked similarly on meats and meat products. These are only a few examples of methods from food-industry laboratories.

Some of the publications that include microbiological methods for the examination of foods are listed at the end of this chapter and should be consulted for details.

Bacteriological examination of eating and drinking utensils

Recommended methods for the bacteriological examination of eating and drinking utensils are described in the "Ordinance and Code Regulating Eating and Drinking Establishments" (U.S. Public Health Service Bull. 280). The swab technique is employed for sampling a significant surface on each utensil, and at least four utensils of each type are tested. Significant surfaces are (1) for cups and glasses, the upper ½ in. of the inner and outer rims; (2) for spoons, the entire inner and outer surfaces of the bowls; (3) for forks, the entire inner and outer surfaces of the tines; and (4) for plates, bowls, and saucers, any 4 sq in. that would come in contact with food when used. The swabs are rinsed in buffered distilled water to which is added sodium thiosulfate solution if utensils are likely to contain residual chlorine. The swab rinsings are plated in standard tryptone glucose extract agar and incubated for 48 hr at 37 C (98.6 F). The average plate count per significant utensil surface examined should not exceed 100 bacterial colonies. Higher counts are presumptive evidence of inadequate cleansing or bactericidal treatment or of recontamination by handling or during storage.

REFERENCES

American Bottlers of Carbonated Beverages. 1952–1953. Suggested bacteriological test methods for granulated and liquid sugar (and tentative standards). American Bottlers of Carbonated Beverages, Washington, D.C.

American Dry Milk Institute, Inc. 1955. The grading of dry whole milk and sanitary and quality standards. Bull. 913 (rev.)

American Dry Milk Institute, Inc. 1954. The grading of nonfat dry milk solids and sanitary and quality standards. Bull. 911 (rev.)

American Public Health Association. 1958. Recommended methods for the microbiological examination of foods. American Public Health Association, New York.

American Public Health Association. 1960. Standard methods for the examination of dairy products. 11th ed. American Public Health Association, New York.

American Public Health Association. 1960. Standard methods for the examination of water and wastewater. 11th ed. American Public Health Association, New York.

American Public Health Association. 1962. Recommended procedures for the bacteriological examination of sea water and shellfish. 3rd ed. American Public Health Association, New York.

ANGELOTTI, R., M. J. FOTER, K. A. BUSCH, and K. H. LEWIS. 1958. A comparative evaluation of methods for determining the bacterial contamination of surfaces. Food Res. 23:175–185.

Association of Official Agricultural Chemists. Official (and tentative) methods of analysis. Various editions. Association of Official Agricultural Chemists, Washington, D.C.

AYRES, J. C. 1949. A procedure for the quantitative estimation of *Salmonella* in dried egg products. Food Technol. 3:172–176.

AYRES, J. C., A. A. KRAFT, H. E. SNYDER, and H. W. WALKER (*Eds.*) 1962. Chemical and biological hazards in food, pp. 157–201, 224–247, 279–302. The Iowa State University Press, Ames, Iowa.

BORGSTROM, G. 1955. Microbiological problems of frozen food products. Advances Food Res. 6:163–230.

CAPPS, BERYL F., MARY K. WOLLAM, and N. L. HOBBS. 1949. A method for the bacteriological examination of edible fat preparations. Food Technol. 3:260–263.

EHRLICH, R. 1960. Applications of membrane filters. *In* W. W. Umbreit (*Ed.*) Advances Appl. Microbiol. 2:95–112.

FOSTER, E. M., and W. C. FRAZIER. 1961. Laboratory manual for dairy microbiology. 3rd ed. Burgess Publishing Company, Minneapolis.

FRAZIER, W. C., and E. M. FOSTER. 1959. Laboratory manual for food microbiology. 3rd ed. Burgess Publishing Company, Minneapolis.

GREENE, V. W., D. VESLEY, and K. M. KEENAN. 1962. New method for microbiological sampling of surfaces. J. Bacteriol. 84:188–189.

LEWIS, K. H., and R. ANGELOTTI (*Eds.*) 1963. Examination of foods for enteropathogenic and indicator bacteria; review of methodology and manual of selected procedures. Milk, Food Research Section. Milk, Food Branch, Div. Environmental Eng., Food Protection, Public Health Serv., U.S. Dep. Health, Educ., Welfare.

U.S. Department of Health, Education, and Welfare, Public Health Service. 1958. Frozen desserts ordinance and code. [Reprinted March, 1958]

U.S. Public Health Service. 1943. Ordinance and code regulating eating and drinking establishments. Bull. 280.

U.S. Public Health Service. 1964. Training course manual—food microbiology. Robert A. Taft Sanitary Engineering Center, Cincinnati, Ohio.

WALTER, W. G. 1955. Symposium on methods for determining bacterial contamination on surfaces. Bacteriol. Revs. 19:284–287.

CHAPTER THIRTY

FOOD CONTROL

The objectives of the control, regulation, and inspection of food are primarily to give assurance that foods, as received by the consumer, will be pure, healthful, and of the quality claimed.

ENFORCEMENT AND CONTROL AGENCIES

Enforcement and control agencies range from Federal to private, as the following brief outline will illustrate:

1. *Federal agencies.* The authority of Federal enforcement agencies is confined to foods shipped interstate or foods produced in or shipped into territories. Agencies include (*a*) The Food and Drug Administration of the Department of Health, Education, and Welfare, which enforces The Federal Food, Drug, and Cosmetic Act, The Tea Act, and The Import Milk Act; (*b*) The Meat Inspection Branch, Agricultural Research Service, United States Department of Agriculture, which enforces The Meat Inspection Act, The Imported Meat Act, and The Horse Meat Act; (*c*) The Agricultural Marketing Service of the United States Department of Agriculture, which is concerned primarily with the inspection and grading of fruits, vegetables, poultry, and dairy products; (*d*) The United States Public Health Service, which is not an enforcement agency, but which makes recommendations, gives publicity, and otherwise aids other agencies; (*e*) The Armed Services, which may set their own standards and recommend special laboratory procedures.

    The Federal Register, published almost daily, contains new pronouncements of agencies of the Federal Government pertaining to food standards and inspection. The Code of Federal Regulations summarizes and groups such information from the Federal Register. Periodically the Agricultural Marketing Service publishes separates which bring up to date regulations and recommendations regarding the inspection, standards, and grade of various foods. Some of these are listed at the end of the chapter.

2. *State agencies.* State food laws usually are enforced through the state Department of Public Health, Agriculture, or Sanitary Engineering.

3. *Municipal agencies.* City or town food laws usually are enforced through the local board of health.

4. *Commercial agencies.* Trade associations or institutes may make recommendations or even attempt regulation within their own industries. Thus the National Canners Association has set bacteriological standards for sugar and starch for canning; the American Dry Milk Institute has established bacterial standards for dry milk; the American Bottlers of Carbonated Beverages has bacterial standards for sweetening agents used in soft drinks, etc.

5. *Professional societies.* The American Public Health Association has published many of the recommended methods and official methods for the bacteriological examination of foods; the International Association of Milk and Food Sanitarians has published recommended methods for the investigation of food-poisoning outbreaks, etc.

6. *Private agencies.* Several private agencies approve and list tested foods, for example, the Good Housekeeping Institute.

THE FEDERAL FOOD, DRUG, AND COSMETIC ACT

The Federal Food, Drug, and Cosmetic Act of 1938, as amended to date, will be discussed primarily from the bacteriological viewpoint. The act applies to interstate shipment of foods and foods produced in or shipped into territories and the District of Columbia. The act includes (1) authorization for standards of quality for foods, except fresh or dried fruits (with some exceptions), vegetables, and butter; and (2) classification of violations into (*a*) *adulteration* and (*b*) *misbranding*.

Adulteration

The following descriptions of adulteration may have bacteriological implications:

(1) If it [the food] bears or contains any poisonous or deleterious substance which may render it injurious to health . . . [unless naturally there at less than a harmful level]; (2) if it bears or contains any added poisonous or added deleterious substance which is unsafe . . . [i.e., above tolerances set]; or (3) if it consists in whole or in part of any filthy, putrid, or decomposed substance, or if it is otherwise unfit for food; or (4) if it has been prepared, packed, or held under insanitary conditions whereby it may have become contaminated with filth, or whereby it may have been rendered injurious to health; or (5) if it is, in whole or in part, the product of a diseased animal or of an animal which has died otherwise than by slaughter; or (6) if its container is composed, in whole or in part, of any poisonous or deleterious substance which may render the contents injurious to health.

Among the things "injurious to health" might be microorganisms causing infections or poisoning, or their products. The microorganisms implied in Section (3) would be those causing the decomposition of foods. Section (4) would apply to the cleaning and sanitizing of equipment that comes in contact with foods, and Section (6) to the bacteriological condition of containers for packaging foods. It should be emphasized that the presence of disease organisms or their products, i.e., the hazard to health, the presence of spoilage organisms or of the results of their action, and the evidence of insanitary practices, do not have to be *proved*, but merely shown to be possible, to lead to the condemnation of a food.

Misbranding

A food is misbranded if its labeling is false or misleading in any particular. A food is deemed to be misbranded if it contains a chemical preservative, unless this content is stated on the label. The act states that the term **chemical preservative** "means any chemical which, when added to food, tends to prevent or retard deterioration thereof; but does not include common salt, sugars, vinegars, spices, or oils extracted from spices, or substances added by . . . wood smoke."

Emergency permit

The act also states that: "Whenever . . . the distribution in interstate commerce of any class of food may, by reason of contamination with microorganisms during manufacture, processing, or packing thereof in any locality, be injurious to health, and . . . such injurious nature cannot be adequately determined after such articles have entered interstate commerce," then an emergency permit may be issued to specify conditions that will protect public health; and all of the food must be produced and handled under these conditions.

FOOD ADDITIVES AMENDMENT OF 1958

This amendment has been discussed in Chapter 9. It defines food additives, covers food additives not generally recognized as safe, requires that complete information be submitted and approval obtained from the Food and Drug Administration prior to commercial use of a new food additive, and provides for tolerances and for safety in regard to the health of man or animal. A special amendment states that no additive shall be deemed safe if when ingested it induces cancer in man or animal.

THE MEAT INSPECTION ACT OF 1906

The Meat Inspection Act and regulations governing meat inspection are enforced by the Meat Inspection Branch, Agricultural Research Service, United States Department of Agriculture. It is required that all meat and meat products that move in interstate commerce, are produced in the territories or the District of Columbia, or are imported into this country be packed under license and only after inspection as specified by the act and regulations. Government inspectors make ante- and post-mortem examinations of animals, inspect the sanitation of the plant and its facilities, supervise slaughter of the animals and dressing of the meat, and check on methods of handling, processing, and labeling of meats and other animal parts and products made from them. From a bacteriological standpoint, of chief interest are (1) ante- and post-mortem inspection of animals for disease; (2) inspection of sanitation throughout the plant; (3) description of procedures of heating, refrigeration, freezing, and curing that guarantee the destruction of *Trichinella spiralis* in pork; (4) authorization of the use of lactic acid starter bacteria in the preparation of certain types of cured sausage, such as Thuringer or salami; and (5) testing of canned meat or meat products for keeping quality.

THE POULTRY PRODUCTS INSPECTION ACT OF 1957

Compulsory inspection of processed poultry (for interstate shipment) by the Poultry Inspection Service, Poultry Division, Agricultural Marketing Service, United States Department of Agriculture, involves ante- and post-mortem inspection of fowls and inspection of plant sanitation. The act went into effect on a voluntary basis on May 1, 1958, and was mandatory as of January 1, 1959.

INSPECTION AND GRADING BY AGRICULTURAL MARKETING SERVICE

Application may be made by any interested party or his authorized agent for the investigation and certification by a licensed inspector from the Agricultural Marketing Service of the class, quality, and condition of any agricultural commodity or food product, whether raw, dried, canned, or otherwise processed, and any product containing an agricultural commodity or derivative thereof when offered for interstate shipment. The applicant pays the cost of the services of the government inspector. Bacteriological analyses, such as plate count, direct microscopic count, coliform tests, and yeast and mold counts are made for appropriate fees. When dairy plants are operating under government (USDA) inspection, bacteriological tests are made on the water for potability and on raw milk and cream for grading purposes. Recommendations are made regarding methods for cooling and handling the milk or cream and for adequate cleansing and sanitizing of equipment and utensils.

STATE AND MUNICIPAL FOOD LAWS

States and cities have food laws that usually are modeled on Federal laws or recommendations. Ordinarily, more attention is given to milk and milk products than to other foods. Municipal ordinances may be stricter than state laws but never more lenient. The regulations usually provide for inspection of food-handling operations, for the examination of foods, and sometimes for examination of food handlers. Direct laws may be enforced or a license system may be employed in which revocation of the license may be a penalty for nonconformance. Many of the states and municipalities have laboratories in which to do bacteriological work (Chapter 29) in connection with the enforcement of the regulations.

CONTROL BY FOOD PLANTS

Inspection and sanitation in the plant by plant employees have been discussed in Chapter 28, and some of the laboratory methods employed have been outlined in Chapter 29.

REFERENCES

Federal Food, Drug, and Cosmetic Act and general regulations for its enforcement, April, 1955, U.S. Dep. Health, Education, and Welfare, FDA. SRA, Food, Drug, and Cosmetic No. 1, Rev. with Addenda.

Food Additives Amendment of 1958. Public Law 85-929, 85th Congress, H.R. 13254, Sept. 6, 1958.

GUNDERSON, F. L., HELEN W. GUNDERSON, and E. R. FERGUSON, JR. 1963. Food standards and definitions in the United States. Academic Press Inc., New York.

LACHMANN, A. 1963. International food regulations. Food Eng. 35(8):59–63.

MAHONEY, J. F. 1958. Food additives amendment enacted. Food Technol. 12:637–640.

Regulations of the United States Department of Agriculture, Agricultural Marketing Service*

Regulations of the Secretary of Agriculture for the enforcement of the Perishable Agricultural Commodities Act, 1930, as amended to Jan., 1965. SRA—AMS 121.

Minimum specifications for approved plants manufacturing, processing, and packaging dairy products. (7 CFR, 1946 Supp., 55.101; 20 FR 8444, Dec., 1955.)

Regulations governing grading and inspection of manufactured or processed dairy products. (7, chap. I, pt. 58 CFR, amended Aug. 1, 1959, Jan. 1, 1963.)

Regulations governing the grading, inspection, and supervision of packaging of butter, cheese, and other manufactured or processed dairy products. (SRA—AMS 169, July 1, 1959.)

United States standards for instant nonfat dry milk (FR, Mar. 26, 1963).

United States standards for grades of nonfat dry milk (roller process). (Subpart M, FR, May 23, 1958.)

United States standards for grades of nonfat dry milk (spray process). (Subpart L, FR, May 23, 1958.)

Amendment to United States standards for grades of nonfat dry milk (spray and roller processes). (FR, Dec. 30, 1964.)

Regulations governing the grading of shell eggs and the United States standards, grades, and weight classes for shell eggs. USDA, AMS, Poultry Division. (7 CFR, pt. 56, Jan., 1965.)

Regulations governing the grading and inspection of egg products. USDA, AMS, Poultry Division. (7 CFR, pt. 55, 1964; amendments, May 1, 1965.)

Regulations governing the inspection of poultry and poultry products. USDA, AMS, Poultry Division. (7 CFR, pt. 81, Dec., 1964.)

Regulations governing the inspection and certification of fresh fruits, vegetables, and other products. USDA, AMS, Fruit and Vegetable Division. (SRA—AMS—93, revised; amended Nov. 1, 1962.)

Regulations governing inspection and certification of processed fruits and vegetables, and other products. USDA, AMS, Fruit and Vegetable Division. (SRA—AMS—155, revised; amended Nov. 1, 1962.)

Regulations governing the meat inspection of the United States Department of Agriculture. (USDA, Agricultural Research Service, Meat Inspection Branch, 1960.)

* Key to abbreviations:
AMS = Agricultural Marketing Service
CFR = Code of Federal Regulations
FDA = Food and Drug Administration
FR = Federal Register
USDA = United States Department of Agriculture

APPENDIX

MICROBIOLOGICAL STANDARDS

A. **Standards for Starch and Sugar** (National Canners Association).
   1. *Total thermophilic spore count:* Of the five samples from a lot of sugar or starch none shall contain more than 150 spores per 10 g, and the average for all samples shall not exceed 125 spores per 10 g.
   2. *Flat sour spores:* Of the five samples none shall contain more than 75 spores per 10 g, and the average for all samples shall not exceed 50 spores per 10 g.
   3. *Thermophilic anaerobe spores:* Not more than three (60 percent) of the five samples shall contain these spores, and in any one sample not more than four (65+ percent) of the six tubes shall be positive.
   4. *Sulfide spoilage spores:* Not more than two (40 percent) of the five samples shall contain these spores, and in any one sample there shall be no more than five colonies per 10 g (equivalent to two colonies in the six tubes).
B. **Standard for "Bottlers" Granulated Sugar, Effective July 1, 1953** (American Bottlers of Carbonated Beverages).
   1. *Mesophilic bacteria:* Not more than 200 per 10 g.
   2. *Yeasts:* Not more than 10 per 10 g.
   3. *Molds:* Not more than 10 per 10 g.
C. **Standard for "Bottlers" Liquid Sugar, Effective in 1959** (American Bottlers of Carbonated Beverages). All figures based on dry-sugar equivalent (D.S.E.).
   1. *Mesophilic bacteria:* (*a*) Last 20 samples average 100 organisms or less per 10 g D.S.E.; (*b*) 95 percent of last 20 counts show 200 or less per 10 g; (*c*) 1 of 20 samples may run over 200; other counts as in (*a*) or (*b*).
   2. *Yeasts:* (*a*) Last 20 samples average 10 organisms or less per 10 g D.S.E.; (*b*) 95 percent of last 20 counts show 18 or less per 10 g; (*c*) 1 of 20 samples may run over 18; other counts as in (*a*) and (*b*).
   3. *Molds:* Standards like those for yeasts.
D. **Standards for Dairy Products.**
   1. From 1965 recommendations of the U.S. Public Health Service
      a. *Grade A raw milk for pasteurization:* Not to exceed 100,000 bacteria per milliliter prior to commingling with other producer milk; and not exceeding 300,000 per milliliter as commingled milk prior to pasteurization.
      b. *Grade A pasteurized milk and milk products* (except cultured products): Not over 20,000 bacteria per milliliter, and not over 10 coliforms per milliliter.
      c. *Grade A pasteurized cultured products:* Not over 10 coliforms per milliliter.
      NOTE: Enforcement procedures for *a*, *b*, and *c* require three-out-of-five compliance by samples. Whenever two of four successive samples do not meet the standard, a fifth sample is tested; and if this exceeds any standard, the permit from the health authority may be suspended. It may

be reinstated after compliance by four successive samples has been demonstrated.

2. *Certified milk* (American Association of Medical Milk Commissions, Inc.)
   a. Certified milk (raw): Bacterial plate count not exceeding 10,000 colonies per milliliter; coliform colony count not exceeding 10 per milliliter.
   b. Certified milk (pasteurized): Bacterial plate count not exceeding 10,000 colonies per milliliter before pasteurization and 500 per milliliter in route samples. Milk not exceeding 10 coliforms per milliliter before pasteurization and 1 coliform per milliliter in route samples.

3. *Milk for manufacturing and processing* (U.S. Dep. Agr., 1955)
   a. Class 1: Direct microscopic clump count (DMC) not over 200,000 per milliliter.
   b. Class 2: DMC not over 3 million per milliliter.
   c. Milk for Grade A dry milk products: must comply with requirements for Grade A raw milk for pasteurization (see above).

4. *Dry milk*
   a. Grade A dry milk products: at no time a standard plate count over 30,000 per gram, or coliform count over 90 per gram (U.S. Public Health Service).
   b. Standards of Agricultural Marketing Service (U.S. Dep. Agr.):
      (1) Instant nonfat: U.S. Extra Grade, a standard plate count not over 35,000 per gram, and coliform count not over 90 per gram.
      (2) Nonfat (roller or spray): U.S. Extra Grade, a standard plate count not over 50,000 per gram; U.S. Standard Grade, not over 100,000 per gram.
      (3) Nonfat (roller or spray): Direct microscopic clump count not over 200 million per gram; and must meet the requirements of U.S. Standard Grade. U.S. Extra Grade, such as used for school lunches, has an upper limit of 75 million per gram.
   c. Standards of American Dry Milk Institute, Inc.

Kind of dry milk	Grade	Process	Maximal SPC*, nos/g
Nonfat	Extra	Spray or roller†	50,000
Nonfat	Standard	Spray or roller	100,000
Nonfat, instant	Extra		35,000
Whole milk	Premium	Gas-packed, spray	30,000
Whole milk	Extra	Bulk spray or roller	50,000
Whole milk	Standard	Bulk spray or roller	100,000
Buttermilk	Extra	Spray or roller	50,000
Buttermilk	Standard	Spray or roller	200,000

* Standard plate count.
† Atmospheric roller throughout table.

5. *Frozen desserts*
   States and cities that have bacterial standards usually specify a maximal count of 50,000 to 100,000 per milliliter or gram. The U.S. Public

Health Ordinance and Code sets the limit at 50,000 and recommends bacteriological standards for cream and milk used as ingredients. Few localities have coliform standards.

E. **Standard for Tomato Juice and Tomato Products—Mold-count Tolerances** (Food and Drug Administration).

The percentage of positive fields tolerated is 20 percent for tomato juice and 40 percent for other comminuted tomato products, such as catchup, purée, paste, etc. A microscopic field is considered positive when aggregate length of not more than three mold filaments present exceeds one-sixth of the diameter of the field (Howard mold count method). This method also has been applied to raw and frozen fruits of various kinds, especially to berries.

F. **Standards for Precooked and Partially Cooked Frozen Foods.**

Massachusetts requires that these foods shall have not more than 50,000 colonies per gram as a standard plate count and not over 10 colonies of coliform bacteria per gram as a coliform plate count. They shall contain no coagulase-positive staphylococci and no organisms of the salmonella-shigella-typhoid group.

G. **Standards for Crab Meat.**

New York City standards for crab meat are: not more than 100 per gram of hemolytic *Staphylococcus aureus* or coliform bacteria, or not more than 1,000 per gram of enterococci, or not more than 100,000 per gram by the total bacteria plate count.

H. **Standards for Custard-filled Items.**

New York City standards for custard-filled items are a standard plate count of not over 100,000 per gram and a coliform count of not over 100 per gram.

SOME SUGGESTED MICROBIOLOGICAL STANDARDS

A. **Standard for Sugar Used In Meat Packing** (Owen).

When five samples of sugar are examined, not more than two samples may show more than two (out of five) tubes positive for gas, odor, or acidity at any of the three temperatures of incubation, and no samples may show more than four (out of five) tubes positive for gas, odor, or acidity at any one of the three temperatures. Tubes of nutrient liver broth are inoculated with 1 ml of a solution of 10 g of sugar per 100 g of sterile water, and incubation is for 3 days at 98 and 131 F, and 5 days at 80 F.

B. **Standards for Drinking Water.**

"Standard Methods for the Examination of Water and Wastewater" does not stipulate any bacteriological standards for water. The U.S. Department of Health, Education, and Welfare in its Public Health Service Drinking Water Standards of 1962 states that the presence of organisms of the coliform group . . . shall not exceed the following limits:

1. When 10-ml standard portions are examined, not more than 10 percent in any month shall show the presence of the coliform group. The presence of the coliform group in three or more 10-ml portions of a standard sample shall not be allowable if this occurs:

a. In two consecutive samples
b. In more than one sample per month when less than 20 are examined per month
c. In more than 5 percent of the samples when 20 or more are examined per month

When organisms of the coliform group occur in three or more of the 10-ml portions of a single standard sample, daily samples from the same sampling point shall be collected promptly and examined until the results obtained from at least two consecutive samples show the water to be of satisfactory quality.

2. When 100-ml standard portions are examined, not more than 60 percent in any month shall show the presence of the coliform group. The presence of the coliform group in all five of the 100-ml portions of a standard sample shall not be allowable if this occurs:
a. In two consecutive samples
b. In more than one sample per month when less than five are examined per month
c. In more than 20 percent of the samples when five or more are examined per month

When organisms of the coliform group occur in all five of the 100-ml portions of a single standard sample, daily samples from the same sampling point shall be collected promptly and examined until the results obtained from at least two consecutive samples show the water to be of satisfactory quality.

3. When the membrane filter technique is used, the arithmetic mean coliform density of all standard samples examined per month shall not exceed one per 100 ml. Coliform colonies per standard sample shall not exceed three per 50 ml, four per 100 ml, seven per 200 ml or thirteen per 500 ml in:
a. Two consecutive samples
b. More than one standard sample when less than 20 are examined per month
c. More than 5 percent of the standard samples when 20 or more are examined per month

When coliform colonies in a single standard sample exceed the above values, daily samples from the same sampling point shall be collected promptly and examined until the results obtained from at least two consecutive samples show the water to be of satisfactory quality.

C. **Standards for Eggs.**

1. *Liquid*
Not more than 5 million organisms per gram when plates are incubated at 22 C (71.6 F).

2. *Dried*
For purchase of whole, stabilized egg solids: Before drying the liquid egg must have been heated at 140 to 142 F (60 to 61.1 C) for from 3 to 3.5 min. The standard plate count of the dry egg shall not exceed an average of 50,000 per gram for a lot. No sample unit shall contain more than 75,000 per gram. The coliform plate count shall not exceed an average of 50 per gram for a lot, or 100 per gram for a sample unit. The yeast and mold count shall not exceed an average of 20 per gram

for a lot, or 50 per gram for a sample unit (Quartermaster Corps, U.S. Army).

3. *Frozen* (whole)

For purchase of whole, frozen egg: The liquid egg must have been flash-heated to 140 to 142 F (60 to 61.1 C) and held 3 to 3.5 min and cooled rapidly to 45 F (7.2 C). The standard plate count on the frozen egg shall not exceed an average of 10,000 per gram, and no lot shall exceed an average of 15,000 per gram. The yeast and mold count shall not exceed an average of 50 per gram, and in no lot an average of 75 per gram. The coliform plate count shall not exceed an average of 50 per gram, or 100 per gram in any lot. The direct microscopic count shall not exceed an average of 500,000 per gram or 1 million per gram in any lot (Department of Defense).

D. Standards for Fruit Preserves and Jams.

It has been suggested that the percentage of fields positive for mold filaments should not exceed 4 to 40 percent, depending on the fruit concerned.

E. Standard for Apple Juice.

Over 2 million yeasts or 200,000 molds per milliliter indicates that original apples were unfit for use.

F. Standards for Frozen Vegetables.

Vegetables as they enter the freezer should contain not over 50,000 bacteria per gram for peas, not over 60,000 for corn, and not over 100,000 for string beans. Most workers agree that frozen vegetables should contain not more than 100,000 bacteria per gram, and fruits fewer than vegetables.

G. Standard for Precooked Frozen Foods (Meals).

They should contain not over 100,000 bacteria per gram by standard plate count and less than 10 coliform bacteria per gram by coliform plate count. None of the meal components shall contain pathogenic organisms (Department of Defense).

H. Standards for Raw Meats.

Many workers have suggested standards, mostly ranging from 250,000 to 10 million bacteria per gram, for hamburger. Standards for other meats might be 10,000 to 100,000 per square centimeter of surface, or less than 2 million to 5 million per gram. For poultry, suggestions are 5,000 per square centimeter or 100,000 per gram.

I. Standards for Precooked Meats.

For precooked sausage and canned ham (pasteurized) there should be not over 10,000 viable aerobes per gram. Various workers have stated that coliforms should be absent in 0.1 g, or that *Escherichia coli* and *Clostridium perfringens* should be absent in 1 g or totally absent, that pathogens should be absent, and that bacterial spores should not be more than 10 per gram.

J. Standards for Fish and Shellfish.

Fish should contain not over 100,000 aerobes per gram, and breaded shrimp not over 500,000 per gram. Crab meat should contain not over 100,000 bacteria per gram by plate count and not over 100 coliforms. The U.S. Public Health Service has recommended that oysters have a most probable number (MPN) of coliform bacteria not over 16,000 per 100 ml, and/or a standard plate count of not more than 50,000 per milliliter. Under some conditions, standards less strict than these might be applied. Tentative standards suggested in 1960 specified a MPN of coliforms of not more than 78 per milliliter and a plate count of not more than 100,000 per milliliter, with exceptions for occasional samples.

REFERENCES

American Association of Medical Milk Commissions, Inc. 1965. Methods and standards for the production of Certified Milk. American Association of Medical Milk Commissions, Inc., New York.

American Bottlers of Carbonated Beverages. 1962. Standards and test procedures for "bottlers" granulated and liquid sugar. American Bottlers of Carbonated Beverages, Washington, D.C.

American Dry Milk Institute, Inc. 1965. Standards for grades of dry milks including methods of analysis. Bull. 916 (rev). Chicago.

DAHLBERG, A. C., and H. S. ADAMS. 1950. Sanitary milk and ice cream legislation in the United States. Natl. Res. Council (U.S.) Bull. 121.

Department of Defense, Military Specification: egg, whole, frozen (Table Grade), MIL-E-35001 (1954); margarine, canned, MIL-M-10958B (1956); meals, precooked, frozen, MIL-M-13966 (1955).

ELLIOTT, R. P., and H. D. MICHENER. 1961. Microbiological standards and handling codes for chilled and frozen foods; a review. Appl. Microbiol. 9:452–468.

Massachusetts Department of Public Health. 1959. Rules and regulations relative to the storage and distribution of frozen foods.

National Canners Association, 1949. Bacterial standards for sugar (and starch). National Canners Association, Washington, D.C. [mimeographed sheets]

OWEN, W. L. 1956. Standards for sugar used in meat packing. Sugar 51(7):28–30.

Quartermaster Corps (U.S. Army). 1957. Purchase description for egg solids, whole, stabilized. IP/DES S-10-7.

TROY, V. S. 1956. Mold counting of tomato products. Continental Can Co., Research Department Bulletin.

United States Department of Health, Education, and Welfare. 1962. Public Health Service drinking water standards (revised, 1962). Public Health Serv. Pub. 956.

U.S. Department of Health, Education, and Welfare, Public Health Service. 1958. Frozen desserts ordinance and code.

U.S. Department of Health, Education, and Welfare, Public Health Service. 1959. Grade A dry milk products (ordinance and code).

U.S. Department of Health, Education, and Welfare. 1965. Grade "A" pasteurized milk ordinance. U.S. Government Printing Office, Washington, D.C.

United States standards for instant nonfat dry milk (FR, Mar. 26, 1963); for nonfat dry milk—roller process (Subpart M, FR, May 23, 1958); for nonfat dry milk—spray process (Subpart L, FR, May 23, 1958); and amendment (FR, Dec. 30, 1964). (See references at end of Chapter 30.)

INDEX

Absidia, 13
Accessory food substances, 169
Acetic acid, 137
  (*See also* Vinegar)
Acetic acid bacteria (acetics), 58, 375, 388, 399
Acetification, 388, 395
Acetobacter, 44, 58, 184, 388, 399
  aceti, 395, 400
  acetigenum, 403
  capsulatum, 388
  oxydans, 395
  rancens, 400
  suboxydans, 375, 422
  turbidans, 388
  viscosum, 388
  xylinum, 44, 400, 406
Acetoin, 52
Acetomonas, 44, 58
Achromobacter, 45, 232, 271, 293, 353
  anaerobium, 388
  perolens, 307
Achromobacteraceae, 44
Acid-proteolytic bacteria, 59, 334
Acidity, of cultured buttermilk, 328
  of olives, 218
  of pickles, 214, 217
  of sauerkraut, 211
  of vinegar, 399, 405
  of wines, 392, 394
    total, 394
    volatile, 392
Acidophilus milk, 328
Acids (*see* specific acids)
Actinomycetales, 56
Actinomycetes, 269, 338
  (*See also Streptomyces*)
Additives, 132–143, 208, 224
Adulteration, 483, 503
Aerobacilli, 56, 333, 363, 365
Aerobacter, 45–47, 343
  aerogenes, 45–47, 198, 336
  cloacae, 190, 211, 308, 336
  oxytocum, 338

Aerobes, 41
Aeromonas, 306
Aflatoxin, 450
Aging, of beer, 386
  of meats, 255, 258, 265
  of wines, 392
Agitation of canned foods, 95
Agricultural Marketing Service inspection and grading, 505
Air, 68–71
  bacteriological analysis of, 70
  contamination from, 68–71
  filtration of, 70
  microorganisms in, 68–70
  sampling of, 70
  treatment of, 70
Alcaligenes, 45, 306
  faecalis, 337
  metalcaligenes, 45
  viscolactis (*viscosus*), 45, 61, 325
Alcohol (*see* Ethanol)
Ale, 390
  (*See also* Beer)
Algae, 353, 407, 417
Alkali dipping, 124
Alkalies and alkaline salts, 135, 484
Alternaria, 22, 227, 235, 237, 450
  brassicae, 22
  citri, 22, 236
  radicina, 230
  tenuis, 22, 238
Alternaria rot, 227
Amertume (wines), 395
Amino acids, decomposition of, 177
  manufacture of, 375, 422
Amylases, 59, 381, 425–427
Amyloglucosidase, 426
Amylopectin, 425
Amylose, 425
Anaerobes, 41
Anaerobic conditions in preservation, of canned foods, 80
  of molasses and sirups, 197
Ang-khak, 411

515

Anguillula aceti, 406
Animals, contamination from, 64
  humane slaughter of, 252
Anthracnose, 227
Antibiotics, 139
  antifungal, 143
  in ice, 140, 284
  plus heat, 105, 139
  as preservatives, 139
    in meat, 265
    in poultry, 314, 316
    in seafood, 288
Appert, Nicolas, 101
Applejack (brandy), 398
Apple juice (cider), pasteurization of,
  100
  standards for, microbiological, 513
Apple wine or hard cider, 396
Apples, gas storage of, 221
  spoilage of, 240
Apricots, spoilage of, 239
Arizona group, 451
Arthrobacter, 55
Arthrospores, 4
Artichokes, spoilage of, 232
Ascocarp, 5
Ascomycetes, 23
Ascospores, 5
Asepsis in preservation, 78
  (*See also* specific foods)
Ashbya gossypii, 421
Asparagus, spoilage of, 228
Aspergillus, 14, 227, 342
  characteristics of, 14
  clavatus, 424
  fischeri, 421
  flavus-oryzae group, 14
  glaucus group, 14
  niger, 14, 185, 227, 238, 424, 430
  oryzae, 378, 390, 409, 410, 426, 429
  repens, 14
  tamarii, 410
  wentii, 424
Associative growth, microbial, 163
Atmosphere, gases in, 114, 136, 221, 257,
  303
  irradiation of, 115
  relative humidity of, 113
  storage, 114
  temperature of, 112

Atmosphere, ventilation of, 114
  (*See also* Air)
Autosterilization, 366
Avidin, 297
Avocados, spoilage of, 238
a$_w$ (*see* Water activity)

Bacillaceae, 55
Bacillus, 56, 247, 273, 333, 341, 354,
  363, 365
  betanigrificans, 366
  calidolactis, 332, 341
  cereus, 334, 450
  characteristics of, 56
  citri medicae, 408
  coagulans, 341, 361, 365, 423
  kefir, 328
  licheniformis, 187
  macerans, 363
  megaterium, 341, 352
  mesentericus, 363
  natto, 412
  nigrificans, 246
  pepo, 361
  polymyxa, 345, 363
  pumilus, 248
  stearothermophilus, 361
  subtilis, 187, 363, 375, 426, 429, 450
  thermoacidurans (*see coagulans
    above*)
  vulgatus, 352
Bacon, 260–264, 277–279
  curing of, 260–264
  spoilage of, 277–279
  Wiltshire, 278
Bacteria, 36–62
  acetic acid-forming, 58, 375, 388, 399
  acid-proteolytic, 59, 334
  aerobic, 41
  alkali-forming, 337
  anaerobic, 41
  butyric acid-forming, 58, 362
  cell aggregates of, 38
  coliform (*see* Coliform bacteria)
  cultural characteristics of, 38
  encapsulation of, 36
  factors influencing growth of, 39–41
  facultative, 41
  of food-poisoning, 61, 434–450
  formation of endospores, 36–38

Bacteria, gas-forming, 61
  generation time of, 77
  groups of, in foods, 57–62
  growth curve of, 76–78
  halophilic, 60
  infections caused by, 61, 435, 450–459
  intestinal, 60
  iron, 57
  lactic acid, 49–54, 58, 332, 373
  lipolytic, 59
  mesophilic, 41
  microaerophilic, 41
  morphology of, 36–38
  osmophilic, 60
  pathogenic, 61, 435–457, 463, 467–470
  pectolytic, 59
  phosphorescent, 44
  physiological characteristics of, 39–41,
      163–174
    foods, 39, 167–170
    inhibitors, 41, 170
    moisture requirements, 40, 164
    O-R potential, 41, 171
    pH, 41, 170
    temperature requirements, 40,
      173
  pigmented, 61
  propionic acid-forming, 54, 58
  proteolytic, 58, 334
  psychrophilic, 40, 60, 110
  psychrotrophic, 41
  rope-forming, 61, 187–189, 198, 243,
      247
  saccharolytic anaerobes, 59
  salt-tolerant, 60
  spores, 36–38
  sugar-tolerant, 60
  thermoduric, 322, 363
  thermophilic, 41, 60, 360–362
Bacterial soft rot, 226
Bacteriophages, 42
Bacterium curvum, 400
  orleanense, 400
  proteolyticum, 347
  scheutzenbachii, 400
  termo, 388
  (See also Brevibacterium)
Baking (see Bread; Cake)
Bananas, spoilage of, 238
Beans (geen, lima, wax), spoilage of,
    228

Beef, dried, 259, 273
  fresh, spoilage of, 272
  ground, 272, 513
Beer, 384–389
  aging of, 386
  continuous brewing of, 386
  filtration of, 386
  ginger, 390
  manufacture of, 384–386
Beer defects and diseases, 387–389
Beer fermentation, 385
Beer microbiology, 386–389
Beets, spoilage of, canned, 366
    raw, 232
Berries, spoilage of, 235
Berry wines, 396
Beta rays (see Cathode rays)
Beverages (see specific beverages)
Biochemical oxygen demand, 478
Bitterness, of cheese, 347
  of milk, 334, 338
  of wine, 395
Black mold rot, 227
Black rot, 227, 306
Black spot, 270, 280, 307
Blanching, 103, 116, 124, 208
Blastospores, 20
Bleeding, 119
Bloaters, 244
Bloom, on eggs, 297
  on meat, 257, 269
Blue mold rot, 227
BOD (biochemical oxygen demand),
    478
Boiler water additives, 143
Boric acid and borates, 135
Botrytis, 19, 226
  cinerea, 19, 226, 393
Botulism, 436–444
  cause of, 436
  as disease, 443
  foods involved, 442
  incidence of, 441
  mortality arising from, 443
  outbreaks of, 443
  prevention of, 444
  tests for, 467
  toxin production in, 438–440
  (See also Clostridium, botulinum)
Brandy, 398

Bread, 181, 184–189, 380–384
  baking of, 383
  chalky, 189
  continuous making of, 382
  flavor of, 382
  leavening in, 380–382
  liquid-ferment process, 382
  moldy, 184–187
  red, 189
  ropy, 187–189
  rye, 383
  sour, 184
  spoilage of, 184–189
  sponge method of making, 381
Breather, 359
Bremia, 227
Brettanomyces, 32, 215
Brevibacteriaceae, 49
Brevibacterium, 49
  erythrogenes, 339
  glutamicum, 375, 422, 425
  linens, 348, 375
Brines, 213, 217, 218, 263, 289
  curing, 263, 280
  fish, 289
  meat, 263
  olive, 218
  pickle, 213, 217
  vegetable, 208
Broccoli, spoilage of, 233
Brucellaceae, 47
Brussels sprouts, spoilage of, 233
Buffers, 170
Bulgarian buttermilk, 328
2, 3-Butanediol, 52, 176
Butter, 342–344
  cream for, 342
  spoilage of, 342–344
  storage of, 325, 343
Buttermilks, 328, 344
  (See also specific fermented milks)
Buttons, 366
Butyric acid, 58, 333
Butyric acid bacteria (butyrics), 58, 176,
    270, 333, 362, 387
Byssochlamys fulva, 88, 365

Cabbage, spoilage of, 233
Cacao, 408
Cake, 189

Cake, irradiation of, 189
  spoilage of, 189
Candida, 32, 215, 353
  arborea, 417
  krusei, 32, 408
  lipolytica, 32
  pulcherrima, 417, 419
  utilis, 32, 417
Candling of eggs, 299, 304
Candy, 195, 199
  contamination of, 195
  spoilage of, 199
Canned foods, agitation of, 95
  spoilage of, 357–368
    biological, 357–367
    chemical, 357
    diagnosis of cause of, 367
    fish, 365
    flat sour, 360
    grouping on pH, 85, 359, 365
    grouping on solids content, 86
    through leakage, 357, 364
    meats, 365
    mesophilic, 362–366
    by molds, 364
    through putrefaction, 363, 365
    sulfide, 361
    swelling of cans, 359
    T.A., 361
    thermophilic, 360–362
    by yeasts, 364
Canning, 101–107
  containers used in, 102
  of fish, 285
  of fruits, 106
  history of, 101–103
  home, 106
  of meats, 255
  of milk, 323, 326
  new methods of, 105
  procedure for, 103
  of vegetables, 104
Cantaloupe (muskmelon), spoilage of,
    233
Capsules, 36
Carbohydrates, changes in, 176
Carbon dioxide, preservation with, 114,
    136, 157, 205, 221, 257, 303, 353
  of carbonated beverages, 157, 353
  of fruits (gas storage), 221
  of meats (gas storage), 257

Carbon dioxide, under pressure, 157
Carrots, spoilage of, 228
Casehardening, 124
Catadyn process, 68
Catalase, 136, 430
Catering, 487
Cathode rays, 152–157
    applications of, 156
    effect of, on foods, 155
        on microorganisms, 154
Cauliflower, spoilage of, 233
Celery, spoilage of, 230
Cellulase, 430
Centrifugation, 80
Cephalosporium acremonium, 19
Ceratostomella, 227
    fimbriata, 233
Cereals and cereal products, 180–191
    contamination of, 180
    preservation of, 181
    spoilage of, 183–190
Cheese, 329, 345–348
    discoloration of, 347
    preservative factors of, 329
    ripening of, 329
    spoilage of, 345–348
    starters for, 373, 375
Chemical changes by microorganisms,
        175–178
Cherries, spoilage of, 239
Chicken (see Poultry)
Chilling, preservation by, 112–115, 205,
        221, 256, 286, 299, 313, 324, 330
    of cheese, 330
    of eggs, 299
    factors in, 112–115
    of fish, 286
    of fruits, 221
    of meats, 256
    of milk, 324
    of poultry, 313
    temperatures involved in, 112
    of vegetables, 205
Chlamydobacteriales, 57
Chlamydospores, 4, 27
Chlorination, in-plant, 477
    of water, 67, 135, 477
Chlorine (sanitizer), 135, 485
Chlorohydrins, 136
Chlortetracycline (see Antibiotics)
Chou, 409

Chromobacterium lividum, 269
Cider (see Apple juice)
CIP (cleaned-in-place) systems, 485
Citric acid production, 424
Citron, 408
Citrus fruits, spoilage of, grapefruit,
        235–237
    lemons, 235–237
    oranges, 235–237
Cladosporium, 21, 227, 235, 270, 306,
        348, 450
    black spot, 270, 307
    green mold rot, 227
    herbarum, 21, 238, 270
Classification, of bacteria of foods, 42–57
    of molds of foods, 8–23
    of yeasts of foods, 28–33
Cleansing, 142, 483
Cleistothecium, 16
Clonothrix, 57
Clostridium, 56, 199, 247, 270, 333, 341,
        362, 365, 436–444
    acetobutylicum, 421
    botulinum, 207, 258, 362, 436–444
    butyricum, 197, 247, 362
    characteristics of, 56
    lentoputrescens, 347
    nigrificans, 361
    parabotulinum, 436
    pasteurianum, 362, 365
    perfringens (welchii), 449, 470
    putrefaciens, 279, 362
    sporogenes, 347, 362
    thermosaccharolyticum, 361, 497
    tyrobutyricum, 347
Cocoa butter, spoilage of, 352
Coffee, 407
Cold storage (see Freezing)
"Cold-storage flavor," 269, 308
Coliform bacteria, 45–47, 163, 247, 332,
        477
    in food poisoning, 450
    in fruit juices, 223, 495
    in milk, 332, 495
    in oysters, 495
    in soft drinks, 495
    sources of, 46
    tests for, 66, 495
    in water, 66
Colletotrichum, 23, 227
    lindemuthianum, 227

Colorless rots of eggs, 305
Common (cellar) storage, 111
Conalbumin, 297
Condensed milk, 325, 340
Conidia, 4
Conidiophores, 4
Contamination, of foods, 63–72, 162, 456
    in disease outbreaks, 456
    sources of, air, 68–71
        animals, 64
        carriers, 446, 452, 467, 469
        equipment, 71, 318
        food handlers, 456, 467, 469
        handling and processing, 71
        natural, 63–70
        plants, 63
        salt, 354
        sewage, 65
        soil, 65
        spices, 354
        water, 66–68
    specific foods, cereals and cereal prod-
        ucts, 180
        eggs, 296
        fish and seafood, 283
        fruits and vegetables, 63, 201–203,
            222
        meats and meat products, 252
        milk, 318
        poultry, 310
        soft drinks, 353
        sugars, 192
Control, food, 502–507
Control agencies, 502
Cooling, of canned foods, 95, 106, 477
    of eggs, 296
    of meats, 256
    of milk, 323
    of poultry, 313
Copra, spoilage of, 352
Coremia, 16
Corynebacteriaceae, 55
Corynebacterium bovis, 55
    diphtheriae, 55
    pyogenes, 55
Coxiella burnetii, 99, 456
Crab meat, standards for, 511, 513
Crabs, spoilage of, 294
Crackers, 382
Crenothrix, 57
Cryptococcaceae, 23

Cucumbers, spoilage of, 233
Cultured buttermilk, 328
Cultured sour cream, 328
Cultures, 370–378
    acetic acid, 375
    activity of, 372
    bacterial, 373–375
    lactic acid, 373
    maintenance of, 371
    mixed, 372
    mold, 377
    preparation of, 371
    propionic acid, 374
    starter (see Starters)
    yeast, 375–377
Curing of meats, 260–264
Curing solutions, spoilage of, for fish,
        289
    for meats, 280
Custard-filled items, 511

D value, 91
Dairy products (see specific products)
Dates, spoilage of, 239
Debaryomyces, 32, 215
    kloeckeri, 32
Dehydration (see Drying)
Dehydrocanning, 105
Dehydrofreezing, 116
Dematiaceae, 21–22
DEPC (diethyl pyrocarbonate), 143
Desiccation (see Drying)
Detergents, 142, 483
Dextran, production of, 375, 422
Dextransucrase, 430
Diacetyl, 52, 211
Diaporthe batatatis, 233
Dielectric heating, 147
Diethyl pyrocarbonate (DEPC), 143
Dill pickles, 216
Diplodia, 23, 227, 236
    tubericola, 233
Diseases, food-borne, 433–473
    food infections, 450–459
    food poisonings, 436–450
    investigation of outbreaks of,
        463–473
    kinds of, 463
    prevention of outbreaks of, 472
Dithermal processing, 147

Dole process, 105
Downy mildew, 227
Dried foods, microbiology of, 125–129
  rehydration of, 127
  (*See also* specific foods)
Drip, 119
Drying, 122–130
  control of, 124
  of cultures, 371
  foam-mat, 124
  freeze-, 123, 127
  methods of, 122
  pressure-gun, 124
  pretreatments for, 124
  procedures after, 125
  sun, 123
  tower, 124
  (*See also* specific foods)
Dysentery, amebic, 456, 463
  bacillary, 456, 463, 469

Eggplants, spoilage of, 235
Eggs, 296–309, 512
  bacterial genera on, 296
  candling of, 299, 304
  contamination of, 296
  defects in, 304
  dried, microbiology of, 128
  pasteurization of, 299
  preservation of, 296–304
    by chilling, 299
    by drying, 301
    by freezing, 300
    by heating, 298
    by oiling, 300
    by rays, 304
    by use of preservatives, 303
  preserved, 413
  pretreatments for, 300
  spoilage of, 304–308
    bacterial, 305–308
    fungal, 306
    mustiness, 306
    nonmicrobial changes in, 304
    pin-spot, 306
  spoiled, bacterial genera in, 305–308
    mold genera in, 306
  standards for, microbiological, 512
  washing of, 298

Electric currents in preservation, 146, 331
Electron-volt (ev), 152
Electrons, 151
Employee health, 489
Endamoeba histolytica, 456, 463
Endives, spoilage of, 232
Endomyces, 23
Endomycopsis, 189, 215
  fibuliger, 189
Endosepsis, 238
Endospores (bacterial), 36–38, 89
  dormancy of, 38
  formation of, 36
  germination of, 37
  heat resistance of, 89
Enterobacter aerogenes (*see Aerobacter, aerogenes*)
Enterobacteriaceae, 45–47
Enterococci, 50, 224, 332, 363
  tests for, 469
Enterotoxin, 446
Enzymes, heat resistance of, 89
  manufacture and uses of, 425–431
    amylase, 381, 425–427
    catalase, 430
    cellulase, 430
    dextransucrase, 430
    glucose-oxidase, 430
    invertase, 427
    lactase, 430
    lipase, 430
    pectolytic, 428
    proteolytic, 381, 429
Eremothecium ashbyii, 421
Erwinia carotovora, 47, 226
Escherichia, 45–47
  coli, 45–47, 308, 336, 354, 422
Ethanol, content of wines, 393
  as preservative, 138
Eumycetes, 8
Evaporated milk, 325, 341, 366
  heat-treatment for, 323
  spoilage of, 341, 366
Evaporation, 123

F value, 91
Fats (and oils), decomposition of, in meats, 269–270
  as energy food, 168

Fats (and oils), from microorganisms, 419
  oxidation of, 351
  spoilage of, 337, 350–352
Federal Food, Drug, and Cosmetic Act, 132, 503
Fermentations, 144, 176, 209–220, 380–415
  acetic acid (vinegar), 399
  alcoholic, 176, 242
    in beer (ale), 385
    in bread, 380
    in distilled liquors, 398
    in ginger beer, 390
    in mead, 398
    in perry, 396
    in pulque, 390
    in sake, 390
    in wines, 391
  butyric acid, 176, 362
  citric acid, 424
  of egg albumin, 301
  lactic acid, 49–54, 176, 211–220, 242, 407
    in cheese, 329, 345–348
    in fermented milks, 328, 344
    in olives, green, 217
    in pickles, 212–217
    production of, 422
    in sauerkraut, 209
    in sauerrüben, 219
    in sausage, 261, 271, 505
  leavening (bread), 380–382
  malolactic (wines), 395
  mannitic (wines), 395
  mixed acid, in cacao, 408
    in citron, 408
    in coffee, 407
    in poi, 413
    in vanilla, 408
  in Oriental foods, 409–414
    in ang-khak, 411
    in eggs, preserved, 413
    in fish, 413
    in idli, 412
    in minchin, 412
    in miso, 410
    in natto, 411
    in soy sauce, 409
    in soybean cheese, 412

Fermentations, in tamari sauce, 410
  in tempeh, 411
  propionic acid, 176
Fermented milks (buttermilks) (see specific kinds)
Field investigation of food-borne diseases, 465–467
Figs, spoilage of, 238
Film yeasts (see Yeasts)
Films, plastic, 255, 280
Filtration, 68, 70, 80
  of air, 70
  of beer, 386
  of beverages, 80
  of water, 68
Fish, 283–295, 413
  bacterial genera in, 283
  contamination of, 283
  microbiological standards for, 513
  preservation of, 284–290
    by asepsis, 284
    by chilling, 286
    by drying, 287
    fermented, 413
    by freezing, 286
    by heat, 285
    by use of antibiotics, 288
    by use of antioxidants, 290
    by use of preservatives, 287–290
  sausage made from, 294
  spoilage of, 290–294
    bacteria causing, 292
    canned, 365
    evidences of, 292
    factors influencing, 291
    special kinds of, 293
  (See also Seafoods)
Fish brines, 289
Fisheye, 247, 248
Fishiness, 308, 338, 343
  in butter, 343
  in eggs, 308
  in milk, 338
Fitness of food, 160
"Flash 18" process, 105
Flat sour spoilage, 360, 365
Flavobacterium, 45, 243, 293, 306
  proteus, 388
  rhenanus, 211
Flavor reversion, 351

Flipper, 359
Floaters, 244, 247
Flour, 180, 183
  contamination of, 180
  spoilage of, 183
  spores in, 189
Flour flora, 180
Fluidized bed process, 105
Foam-mat drying, 124
Food additives, 132–143, 224
Food Additives Amendment, 132, 504
Food handlers, 456, 467, 469
Food infections, 435, 450–459
  miscellaneous, 435, 456
  prevention of, 456
  Salmonella, 451–455
  Streptococcus, 455
Food laws, 503–506
Food poisoning, 434–450, 463–473
  aflatoxin, 450
  animal sources of, 435
  Bacillus cereus, 450
  Bacillus subtilis, 450
  botulism, 436–444
  chemical, 434
  Clostridium perfringens, 449
  coliform bacteria in, 450
  fungal, 450
  investigation of, 463–473
  plant sources of, 434
  Proteus, 450
  Staphylococcus, 444–449
  tests for, 467
  true, 436–449
  Vibrio parahemolyticus, 450
  virus in, 435, 456
Foods, bacteria in, 36–62
  biological structure of, 167
  chemical properties of, 167
  colloidal constituents of, 167
  contamination of, 63–72, 162, 456
  control of, 502–507
  cultures for, 370–378
  fermentations of, 144, 176, 209–220
  fitness of, 160
  groups of, 74
  heated canned, 101–107
  infections from (*see* Food infections)
  ingredients of, 482
  laboratory methods for testing samples
    of, 467–470, 492–501

Foods, from microorganisms, 416–422
  molds on, 2–24
  nutrients for microorganisms, 167–170
  perishable, 161
  physical state and structure of,
    164–167
  poisoning from (*see* Food poisoning)
  preservation principles for, 73–78
  preservatives for, 131–145
  sanitation of, 475–491
  semiperishable, 161
  spoilage principles, 160–174
  stable, 161
  standards for, 509–514
  washing of, 80, 204, 211, 220
  yeasts in, 25–35
  (*See also* specific foods)
Formaldehyde, 138
Formula method, 97
Frankfurters, spoilage of, 274
Freeze-drying, 123, 127, 259
Freezerburn, 118
Freezing, 115–120
  changes during, 118
  effect on microbial growth, 166
  lethal effects of, 110
  methods, 116
  preparation for, 115
    changes during, 117
  quick, sharp, slow, 116
  (*See also* specific foods)
Frozen desserts (*see* Ice cream)
Frozen foods, precooked, 120, 511
  bacteriological standards for, 511, 513
  (*See also* specific frozen foods)
Fruit juices, 223, 241
  contamination of, 223
  pasteurization of, 100
  preservation of, 221
  spoilage of, 241
Fruits, 63, 201–203, 220–241
  bacterial genera in, 204
  contamination of, 63, 201–203, 222
  dried, microbiology of, 128
  frozen, mold genera in, 223
  preservation of, 220–225
    by canning, 221
    by chilling, 221
    by drying, 124, 224
    by freezing, 222
    by gas storage, 221

Fruits, preservation of, by heat, 221
    by rays, 220, 225
    by use of preservatives, 224
    spoilage of, 225–241
    types of, 226
    washing of, 220
    (*See also* specific fruits)
Fungal red rot (eggs), 307
Fungal spoilage (*see* Molds)
Fungi Imperfecti, 14–23, 27, 28
Fungicides and fungistats, 8, 142, 224,
    486
Fusarium, 22, 227, 238, 450
    culmorum, 344
    moniliforme, 239
Fusarium rots, 227

Gallionella, 57
Gamma rays, 151, 153–157
    applications of, 156
    effect of, on foods, 155
        on microorganisms, 154
Garlic, spoilage of, 228
Gases in food preservation, 114, 136,
    221, 257, 303
Gassiness, in candy, 199
    in canned foods, 357, 359, 361–364
    in cheese, 345
    in hams, 279
    in milk, 333
    in molasses, 197
    in olives, 247
    in pickles, 244
    in sausage, 277
Generation times, 77
Geotrichum (*Oöspora, Oidium*), 17, 25,
    270, 348
    aurianticum, 189
    candidum (*O. lactis*), 17, 342, 417,
        419
    caseovorans, 348
    citri-aurantii, 237
    crustacea, 348
    rubrum, 348
Ginger beer, 390
Gleosporium, 23, 238
Glucose-oxidase, 430
Glycerol as preservative, 138
Grains, flora of, 180
    spoilage of, 183

Grapefruit, spoilage of, 235–237
Grapes, juice from, pasteurization of, 100
    spoilage of, 235
    for wine, 391
Graphical method, 97
GRAS chemicals, 132
Gray mold rot, 226
Green mold rot (fruits, vegetables), 227
Green rot (eggs), 305
Growth of microorganisms, associative,
    163
    delay, 76–78
    factors affecting, 163–174
    in fish, 290–294
    in meat, 266
    prevention, 78
Growth curve (history), 76–78

Halobacterium, 44, 354
    cutirubrum, 273
    salinarium, 44
Halogens, 135, 485
Halophiles, 60, 263, 288, 293
Hamburger, spoilage of, 272
    standards for, bacteriological, 513
Hams, canned, 255
    curing of, 260–264
    smoking of, 264
    spoilage of, 279
    tenderized, 256
Hands, washing of, 490
Hanseniaspora, 32
Hansenula, 32, 215
    anomala, 388
Hard swell, 359
Heat (in preservation), 82–108
    boiling, 100
    canning, 101–107
    pasteurization, 99
    penetration of, 93–95
    resistance to, 82–93
    in thermal processes (*see* Thermal
        processes)
Heat penetration, 93–95
Heat resistance, 82–93
    of bacteria and spores, 88
    determination of, 89–91
    of enzymes, 89
    factors affecting, 82–87

Heat resistance, of molds and spores, 87
  of yeasts and spores, 87
Helminthosporium, 22
Hepatitis, 456
Honey, contamination of, 194
  pasteurization of, 196
  spoilage of, 198
Horse-radish, 142, 354
Hromatka and Ebner method (vinegar), 403
Humidity, relative, and bread storage, 185
    and chilling storage, 113
    and drying, 124
    and gas storage, 114
    and ultraviolet irradiation, 115
    and water activity, 6, 164
Hydrogen-ion concentration (*see* pH)
Hydrogen peroxide, 136, 195, 328
Hydrogen swells, 357
Hydrostatic sterilizer, 105
Hyphae, 3
Hypochlorites, 134, 485

Ice, 67, 478
  bacteriology of, 67
  germicidal, 284, 288–290, 478
Ice cream, 99, 324, 342, 510
  ingredients of, 319, 342
  standards for, bacteriological, 510
Ice-cream mix, pasteurization of, 99
Idli, 412
Induction heating, 147
Infections (*see* Food infections)
Inhibitors, 8, 40, 170
Inspection and grading, 505
Invertase, 427
Iodine, 135, 224
Iodophors, 135, 485
Ionizing radiations, 151–157
Iron bacteria, 57
Irradiation (*see* Radiations)
Isoamylase, 426

Jams, 364, 513
  spoilage of, 364
  standards for, microbiological, 513
Jellies, spoilage of, 364

Keeping-quality tests, 498
Kefir, 328
Klebsiella aerogenes (*see Aerobacter aerogenes*)
Kloeckera, 33, 215
  apiculata, 33
Koji, 409, 426
Kumiss, 328

Laboratory methods, microbiological, 467–470, 492–501
    agencies recommending, 498
    for eating and drinking utensils, 500
    for food-borne disease, 467–470
    qualitative, 497
    quantitative, 496
    samples tested by, 494
    sampling, 493
    tests, keeping-quality, 498
      kinds of, 495
      purposes of, 492
      for sterility, 498
Lactase, 430
Lactenins, 170
Lactic acid, fermentations (*see* Fermentations)
  manufacture and uses, 422–424
  as preservative, 137
Lactic acid bacteria (lactics), 49–54, 58, 332, 373
  heterofermentative, 49, 332
  homofermentative, 49, 332
  starters, 344, 373
  (*See also* Fermentations)
Lactobacillaceae, 49–54
Lactobacillus, 52–54, 242, 275, 280, 363, 406
  acidophilus, 328
  brevis, 212, 214, 217, 218, 243, 244, 395, 396, 423
    var. *rudensis,* 347
  buchneri, 395
  bulgaricus, 328, 332, 336, 374, 423
  casei, 374
  delbrueckii, 378, 410, 423
  fermenti, 395
  helveticus, 374
  hilgardii, 395, 396
  lactis, 374
  pastorianus, 388

Lactobacillus, plantarum, 211, 217, 218, 243, 395, 423
    var. *rudensis,* 347
  salimandus, 277
  thermophilus, 332
  trichodes, 395, 396
  vermiformis, 390
Lactoperoxidase, 332
Lag phase of growth, 76
Laws, food, 503–506
Leakage, 119, 357, 364
Leaks in canned foods, 357, 363
Leavening, 380–382
Lemons, spoilage of, 235–237
Lettuce, spoilage of, 232
Leuconostoc, 52, 196, 198, 242, 275, 280, 363, 406
  citrovorum, 328, 373
  dextranicum, 197, 328, 396
  mesenteroides, 197, 211, 214, 217, 218, 242, 375, 396, 422
Lipase, 59, 337, 430
Lipids, changes in, 177
Lipolysis, 59, 269, 337
  bacterial genera in, 337, 351
  mold genera in, 352
Liquors, distilled, 398
Logarithmic phase of growth, 76–78
Low temperatures, 109–121
  growth at, 110
  lethal effect of, 110
  use of, 111–120
    in chilling, 112–115, 256
    in common or cellar storage, 111
    in freezing, 115–120
    principles of, 109
    temperatures employed, 111–116
  (*See also* Chilling; Freezing)
Lyophilization, 123, 127, 371
Lysozyme, 170, 297

Macaroni, spoilage of, 190
Malt, 384
  continuous malting, 386
Malt beverages (*see* Ale; Beer)
Maltase, 426
Mannitic fermentation (wines), 395
Manure, 65
Maple sap, 193, 198

Maple sap, contamination of, 193
  spoilage of, 198
Maple sirup, 193, 196, 198
Martin process, 105
Mayonnaise, 352
Mead, 398
Meals, precooked, standards for, 513
Meat Inspection Act, 505
Meat products, bacon, 259–264, 277–279
  canned, 255, 365
  dried beef, 259, 273
  ham, 255, 260–264, 279
  pork sausage, 272
  sausage, 261, 273–277
Meats, 252–282
  bacterial genera in, 253
  contamination of, 252
  cured, 259–264, 273–280, 366
  growth of microorganisms in, 266
  mold genera in, 253
  packaging of, 255
  pigments in, 261, 269
  precooked, bacteriological standards for, 513
  preservation of, 253–265
    by use of antibiotics, 265
    by asepsis, 254
    by chilling, 256
    by curing, 260–264
    by drying, 259
    by fermentation, 261
    by freezing, 257
    by heat, 255
    by irradiation, 258
    by smoking, 264
    by use of preservatives, 260–265
  spoilage of, 257, 265–280
    aerobic, 268–270
    anaerobic, 270
    bacterial, 268, 270–280
    canned, 365
    cured, 273–280, 366
    discoloration in, 269, 272, 275
    fresh, 271–273
    greening in, 270, 272, 275
    by lactics, 271
    by molds, 269
    off-odors and off-tastes in, 265, 269–280
    packaged, 280
    pigment changes in, 269

Meats, spoilage of, putrefaction in, 270
    slime in, 268, 272, 274
    souring in, 265, 269, 270, 272
    tissue invasion in, 266
    types of, 268–271
    by yeasts, 269
    (*See also* specific meats)
Melanconiales, 23
Melons, spoilage of, 233
Membrane filters, 79, 386
Mesophiles, 41
Metabiosis, 163
Metacryotic liquid, 118
Metals as preservatives, 135
Microaerophilic bacteria, 41
Microbacterium, 55, 363
    lacticum, 55
Micrococcaceae, 48
Micrococcus, 48, 332, 354, 363, 366
    candidus, 275
    caseolyticus, 59
    freudenreichii, 48, 335
    lipolyticus, 280
    lysodeikticus, 430
    roseus, 198, 339
    ureae, 337
Microwave heating, 148
Milk, 65, 318–340, 509
    certified, 510
    contamination of, 318
    preservation of, 320–331
        by asepsis, 320
        by chilling, 324
        by condensing, 325
        by drying, 325–327
        by fermentation, 328–330, 344
        by freezing, 324
        by heat, 321–323
        by irradiation, 330
        by pasteurization, 321
        by removal of organisms, 321
        by use of preservatives, 327–330
    smothered, 338
    spoilage of, 331–340
        alkali production in, 337
        color changes in, 339
        flavor changes in, 337
        gassiness in, 333
        lipolysis in, 337
        proteolysis in, 334

Milk, spoilage of, ropiness in, 335–337
    souring in, 332
    sweet curdling in, 56, 334
    standards for, bacteriological, 509
Milk products, 322–331, 340–348
    butter, 342
    buttermilks, 328, 344
    cheese, 329, 345–348
    condensed milk, 325, 340
    dry milk, 129, 326, 340, 510
        standards for, bacteriological, 129,
            327, 510
    evaporated milk, 325, 341
    frozen desserts, 324, 342, 510
        standards for, bacteriological, 510
    preservation of, 322–331
    spoilage of, 340–348
    sweetened condensed milk, 326, 341,
        366
Minchin, 412
Misbranding, 503
Miso, 410
Moisture (*see* Water)
Molasses, 195, 197
    Barbados, 193
    preservation of, 195
    spoilage of, 197
Molds, 2–24
    asexual spores of, 4
    classification and identification of, 8–23
    counts of filaments in, 223, 495, 497
    cultural characteristics of, 5
    fungistatic paints against, 486
    fungistats and fungicides against, 8,
        142, 224, 486
    heat resistance of, 87
    industrial, 10–23
    morphological characteristics of, 2–5
    physiological characteristics of, 5–8
    requirements of, foods, 8
        moisture, 5, 165
        oxygen, 8
        pH, 8, 170
        temperature, 8
    sexual spores of, 4
    toxic, 450
Monascus purpureus, 23, 411
Monilia (*see Neurospora*)
Moniliaceae, 14–21
Moniliales, 14–23

Monosodium glutamate, 425
Morphology, of bacteria, 36–38
  of molds, 2–5
  of yeasts, 25
Mucor, 10, 198, 269, 412, 450
  lusitanicus, 269
  mucedo, 269
  pyriformis, 424
  racemosus, 10, 269
  rouxii, 10, 426
Mucorales, 10–13
Mustiness, in eggs, 306
  in meat, 269, 270
  in milk, 338
Mycelium, 3
Mycobacterium, 57
Mycostats (fungistats), 8, 142, 224

Nadsonia, 32
Natto, 411
Neurospora (Monilia), 18, 190
  nigra, 348
  sitophila, 18, 185, 189
Nisin, 139
Nitrates, 134, 260
Nitrites, 134, 260
Nitrogenous compounds, changes in, 175
Nomogram method, 98
Nutmeats, 354

O-R (oxidation-reduction) potential,
    41, 171
Odor, time of appearance of, 268
Oidia, 4
Oils, essential, 352
Olives, 217–219, 247
  green, 217, 247
    manufacture of, 217
    spoilage of, 247
  ripe, 219, 248
Onions, spoilage of, 228
Oömycetes, 10
Oöspora (see Geotrichum)
Oöspores, 5
Oranges, spoilage of, 235–237
  (*See also* Fruit juices)
Oriental fermented foods, 409–414
Osmophiles, 60
Ovomucoid, 297

Oxidation-reduction (O-R) potential,
    41, 171
Oxidizing agents, 142
Oxygen requirements, 4, 8, 27, 171
  of bacteria, 41, 171
  of molds, 8, 171
  of yeasts, 27, 171
Oxytetracycline (*see* Antibiotics)
Oysters, spoilage of, 294, 513
  standards for, bacteriological, 513
Ozone, 114, 136, 205, 222, 257, 303

Packaging, materials for, 482
  of meats, 255, 280
Paints, fungistatic, 486
Papain, 429
Parabens, 137
Paracolobactrum, 45, 306
Parsley, spoilage of, 230
Parsnips, spoilage of, 230
Pasteurization, 99
  of beer, 100
  of cream, 322
  of cucumber pickles, 212
  of dried fruits, 100, 125
  of eggs, liquid, 299
  of fruit juices, 100
  high-temperature–short-time, 99
  holding method for, 99
  of honey, 196
  of ice-cream mix, 99
  of market milk, 99
  of milk for cheese, 99
  of pickles, 212
  of vinegar, 100
  of wine, 100
Pathogens, 61, 225, 435–457, 463,
    467–470
  food-borne, 61, 435–457, 463
  groups of, 463
  plant, 225
  tests for, 467–470
Peaches, spoilage of, 239
Pears, gas storage of, 222
  spoilage of, 240
Peas, spoilage of, 228
Pectin, changes in, 177, 428
Pectinesterase, 428
Pectolytic enzymes, 59, 177, 428

Pediococcus, 51
  cerevisiae, 51, 212, 214, 217, 243, 374, 388
  soyae, 410
Penicillium, 15, 198, 227, 270, 306, 450
  asperulum, 270
  aurantio-virens, 348
  camemberti, 16, 378
  casei, 348
  citrinum, 424
  digitatum, 16, 236
  expansum, 16, 185, 240, 270
  italicum, 16, 236
  luteum, 424
  oxalicum, 270
  puberulum, 348
  roqueforti, 16, 378
  spinulosum, 273
  stoloniferum, 185
Peppers, spoilage of, 235
Peronosporales, 10
Peroxides, 136
Perry, 396
Pestalozzia, 23, 238
pH, 170
  for bacterial growth, 41, 170
  canned foods grouped on, 85, 359
  effect on heat-killing, 85
  for mold growth, 8, 170
  for yeast growth, 28, 170
Phialides (*see* Sterigmata)
Phoma, 23
Phomopsis, 227, 236
  vexans, 235
Phosphorescence, 44, 269
Phosphoric acid, 134
Phosphorylase, 426
Photobacterium, 44, 269
Phycomycetes, 10–13
Physalospora, 227
  malorum, 240
Physiological characteristics, of bacteria, 39–41, 164–171
  of molds, 5–8, 165, 170
  of yeasts, 27, 165, 168–170
Phytophthora, 227, 236
  cactorum, 235
Pichia, 32
  membranaefaciens, 32
Pickles (cucumber), 212–217, 244–247
  dill, 216

Pickles (cucumber), hollow, 244
  pasteurization of, 212
  salt-stock, 213–216
  spoilage of, 244–247
    black, 246
    bloaters or floaters in, 244
    slipperiness in, 244
    softness in, 244
  unfermented, 212
  yeast genera in, 215
Pidan, 413
Pigments, bacterial, 61, 269
  of meat, 269
  of mold, 5
  of yeast, 33
Pink mold rot (fruits, vegetables), 227
Pink rot (eggs), 306
Pink-spot (eggs), 306
Plants (green), bacterial genera on, 63
Plenodomus destruens, 233
Plums, spoilage of, 239
Podosphaera leucotricha, 240
Poi, 413
Poising capacity (O-R), 171
Poisoning (*see* Food poisoning)
Polygalacturonase, 428
Potatoes, spoilage of, 233
  storage of, 205
Pouches, plastic, 103
Poultry, 310–317
  bacterial genera in, 310
  contamination of, 310
  plucking and dressing of, 311
  preservation of, 311–315
    by chilling, 313
    by freezing, 313
    by heat, 313
    by irradiation, 315
    by use of antibiotics, 314
    by use of preservatives, 314
  spoilage of, 315
    bacterial genera in, 315
  standard for, bacteriological, 513
Poultry Products Inspection Act, 505
Pousse (wines), 395
Preservation of foods, by anaerobic conditions, 80
  by asepsis, 78
  by cold, 109–121
  by drying, 122–130
  electric currents in, 146, 331

Preservation of foods, general methods
    for, 74
    general principles for, 75–78
    by heat, 82–108
    by radiations, 146–157
    by removal of microorganisms, 79
    by use of preservatives, 131–145
    (*See also* specific foods)
Preservatives, 131–145, 208–220, 224, 504
    added, 132–143, 208, 224
    alcohols, 138
    alkalies and alkaline salts, 135
    antibiotics, 139
    definition, 131, 132, 504
    developed, 144, 209–220
    formaldehyde, 138
    fungistatic and fungicidal, 142
    gases, 114, 136, 221, 257
    halogens, 135
    inorganic, 133–136
    inorganic acids and salts, 133
    metals, 135
    organic, 136–142
        acids and salts, 137
    peroxides, 136
    plant extracts, 140
    selection of, 132
    sorbic acid and sorbates, 138
    spices, 141
    sugars, 138
    tests for, 132
    tolerances of, 143
    wood smoke, 123, 140, 264, 289, 315
    (*See also* specific foods)
Pressure, mechanical, 157
Pressure-gun drying, 124
Pressurized foods, 105
Pretreatments, effect on microorganisms,
    162
Process time (thermal), calculation of,
    96
Propionibacteriaceae, 54
Propionibacterium, 54, 58, 347
    rubrum, 347
    shermanii, 374
    thoenii, 347
    zeae, 347
Propionic acid and propionates, 137, 176
Propionic acid bacteria (propionics), 54,
    58, 176
Propylene oxide, 136, 354

Proteases (proteinases, peptidases), 429
Proteolysis, 58, 334
    bacterial genera causing, 58
Proteolytic enzymes, production of, 429
Proteus, 47, 306, 450
    melanovogenes, 306
    mirabilis, 450
    vulgaris, 47, 450
Pseudomonadaceae, 42–44
Pseudomonas, 42, 271, 280, 293,
        305–307 315, 339
    fluorescens, 42, 198, 273, 293, 305,
        337, 422
    fragi, 343, 346
    graveolens, 307
    ichthyosmia, 338, 343
    mephitica, 343
    mucidolens, 307
    nigrifaciens, 344
    putrefaciens, 343
    sapolactica, 338
    syncyanea, 163, 273, 339
    synxantha, 339
    viscosa, 346
Psychrophiles, 40, 60, 110, 257, 293, 339,
    477
Psychrotrophs (*see* Psychrophiles), 41
Pullularia, 20
    pullulans, 20
Pulque, 390
Pumpkins, spoilage of, 233
Putrefaction, 56, 59, 175, 270, 279, 293,
    347, 363, 365
Pythium, 10
    butleri, 228
    debaryanum, 233

Q fever, 99, 456
Quaternary ammonium compounds, 485
Quick freezing, 116
Quick methods for vinegar, 401–405
Quinces, spoilage of, 240

Rad, 152
Radiations, 115, 146–157
    applications of, to foods, 151, 156
    beta rays, 151
    cathode rays, 152–157

Radiations, corpuscular, 146
  electric currents, 146
  electromagnetic, 146
  electrons, 151
  gamma rays, 151–157
  heating, 146–148
  ionizing, 151–157, 259
  microwaves, 148
  radiofrequency, 147
  ultraviolet, 149–151
  X-rays, 151, 152
Radishes, spoilage of, 233
Rancidity, 269, 343, 351
  flavor reversion in, 351
  hydrolytic, 351
  ketonic, 351
  oxidative, 269, 351
  tallowiness in, 344, 351
Red rot (eggs), 306
Red spot (meat), 269
Removal of microorganisms, 79
Rep (roentgen-equivalent-physical), 152
Rhizoctonia crown rot, 230
Rhizopus, 11, 226, 269
  arrhizus, 411
  delemar, 426
  nigricans, 11, 185, 226
  oligosporous, 411
  oryzae, 411, 423
  sonti, 390
  stolonifer, 411
Rhizopus soft rot, 226
Rhodotorula, 27, 33, 273
  glutinis, 419
Rhodotorulaceae, 23
Rhubarb, spoilage of, 232
Rigor mortis, 284, 286, 290
Roasting, 100
Roentgen (r), 152
Roentgen-equivalent-physical (rep), 152
Ropiness, 61, 187–189, 198, 243, 247,
      280, 335–337, 353, 396
  in bread, 187–189
  in maple sap, sirup, 198
  in meat-curing pickle, 280
  in milk, 335–337
  in pickle brine, 247
  in sauerkraut, 243
  in soft drinks, 353
  in wines, 396

Rots, egg, 305–308
  black, 306
  colorless, 305
  fungal, 306
  green, 305
  pink, 306
  red, 306
  fruit and vegetable, 226–241
  alternaria, 227
  anthracnose, 227
  bacterial soft rot, 226
  black, 227
  black mold, 227
  blue mold, 227
  brown, 227
  downy mildew, 227
  dry, 227
  finger, 238
  fusarium, 227
  gray mold, 226
  green mold, 227
  pink mold, 227
  rhizoctonia crown, 230
  rhizopus soft, 226
  stalk, 238
  stem-end, 227
  watery soft, 227
Rum, 398
Rutabagas, spoilage of, 233

Saccharomyces, 31, 198
  carlsbergensis, 31, 376, 385, 421
  cerevisiae, 31, 375, 376, 390, 419, 427
    var. *ellipsoideus,* 31, 377, 391, 398,
      399
  citri medicae, 408
  fragilis, 31, 431
  lactis, 32
  mellis, 32
  pastorianus, 387
  pyriformis, 390
  rouxii, 32, 410
Saccharomycodes, 34
Sake, 390
Salad dressings, spoilage of, 352
    French, 352
    mayonnaise, 352
    Thousand Island, 352
Salicylates, 137

Salmonella, 47, 258, 301
  enteritidis, 451
  pullorum, 452
  species of, 451
  typhimurium, 451
Salmonella infections, 451–455
  cause of, 451
  disease from, 454
  foods involved in, 453
  incidence of, 453
  outbreaks of, 454
  prevention of, 455
  sources of salmonellae, 452
  tests for, 469
Salt (sodium chloride), 133, 208, 210,
    213, 217, 218, 260, 288, 354
  effect on heat resistance, 86
Salt-stock pickles, 213–216
Sampling, for investigating disease out-
    breaks, 466
  for microbiological examination, 493
Sandwiches, 487
Sanitarian, duties of, 476
Sanitation, 475–491
  of eating and drinking utensils,
    489
  of plant equipment, 478, 483–485
  employee health, 489
  waste (sewage) treatment and dis-
    posal, 478–481
  water supplies, 476–478
Sanitizing (sanitizers), 71, 142, 179, 484
Saprolegnia parasitica, 10
Sarcina, 48, 293, 354
Sarcina sickness (beer), 388
Sauerkraut, 209–212, 243
  canning of, 212
  definition, 209
  fermentation, 211
  manufacture of, 210
  spoilage of, 243
Sauerrüben, 219
Sausage, 261, 272–277
  fish, 294
  greening of, 275
  lactic acid fermentation of, 261, 505
  pork, spoilage of, 272
  souring of, 275
  spoilage of, 273–277
Schizosaccharomyces, 30, 197
Sclerotia, 3

Sclerotinia, 23, 227, 237
  sclerotiorum, 227, 237
Sclerotium bataticola, 233
Scopulariopsis, 19, 348
  brevicaulis, 19
Seafoods, crabs, spoilage of, 294
  mussel and clam poisoning, 435
  oysters, 294, 513
    coliform test for, 495
    spoilage of, 294
  shrimp, spoilage of, 294
  standards for, bacteriological, 513
  (*See also* Fish; Shellfish)
Sedimentation, 80
Serratia, 47, 306, 339
  marcescens, 47, 189, 339
  salinaria, 354
Sewage and wastes, 65, 478–481
  BOD values, 478
  treatment and disposal of, 478–481
  types of, 481
Shellfish, bacterial genera in, 284
  coliform test for, 495
  contamination from sewage in, 65
  spoilage of, 294
  standards for, bacteriological, 511, 513
Shigella, 47, 463, 469
  tests for, 469
Shrimp, spoilage of, 294
Sirups, preservation of, 195
  spoilage of, 197
Slaughter, humane, 252
Sliminess, 61, 227, 243, 268, 272, 274,
    316
  in meat, 268, 272, 274
  in poultry, 316
  in sauerkraut, 243
  time of appearance of, 268
  in vegetables, 227
  in vinegar, 406
Smoking (wood smoke), 123, 140, 264,
    289, 315
Smudge (onion), 228
Smut, 227, 238
Sodium benzoate, 137
Sodium chloride (*see* Salt)
Soft drinks, 353
  contamination of, 353
  microbiological tests for, 495
  spoilage of, 353
Soft swell, 359

Soil, bacterial genera in, 66
contamination from, 65
Sonic (sound) waves, 148, 330
Sonti, 390
Sorbic acid and sorbates, 138
Sour, for rye bread, 374, 383
Souring, 227, 238, 265, 269, 270, 272,
279, 294, 332, 338, 365
of canned fish, 365
of canned meat, 365
of figs and dates, 238
of meats, 265, 269, 270, 272, 279
of milk and cream, 332
of oysters, 294
Soy sauce, 409
Soybean cheese, 412
Sphaeropsidales, 23
Sphaerotheca mors-uvae, 235
Spices and other condiments, 141, 264,
354
Spinach, spoilage of, 232
Spirillaceae, 44
Spoilage, causes of, 161
ease of, classification on, 161
fitness for consumption, 160
general principles of, 160–174
growth of microorganisms in, 162–174
(*See also* specific foods)
Sporangia, 4
Sporangiophores, 4
Sporangiospores, 4
Spores, arthrospores (oidia), 4
ascospores, 5
asexual (mold), 4
bacterial endospores, 36–38, 88
in air, 69
dormancy of, 38
formation of, 37
germination of, 37
heat resistance of, 88, 440
chlamydospores, 4, 27
conidia, 4
mold, 4
in air, 69
heat resistance of, 87
oöspores, 5
sexual (mold), 4
sporangiospores, 4
yeast, 26, 30, 87
heat resistance of, 87
zygospores, 5

Sporotrichum, 18, 270, 306
carnis, 18, 270
Springer, 359
Squash, spoilage of, 233
Stamping-ink discoloration, 269, 351
Standards, microbiological, 488, 509–514
for apple juice, 513
for crab meat, 511, 513
for custard-filled items, 511
for dry milk, 129, 327, 510
for eggs (dried, frozen, liquid), 512
for fish and shellfish, 511, 513
for frozen desserts, 510
for frozen vegetables, 513
for fruit preserves and jams, 513
for hamburger, 513
for market milk, 509
for poultry, 513
for precooked frozen foods, 511
for precooked meals, 513
for precooked meats, 513
for starch, 509
for sugar, 509, 511
for tomatoes and tomato products,
511
for water, 511
Staphylococcus, 49
aureus, 49, 280, 444–449, 468
epidermidis, 49
lactis, 49
Staphylococcus food poisoning, 444–449,
468
cause of, 444
disease from, 448
foods involved in, 447
incidence of, 447
outbreaks of, 448
prevention of, 448
tests for, 468
toxin, 446
Starch, amylases, 59, 381, 425–427
amylopectin, 425
amylose, 425
standards for, bacteriological, 509
Starters, 344, 372–374
for butter, 373
for buttermilks, 344, 373
for cheese, 345, 373
for rye bread, 374, 383
for sausage, 374
for soy sauce (koji), 409, 426

Stem-end rots (fruits, vegetables), 227
Stemphylium, 22, 197
Sterigmata, 4
Sterigmatocystis, 197
Sterility, commercial, 103, 255, 285
  tests for, 498
Sterilization, cold, 146
Stickiness, 269
Storage, 111–120, 173
  chilling, 112–115
  common or cellar, 111
  dry, 181
  freezing (cold storage), 115–120
  gas, 114, 136, 221, 257, 303
Streptococcus, 50
  agalactiae, bovis, durans, pyogenes, 50
  cremoris, 50, 328, 336
  diacetilactis, 52, 328
  faecalis, 50, 211, 214, 217, 224, 277,
      332, 364, 366, 455, 469
    var. *liquefaciens,* 51, 59, 334
    var. *zymogenes,* 51
  faecium, 50, 364, 366
  groups of, 50
  lactis, 50, 163, 328, 332
    var. *hollandicus,* 336
    var. *maltigenes,* 338, 343
  mucilaginosus, 388
  thermophilus, 50, 328, 332, 363
Streptococcus infections, 455
  tests for, 469
Streptomyces, 56, 293, 307, 421
Subtilin, 139
Sucrose, spoilage of, 196
Sugar products, candy, 195, 199
  honey, 194, 196, 198
  maple, 193, 197
    sap, contamination of, 193
      spoilage of, 198
    sirup, preservation of, 196
      spoilage of, 198
    sugar, 198
  molasses, sirups, 195, 197
Sugars, 192, 196
  beet and cane, 192
  contamination of, 192
  effect of, on heat resistance, 86
  invert, 427
  liquid, bacteriological standards for,
      509
  preservation by, 138

Sugars, preservation of, 195
  spoilage of, 196
  standards for, microbiological, 509
  sucrose, spoilage of, 196
Sulfide spoilage, 361
Sulfur dioxide (sulfuring), 124, 134,
      208, 224
  in dried fruits, 124, 208
  in wines, 391
Sun drying, 123
Surface taint (butter), 343
Survivor curves, 92
Sweating, 125
Sweet curdling, 56, 334
Sweet potatoes, 112, 232
  spoilage of, 232
  storage of, 112, 232
Symbiosis, 163
Synergism, 163

T.A. spoilage, 361
Taette, 328
Taints in meat, 269, 271
Tallowiness (fats), 344, 351
Tamari sauce, 410
Tapioca, spoilage of, 190
Tea, 407
Tempeh, 411
Temperature-time relationship, 83
Temperatures, high, preservation by,
      82–108
  low, growth at, 40, 60, 110
    preservation by, 109–121
  maximal, minimal, optimal for growth,
      40
  for storage, 111–120, 173
  (*See also* Thermal processes)
Temperature requirements, 8, 28, 40, 172
  for bacteria, 40, 173
  for mesophiles, 41
  for molds, 8, 172
  for psychrophiles, 40
  for thermophiles, 41
  for yeasts, 28, 172
Tetracyclines, 265
  (*See also* Antibiotics)
Thamnidium, 13, 269
  chaetocladioides, 269
  elegans, 13, 269
Thamnidium taint, 270

Thawing, 119
thawed food, disposal of, 119
Thermal death point, 87
Thermal death rate, 87
Thermal death time, 82–93
absolute, 87
curves, 91
majority, 87
(*See also* Heat resistance)
Thermal processes, 96–107
for canned fish, 285
for canned fruits, 221
for canned meats, 255
for canned vegetables, 104
determination of, 96–98
effect on microbial growth, 166
equivalent, 96
heat-treatments employed in, 98–107
Thermodurics, 322, 363
Thermophiles, 60, 322
Tomatoes, spoilage of, 235
standards for, microbiological, 511
Torula globula, 342
glutinis, 339
lactis-condensi, 342
mellis, 198
Torulaspora, 215
Torulopsis, 32, 215, 353
caroliniana, 216
lipofera, 419
sphaerica, 32
Tou-fu-ru, 412
Tourne (wines), 395
Toxin, in botulism, 438–440
mold, 450
staphylococcal, 446
Trichinella spiralis, 457, 505
Trichinosis, 457–459
Trichoderma, 19, 227
viride, 19, 237
Trichosporon, 33, 189
pullulans, 33, 419
variable, 189
Trichothecium, 17, 227
roseum, 17, 227
Trimethylamine, 291–293
Trimming of foods, 80
Tuberculariaceae, 22
Turnips, spoilage of, 233
Tylosin, 139

Udder, contamination from, 65
Ultraviolet rays, 68, 149–151, 182, 195, 258, 330
on bread, cakes, 151, 182
on dairy products, 330
on meats, 258
on sugar, 151, 195
on water, 68, 151
Utensils, cleaning, 483
dairy, contamination from, 318
eating and drinking, 489, 500
bacteriological examination of, 500
cleaning and sanitizing of, 490
standards for, bacteriological, 500
microbiological examination for, 493
sanitizing of, 490

Vanilla, 408
Vegetables, 112, 124, 128, 201–217, 242–247, 513
bacterial genera in, 204
contamination of, 201–203
dried, microbiology of, 128
frozen, bacteriological standards for, 513
preservation of, 204–217
by asepsis, 204
by chilling, 205
by common storage, 112, 205
by drying, 124, 207
by fermentation, 209–217
by freezing, 206
by heat, 205
by preservatives, 208–217
by salting, 208
spoilage, 226–235
causes of, 226
of fermented, 243–247
of vegetable juices, 242
types of, 226
(*See also* specific vegetables)
Vending of foods, 486
from commissaries, 487
from machines, 486
sandwiches, 487
Ventilation, 114
Vibrio, 44, 280
parahemolyticus, 450
Vinegar, 100, 398–407
bacteria, 44, 58, 399

Vinegar, defects in, 405
  diseases of, 406
  eels, 406
  fermentation of, 399
  kinds of, 399
  manufacture of, 400–405
  pasteurization of, 100
  quick methods of manufacture,
      401–405
  slow methods of manufacture, 400
  submerged method of manufacture,
      403–405
Vitamins, 169
  for microorganisms, 169
  from microorganisms, 420–422

Washing, equipment for, 483
  of foods, 80, 204, 211, 220
  of hands, 490
Waste treatment and disposal, 478–481
Water, 5, 27, 40, 66–68, 164–166
  availability of, 5, 27, 40, 164
  bacterial genera in, 66
  bacteriology of, 66–68, 476–478
  chemical properties of, 67
  chlorination of, 67, 135, 477
  contamination from, 66–68
  cooling of, 67, 106
  plant, 67, 477
  potable, 66, 477, 511
  requirements for, (see Water activity)
  reused, 477
  standards for, bacteriological, 511
  treatment of, 68
  ultraviolet irradiation of, 68, 151
Water activity (a_w), 5–7, 27, 40, 164–166
  for bacteria, 40
  for molds, 7
  for yeasts, 27
Watermelons, spoilage of, 233
Watery soft rot (fruits, vegetables), 227
Whiskers, on eggs, 307
  on meats, 269
Whiskey, 398
White spot (meats), 270
Wieners, spoilage of, 274
Wines, 390–398
  aging of, 392
  defects in, 393–396
  fermentation of, 391

Wines, flowers in, 394, 407
  kinds of, 393
  manufacture of, 391
  microbiology of, 392
  spoilage of, 394–396
Wood smoke (smoking), 123, 140, 264,
      289, 315
Wort, 385

X-rays, 151, 152
Xerophilic fungi, 165

Yeast and mold counts, 495
Yeasts, 25–35, 375–377
  apiculate, 34
  bakers', 375, 381
  for beer, 376, 385
  bottom, 31, 385
  classification and identification of, 28
  cultural characteristics of, 27
  definition, 25
  distillers', 377
  false, 27, 32
  fats from, 419
  fermentative, 25, 27, 28, 30, 242
  film (oxidative), 27, 30, 33, 243, 244,
      247
  as food, 416–419
  industrial, 30–34
  morphology of, 25
  nutritive value of, 418
  osmophilic, 27, 32, 34, 196, 198
  physiological characteristics of, 27
    foods, 28
    requirements, moisture, 27, 164
      oxygen, 27
      pH, 28, 170
      temperature, 28
  primary, 417
  salt-tolerant, 34
  secondary, 416
  top, 31, 390
  torula, 417
  true, 26, 30–32
  wild, 30
  wine, 377
Yoghurt, 328

z value, 91
Zapatera spoilage of olives, 247
Zygomycetes, 10–13
Zygorrhynchus, 11
Zygosaccharomyces, 32, 193, 196–198,
    215, 410

Zygosaccharomyces, mellis (*see Saccha-*
    romyces mellis), 198
    nussbaumeri, 32, 198
    richteri, 198
    soyae, 410
Zygospores, 5